Lecture Notes in Artificial Intelligence   1532

Subseries of Lecture Notes in Computer Science
Edited by J. G. Carbonell and J. Siekmann

Lecture Notes in Computer Science

Edited by G. Goos, J. Hartmanis and J. van Leeuwen

T0223203

Springer
Berlin
Heidelberg
New York
Barcelona
Hong Kong
London
Milan
Paris
Singapore
Tokyo

Setsuo Arikawa Hiroshi Motoda (Eds.)

Discovery Science

First International Conference, DS'98
Fukuoka, Japan, December 14-16, 1998
Proceedings

Springer

Series Editors

Jaime G. Carbonell, Carnegie Mellon University, Pittsburgh, PA, USA
Jörg Siekmann, University of Saarland, Saarbrücken, Germany

Volume Editors

Setsuo Arikawa
Kyushu University, Department of Informatics
Fukuoka 812-8581, Japan
E-mail: arikawa@i.kyushu-u.ac.jp

Hiroshi Motoda
Osaka University, Institute of Scientific and Industrial Research
Division of Intelligent Systems Science
8-1 Mihogaoka, Ibaraki, Osaka 567-0047, Japan
E-mail: motoda@sanken.osaka-u.ac.jp

Cataloging-in-Publication Data applied for

Die Deutsche Bibliothek - CIP-Einheitsaufnahme

Discovery science : first international conference ; proceedings / DS '98,
Fukuoka, Japan, December 14 - 16, 1998 / Setsuo Arikawa ; Hiroshi Motoda
(ed.). - Berlin ; Heidelberg ; New York ; Barcelona ; Hong Kong ; London ;
Milan ; Paris ; Singapore ; Tokyo : Springer, 1998
  (Lecture notes in computer science ; Vol. 1532 : Lecture notes in artificial
  intelligence)
  ISBN 3-540-65390-2

CR Subject Classification (1998): I.2, H.2.8, H.3, J.1, J.2

ISBN 3-540-65390-2 Springer-Verlag Berlin Heidelberg New York

© Springer-Verlag Berlin Heidelberg 1998
Printed in Germany

Typesetting: Camera ready by author
SPIN 10692930   06/3142 – 5 4 3 2 1 0   Printed on acid-free paper

Preface

This volume contains all the papers presented at the First International Conference on Discovery Science (DS'98), held at the Hotel Uminonakamichi, Fukuoka, Japan, during December 14–16, 1998.

This conference was organized as a part of activities of the Discovery Science Project sponsored by Grant-in-Aid for Scientific Research on Priority Area from the Ministry of Education, Science, Sports and Culture (MESSC) of Japan. This is a three-year project starting from 1998 that aims to (1) develop new methods for knowledge discovery, (2) install network environments for knowledge discovery, and (3) establish Discovery Science as a new area of computer science.

The DS'98 program committee selected 28 papers and 34 posters/demos from 76 submissions. Additionally, five invited talks by Pat Langley of the Institute for the Study of Learning & Expertise, Heikki Mannila of University of Helsinki, Shinichi Morishita of University of Tokyo, Stephen Muggleton of University of York, and Keiichi Noe of Tohoku University were featured at the conference.

DS'98 series provide an open forum for intensive discussions and interchange of new information, be it academic or business, among researchers working in the new area of Discovery Science. It focuses on all areas related to discovery including, but not limited to, the following areas: logic for/of knowledge discovery, knowledge discovery by inferences, knowledge discovery by learning algorithms, knowledge discovery by heuristic search, scientific discovery, knowledge discovery in databases, data mining, knowledge discovery in network environments, inductive logic programming, abductive reasoning, machine learning, constructive programming as discovery, intelligent network agents, knowledge discovery from unstructured and multimedia data, statistical methods for knowledge discovery, data and knowledge visualization, knowledge discovery and human interaction, and human factors in knowledge discovery.

Continuation of the DS series is supervised by its Steering Committee consisting of Setsuo Arikawa (Chair, Kyushu Univ.), Yasumasa Kanada (Univ. of Tokyo), Akira Maruoka (Tohoku Univ.), Satoru Miyano (Univ. of Tokyo), Masahiko Sato (Kyoto Univ.), and Taisuke Sato (Tokyo Inst. of Tech.).

DS'98 was chaired by Setsuo Arikawa (Kyushu Univ.), and assisted by the Local Arrangement Committee: Takeshi Shinohara (Chair, Kyushu Inst. of Tech.), Hiroki Arimura (Kyushu Univ.) and Ayumi Shinohara (Kyushu Univ.), It was cooperated by SIG of Data Mining, Japan Society for Software Science and Technology.

We would like to express our immense gratitude to all the members of the Program Committee, which consists of:

Hiroshi Motoda (Chair, Osaka Univ., Japan)
Peter Flach (Univ. of Bristol, UK)
Koichi Furukawa (Keio Univ., Japan)
Randy Goebel (Univ. of Alberta, Canada)
Ross King (Univ. of Wales, UK)
Yves Kodratoff (Univ. de Paris-Sud, France)
Nada Lavrac (Jozef Stefan Inst., Slovenia)
Wolfgang Maass (Tech. Univ. of Graz, Austria)
Katharina Morik (Univ. of Dortmund, Germany)
Shinichi Morishita (Univ. of Tokyo, Japan)
Koichi Niijima (Kyushu Univ., Japan)
Toyoaki Nishida (Naist, Japan)
Hiroakira Ono (Jaist, Japan)
Claude Sammut (Univ. of NSW, Australia)
Carl Smith (Univ. of Maryland, USA)
Yuzuru Tanaka (Hokkaido Univ., Japan)
Esko Ukkonen (Univ. of Helsinki, Finland)
Raul Valdes-Perez (CMU, USA)
Thomas Zeugmann (Kyushu Univ., Japan)

The program committee members, the steering committee members, the local arrangement members, and subreferees enlisted by them all put a huge amount of work into reviewing the submissions and judging their importance and significance, ensuring that the conference had high technical quality. We would like to express our sincere gratitude to the following subreferees.

Sebastien Augier	Peter Brockhausen	Xiao-ping Du
Toshiaki Ejima	Tapio Elomaa	Russ Greiner
Toshinori Hayashi	Koichi Hirata	Eiju Hirowatari
Daisuke Ikeda	Thorsten Joachims	Yukiyoshi Kameyama
Juha Karkkainen	Satoshi Matsumoto	Tetsuhiro Miyahara
Paul Munteanu	Teigo Nakamura	Claire Nedellec
Masayuki Numao	Seishi Okamoto	Celine Rouveirol
Hiroshi Sakamoto	Michele Sebag	Shinichi Shimozono
Noriko Sugimoto	Masayuki Takeda	Izumi Takeuchi
Shusaku Tsumoto	Tomoyuki Uchida	Takashi Washio
Takaichi Yoshida		

We would also like to thank everyone who made this meeting possible: authors for submitting papers, the invited speakers for accepting our invitation, the steering committee, and the sponsors for their support. Special thanks are due to Takeshi Shinohara, the Local Arrangement Chair, for his untiring effort in organization, and to Ayumi Shinohara for his assistance with the preparation for the proceedings.

Fukuoka, December 1998
Osaka, December 1998

Setsuo Arikawa
Hiroshi Motoda

Table of Contents

Philosophical Aspects of Scientific Discovery : A Historical Survey

Keiichi Noé

Tohoku University, Faculty of Arts and Letters,
Kawauchi, Aoba-ku, Sendai 980-8576, Japan
Noe@sal.tohoku.ac.jp

Abstract. This paper is intended as an investigation of scientific discoveries from a philosophical point of view. In the first half of this century, most of philosophers rather concentrated on the logical analysis of science and set the problem of discovery aside. Logical positivists distinguished a context of discovery from a context of justification and excluded the former from their analysis of scientific theories. Although Popper criticized their inductivism and suggested methodological falsificationism, he also left the elucidation of discovery to psychological inquiries. The problem of scientific discovery was proprely treated for the first time by the "New Philosophy of Science" school in the 1960's. Hanson reevaluated the method of "retroduction" on the basis of Peirce's logical theory and analysed Kepler's astronomical discovery in detail as a typical application of it. Kuhn's theory of paradigm located discoveries in the context of scientific revolutions. Moreover, he paid attention to the function of metaphor in scientific discoveries. Metaphorical use of existing terms and concepts to overcome theoretical difficulties often plays a decisive role in developping new ideas. In the period of scientific revolution, theoretical metaphors can open up new horizons of scientific research by way of juxtapositions of heterogeneous concepts. To explicate such a complicated situation we need the "rhetoric" of science rather than the "logic" of science.

1 Introduction

In the field of philosophy of science, scientific "discovery" has been a problematic concept because it is too ambiguous to formulate in terms of methodology of science. If the procedure of discovery can be reduced to a certain logical algorithm, it is not a philosophical but a mathematical problem. On the other hand, if it concerns with flashes of genius, it amounts to a kind of psychological problem. As a result, philosophers of science in the twentieth century almost ignored the problem of discovery and excluded it from their main concern.

However, in the 1960's, the trend of philosophy of science was radically changed. At that time the so-called "New Philosophy of Science" school was on the rise in opposition to Logical Positivism. Representative figures of this school were Norwood Hanson, Thomas Kuhn, Stephen Toulmin and Paul Feyerabend. They tackled, more or less, the problem of scientific discovery from philosophical as well as sociological viewpoints. It is no exaggeration to say that the concept of discovery was properly treated as a main subject of the philosophy of science for the first time in this movement.

The purpose of this paper is to examine how the notion of "discovery" has been discussed and analysed along the historical development of the philosophy of science in this century. Furthermore, I would like to consider the role of rhetorical moments, especially the explanatory function of metaphor, in scientific discoveries. It is not my present concern to investigate the logical structure of discovery. For many participants of this conference will probably make the problem clear.

2   A context of justification vs. a context of discovery

In the broad sense of the word, the origin of philosophy of science goes back to Aristotle's *Analytica Posteriora*. There he elucidated the structure of scientific reasoning, e.g., deduction, induction and abduction. But his analysis mainly dealt with demonstrative knowledge and did not step into the problem of discovery in empirical sciences. For the process of discovery is contingent and can not be explicitly formulated. In the last chapter of *Analytica Posteriora* Aristotle characterized the ability of discovery as "νοῦς", i.e. "insight of reason" or "comprehension."[1] As is well known, in the medieval scholastic philosophy this ability was called *lumen naturale*, which stood for natural light. Even in the modern era, such tradition has dominated the main stream of the philosophy of science, so that philosophers have been considered discovery to be an intuitive mental process irrelevant to their logical analysis.

One exception was C. S. Peirce, who tried to construct the logic of discovery, i.e. how to find a hypothesis to solve a problem, on the basis of Aristotle's concept of abduction. This is what Peirce has to say on the matter : "All the ideas of science come to it by the way of Abduction. Abduction consists in studying facts and devising a theory to explain them. Its only justification is that if we are ever to understand things at all, it must be in that way."[14] He made unique functions of abduction in scientific research clear by contrast with the method of deduction and induction. Unfortunately his attempt was regarded as a deviation from orthodoxy and almost forgotten for a long time. It was in the end of 1950's that Hanson reevaluated Peirce's theory of "abduction" or rather "retroduction" in the context of contemporary philosophy of science.

In the narrow sense of the word, the philosophy of science began with a booklet entitled *The Scientific Conception of the World*, which was a platform of the Vienna Circle in 1929. This group was founded by Austrian philosopher Moritz Schlick and consisted of philosophers, physicists and mathematicians. Their outlook is generally described as *logical positivism* and symbolized by their slogan "elimination of metaphysics" as well as "the unity of science." Achieving their purpose, the Vienna Circle developped the method of logical analysis that was based on mathematical logic constructed by Frege and Russell at that time. Then they adopted a "verifiability principle of meaning" as a criterion of demarcation, which rejected metaphysical statements as cognitively meaningless. Generally speaking, logical positivists regarded the task of philosophy as the logical clarification of scientific concepts and aimed at "scientific philosophy" rather than "philosophy of science."

So far as the logical structure of scientific research is concerned, analytic approaches of the Vienna Circle achieved a brilliant success. However, they could not help excluding the problem of scientific discovery from their consideration so as to maintain the logical clarity of analysis. Take Hans Reichenbach's argument for example. After pointing out that the role of inductive inference is not to find a theory but to justify it, he continues as follows :

> The mystical interpretation of the hypothetico-deductive method as an irrational guessing springs from a confusion of *context of discovery* and *context of justification*. The act of discovery escapes logical analysis; there are no logical rules in terms of which a "discovery machine" could be constructed that would take over the creative function of the genius. But it is not the logician's task to account for scientific discoveries; all he can do is to analyze the relation between given facts and a theory presented to him with the claim that it explains those facts. In other words, logic is concerned only with the context of justification. And the justification of a theory in terms of observational data is the subject of the theory of induction.[20]

What is immediately apparent in this extract is that Reichenbach identifies a discovery with a casual idea or a flash. The hypothetico–deductive method, which is a kernel of methodology for modern science, is usually devided into four steps: (1) observations, (2) postulating a general hypohthesis, (3) deducing a test statement from it, (4) testing the statement by experiments. In general, the transition from the first step to the second step has been thought of the process of discovery that is closely connected to the procedure of induction. But Reichenbach maintains that the central subject of the theory of induction is not the second step but the fourth step, namely justification of theories. The second step is merely an activity of scientific guess. Thus there is no logical rules for promoting scientific discoveries. He leaves an explanation of mental mechanism of discovery to the psychological or sociological studies. The strict distinction between a context of discovery and a context of justification became an indispensable thesis of logical positivism. After that, philosophers of science have devoted themselves to investigate the latter and almost ignored the former problem.

3 Popper's falsificationism

Karl Popper was one of the most influential philosopher of science in the twentieth century. Though he associated with members of the Vienna Circle, he did not commit himself to the movement of logical positivism. Rather, he criticized their basic premise of the verifiability principle of meaning, and proposed the concept of "falsifiability" instead. For the procedure of verification presupposes the legitimacy of induction which justifies a hypothesis on the basis of experimental data. For that purpose, logical positivists were eager in constructing the system of inductive logic to calculate the probability of verification. They wanted to give a firm basis to scientific theories by appealing to inductive logic.

However, for Popper, it is evident that inductive inferences are logically invalid. Induction is simply characterized as a transition from particular observations to a general law. Logically speaking, since this transition is nothing but a leap from a conjunction of singular statements to a universal statement, it needs to be justified by some other principles which make possible to bridge the gap, e.g. "the principle of uniformity of nature" by J. S. Mill. Then, where can we obtain such a principle? If it is not a metaphysical principle but an empirical one, we have to gain it by means of induction again. Obviously this argument contains a circular reasoning. Therefore it is impossible to justify inductive inference in a proper sense, and we cannot reach any indubitable knowledge by means of it.

Such difficulties were nearly pointed out by David Hume in the middle of the seventeenth century. As is generally known, he headed straight for skepticism concerning scientific knowledge through the distrust of inductive method, and his skeptical conclusion was refuted by Kant's transcendental argument. Reexamining this controversy, Popper criticizes Hume for abandoning rationality. On the other hand, he is not necessarily satisfied with Kant's solution. According to Kant's *apriorism*, natural laws are not discovered but invented by human intellect, as can be seen in the following quotation: "The understanding does not derive its laws (*a priori*) from, but prescribes to, nature."[3] Popper basically accepts this argument, but rejects Kant's thesis of the *apriority* of natural laws. Because every scientific statement about natural laws is no other than a tentative hypothesis for him. As far as it belongs to empirical sciences, it must be always exposed to the risk of falsification.

Consequently Popper suggests a negative solution to the foregoing "Hume's Paradox." He asserts that whereas a scientific hypothesis cannot be verified, it can only be falsified. His argument depends upon the asymmetry between verification and falsification. Even a vast collection of examples cannot completely verify a universal statement. On the contrary, only one counter-example is enough to falsify it. If induction is a procedure of verifying a hypothesis, it is neither necessary for nor relevant to scientific researches. The essence of scientific inquiry consists in the process of strict falsification. We are merely able to maintain a hypothesis as a valid theory until it will be falsified in the future. In this respect we have to accept the fallible character of knowledge. This is the central point of Popper's thesis of "falsificationism."

From such a viewpoint, Popper characterizes the methodology of science as a series of "conjecture and refutation" and reformulates the structure of hypothetico-deductive method. Since induction is a white elephant, it would be useless to start with observations. What is worse, observations are complicated activities which should be described as "theory-impregnated". There is no pure observation without theoretical background. Therefore, to begin with, scientists have to start from a problem (P_1) with which they are confronted. Then they try to propose a tentative theory (TT) to settle the problem by means of imaginative conjectures. Finally, they proceed to the procedure of error-elimination (EE) by a rigorous critical examination of the conjecture. If the tentative theory is refuted, they have to grope for some other conjecture. If not, they will grapple with a new problem (P_2). Popper summarizes the general schema of "the method of conjecture and refutation" as follows[15]:

$$P_1 \rightarrow TT \rightarrow EE \rightarrow P_2.$$

In this schema, a step toward a scientific discovery corresponds to the process of conjecturing a tentative theory to solve the problem. The word "conjecture" sounds a little illogical or contingent. Then, is there any logic of scientific discovery? Popper's answer is negative. First of all, he points out that we should distinguish sharply "between the process of conceiving a new idea, and the methods and results of examining it logically"[16]. He goes on to say:

> Some might object that it would be more to the purpose to regard it as the business of epistemology to produce what has been called a *rational reconstruction* of the steps that have led the scientists to a discovery — to the finding of some new truth. But the question is: what, precisely, do we want to reconstruct? If it is the process involved in the stimulation and release of an inspiration which are to be reconstructed, then I should refuse to take it as the task of the logic of knowledge. Such processes are the concern of empirical psychology but hardly of logic[17].

What the passage makes clear at once is that Popper distinguishes the logic of knowledge from the psychology of knowledge, and confines the task of philosophy of science to the former. Thus, the problem of scientific discovery is thrown out of it and left to the latter. There is no such thing as logical rules for discovery. Moreover, he adds that "every discovery contains 'an irrational element', or 'a creative intuition', in Bergson's sense."[18] In such a way of thinking, Popper undoutedly succeeds to the logical positivists' distinction between a context of justification and a context of discovery.

According to Popper's falsificationism, the maxim of conjecture is, to borrow Feyerabend's phrase, "anything goes." It is all right if we use mystical intuition or oracle to conjecture a hypothesis. In conjecturing, the bolder the better. Such a view underlies the following remarks by him: "And looking at the matter from the psychological angle, I am inclined to think that scientific discovery is impossible without faith in ideas which are of a purely speculative kind, and sometimes even quite hazy; a faith which is completely unwarranted from the point of view of science, and which, to that extent, is 'metaphysical'."[19] It is for this reason that Popper regards the elucidation of discovery as business of the psychology of knowledge. In this respect, the title of his first work *The Logic of Scientific Discovery* was very misleading. For it did not include any discussion about the logic of discovery.

4 The concept of discovery in the New Philosophy of Science

From the late 1950's to the early 1960's, some philosophers of science took a critical attitude toward the dominant views of logical positivism and developped historical and sociological analyses of science. This new trend was later called the school of "New Philosophy of Science." In this stream of scientific thought, they cast doubt on the distinction between a context of justification and a context of discovery, and proved that both of them were closely interrelated and inseparable. After that, the concept of discovery became a proper subject of the philosophy of science.

Norwood Hanson's book *Patterns of Discovery* laid the foundations of this movement. In the "introduction" he clearly says that his main concern in this book is not with the testing of hypotheses, but with their discovery. His arguments begins with reexamination of the concept of observation. Needless to say, Hanson is known as an adovocator of the thesis of "the theory-ladenness of observation." It means that we cannot observe any scientific fact without making any theoretical assumptions. Observations must reflect these assumptions. Thus he concludes that "There is a sense, then, in which seeing is a 'theory-laden' undertaking. Observation of x is shaped by prior knowledge of x."[4]

The observation is not a simple experience of accepting sense-data. It presupposes a complicated theoretical background. Excellent scientists can observe in familiar objects what no one else has observed before in the light of a new theoretical hypothesis. In this sense, making observations and forming a hypothesis are one and inseparable. We should rather suggest that an observational discovery is usually preceded by a theoretical discovery. The discovery of pi-meson predicted by Hideki Yukawa's physical theory was a notable example. He intended to explain the phenomena of nuclear force and finally introduced a new particle in terms of analogy with the photon's role in medeating electromagnetic interactions. Hanson calls such a process of inference "retroduction" on the basis of Peirce's logical theory. Relevant to this point is his following remark:

> Physical theories provide patterns within which data appear intelligible. They constitute a 'conceptual Gestalt'. A theory is not pieced together from observed phenomena; it is rather what makes it possible to observe phenomena as being of a certain sort, and as related to other phenomena. Theories put phenomena into systems. They are built up 'in reverse' — retroductively. A theory is a cluster of conclusions in search of a premiss. From the observed properties of phenomena the physicist reasons his way towards a keystone idea from which the properties are explicable as a matter of course. The physicist seeks not a set of possible objects, but a set of possible explanations.[5]

Formally speaking, the retroductive inference is simply formulated as follows:

$$P \rightarrow Q, Q \vdash P.$$

It is evident that this is a logical mistake called "fallacy of affirming the consequent." The procedure of hypothetico-deductive method provides no room for such a fallacy to play a positive role. Therefore logical positivists excluded retroduction from their logical analysis, and left it to the task of psychology. Hanson, on the other hand, maintains that although retroduction is logically invalid, it works as reasonable inference in scientists' actual thinking. Because retroduction can bring about the change of a "conceptual Gestalt" or "conceptual organization." Such changes are nothing but a symptom of scientific discovery.

However, it is rather difficult to describe the "conceptual Gestalt" in detail because of its vagueness. Hanson did no more than explain it figuratively by appealing to examples of optical Gestalts. In the meantime, Thomas Kuhn characterized its function as "paradigm" from a historical and sociological point of view. His main work *The Structure of Scientific Revolution* was published in 1962 and radically transformed the trend of philosophy of science in the twentieth century. In this book Kuhn objected to logical positivists' basic assumptions, to be precise, the accumulation of scientific knowledge and the the Whiggish historiography of science. Instead of the continuous progress of science, he proposed a new image of scientific development, namely the intermittent transformation of scientific theories. To use Kuhn's famous term, it should be calld "paradigm-shift" or "paradigm-change."

The very concept of paradigm is, to echo him, "universally recognized scientific achievements that for a time provide model problems and solutions to a community of practitioners."[7] The scientific research conducted by a certain paradigm is called "normal science," which is compared to puzzle-solving activities following preestablished rules. Anomalies occur from time to time in normal scientific inquiries. If there seems no way to cope with them, the present paradigm of a discipline loses scientists' confidence. Thus a crisis arises. In order to overcome the crisis, scientists try to change their conceptual framework from the old paradigm to the new one. This is the so-called scientific revolutions. The historical process of scientific development, according to Kuhn's picture, can be summarized in the following sequence:

$$paradigm \rightarrow normal\ science \rightarrow anomalies \rightarrow crisis$$
$$\rightarrow scientific\ revolution \rightarrow new\ paradigm \rightarrow new\ normal\ science.$$

The scientific revolution involves a total reconstruction of materials and a new way of looking at things. Kuhn considers it as a kind of "Gestalt switch" in scientific thinking. From a viewpoint of his historiography, we have to distinguish two kinds of scientific discovery, that is to say, discoveries in the stage of normal science and ones in the stage of scientific revolution. To borrow Kuhn's phrase, it is a distinction between "discoveries, or novelties of fact" and "inventions, or novelties of theory."[8] A simple example of the former is the discoveries of new chemical elements on the basis of the periodic table in the nineteenth century. By contrast, the discoveries of Einstein's relativity theory and Planck's quantum theory are typical examples of the latter. On this point the method of retroduction suggested by Hanson mainly concerns the invention of a new theory. But Kuhn himself thinks that there is no logical algorithm to create a new paradigm and entrusts the problem of paradigm-shift to some psychological and sociological considerations. He only suggests that: "almost always the men who achieve these fundamental inventions of a new paradigm have been either very young or very new to the field whose paradigm they change."[9]

5   The function of metaphor in scientific discoveries

Although Kuhn takes a negative attitude to the logic of discovery, his theory of paradigm gives some hints to explicate the essence of scientific discovery. In the period of normal science most of discoveries have to do with novelties of fact and partial revision of theories. Therefore normal scientific research is a rather cumulative enterprise to extend the scope and precision of scientific knowledge. On the other hand, in the period of scientific revolution, discoveries are usually accompanied by a radical change of fact as well as theory. Both fact and assimilation to theory, observation and conceptualization, are closely interrelated in this kind of discovery. It consists of a complicated series of events. Kuhn makes several important statements on this process.

> Discovery commences with the awareness of anomaly, i.e., with the recognition that nature has somehow violated the paradigm-induced expectations that govern normal science. It then continues with a more or less extended exploration of the area of anomaly. And it closes only when the paradigm theory has been adjusted so that the anomalous has become the expected. Assimilating a new sort of fact demands a more than additive adjustment of theory, and until that adjustment is completed — until the scientist has learned to see nature in a different way — the new fact is not quite a scientific fact at all.[10]

The phrase "the paradigm-induced expectation" is suggestive in this context, because to aware anomalies scientists must commit themselves to a particular paradigm. Curiously enough, scientific discovery leading to a new paradigm-change presupposes a committment to the former paradigm. Kuhn called such a paradoxical situation "the essential tension" in scientific researches. In other words, the dialectic of tradition and novelty is working during scientific revolutions. It is a process of trial and error, and takes time. One may say, as Kuhn in fact does, that "only as experiment and tentative theory are together articulated to a match does the discovery emerge and the theory become a paradigm."[11] As a rule, a conspicuous discovery is recognized as such after the concerned scientific revolution. The title of "discovery" might be awarded only with hindsight by the scientific community. It is for this reason that the multiple aspects of discovery require not only epistemological but also sociological analyses.

There is one further characteristic of discovery that we must not ignore. It is a metaphorical usage of scientific concepts and terms in inventing a new theory. Scientists, as mentioned above, have to devote themselves to normal scientific research in order to aware anomalies. Even when they deal with anomalies from a revolutionary viewpoint, they cannot help using the existing concepts and terms.

In such a case it is very useful for them to employ metaphor and analogy. Scientists involved in scientific revolution try to develop their new ideas by using old terms metaphoricaly. One may say from what has been said that an important symptom of theoretical discovery is an appearance of metaphors in the context of scientific explanation.

Kuhn cites many examples which seem to support this point in his article "What Are Scientific Revolutions?"[12] The conspicuous one is Planck's discovery of the quantum theory. In the end of the nineteenth century, he was struggling with the problem of black-body radiation which was one of representative anomalies in modern physics. The difficulty was how to explain the way in which the spectrum distribution of a heated body changes with temperature.

Planck first settled the problem in 1900 by means of a classical method, namely the probability theory of entropy proposed by Ludwig Boltzmann. Planck supposed that a cavity filled with radiation contained a lot of "resonators." This was obviously a metaphorical application of acoustic concept to the problem of thermodynamics. In the meantime, he noticed that a resonator could not have energy except a multiple of the energy "element." As a result, to quote Kuhn's expression, "the resonator has been transformed from a familiar sort of entity governed by standard classical laws to a strange creature the very existence of which is incompatible with traditional ways of doing physics."[13] Later in 1909, Planck introduced the new concept "quantum" instead of the energy "element." At the same time it turned out that the acoustic metaphor was not appropriate because of the discontinuous change of energy, and the entities called "resonators" now became "oscillators" in the expaneded context. Finally Planck was loaded with honors of discovering the quantum theory.

With regard to the functions of metaphor Max Black's interaction theory is broadly accepted. To take a simple example, "Man is a wolf" is a well-known metaphor. One can understand this by recognizing a certain similarity between two terms. Its content is easily transformed into a form of simile "Man is like a wolf." Another, less typical, example is "Time is a beggar." In this case, it must be difficult to find some similarity betweeen them. Therefore it cannot have any corresponding simile. Here although there is no similarity beforehand, a new sort of similarity is formed by the juxtaposition of two terms. A poet often employs such kind of metaphors to look at things from a different angle. According to Black's account, two heterogeneous concepts interact with each other in metaphor, and a meaning of new dimensions emerges out of our imagination. He suggests that "It would be more illuminating in some of these cases to say that the metaphor creates the similarity than to say that it formulates some simirality antecedently existing."[2] In short, the essence of metaphor consists in creating new similarities and reorganizing our views of the world.

One can see the creative function of metaphor not only in literary works but also in scientific discourses. But it must be noted that metaphorical expressions in scientific contexts are highly mediated by networks of scientific theories in comparison with literary or ordinary contexts, cf [6]. For example, the following statements cannot play a role of metaphor without some theoretical background;

"Sound is propagated by wave motion" and "Light consists of particles." These opened up new ways of articulating the nature in cooperation with physical theories. In this sense introducing metaphor and constructing a new theory are interdependent and, as it were, two sides of the same coin. Of course Planck's introduction of the acoustic concept of "resonators" into the black-body theory is another illustration of the same point.

Even the statement "The earth is a planet" was a kind of metaphor before the Copernican Revolution. For, in the sixteenth century, people who was tied to Aristoterian world-view could not find any similarity between the earth and a planet. The heliocentric theory of Copernicus made possible to see a new similarity between them. Needless to say, nowadays we do not regard that statement as metaphor. We can understand it as a literal description without any difficulties. Thus metaphorical expressions gradually switch over to trivial ones and lose their initial inpact. Matters are quite similar in the case of ordinary language. Such a process overlaps with the transition from scientific revolution to normal science in Kuhnian picture of scientific development. It might be of no use employing metaphors in normal science. On the contrary, in the midst of scientific revolutions, making use of metaphors is indispensable means to renew our theoretical perspective.

6   Conclusion

It follows from what has been said that the metaphorical use of existing terms and concepts to deal with the anomalies takes an important role in the initial stage of scientific discoveries. One can find a number of examples in the history of science. Especially at times of scientific revolution, as Kuhn pointed out, there is no explicit logical rules to invent a new theory. Because scientific revolutions include drastic changes of our basic premisses of scientific knowledge. Logical procedures are certainly useful for normal scientific research. By contrast, in the period of paradigm-shift, scientists cast doubt on the existing conceptual scheme and grope for a new categorization of phenomena. At this poit, metaphor shows its ability fully to alter the criteria by which terms attach to nature. It opens up a new sight of research through juxtapositions of different kind of ideas.

However, formal logical analysis is not necessarily applicable to the investigation of metaphor. In this sense, it was unavoidable for logical positivists to exclude a context of discovery from their scientific philosophy. The important role of model, metaphor, and analogy in scientific discoveries was recognized for the first time in the discussions of New Philosophy of Science. Traditionally the study of figurative expressions has been called "rhetoric." Logic and rhetoric are not contradictory but complementary disciplines. Whereas logical structures of scientific theories have been analyzed in detail, rhetorical aspects of scientific thought have been set aside. It seems reasonable to conclude that if we are to search for the many-sided structure of scientific discoveries, what we need is the rhetoric of science rather than the logic of science.

References

1. Aristotle, *Analytica Posteriora*, 100b1–15.
2. Black, M., *Models and Metaphors*, Ithaca, 1962, p. 37.
3. Carus, P. (ed.), *Kant's Proregomena to Any Future Metaphysics*, Chicago, 1909, p.82.
4. Hanson, N., *Patterns of Discovery*, Cambridge, 1969, p.19.
5. Hanson, N., *Ibid.*, p. 90.
6. Hesse, M., *Models and Analogies in Science*, Notre Dame, 1970, p.157ff.
7. Kuhn, T., *The Structure of Scientific Revolutions*, Chicago, 1970, p. viii.
8. Kuhn, T., *Ibid.*, p.52.
9. Kuhn, T., *Ibid.*, p.90.
10. Kuhn, T., *Ibid.*, pp. 52–53.
11. Kuhn, T., *Ibid.*, p. 61.
12. Kuhn, T., "What Are Scientific Revolutions?", in Lorenz Krueger et al. (eds.), *The Probabilistic Revolution*, Vol. 1, Cambridge, Mass., 1987.
13. Kuhn, T., *Ibid.*, p. 18.
14. Peirce, C. S., *Collected Papers*, Vol. V, §146, Cambridge/Mass., 1958.
15. Popper, K., *Objective Knowledge*, Oxford, 1972, p.164.
16. Popper, K., *The Logic of Scientific Discovery*, London, 1958, p.31.
17. Popper, K., *Ibid.*
18. Popper, K., *Ibid.*, p.32.
19. Popper, K., *Ibid.*, p.38.
20. Reichenbach, H., *The Rise of Scientific Philosophy*, Berkeley, 1951, 1973, p.231.

Learning, Mining, or Modeling?
A Case Study from Paleoecology

Heikki Mannila[1], Hannu Toivonen[2], Atte Korhola[3], and Heikki Olander[3]

[1] Department of Computer Science, PO Box 26,
FIN-00014 University of Helsinki, Finland
Heikki.Mannila@cs.helsinki.fi
[2] Rolf Nevanlinna Institute and Department of Computer Science,
PO Box 4, FIN-00014 University of Helsinki, Finland
Hannu.Toivonen@rni.helsinki.fi
[3] Laboratory of Physical Geography, PO Box 9,
FIN-00014 University of Helsinki, Finland
{Atte.Korhola, Heikki.Olander}@helsinki.fi

Abstract. Exploratory data mining, machine learning, and statistical modeling all have a role in discovery science. We describe a paleoecological reconstruction problem where Bayesian methods are useful and allow plausible inferences from the small and vague data sets available.

Paleoecological reconstruction aims at estimating temperatures in the past. Knowledge about present day abundances of certain species are combined with data about the same species in fossil assemblages (e.g., lake sediments). Stated formally, the reconstruction task has the form of a typical machine learning problem. However, to obtain useful predictions, a lot of background knowledge about ecological variation is needed.

In paleoecological literature the statistical methods are involved variations of regression. We compare these methods with regression trees, nearest neighbor methods, and Bayesian hierarchical models. All the methods achieve about the same prediction accuracy on modern specimens, but the Bayesian methods and the involved regression methods seem to yield the best reconstructions. The advantage of the Bayesian methods is that they also give good estimates on the variability of the reconstructions.

Keywords Bayesian modeling, regression trees, machine learning, statistics, paleoecology

1 Introduction

Exploratory data mining, machine learning, and statistical modeling all have a role in discovery science, but it is not at all clear of what the applicability of different techniques is. In this paper we present a case study from paleoecological reconstruction, discussing how the nature of the problem affects the selection of the techniques. Our emphasis is on the methodology: we try to discuss the motivation between different choices of techniques. For this particular problem, the application of Bayesian methods seems to be useful: background knowledge can be encoded in the model considered, and this allows making more plausible inferences than, e.g., by 'blind' machine learning methods.

The application domain is paleoecological reconstruction, which aims at estimating temperatures in the past. Such reconstructions are crucial, e.g., in trying to find out what part of the global warming is due to natural variation and how significant is the human influence. In paleoecological reconstruction, knowledge about present day abundances of certain species are combined with data about the same species in fossil assemblages (e.g., lake sediments). The present day abundances and present day temperatures can be viewed as the training set, and the abundances of the taxa in sediments are the test set, for which the temperatures are not known. Stated formally, the reconstruction task thus has the form of a typical machine learning problem. However, to obtain useful predictions, a lot of background knowledge about ecological variation is needed.

In paleoecological literature the statistical methods are involved variations of regression. We compare these methods with regression trees, nearest neighbor methods, and Bayesian hierarchical models. All the methods achieve about the same prediction accuracy on modern specimens, but the Bayesian methods and the involved regression methods seem to yield the best reconstructions. The advantage of the Bayesian methods is that they also give good estimates on the variability of the reconstructions. While exploratory pattern discovery methods do not have a direct use in the actual reconstruction task, they can be used to search for unexpected dependencies between species.

Data The data we analyze in this paper consists of surface-sediment and limnological data from 53 subarctic lakes in northern Fennoscandia [6, 7]. The data contains abundances of 36 taxa and measurements of 25 environmental variables such as mean July temperature, alcalinity, or the size of the catch area. The fossil assemblage available consists of abundances of the 36 taxa in 147 different depths in the sediment of one of the lakes. The deepest layer has accumulated approximately 10 000 years ago.

2   The reconstruction problem

In this section we describe a simplified version of the paleoecological reconstruction problem and introduce the notation needed for the description of the models. We mainly use the notational conventions of [1]; a summary of notations is given in Table 1. Suppose we have n sites and m taxa, and let y_{ik} be the abundance of taxon k at site i. Let $Y = (y_{ik})$ be the $n \times m$ matrix of abundances.

Suppose that for the n sites we have measured p physico-chemical variables of the environment; denote by x_{ij} the value of variable j on site i, and by $X = (x_{ij})$ the $n \times p$ matrix of these variables. The environment at site i is thus described by the vector $x_i = (x_{i1}, x_{i2}, \ldots, x_{im})$. The problem in quantitative reconstruction is to obtain good estimates of the environmental variables in the past, such as temperature, based on fossil assemblages of abundances.

Table 1. Notations.

Symbol	Meaning
n	number of sites
m	number of taxa
p	number of environmental variables
x_{ij}	value of environmental variable j on site i
$X = (x_{ij})$	matrix of values of environmental variables
y_{ik}	abundance of taxon k at site i
$Y = (y_{ik})$	matrix of abundances
$y_0 = (y_{01}, y_{02}, \ldots, y_{0m})$	the fossil assemblage for which the environmental variables are to be inferred
$x_0 = (x_{01}, x_{02}, \ldots, x_{0p})$	values of environmental variables for the fossil assemblage; to be inferred

Definition 1 (Quantitative reconstruction problem) *Let $Y = (y_{ik})$ be the matrix of abundances and $X = (x_{ij})$ the matrix of environmental variables. Let $y_0 = (y_{01}, y_{02}, \ldots, y_{0m})$ be a fossil assemblage, and let $x_0 = (x_{01}, x_{02}, \ldots, x_{0p})$ be the set of values of environmental variables for the fossil assemblage. Given Y, X, and y_0, reconstruct the environmental variables x_0.*

In machine learning terms, the input consists of pairs (y_i, x_i) of abundances and the corresponging environment. The task is to learn to predict x_i from y_i, and to apply the learned function in reconstructing x_0 from y_0.

Problems and requirements The data is typically not consistent and does not display obvious response functions for the organisms. Figure 1 shows the observed abundances of a typical chironomid species as the function of lake temperature [6]. The large variation in abundances is due to the large number of both measured and unmeasured ecological variables.

There are two important requirements for any reconstruction method [1]. First, it is essential to obtain information about the reliability of the estimates of environmental variables. For instance, if the estimates are known to be rough, then definitive conclusions against of for global warming can hardly be drawn. It seems obvious that the estimates cannot be exact for any method, due to the vagueness of data. This makes the accuracy estimation even more important.

The other requirement is the capability to extrapolate from the training examples to unknown areas. The temperatures of the modern training set, for instance, range from 9 to 15°C with three exceptions (Figure 1), but it is not reasonable to assume that the temperatures to be reconstructed have been within this range.

This second requirement leads to a more fundamental question about how the phenemenon under consideration is described or modeled by the reconstruction method. There are two major approaches: (1) estimate (e.g., by learning or modeling) the modern response function U of each species by $\hat{U}$ such that

$$y_i \approx \hat{U}(x_i),$$

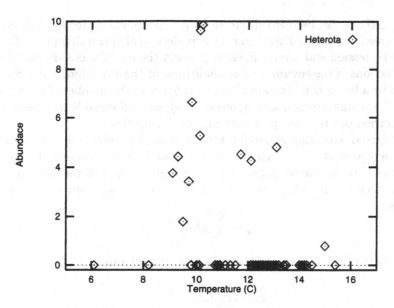

Fig. 1. Abundance of a chironomid taxon (Heterotanytarsus sp.) in 53 subarctic lakes.

and reconstruct an estimate $\hat{x}_0 = \hat{U}^{-1}(y_o)$, or (2) directly construct an estimate $\hat{U}^{-1}$ such that

$$x_i \approx \hat{U}^{-1}(y_i).$$

The latter approach is typical for most machine learning techniques. However, modelling $\hat{U}^{-1}$ directly, e.g., by a regression tree, has the weakness that there is no guarantee at all of the form of the corresponding response function $\hat{U}$; the response functions might be ecologically impossible. This means that even if the model works well for the training set and shows good accuracy in cross validation tests, its performance in extrapolation in real reconstruction might be unpredictable and intolerable.

The data set available for training is not large, only 53 lakes, but it has 36 dimensions (taxa abundances). Whatever the method, learning appropriate generalizations necessary for extrapolation can obviously be difficult. In machine learning terms this can be seen as a question of using learning bias to expedite models with plausible extrapolation.

In the following sections we consider these issues in the context of statistical state-of-the-art methods for paleoecological reconstruction, regression tree learning and nearest neighbor prediction, and finally Bayesian modelling and inference.

3   Traditional statistical methods

In this section we briefly describe some properties of the methods that are typically used for reconstructions. We refer to [1] for an in-depth survey.

The methods can be divided into three groups: linear regression and inverse linear regression, weighted averaging, and modern analogue techniques. Of these, in linear regression and inverse linear regression the abundances are modeled as linear functions of the environmental conditions, or the environmental variable is modeled by a linear combination of the abundances. The number of taxa can be decreased by using a principal components analysis, and variable transformations can be carried out to avoid problems with nonlinearities.

In weighted averaging regression the concept of a unimodal response curve of a taxon is present. The species' optimum value for the environmental variable x is assumed to be the weighted average of the x values of the sites in which the taxon occurs; as weight, the abundance of the taxon is used. That is, the estimate is

$$u_k = \frac{\sum_{i=1}^n y_{ik} x_i}{\sum_{i=1}^n y_{ik}}.$$

The estimate of the variance of the optimum, the tolerance, is

$$\frac{\sum_{i=1}^n y_{ik}(x_i - u_k)^2}{\sum_{i=1}^n y_{ik}}.$$

In this paper, we present results that have been obtained using the partial least squares addition to weighted averaging: the method is abbreviated WA-PLS [1]. The top panel of Figure 3 displays the reconstruction obtained with WA-PLS.

Several other methods have also been used in reconstructions. We mention only canonical correspondence analysis (CCA) and modern anologue methods (MAT), which actually are nearest neigbour techniques.

Remarks Space does not permit us to present a detailed critique of the existing methods, thus we only mention two points.

Many of these methods have a weakness in extrapolation: the predictions tend to regress towards the mean. This is in many cases corrected by "deshrinking", which basically just spreads the predictions farther away from the mean than they originally were.

Another problem is the rigid structure of the response curves: they are often assumed to be symmetric, and the accuracy is assumed to be the same at all locations along the temperature dimension.

4   Machine learning methods

In this section we present results with two machine learning approaches: K nearest neighbors and regression tree learning.

K nearest neighbors In instance or case-based learning the idea is not to learn any function between the input and output attributes, but rather to compare new cases directly to the ones already seen. The methods store the training examples, and base their prediction for a new case on the most similar examples recorded during the "training". As mentioned above, in paleoecology nearest neighbor methods have been used under the name "modern analogue techniques".

We implemented a simple K nearest neighbor method. Given y_0, the method finds I such that $\{y_i \mid i \in I\}$ is the set of K modern environments most similar to y_0. The method then returns the mean of the corresponding environmental variables, i.e., the mean of $x_i, i \in I$. We used Euclidian distance as the (dis)similarity measure and gave all the K neighbors the same weight.

The only parameter for this method is K, the number of nearest neighbors considered. The best cross validation standard error $1.42°C$ was obtained with $K = 6$; values of K in $[2, 8]$ yielded almost equally good standard errors, at most $1.45°C$ in all cases. The second panel from the top in Figure 3 shows the reconstruction with 6 nearest neighbors.

Regression tree learning Regression tree learning is closely related to the better known decision tree learning. In both cases, the idea is to recursively subdivide the problem space into disjoint, more homogeneous areas. Whereas decision trees predict the classification of a new case, regression trees make a numerical prediction based, e.g., on a linear formula on the input parameters [2].

We ran experiments with the publicly available demonstration version of Cubist (*http://www.rulequest.com/*). The best cross validation standard error, $1.80°C$, was obtained with parameter -*i*, directing Cubist to actually combine the regression tree technique with a nearest neighbor component [8, 5]. Without the nearest neighbor component, the best standard error of $1.90°C$ was obtained with the default settings of Cubist. This result is close to the standard deviation $1.92°C$ of the temperatures in the training set, and it is reasonable to say that in this case Cubist was not able learn from the data. The third panel from top in Figure 3 shows the reconstruction made by the regression tree learned by Cubist with option -*i*.

Remarks The nearest neighbor method is attractive as it is simple and it has good cross validation accuracy. Its weakness is that, by definition, it cannot extrapolate outside the space it has been trained with. Some estimation of the accuracy could be derived from the homogeneity of the K nearest neighbors, in addition to the cross validation standard error. The regression tree method we tested gives poor cross validation accuracy and is unreliable in extrapolation, so the reconstruction result is questionable. The reconstruction results do not match with the results obtained with the state of the art methods of paleoecological reconstruction such as WA-PLS.

5   Bayesian approach

We give a brief introduction to the Bayesian approach to data analysis; see [3, 4] for good expository treatments. We then give a statistical model for the reconstruction problem and present experimental results.

Bayesian reconstruction One of the major advantages of the Bayesian approach is that problems can often be described in natural terms. In the case of paleoecology, for instance, a Bayesian model can specify what the plausible response functions $\hat{U}$ are, and also which functions are more likely than others.

In the reconstruction problem, we apply the Bayesian approach as follows. For simplicity, consider only one environmental variable z. Assume the response of a taxon k to variable z is defined by the set $\{w_{k1}, w_{k2}, \ldots, w_{kh}\}$ of parameters, and assume that the taxa are independent. The parameters of the model thus include for each taxon k the vector $w_k = (w_{k1}, w_{k2}, \ldots, w_{kh})$. We can now, in principle, express the abundance of species k at site i as a function of the environmental variables at site i and the parameters of the species k: $y_{ik} = \hat{U}(x_i, w_k)$. We write $W = (w_1, w_2, \ldots, w_m)$.

For the general reconstruction problem, we are interested in obtaining the *posterior distribution* $Pr(x_0 \mid X, Y, y_0)$ of the unknown x_0 given the data X, Y, and y_0. This can be written as

$$Pr(x_0 \mid X, Y, y_0) = \int_W Pr(x_0, W \mid X, Y, y_0).$$

That is, the marginal probability of x_0 is obtained by integrating over all possible parameter values W for the response functions. This equation indicates an important property of the Bayesian approach: we do not use just the "best" response function for each taxon; rather, all possible response functions are considered (in principle) and their importance for the final results is determined by their likelihood and prior probability.

According to Bayes' theorem, we can write

$$Pr(x_0, W \mid X, Y, y_0) = \frac{Pr(x_0, W)Pr(Y, y_0 \mid X, x_0, W)}{Pr(Y, y_0 \mid X)}$$
$$\propto Pr(x_0, W)Pr(Y, y_0 \mid X, x_0, W).$$

Model specification In the case of one environmental variable z (such as temperature), a typical ecological response function might be a smooth unimodal function such as a bell-shaped curve. We use a simple model where the response of each taxon k is specified by such a curve

$$r(z, \alpha_k, \beta_k, \gamma_k) = \alpha \exp\left(\frac{-(z - \beta)^2}{\gamma}\right), \tag{1}$$

where $w_k = (\alpha_k, \beta_k, \gamma_k)$ are the model parameters for taxon k. Given a value of the environmental variable z, the function specifies the suitability of the environment for taxon k.

In the literature the abundances are typically represented as fractions; for ease of modeling, we assume here that they are actual counts, i.e., y_{ik} indicates how many representatives of taxon k were observed on site i. In the model we assume that the abundances y_{ik} are Poisson distributed with parameter $r(z_i, \alpha_k, \beta_k, \gamma_k)$, i.e.,

$$Pr(y_{ik} \mid z_i, \alpha_k, \beta_k, \gamma_k) = (y_{ik}!)^{-1} r(z_i, \alpha_k, \beta_k, \gamma_k)^{y_{ik}} \exp(-r(z_i, \alpha_k, \beta_k, \gamma_k)). \tag{2}$$

We also assume that given the parameters, the observed abundances of the taxa are conditionally independent, although this is not exactly true. Thus the likelihood $Pr(Y, y_0 \mid z, \bar{\alpha}, \bar{\beta}, \bar{\gamma})$ can be written in product form:

$$Pr(Y, y_0 \mid z, \bar{\alpha}, \bar{\beta}, \bar{\gamma}) = \prod_{i=0}^{n} Pr(y_i \mid z_i, \bar{\alpha}, \bar{\beta}, \bar{\gamma}).$$

For individual site i, assuming that the taxa are independent, we can write

$$Pr(y_i \mid z_i, \bar{\alpha}, \bar{\beta}, \bar{\gamma})) = \prod_{k=1}^{m} Pr(y_{ik} \mid z_i, \alpha_k, \beta_k, \gamma_k).$$

Here $Pr(y_{ik} \mid z_i, \alpha_k, \beta_k, \gamma_k)$ is the probability of observing abundance y_{ik} at site i for taxon k, whose response curve is described by the three parameters α_k, β_k, and γ_k.

To complete the model specification, we need to give the prior probabilities $Pr(x_0, W)$. For the priors of the parameters we write

$$Pr(x_0, \bar{\alpha}, \bar{\beta}, \bar{\gamma}) = Pr(x_0) \prod_{k=1}^{m} Pr(\alpha_k) Pr(\beta_k) Pr(\gamma_k),$$

i.e., the priors for the unknown temperature and the individual parameters for the k response functions are independent.

The prior for the temperature x_0 to be reconstructed is normal distribution with mean 12.1 and variance 0.37. The mean has been chosen to be the same as the mean of the modern training set. The variance is only one tenth of the variance of the modern training set, in order to rather make conservative than radical reconstructions.

For the parameters α_k, β_k, and γ_k we use the following priors: α_k has uniform distribution in $[0.1, 50]$, β_k has normal distribution with parameters 12.1 and 3.7, and γ_k is assumed to be gamma distributed with parameters 50 and 5. The prior of α_k allows any possible value with equal probability. The optimum temperatures β_k are assumed to have roughly the same distribution as the temperatures. Finally, the prior of γ_k is defined to encourage conservative, gently sloping response curves.

Computing the posterior distribution of the unknown variables in a closed form is typically not possible. Stochastic simulation techniques called Markov chain Monte Carlo (MCMC) methods, can be used to computationally approximate almost arbitrary distributions. For an overview of MCMC methods, see, e.g., [4]. The following results have been obtained with Bassist [9], a general purpose MCMC tool under development at the Department of Computer Science at the University of Helsinki.

Results Figure 2 shows the posterior mean of the response curve for the species of Figure 1. According to the model and the data (and the simulation), the

response curve is within the broken lines with 95 % probability. This demonstrates a useful aspect of Bayesian reasoning: in the colder areas on the left,
where there is not much evidence for fitting the response curve, there is also
more variation in the posterior. On the right, where it is more obvious from
Figure 1 that the temperatures are not suitable for this species, the response
curve is quite certainly quite low. This should be contrasted with the weighted
averaging methods, where the response curve is always a symmetric unimodal
curve.

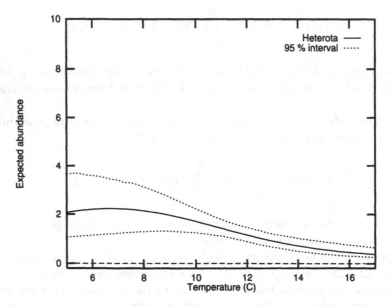

Fig. 2. A response model for the chironomid taxon of Figure 2.

Finally, the bottom panel of Figure 3 shows the reconstruction obtained with
the above model. Again, the solid line is the posterior mean, and the area between the broken lines contains 95 % of the variation. The reconstruction closely
matches the one obtained by the statistical method used by paleoecologists. The
cross validation standard error of the model is 1.49°C.

Remarks A strength of full probability models is that uncertainty is represented
naturally. Here we obtain information about the credibility of the reconstruction:
the uncertainty is carried throughout the process from the vagueness of the data
to the different probabilities of response functions, and then to the probabilities
of different reconstructions.

Bayesian modeling is, in principle, simple: what one needs to describe is (a
suitable abstraction of) the natural effects and causes. The model is essentially
defined by equations 1 and 2 plus the prior distributions. The bias necessary for
the task, that the constructed models are ecologically plausible, was easily built
into the model, in this example, e.g., as the assumption that response curves are

bell-shaped. Consequently the method can be assumed to function reasonably also for fossil data, which may require remarkable extrapolation.

On the down-side is the fact that such models are slow to fit. Running the above model for the above data until convergence takes currently several hours on a PC. This is, however, cheap compared to the man months or year spent on obtaining the data in the first place.

6   Discussion

The role of machine learning, exploratory rule finding methods, and statistical methods is one of the key issues in developing the data mining area. We have presented in a case study from paleoecology how these different methods can be applied to a relatively difficult scientific data analysis and modeling problem.

Experience with this case study shows that the specialized statistical methods for reconstructions developed within the paleoecological community have their strengths and weaknesses. They typically try to use only biologically viable assumptions about the taxa, but the treatment of nonlinearities and the approach to uncertainty in the data are relatively ad hoc.

Machine learning methods are in principle directly applicable to the reconstruction task. However, the background knowledge about how taxa respond to changes in the environmental variable, temperature, is very difficult to incorporate into the approaches. Therefore these methods produce reconstructions whose accuracy is difficult to estimate.

Exploratory data mining methods might seem to be useless for the reconstruction process. This is partly true: once the taxa have been selected, other approaches are more useful. However, in the preliminary phase of data collection, understanding the dependencies between different taxa can well be aided by looking exploratively for rules relating changes in the abundance of one taxon to changes in the abundances of other taxa.

The Bayesian approach to the reconstruction problem gives a natural way of embedding the biological background knowledge into the model. The conceptual structure of the full probability model can actually be obtained directly from the structure of the data: the structure of the probability model is very similar to the ER-diagram of the data. It seems that this correspondence holds for a fairly large class of examples: stripping details of distributions from a bayesian models gives a description of the data items and their relationships, i.e., an ER-diagram. This is one of the reasons why Bayesian hierarchical models seem so well suited for data mining tasks.

In the Bayesian approach, estimating the uncertainty in the reconstruction is also simple, following immediately from the basic principles of the approach. The problem with the Bayesian approach is that it is relatively computer-intensive: the MCMC simulations can take a fair amount of time.

In Table 2 we show the cross validation prediction accuracies of the different approaches. As can be seen, simple nearest neighbor methods achieve the best accuracy, closely followed by the Bayesian method.

Table 2. Summary of cross validation prediction accuracies of different approaches. Results for the methods WA — MAT are from [6].For brevity, these methods are not described in detail in the text; see [1].

Method	Cross validation accuracy (°C)
WA (inverse)	1.56
WAtol (inverse)	1.76
WA (classical)	2.06
WAtol (classical)	2.49
PLS (1 component)	1.57
WA-PLS (1 component)	1.53
GLM	2.16
MAT (6 neighbors)	1.48
6 nearest neighbors	1.42
regression tree + NN	1.80
regression tree	1.92
a Bayesian model	1.49

Prediction accuracy is not by itself the goal. Comparing the reconstructions in Figure 3, we note that the statistical state of the art method WA-PLS and the Bayesian method produce similar results: there is a slight increase in temperature over the last 5000 years, with sudden short episodes during which the temperature falls. This fits rather well with the results obtained from other datasets. The reconstruction produced by Cubist has no trend at all, and the variability in the nearest neighbor result is quite large.

It is somewhat surprising that a reasonably simple Bayesian model can produce results similar to or better than those obtained by using statistical approaches tailored to the reconstruction problem. One explanation for this is that the biological background knowledge is naturally represented. The other is the somewhat nonparametric nature of the method, allowing in practice many different forms of posterior response curves.

Preliminary experiments indicate also that small perturbations in the data do not change the Bayesian results, whereas such changes can have a drastic effect on the results of other methods.

References

1. H. J. B. Birks. Quantitative palaeoenvironmental reconstructions. In D. Maddy and J. Brew, editors, *Statistical modelling quaternary science data*, pages 161 – 254. Quaternary Research Association, Cambridge, 1995.
2. L. Breiman, J. Friedman, R. Olshen, and C. Stone. *Classification and Regression Trees*. Wadsworth International Group, Belmont, CA, 1984.
3. A. Gelman, J. B. Carlin, H. S. Stern, and D. B. Rubin. *Bayesian Data Analysis*. Chapman & Hall, New York, 1995.
4. W. R. Gilks, S. Richardson, and D. J. Spiegelhalter. *Markov Chain Monte Carlo in Practice*. Chapman & Hall, London, 1996.

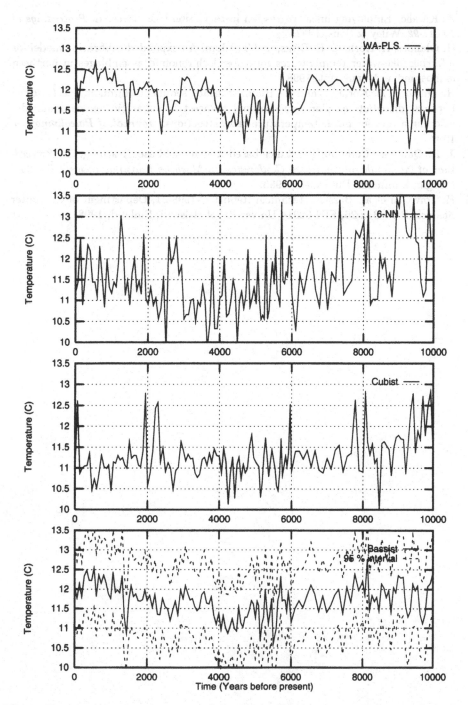

Fig. 3. Temperature reconstructions from top to bottom: statistical state of the art method WA-PLS, 6 nearest neighbors, regression tree (Cubist), and a Bayesian model (Bassist).

5. A. Karalic. Employing linear regression in regression tree leaves. In *Proceedings of ECAI-92*. Wiley & Sons, 1992.
6. H. Olander, A. Korhola, H. Birks, and T. Blom. An expanded calibration model for inferring lake-water temperatures form fossil chironomid assemblages in northern fennoscandia. Manuscript, 1998.
7. H. Olander, A. Korhola, and T. Blom. Surface sediment chironomidae (insecta: Diptera) distributions along an ecotonal transect in subarctic fennoscandia: developing a tool for palaeotemperature reconstructions. *Journal of Paleolimnology*, 1997.
8. J. R. Quinlan. Combining instance-based and model-based learning. In *Proceedings of the Tenth International Conference on Machine Learning*, pages 236 – 243. Morgan Kaufmann Publishers, 1993.
9. H. Toivonen et al. Bassist. Technical Report C-1998-31, Department of Computer Science, P.O. Box 26, FIN-00014 University of Helsinki, Finland, 1998.

The Computer-Aided Discovery of Scientific Knowledge

Pat Langley

Adaptive Systems Group
Daimler-Benz Research and Technology Center
1510 Page Mill Road, Palo Alto, CA 94304 USA
LANGLEY@RTNA.DAIMLERBENZ.COM

Abstract. In this paper, we review AI research on computational discovery and its recent application to the discovery of new scientific knowledge. We characterize five historical stages of the scientific discovery process, which we use as an organizational framework in describing applications. We also identify five distinct steps during which developers or users can influence the behavior of a computational discovery system. Rather than criticizing such intervention, as done in the past, we recommend it as the preferred approach to using discovery software. As evidence for the advantages of such human-computer cooperation, we report seven examples of novel, computer-aided discoveries that have appeared in the scientific literature, along with the role that humans played in each case. We close by recommending that future systems provide more explicit support for human intervention in the discovery process.

1 Introduction

The process of scientific discovery has long been viewed as the pinnacle of creative thought. Thus, to many people, including some scientists themselves, it seems an unlikely candidate for automation by computer. However, over the past two decades, researchers in artificial intelligence have repeatedly questioned this attitude and attempted to develop intelligent artifacts that replicate the act of discovery. The computational study of scientific discovery has made important strides in its short history, some of which we review in this paper.

Artificial intelligence often gets its initial ideas from observing human behavior and attempting to model these activities. Computational scientific discovery is no exception, as early research focused on replicating discoveries from the history of disciplines as diverse as mathematics (Lenat, 1977), physics (Langley, 1981), chemistry (Żytkow & Simon, 1986), and biology (Kulkarni & Simon, 1990). As the collection by Shrager and Langley (1990) reveals, these efforts also had considerable breadth in the range of scientific activities they attempted to model, though most work aimed to replicate the historical record only at the most abstract level. Despite the explicit goals of this early research, some critics (e.g., Gillies, 1996) have questioned progress in the area because it dealt with scientific laws and theories already known to the developers.

Although many researchers have continued their attempts to reproduce historical discoveries, others have turned their energies toward the computational discovery of new scientific knowledge. As with the historical research, this applied work covers a broad range of disciplines, including mathematics, astronomy, metallurgy, physical chemistry, biochemistry, medicine, and ecology. Many of these efforts have led to refereed publications in the relevant scientific literature, which seems a convincing measure of their accomplishment.

Our aim here is to examine some recent applications of computational scientific discovery and to analyze the reasons for their success. We set the background by reviewing the major forms that discovery takes in scientific domains, giving a framework to organize the later discussion. After this, we consider steps in the larger discovery process at which humans can influence the behavior of a computational discovery system. We then turn to seven examples of computer-aided discoveries that have produced scientific publications, in each case considering the role played by the developer or user. In closing, we consider directions for future work, emphasizing the need for discovery aids that explicitly encourage interaction with humans.

2   Stages of the Discovery Process

The history of science reveals a variety of distinct types of discovery activity, ranging from the detection of empirical regularities to the formation of deeper theoretical accounts. Generally speaking, these activities tend to occur in a given order within a field, in that the products of one process influence or constrain the behavior of successors. Of course, science is not a strictly linear process, so that earlier stages may be revisited in the light of results from a later stage, but the logical relation provides a convenient framework for discussion.

Perhaps the earliest discovery activity involves the formation of *taxonomies*. Before one can formulate laws or theories, one must first establish the basic concepts or categories one hopes to relate. An example comes from the early history of chemistry, when scientists agreed to classify some chemicals as acids, some as alkalis, and still others as salts based on observable properties like taste. Similar groupings have emerged in other fields like astronomy and physics, but the best known taxonomies come from biology, which groups living entities into categories and subcategories in a hierarchical manner.

Once they have identified a set of entities, scientists can begin to discover *qualitative laws* that characterize their behavior or that relate them to each other. For example, early chemists found that acids tended to react with alkalis to form salts, along with similar connections among other classes of chemicals. Some qualitative laws describe static relations, whereas others summarize events like reactions that happen over time. Again, this process can occur only after a field has settled on the basic classes of entities under consideration.

A third scientific activity aims to discover *quantitative laws* that state mathematical relations among numeric variables. For instance, early chemists identified the relative masses of hydrochloric acid and sodium hydrochloride that combine to form a unit mass of sodium chloride. This process can also involve postulating the existence of an *intrinsic property* like density or specific heat, as well as estimating the property's value for specific entities. Such numeric laws are typically stated in the context of some qualitative relationship that places constraints on their operation.

Scientists in most fields are not content with empirical summaries and so try to explain such regularities, with the most typical first step involving the creation of *structural models* that incorporate unobserved entities. Thus, nineteenth century chemists like Dalton and Avogadro postulated atomic and molecular models of chemicals to account for the numeric proportions observed in reactions. Initial models of this sort are typically qualitative in nature, stating only the components and their generic relations, but later models often incorporate numeric descriptions that provide further constraints. Both types of models are closely tied to the empirical phenomena they are designed to explain.

Eventually, most scientific disciplines move beyond structural models to *process models*, which explain phenomena in terms of hypothesized mechanisms that involve change over time. One well-known process account is the kinetic theory of gases, which explains the empirical relations among gas volume, pressure, and temperature in terms of interactions among molecules. Again, some process models (like those in geology) are mainly qualitative, while others (like the kinetic theory) include numeric components, but both types make contact with empirical laws that one can derive from them.

In the past two decades, research in automated scientific discovery has addressed each of these five stages. Clustering systems like CLUSTER/2 (Michalski & Stepp, 1983), AUTOCLASS (Cheeseman et al., 1988), and others deal with the task of taxonomy formation, whereas systems like NGLAUBER (Jones, 1986) search for qualitative relations. Starting with BACON (Langley, 1981; Langley, Simon, Bradshaw, & Żytkow, 1987), researchers have developed a great variety of systems that discover numeric laws. Systems like DALTON (Langley et al., 1987), STAHLP (Rose & Langley, 1987), and GELL-MANN (Żytkow, 1996) formulate structural models, whereas a smaller group, like MECHEM (Valdés-Pérez, 1995) and ASTRA (Kocabas & Langley, 1998), instead construct process models.

A few systems, such as Lenat's (1977) AM, Nordhausen and Langley's IDS (1993), and Kulkarni and Simon's (1990) KEKADA, deal with more than one of these facets, but most contributions have focused on one stage to the exclusion of others. Although the work to date has emphasized rediscovering laws and models from the history of science, we will see that a similar bias holds for efforts at finding new scientific knowledge. We suspect that integrated discovery applications will be developed, but only once the more focused efforts that already exist have become more widely known.

This framework is not the only way to categorize scientific activity, but it appears to have general applicability across different fields, so we will use it to organize our presentation of applied discovery work. The scheme does favor methods that generate the types of formalisms reported in the scientific literature, and thus downplays the role of mainstream techniques from machine learning. For example, decision-tree induction, neural networks, and nearest neighbor have produced quite accurate predictors in scientific domains like molecular biology (Hunter, 1993), but they employ quite different notations from those used normally to characterize scientific laws and models. For this reason, we will not focus on their application to scientific problems here.

3   The Developer's Role in Computational Discovery

Although the term *computational discovery* suggests an automated process, close inspection of the literature reveals that the human developer or user plays an important role in any successful project. Early computational research on scientific discovery downplayed this fact and emphasized the automation aspect, in general keeping with the goals of artificial intelligence at the time. However, the new climate in AI favors systems that advise humans rather than replace them, and recent analyses of machine learning applications (e.g., Langley & Simon, 1995) suggest an important role for the developer. Such analyses carry over directly to discovery in scientific domains, and here we review the major ways in which developers can influence the behavior of discovery systems.

The first step in using computational discovery methods is to formulate the discovery problem in terms that can be solved using existing techniques. The developer must first cast the task as one that involves forming taxonomies, finding qualitative laws, detecting numeric relations, forming structural models, or constructing process accounts. For most methods, he must also specify the dependent variables that laws should predict or indicate the phenomena that models should explain. Informed and careful *problem formulation* can greatly increase the chances of a successful discovery application.

The second step in applying discovery techniques is to settle on an effective representation.[1] The developer must state the variables or predicates used to describe the data or phenomena to be explained, along with the output representation used for taxonomies, laws, or models. The latter must include the operations allowed when combining variables into laws and the component structures or processes used in explanatory models. The developer may also need to encode background knowledge about the domain in terms of an initial theory or results from earlier stages of the discovery process. Such *representational engineering* plays an essential role in successful applications of computational discovery.

Another important developer activity concerns preparing the data or phenomena on which the discovery system will operate. Data collected by scientists may be quite sparse, lack certain values, be very noisy, or include outliers, and the system user can improve the quality of these data manually or using techniques for interpolation, inference, or smoothing. Similarly, scientists' statements of empirical phenomena may omit hidden assumptions that the user can make explicit or include irrelevant statements that he can remove. Such *data manipulation* can also improve the results obtained through computational discovery.

---

[1] We are not referring here to the representational formalism, such as decision trees or neural networks, but rather to the domain features encoded in a formalism.

Research papers on machine discovery typically give the algorithm center stage, but they pay little attention to the developer's efforts to modulate the algorithm's behavior for given inputs. This can involve activities like the manual setting of system parameters (e.g., for evidence thresholds, noise tolerance, and halting criteria) and the interactive control of heuristic search by rejecting bad candidates or attending to good ones. Some systems are designed with this interaction in mind, whereas others support the process more surreptitiously. But in either case, such *algorithm manipulation* is another important way that developers and users can improve their chances for successful discoveries.

A final step in the application process involves transforming the discovery system's output into results that are meaningful to the scientific community. This stage can include manual filtering of interesting results from the overall output, recasting these results in comprehensible terms or notations, and interpreting the relevance of these results for the scientific field. Thus, such *postprocessing* subsumes both the human user's evaluation of scientific results and their communication to scientists who will find them interesting. Since evaluation and communication are central activities in science, they play a crucial role in computational discovery as well.

The literature on computational scientific discovery reveals, though often between the lines, that developers' intervention plays an important role even in historical models of discovery. Indeed, early critiques of machine discovery research frowned on these activities, since both developers and critics assumed the aim was to completely automate the discovery process. However, this view has changed in recent years, and the more common perspective, at least in applied circles, is that discovery systems should aid scientists rather than replace them. In this light, human intervention is perfectly acceptable, especially if the goal is to discover new scientific knowledge and not to assign credit.

4   Some Computer-Aided Scientific Discoveries

Now that we have set the stage, we are ready to report some successful applications of AI methods to the discovery of new scientific knowledge. We organize the presentation in terms of the basic scientific activities described earlier, starting with examples of taxonomy formation, then moving on to law discovery and finally to model construction. In each case, we review the basic scientific problem, describe the discovery system, and present the novel discovery that it has produced. We also examine the role that the developer played in each application, drawing on the five steps outlined in the previous section.

Although we have not attempted to be exhaustive, we did select examples that meet certain criteria. Valdés-Pérez (1998) suggests that scientific discovery involves the "generation of novel, interesting, plausible, and intelligible knowledge about objects of scientific study", and reviews four computer-aided discoveries that he argues meet this definition. Rather than repeating his analysis, we have chosen instead to use publication of the result in the relevant scientific literature as our main criterion for success, though we suspect that refereed publication is highly correlated with his factors.

4.1 Stellar Taxonomies from Infrared Spectra

Existing taxonomies of stars are based primarily on characteristics from the visible spectrum. However, artificial satellites provide an opportunity to make measurements of types that are not possible from the Earth's surface, and the resulting data could suggest new groupings of known stellar objects. One such source of new data is the Infrared Astronomical Satellite, which has produced a database describing the intensity of some 5425 stars at 94 wavelengths in the infrared spectrum.

Cheeseman et al. (1988) applied their AUTOCLASS system to these infrared data. They designed this program to form one-level taxonomies, that is, to group objects into meaningful classes or clusters based on similar attribute values. For this domain, they chose to represent each cluster in terms of a mean and variance for each attribute, thus specifying a Gaussian distribution. The system carries out a gradient descent search through the space of such descriptions, starting with random initial descriptions for a specified number of clusters. On each step, the search process uses the current descriptions to probabilistically assign each training object to each class, and then uses the observed values for each object to update class descriptions, repeating this process until only minor changes occur. At a higher level, AUTOCLASS iterates through different numbers of clusters to determine the best taxonomy, starting with a user-specified number of classes and increasing this count until it produces classes with negligible probabilities.

Application of AUTOCLASS to the infrared data on stars produced 77 stellar classes, which the developers organized into nine higher-level clusters by running the system on the cluster descriptions themselves. The resulting taxonomy differed significantly from the one then used in astronomy, and the collaborating astronomers felt that it reflected some important results. These included a new class of blackbody stars with significant infrared excess, presumably due to surrounding dust, and a very weak spectral 'bump' at 13 microns in some classes that was undetectable in individual spectra. Goebel et al. (1989) recount these and other discoveries, along with their physical interpretation; thus, the results were deemed important enough to justify their publication in an refereed astrophysical journal.

Although AUTOCLASS clearly contributed greatly to these discoveries, the developers acknowledge that they also played an important role (Cheeseman & Stutz, 1996). Casting the basic problem in terms of clustering was straightforward, but the team quickly encountered problems with the basic infrared spectra, which had been normalized to ensure that all had the same peak height. To obtain reasonable results, they renormalized the data so that all curves had the same area. They also had to correct for some negative spectral intensities, which earlier software used by the astronomers had caused by subtracting out a background value. The developers' decision to run AUTOCLASS on its own output to produce a two-level taxonomy constituted another intervention. Finally, the collaborating astronomers did considerable interpretation of the system outputs before presenting them to the scientific community.

4.2 Qualitative Factors in Carcinogenesis

Over 80,000 chemicals are available commercially, yet the long-term health effects are known for only about 15 percent of them. Even fewer definitive results are available about whether chemicals cause cancer, since the standard tests for carcinogens involve two-year animal bioassays that cost \$2 million per chemical. As a result, there is great demand for predictive laws that would let one predict carcinogenicity from more rapid and less expensive measurements.

Lee, Buchanan, and Aronis (1998) have applied the rule-induction system RL to the problem of discovering such qualitative laws. The program constructs a set of conjunctive rules, each of which states the conditions under which some result occurs. Like many other rule-induction methods, RL invokes a general-to-specific search to generate each rule, selecting conditions to add that increase the rule's ability to discriminate among classes and halting when there is no improvement in accuracy. The system also lets the user bias this search by specifying desirable properties of the learned rules.

The developers ran RL on three databases for which carcinogenicity results were available, including 301, 108, and 1300 chemical compounds, respectively. Chemicals were described in terms of physical properties, structural features, short-term effects, and values on potency measures produced by another system. Experiments revealed that the induced rules were substantially more accurate than existing prediction schemes, which justified publication in the scientific literature (Lee et al., 1996). They also tested the rules' ability to classify 24 new chemicals for which the status was unknown at development time; these results were also positive and led to another scientific publication (Lee et al., 1995).

The authors recount a number of ways in which they intervened in the discovery process to obtain these results. For example, they reduced the 496 attributes for one database to only 75 features by grouping values about lesions on various organs. The developers also constrained the induction process by specifying that RL should favor some attributes over others when constructing rules and telling it to consider only certain values of a symbolic attribute for a given class, as well as certain types of tests on numeric attributes. These constraints, which they developed through interaction with domain scientists, took precedence over accuracy-oriented measures in deciding what conditions to select, and it seems likely that they helped account for the effort's success.

4.3 Quantitative Laws of Metallic Behavior

A central process in the manufacture of iron and steel involves the removal of impurities from molten slag. Qualitatively, the chemical reactions that are responsible this removal process increase in effectiveness when the slag contains more free oxide (O^{2-}) ions. However, metallurgists have only imperfect quantitative laws that relate the oxide amount, known as the *basicity* of the slag, to dependent variables of interest, such as the slag's sulfur capacity. Moreover, basicity cannot always be measured accurately, so there remains a need for improved ways to estimate this intrinsic property.

Mitchell, Sleeman, Duffy, Ingram, and Young (1997) applied computational discovery techniques to these scientific problems. Their DAVICCAND system includes operations for selecting pairs of numeric variables to relate, specifying qualitative conditions that focus attention on some of the data, and finding numeric laws that relate variables within a given region. The program also includes mechanisms for identifying outliers that violate these numeric laws and for using the laws to infer the values of intrinsic properties when one cannot measure them more directly.

The developers report two new discoveries in which DAVICCAND played a central role. The first involves the quantitative relation between basicity and sulfur capacity. Previous accounts modeled this relation using a single polynomial that held across all temperature ranges. The new results involve three simpler, linear laws that relate these two variables under different temperature ranges. The second contribution concerns improved estimates for the basicity of slags that contain TiO_2 and FeO, which DAVICCAND inferred using the numeric laws it induced from data, and the conclusion that FeO has quite different basicity values for sulphur and phosphorus slags. These results were deemed important enough to appear in a respected metallurgical journal (Mitchell et al., 1997).

Unlike most discovery systems, DAVICCAND encourages users to take part in the search process and provides explicit control points where they can influence choices. Thus, they formulate the problem by specifying what dependent variable the laws should predict and what region of the space to consider. Users also affect representational choices by selecting what independent variables to use when looking for numeric laws, and they can manipulate the data by selecting what points to treat as outliers. DAVICCAND presents its results in terms of graphical displays and functional forms that are familiar to metallurgists, and, given the user's role in the discovery process, there remains little need for postprocessing to filter results.

4.4 Quantitative Conjectures in Graph Theory

A recurring theme in graph theory involves proving theorems about relations among quantitative properties of graphs. However, before a mathematician can prove that such a relation always holds, someone must first formulate it as a conjecture. Although mathematical publications tend to emphasize proofs of theorems, the process of finding interesting conjectures is equally important and has much in common with discovery in the natural sciences.

Fajtlowicz (1988) and colleagues have developed GRAFFITI, a system that generates conjectures in graph theory and other areas of discrete mathematics. The system carries out search through a space of quantitative relations like $\sum x_i \geq \sum y_i$, where each x_i and y_i is some numerical feature of a graph (e.g., its diameter or its largest eigenvalue), the product of such elementary features, or their ratio. GRAFFITI ensures that its conjectures are novel by maintaining a record of previous hypotheses, and filters many uninteresting conjectures by noting that they seem to be implied by earlier, more general, candidates.

GRAFFITI has generated hundreds of novel conjectures in graph theory, many of which have spurred mathematicians in the area to attempt their proof or refutation. In one case, the conjecture that the 'average distance' of a graph is no greater than its 'independence number' resulted in a proof that appeared in the refereed mathematical literature (Chung, 1988). Although GRAFFITI was designed as an automated discovery system, its developers have clearly constrained its behavior by specifying the primitive graph features and the types of relations it should consider. Data manipulation occurs through a file that contains qualitatively different graphs, against which the system tests its conjectures empirically, and postprocessing occurs when mathematicians filter the system output for interesting results.

4.5 Temporal Laws of Ecological Behavior

One major concern in ecology is the effect of pollution on the plant and animal populations. Ecologists regularly develop quantitative models that are stated as sets of differential equations. Each such equation describes changes in one variable (its derivative) as a function of other variables, typically ones that can be directly observed. For example, Lake Glumsoe is a shallow lake in Denmark with high concentrations of nitrogen and phosphorus from waste water, and ecologists would like to model the effect of these variables on the concentration of phytoplankton and zooplankton in the lake.

Todorovski, Džeroski, and Kompare (in press) applied techniques for numeric discovery to this problem. Their LAGRAMGE system carries out search through a space of differential equations, looking for the equation set that gives the smallest error on the observed data. The system uses two constraints to make this search process tractable. First, LAGRAMGE incorporates background knowledge about the domain in the form of a context-free grammar that it uses to generate plausible equations. Second, it places a limit on the allowed depth of the derivations used to produce equations. For each candidate set of equations, the system uses numerical integration to estimate the error and thus the quality of the proposed model.

The developers report a new set of equations, discovered by LAGRAMGE, that model accurately the relation between the pollution and plankton concentrations in Lake Glumsoe. This revealed that phosphorus and temperature are the limiting factors on the growth of phytoplankton in the lake. We can infer Todorovski et al.'s role in the discovery process from their paper. They formulated the problem in terms of the variables to be predicted, and they engineered the representation both by specifying the predictive variables and by providing the grammar used to generate candidate equations. Because the data were sparse (from only 14 time points over two months), they convinced three experts to draw curves that filled in the gaps, used splines to smooth these curves, and sampled from these ten times per day. They also manipulated LAGRAMGE by telling it to consider derivations that were no more than four levels deep. However, little postprocessing or interpretation was needed, since the system produces output in a form familiar to ecologists.

4.6   Chemical Structures of Mutagens

Another area of biochemistry with important social implications aims to understand the factors that determine whether a chemical will cause mutations in genetic material. One data set that contains results of this sort involves 230 aromatic and heteroaromatic nitro compounds, which can be divided into 138 chemicals that have high mutagenicity and 92 chemicals that are low on this dimension. Structural models that characterize these two classes could prove useful in predicting whether new compounds pose a danger of causing mutation.

King, Muggleton, Srinivasan, and Sternberg (1996) report an application of their PROGOL system to this problem. The program operates along lines similar to other rule-induction methods, in that it carries out a general-to-specific search for a conjunctive rule that covers some of the data, then repeats this process to find additional rules that cover the rest. The system also lets the user specify background knowledge, stated in the same form, which it takes into account in measuring the quality of induced rules. Unlike most rule-induction techniques, PROGOL assumes a predicate logic formalism that can represent relations among objects, rather than just attribute values.

This support for relational descriptions led to revealing structural descriptions of mutation factors. For example, for the data set mentioned above, the system found one rule predicting that a compound is mutagenic if it has "a highly aliphatic carbon atom attached by a single bond to a carbon atom that is in a six-membered aromatic ring". Combined with four similar rules, this characterization gave 81% correct predictions, which is comparable to the accuracy of other computational methods. However, alternative techniques do not produce a structural model that one can use to visualize spatial relations and thus to posit the deeper causes of mutation,[2] so that the results justified publication in the chemistry literature (King et al., 1996).

As in other applications, the developers aided the discovery process in a number of ways. They chose to formulate the task in terms of finding a classifier that labels chemicals as causing mutation or not, rather than predicting levels of mutagenicity. King et al. also presented their system with background knowledge about methyl and nitro groups, the length and connectivity of rings, and other concepts. In addition, they manipulated the data by dividing into two groups with different characteristics, as done earlier by others working in the area. Although the induced rules were understandable in that they made clear contact with chemical concepts, the authors aided their interpretation by presenting graphical depictions of their structural claims. Similar interventions have been used by the developers on related scientific problems, including prediction of carcinogenicity (King & Srinivasan, 1996) and pharmacophore discovery (Finn, Muggleton, Page, & Srinivasan, 1998).

---

[2] This task does not actually involve structural modeling in the sense discussed in Section 2, since the structures are generalizations from observed data rather than combinations of unobserved entities posited to explain phenomena. However, applications of such structural modeling do not appear in the literature, and the King et al. work seems the closest approximation.

4.7 Reaction Pathways in Catalytic Chemistry

For a century, chemists have known that many reactions involve, not a single step, but rather a sequence of primitive interactions. Thus, a recurring problem has been to formulate the sequence of steps, known as the *reaction pathway*, for a given chemical reaction. In addition to the reactants and products of the reaction, this inference may also be constrained by information about intermediate products, concentrations over time, relative quantities, and many other factors. Even so, the great number of possible pathways makes it possible that scientists will overlook some viable alternatives, so there exists a need for computational assistance on this task.

Valdés-Pérez (1995) developed MECHEM with this end in mind. The system accepts as input the reactants and products for a chemical reaction, along with other experimental evidence and considerable background knowledge about the domain of catalytic chemistry. MECHEM lets the user specify interactively which of these constraints to incorporate when generating pathways, giving him control over its global behavior. The system carries out a search through the space of reaction pathways, generating the elementary steps from scratch using special graph algorithms. Search always proceeds from simpler pathways (fewer substances and steps) to more complex ones. MECHEM uses its constraints to eliminate pathways that are not viable and also to identify any intermediate products it hypothesizes in the process. The final output is a comprehensive set of simplest pathways that explain the evidence and that are consistent with the background knowledge.

This approach has produced a number of novel reaction pathways that have appeared in the chemical literature. For example, Valdés-Pérez (1994) reports a new explanation for the catalytic reaction *ethane* $+ H_2 \rightarrow 2$ *methane*, which chemists had viewed as largely solved, whereas Zeigarnik et al. (1997) present another novel result on acrylic acid. Bruk et al. (1998) describe a third application of MECHEM that produced 41 novel pathways, which prompted experimental studies that reduced this to a small set consistent with the new data. The human's role in this process is explicit, with users formulating the problem through stating the reaction of interest and manipulating the algorithm's behavior by invoking domain constraints. Because MECHEM produces pathways in a notation familiar to chemists, its outputs require little interpretation.

4.8 Other Computational Aids for Scientific Research

We have focused on the examples above because they cover a broad range of scientific problems and illustrate the importance of human interaction with the discovery system, but they do not exhaust the list of successful applications. For example, Pericliev and Valdés-Pérez (in press) have used their KINSHIP program to generate minimal sets of features that distinguish kinship terms, like *son* and *uncle*, given genealogical and matrimonial relations that hold for each. They have applied their system to characterize kinship terms in both English and

Bulgarian, and the results have found acceptance in anthropological linguistics because they are stated in that field's conventional notation.

There has also been extensive work in molecular biology, where one major goal is to predict the qualitative structure of proteins from their nucleotide sequence, as Fayyad, Haussler, and Stolorz (1996) briefly review. This work has led to many publications in the biology and biochemistry literature, but we have chosen not to focus on it here. One reason is that most studies emphasize predictive accuracy, with low priority given to expressing the predictors in some common scientific notation. More important, many researchers have become concerned less with discovering new knowledge than with showing that their predictors give slight improvements in accuracy over other methods. A similar trend has occurred in work on learning structure-activity relations in biochemistry, and we prefer not to label such efforts as computational scientific discovery.

We also distinguish computer-aided scientific discovery from the equally challenging, but quite different, use of machine learning to aid scientific data analysis. Fayyad et al. (1996) review some impressive examples of the latter approach in astronomy (classifying stars and galaxies in sky photographs), planetology (recognizing volcanoes on Venus), and molecular biology (detecting genes in DNA sequences). But these efforts invoke induction primarily to automate tedious recognition tasks in support of cataloging and statistical analysis, rather than to discover new knowledge that holds scientific interest in its own right. Thus, we have not included them in our examples of computer-aided discovery.

5   Progress and Prospects

As the above examples show, work in computational scientific discovery no longer focuses solely on historical models, but also contributes novel knowledge to a range of scientific disciplines. To date, such applications remain the exception rather than the rule, but the breadth of successful computer-aided discoveries provides convincing evidence that these methods have great potential for aiding the scientific process. The clear influence of humans in each of these applications does not diminish the equally important contribution of the discovery system; each has a role to play in a complex and challenging endeavor.

One recurring theme in applied discovery work has been the difficulty in finding collaborators from the relevant scientific field. Presumably, many scientists are satisfied with their existing methods and see little advantage to moving beyond the statistical aids they currently use. This attitude seems less common in fields like molecular biology, which have taken the computational metaphor to heart, but often there are social obstacles to overcome. The obvious response is to emphasize that we do not intend our computational tools to replace scientists but rather to aid them, just as simpler software already aids them in carrying out statistical analyses.

However, making this argument convincing will require some changes in our systems to better reflect the position. As noted, existing discovery software already supports intervention by humans in a variety of ways, from initial problem formulation to final interpretation. But in most cases this activity happens in spite of the software design rather than because the developer intended it. If we want to encourage synergy between human and artificial scientists, then we must modify our discovery systems to support their interaction more directly. This means we must install interfaces with explicit hooks that let users state or revise their problem formulation and representational choices, manipulate the data and system parameters, and recast outputs in understandable terms. The MECHEM and DAVICCAND systems already include such facilities and thus constitute good role models, but we need more efforts along these lines.

Naturally, explicit inclusion of users in the computational discovery process raises a host of issues that are absent from the autonomous approach. These include questions about which decisions should be automated and which placed under human control, the granularity at which interaction should occur, and the type of interface that is best suited to a particular scientific domain. The discipline of human-computer interaction regularly addresses such matters, and though its lessons and design criteria have not yet been applied to computer-aided discovery, many of them should carry over directly from other domains. Interactive discovery systems also pose challenges in evaluation, since human variability makes experimentation more difficult than for autonomous systems. Yet experimental studies are not essential if one's main goal is to develop computational tools that aid users in discovering new scientific knowledge.

Clearly, we are only beginning to develop effective ways to combine the strengths of human cognition with those of computational discovery systems. But even our initial efforts have produced some convincing examples of computer-aided discovery that have led to publications in the scientific literature. We predict that, as more developers realize the need to provide explicit support for human intervention, we will see even more productive systems and even more impressive discoveries that advance the state of scientific knowledge.

References

1. Bruk, L. G., Gorodskii, S. N., Zeigarnik, A. V., Valdés-Pérez, R. E., & Temkin, O. N. (1998). Oxidative carbonylation of phenylacetylene catalyzed by Pd(II) and Cu(I): Experimental tests of forty-one computer-generated mechanistic hypotheses. *Journal of Molecular Catalysis A: Chemical, 130*, 29–40.
2. Cheeseman, P., Freeman, D., Kelly, J., Self, M., Stutz, J., & Taylor, W. (1988). AUTOCLASS: A Bayesian classificiation system. *Proceedings of the Fifth International Conference on Machine Learning* (pp. 54–64). Ann Arbor, MI: Morgan Kaufmann.
3. Cheeseman, P., Goebel, J., Self, M., Stutz, M., Volk, K., Taylor, W., & Walker, H. (1989). *Automatic classification of the spectra from the infrared astronomical satellite (IRAS)* (Reference Publication 1217). Washington, DC: National Aeronautics and Space Administration.
4. Cheeseman, P., & Stutz, J. (1996). Bayesian classification (AUTOCLASS): Theory and results. In U. M. Fayyad, G. Piatetsky-Shapiro, P. Smyth, & R. Uthurusamy (Eds.), *Advances in knowledge discovery and data mining*. Cambridge, MA: MIT Press.

5. Chung, F. (1988). The average distance is not more than the independence number. *Journal of Graph Theory*, *12*, 229–235.
6. Fajtlowicz, S. (1988). On conjectures of GRAFFITI. *Discrete Mathematics*, *72*, 113–118.
7. Fayyad, U., Haussler, D., & Stolorz, P. (1996). KDD for science data analysis: Issues and examples. *Proceedings of the Second International Conference of Knowledge Discovery and Data Mining* (pp. 50–56). Portland, OR: AAAI Press.
8. Finn, P., Muggleton, S., Page, D., & Srinivasan, A. (1998). Pharmacophore discovery using the inductive logic programming system PROGOL. *Machine Learning*, *30*, 241-270.
9. Gillies, D. (1996). *Artificial intelligence and scientific method*. Oxford: Oxford Univerity Press.
10. Goebel, J., Volk, K., Walker, H., Gerbault, F., Cheeseman, P., Self, M., Stutz, J., & Taylor, W. (1989). A Bayesian classification of the IRAS LRS Atlas. *Astronomy and Astrophysics*, *222*, L5–L8.
11. Hunter, L. (1993). (Ed.). *Artificial intelligence and molecular biology*. Cambridge, MA: MIT Press.
12. Jones, R. (1986). Generating predictions to aid the scientific discovery process. *Proceedings of the Fifth National Conference on Artificial Intelligence* (pp. 513–517). Philadelphia: Morgan Kaufmann.
13. King, R. D., Muggleton, S. H., Srinivasan, A., & Sternberg, M. E. J. (1996). Structure-activity relationships derived by machine learning: The use of atoms and their bond connectives to predict mutagenicity by inductive logic programming. *Proceedings of the National Academy of Sciences*, *93*, 438–442.
14. King, R. D., & Srinivasan, A. (1996). Prediction of rodent carcinogenicity bioassays from molecular structure using inductive logic programming. *Environmental Health Perspectives*, *104* (Supplement 5), 1031–1040.
15. Kocabas, S. (1991). Conflict resolution as discovery in particle physics. *Machine Learning*, *6*, 277–309.
16. Kocabas, S., & Langley, P. (in press). Generating process explanations in nuclear astrophysics. *Proceedings of the ECAI-98 Workshop on Machine Discovery*. Brighton, England.
17. Kulkarni, D., & Simon, H. A. (1990). Experimentation in machine discovery. In J. Shrager & P. Langley (Eds.), *Computational models of scientific discovery and theory formation*. San Mateo, CA: Morgan Kaufmann.
18. Langley, P. (1981). Data-driven discovery of physical laws. *Cognitive Science*, *5*, 31–54.
19. Langley, P., & Simon, H. A. (1995). Applications of machine learning and rule induction. *Communications of the ACM*, *38*, November, 55–64.
20. Langley, P., Simon, H. A., Bradshaw, G. L., & Żytkow, J. M. (1987). *Scientific discovery: Computational explorations of the creative processes*. Cambridge, MA: MIT Press.
21. Lee, Y., Buchanan, B. G., & Aronis, J. M. (1998). Knowledge-based learning in exploratory science: Learning rules to predict rodent carcinogenicity. *Machine Learning*, *30*, 217–240.
22. Lee, Y., Buchanan, B. G., Mattison, D. R., Klopman, G., & Rosenkranz, H. S. (1995). Learning rules to predict rodent carcinogenicity of non-genotoxic chemicals. *Mutation Research*, *328*, 127–149.
23. Lee, Y., Buchanan, B. G., & Rosenkranz, H. S. (1996). Carcinogenicity predictions for a group of 30 chemicals undergoing rodent cancer bioassays based on rules

derived from subchronic organ toxicities. *Environmental Health Perspectives*, *104* (Supplement 5), 1059–1063.

24. Lenat, D. B. (1977). Automated theory formation in mathematics. *Proceedings of the Fifth International Joint Conference on Artificial Intelligence* (pp. 833–842). Cambridge, MA: Morgan Kaufmann.

25. Michalski, R. S., & Stepp, R. (1983). Learning from observation: Conceptual clustering. In R. S. Michalski, J. G. Carbonell, & T. M. Mitchell (Eds.), *Machine learning: An artificial intelligence approach*. San Francisco: Morgan Kaufmann.

26. Mitchell, F., Sleeman, D., Duffy, J. A., Ingram, M. D., & Young, R. W. (1997). Optical basicity of metallurgical slags: A new computer-based system for data visualisation and analysis. *Ironmaking and Steelmaking*, *24*, 306–320.

27. Nordhausen, B., & Langley, P. (1993). An integrated framework for empirical discovery. *Machine Learning*, *12*, 17–47.

28. Pericliev, V., & Valdés-Pérez, R. E. (in press). Automatic componential analysis of kinship semantics with a proposed structural solution to the problem of multiple models. *Anthropological Linguistics*.

29. Rose, D., & Langley, P. (1986). Chemical discovery as belief revision. *Machine Learning*, *1*, 423–451.

30. Shrager, J., & Langley, P. (Eds.) (1990). *Computational models of scientific discovery and theory formation*. San Francisco: Morgan Kaufmann.

31. Todorovski, L., Džeroski, S., & Kompare, B. (in press). Modeling and prediction of phytoplankton growth with equation discovery. *Ecological Modelling*.

32. Valdés-Pérez, R. E. (1994). Human/computer interactive elucidation of reaction mechanisms: Application to catalyzed hydrogenolysis of ethane. *Catalysis Letters*, *28*, 79–87.

33. Valdés-Pérez, R. E. (1995). Machine discovery in chemistry: New results. *Artificial Intelligence*, *74*, 191–201.

34. Valdés-Pérez, R. E. (1998). Why some programs do knowledge discovery well: Experiences from computational scientific discovery. Unpublished manuscript, School of Computer Science, Carnegie Mellon University, Pittsburgh, PA.

35. Zeigarnik, A. V., Valdés-Pérez, R. E., Temkin, O. N., Bruk, L. G., & Shalgunov, S. I. (1997). Computer-aided mechanism elucidation of acetylene hydrocarboxylation to acrylic acid based on a novel union of empirical and formal methods. *Organometallics*, *16*, 3114–3127.

36. Żytkow, J. M. (1996). Incremental discovery of hidden structure: Applications in theory of elementary particles. *Proceedings of the Thirteenth National Conference on Artificial Intelligence* (pp. 750–756). Portland, OR: AAAI Press.

37. Żytkow, J. M., & Simon, H. A. (1986). A theory of historical discovery: The construction of componential models. *Machine Learning*, *1*, 107–137.

On Classification and Regression

Shinichi Morishita

University of Tokyo
moris@ims.u-tokyo.ac.jp
http://platinum.ims.u-tokyo.ac.jp/~moris/

Abstract. We address the problem of computing various types of expressive tests for decision tress and regression trees. Using expressive tests is promising, because it may improve the prediction accuracy of trees. The drawback is that computing an optimal test could be costly. We present a unified framework to approach this problem, and we revisit the design of efficient algorithms for computing important special cases. We also prove that it is intractable to compute an optimal conjunction or disjunction.

1 Introduction

A decision (resp. regression) tree is a rooted binary tree structure for predicting the categorical (numeric) values of the objective attribute. Each internal node has a test on conditional attributes that split data into two classes. A record is recursively tested at internal nodes and eventually reaches a leaf node. A good decision (resp. regression) tree has the property that almost all the records arriving at every node take a single categorical value (a numeric value close to the average) of the objective attribute with a high probability, and hence the single value (the average) could be a good predictor of the objective attribute.

Making decision trees [9, 11, 10] and regression trees [2] has been a traditional research topic in the field of machine learning and artificial intelligence. Recently the efficient construction of decision trees and regression trees from large databases has been addressed and well studied among the KDD community. Computing tests at internal nodes is the most time-consuming step of constructing decision trees and regression trees. In the literature, there have been used simple tests that check if the value of an attribute is equal to (or less than) a specific value.

Using more expressive tests is promising in the sense that it may reduce the size of decision or regression trees while it can retain higher prediction accuracy [3, 8]. The drawback however is that the use of expressive tests could be costly. We consider the following three types of expressive tests for partitioning data into two classes; 1) subsets of categorical values for categorical attributes, 2) ranges and regions for numeric attributes, and 3) conjunctions and disjunctions of tests. We present a unified framework for handling those problems. We then reconstruct efficient algorithms for the former two problems, and we prove the intractability of the third problem.

2   Preliminaries

2.1   Relation Scheme, Attribute and Relation

Let $\mathcal{R}$ denote a relation scheme, which is a set of categorical or numeric attributes. The domain of a categorical attribute is a set of unordered distinct values, while the domain of a numeric attribute is real numbers or integers. We select a Boolean or numeric attribute A as special and call it the *objective* attribute. We call the other attributes in $\mathcal{R}$ *conditional attributes*.

Let B be an attribute in relation scheme $\mathcal{R}$. Let t denote a record (tuple) over $\mathcal{R}$, and let $t[B]$ be the value for attribute B. A set of records over $\mathcal{R}$ is called a *relation* over $\mathcal{R}$.

2.2   Tests on Conditional Attributes

We will consider several types of tests for records in a database. Let B denote an attribute, and let v and v_i be values in the domain of B. $B = v$ is a simple test, and t meets $B = v$ if $t[B] = v$.

When B is a categorical attribute, let $\{v_1, \ldots, v_k\}$ be a subset of values in the domain of B. Then, $B \in \{v_1, \ldots, v_k\}$ is a test, and t satisfies this test if $t[B]$ is equal to one value in $\{v_1, \ldots, v_k\}$. We will call a test of the form $B \in \{v_1, \ldots, v_k\}$ a *test with a subset of categorical values*.

When B is a numeric attribute, $B = v$, $B \leq v$, $B \geq v$, and $v_1 \leq B \leq v_2 (B \in [v_1, v_2])$ are tests, and a record t meets them respectively if $t[B] = v$, $t[B] \leq v$, $t[B] \geq v$, and $v_1 \leq t[B] \leq v_2$. We will call a test of the form $B \in [v_1, v_2]$ a *test with a range*.

The negation of a test T is denoted by $\neg T$. A record t meets $\neg T$ if t does not satisfy T. The negation of $\neg T$ is T.

A conjunction (a disjunction, respectively) of tests $T_1, T_2, \ldots, T_k$ is of the form $T_1 \wedge T_2 \wedge \ldots \wedge T_k$ ($T_1 \vee T_2 \vee \ldots \vee T_k$). A record t meets a conjunction (respectively, a disjunction) of tests, if t satisfies all the tests (some of the tests).

2.3   Splitting Criteria for Boolean Objective Attribute

Splitting Relation in Two Let R be a set of records over $\mathcal{R}$, and let $|R|$ denote the number of records in R. Let $Test$ be a test on conditional attributes. Let R_1 be the set of records that meet $Test$, while let R_2 denote $R - R_1$. In this way, we can use $Test$ to divide R into R_1 and R_2. Suppose that the objective attribute A is Boolean. We call a record whose A's value is true a *positive* record with respect to the objective attribute A. Let R^t denote the set of positive records in R. One the other hand, we call a record whose A's value is false a *negative* record, and let R^f denote the set of negative records in R. The following diagram illustrates how R is partitioned.

$$R = R^t \cup R^f$$

$$R_1 = R_1^t \cup R_1^f \qquad\qquad R_2 = R_2^t \cup R_2^f$$

The splitting by *Test* is effective for characterizing the objective Boolean attribute A if the probability of positive records changes dramatically after the division of R into R_1 and R_2; for instance, $|R^t|/|R| \ll |R_1^t|/|R_1|$, and $|R^t|/|R| \gg |R_2^t|/|R_2|$. On the other hand, the splitting by *Test* is most ineffective if the probability of positive records does not change at all; that is, $|R^t|/|R| = |R_1^t|/|R_1| = |R_2^t|/|R_2|$.

Measuring the Effectiveness of Splitting It is helpful to have a way of measuring the effectiveness of the splitting by a condition. To define the measure, we need to consider $|R|, |R^t|, |R^f|, |R_1|, |R_1^t|, |R_1^f|, |R_2|, |R_2^t|$ and $|R_2^f|$ as parameters, which satisfy the following equations:

$$|R| = |R^t| + |R^f| \quad |R_1| = |R_1^t| + |R_1^f| \quad |R^t| = |R_1^t| + |R_2^t|$$
$$|R| = |R_1| + |R_2| \quad |R_2| = |R_2^t| + |R_2^f| \quad |R^f| = |R_1^f| + |R_2^f|$$

Since R is given and fixed, we can assume that $|R|, |R^t|$, and $|R^f|$ are constants. Let n and m denote $|R|$ and $|R^t|$ respectively, then $|R^f| = n - m$. Furthermore, if we give the values of $|R_1|$ and $|R_1^t|$, for instance, the values of all the other variables are determined. Let x and y denote $|R_1|$ and $|R_1^t|$ respectively. Let $\phi(x, y)$ denote the measurement of the effectiveness of the splitting by condition *Test*. We now discuss some requirements that $\phi(x, y)$ is expected to have.

We first assume that lower value of $\phi(x, y)$ indicates higher effectiveness of the splitting. It does not matter if we select the reverse order. The splitting by *Test* is most ineffective when $|R^t|/|R| = m/n = |R_1^t|/|R_1| = y/x = |R_2^t|/|R_2|$, and hence $\phi(x, y)$ should be maximum when $y/x = m/n$.

Suppose that the probability of positive records in R_1, y/x, is greater than that of positive records in R, m/n. Also suppose that if we divide R by another new test, the number of positive records in R_1 increases by Δ ($0 < \Delta \le x - y$), while $|R_1|$ is the same. Then, the probability of positive records in R_1, $(y+\Delta)/x$, becomes to be greater than y/x, and hence we want to claim that the splitting by the new test is more effective. Thus we expect $\phi(x, y+\Delta) \le \phi(x, y)$. Similarly, since $y/x \le y/(x - \Delta)$ for $0 \le \Delta < x - y$ we also expect $\phi(x - \Delta, y) \le \phi(x, y)$. Figure 1 illustrates points $(x, y), (x, y + \Delta)$, and $(x - \Delta, y)$.

If the probability of positive records in R_1, y/x, is less than the average m/n, then (x, y) is in the lower side of the line connecting the origin and (m, n). See Figure 1. In this case observe that the probability of positive records in R_2, which is $(m - x)/(n - x)$, is greater than m/n. Suppose that the number of positive records in R_1 according to the new test decreases by Δ ($0 < \Delta \le x - y$), while $|R_1|$ is unchanged. Then, the number of positive records in R_2 increases by Δ while $|R_2|$ is the same. Thus the splitting by the new test is more effective, and we expect $\phi(x, y - \Delta) \le \phi(x, y)$. Similarly we also want to require $\phi(x + \Delta, y) \le \phi(x, y)$.

Entropy of Splitting We present an instance of $\phi(x, y)$ that meets all the requirements discussed so far. Let $ent(p) = -p \ln p - (1 - p) \ln(1 - p)$, where p

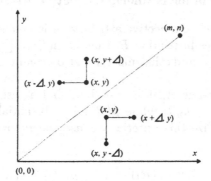

Fig. 1. (x, y), $(x, y + \Delta)$, $(x - \Delta, y)$, $(x, y - \Delta)$, and $(x + \Delta, y)$

means the probability of positive records in a set of records, while $(1 - p)$ implies the probability of negative records. Define the *entropy* $Ent(x, y)$ of the splitting by *Test* as follows:

$$\frac{x}{n} ent(\frac{y}{x}) + \frac{n - x}{n} ent(\frac{m - y}{n - x}),$$

where $\frac{y}{x}$ ($\frac{m-y}{n-x}$, respectively) is the probability of positive records in R_1 (R_2). This function is known as Quinlan's entropy heuristic [9], and it has been traditionally used as a criteria for evaluating the effectiveness of the division of a set of records. $Ent(x, y)$ is an instance of $\phi(x, y)$. We use the following theorem to show that $Ent(x, y)$ satisfies all the requirements on $\phi(x, y)$.

Theorem 1. $Ent(x, y)$ *is a concave function for* $x \geq y \geq 0$; *that is, for any* (x_1, y_1) *and* (x_2, y_2) *in* $\{(x, y) \mid x \geq y \geq 0\}$ *and any* $0 \leq \lambda \leq 1$,

$$\lambda Ent(x_1, y_1) + (1 - \lambda) Ent(x_2, y_2) \leq Ent(\lambda(x_1, y_1) + (1 - \lambda)(x_2, y_2)).$$

$Ent(x, y)$ *is maximum when* $y/x = m/n$.

Proof. See Appendix.   □

We immediately obtain the following corollary.

Corollary 1. *Let* (x_3, y_3) *be an arbitrary dividing point of* (x_1, y_1) *and* (x_2, y_2) *in* $\{(x, y) \mid x \geq y \geq 0\}$. *Then*, $\min(Ent(x_1, y_1), Ent(x_2, y_2)) \leq Ent(x_3, y_3)$.

For any (x, y) such that $y/x > m/n$ and any $0 < \Delta \leq x - y$, from the above corollary we have

$$\min(Ent(x, y + \Delta), Ent(x, (m/n)x)) \leq Ent(x, y),$$

because (x, y) is a dividing point of $(x, y+\Delta)$ and $(x, (m/n)x)$. Since $Ent(x, (m/n)x)$ is maximum,

$$Ent(x, y + \Delta) \leq Ent(x, y),$$

and hence $Ent(x, y)$ satisfies the requirement $\phi(x, y + \Delta) \leq \phi(x, y)$. In the same way we can show that $Ent(x, y)$ meets all the requirements on $\phi(x, y)$.

2.4   Splitting Criteria for Numeric Objective Attribute

Consider the case when the objective attribute A is numeric. Let $\mu(R)$ denote the average of A's values in relation R; that is, $\mu(R) = \sum_{t \in R} t[A]/|R|$. Let R_1 denote again the set of records that meet a test on conditional attributes, while let R_2 denote $R - R_1$.

In order to characterize A, it is useful to find a test such that $\mu(R_1)$ is considerably higher than $\mu(R)$ while $\mu(R_2)$ is substantially lower than $\mu(R)$ simultaneously. To realize this criteria, we use the *interclass variance* of the splitting by the test:

$$|R_1|(\mu(R_1) - \mu(R))^2 + |R_2|(\mu(R_2) - \mu(R))^2.$$

A test is more interesting if the interclass variance of the splitting by the test is larger. We also expect that the variance of A's values in R_1 (resp., R_2) should be small, which lets us approximate A's values in R_1 (R_2) at $\mu(R_1)$ ($\mu(R_2)$). To measure this property, we employ the *intraclass variance* of the splitting by the test:

$$\frac{\sum_{t \in R_1} (t[A] - \mu(R_1))^2 + \sum_{t \in R_2} (t[A] - \mu(R_2))^2}{|R|}.$$

We are interested in a test that maximizes the interclass variance and also minimizes the intraclass variance at the same time. Actually the maximization of the interclass variance coincides with the minimization of the intraclass variance.

Theorem 2. *Given a set of tests on conditional attributes, the test that maximizes the interclass variance also minimizes the intraclass variance.*

Proof. See Appendix. □

In what follows, we will focus on the maximization of the interclass variance. When R is given and fixed, $|R|(= |R_1| + |R_2|)$ and $\sum_{t \in R} t[A]$ can be regarded as constants, and let n and m denote $|R|$ and $\sum_{t \in R} t[A]$ respectively. If we denote $|R_1|$ and $\sum_{t \in R_1} t[A]$ by x and y, the interclass variance is determined by x and y as follows:

$$x(\frac{y}{x} - \frac{m}{n})^2 + (n - x)(\frac{m - y}{n - x} - \frac{m}{n})^2,$$

which will be denoted by $Var(x, y)$. We then have the following property of $Var(x, y)$, which is similar to Theorem 1 for the entropy function.

Theorem 3. *$Var(x, y)$ is a convex function for $0 < x < n$; that is, for any (x_1, y_1) and (x_2, y_2) such that $n > x_1, x_2 > 0$ and any $0 \leq \lambda \leq 1$,*

$$\lambda Var(x_1, y_1) + (1 - \lambda)Var(x_2, y_2) \geq Var(\lambda(x_1, y_1) + (1 - \lambda)(x_2, y_2)).$$

$Var(x, y)$ is minimum when $y/x = m/n$.

Proof. See Appendix. □

Corollary 2. *If (x_3, y_3) be an arbitrary dividing point of (x_1, y_1) and (x_2, y_2) such that $n > x_1, x_2 > 0$, then $\max(Var(x_1, y_1), Var(x_2, y_2)) \geq Var(x_3, y_3)$.*

Since the interclass variance has the property similar to the entropy function, in the following sections, we will present how to compute the optimal test that minimizes the entropy, but all arguments directly carry over to the case of finding the test maximizing the interclass variance.

2.5   Positive Tests and Negative Tests

Let R be a given relation, and let R_1 be the set of records in R that meet a given test. If the objective attribute A is Boolean, we treat "true" and "false" as numbers "1" and "0" respectively. We call the test *positive* if the average of A's values in R_1 is greater than or equal to the average of A's values in R; that is, $(\sum_{t \in R_1} t[A])/|R_1| \geq (\sum_{t \in R} t[A])/|R|$. Otherwise the test is called *negative*. Thus, when A is Boolean, the probability of positive records in R_1 is greater than or equal to the probability of positive records in R.

The test that minimizes the entropy could be either positive or negative. In what follows, we will focus on computing the positive test that minimizes the entropy of the splitting by the positive test among all the positive tests. This is because the algorithm for computing the optimal positive test can be used to calculate the optimal negative test by exchanging "true" and "false" of the objective Boolean attribute value (or reversing the order of the objective numeric attribute value) in each record.

3   Computing Optimal Tests with Subsets of Categorical Values

Let C be a conditional categorical attribute, and let $\{c_1, c_2, \ldots, c_k\}$ be the domain of C. Among all the positive tests of the form $C \in S$ where S is a subset of $\{c_1, c_2, \ldots, c_k\}$, we want to compute the positive test that minimizes the entropy of the splitting. A naive solution would consider all the possible subsets of $\{c_1, c_2, \ldots, c_k\}$ and select the one that minimizes the entropy. Instead of investigating all 2^k subsets, there is an efficient way of checking only k subsets.

We first treat "true" and "false" as real numbers "1" and "0" respectively. For each c_i, let μ_i denote the average of A's values of all the records whose C's values are c_i; that is,

$$\mu_i = \frac{\sum_{t[C]=c_i} t[A]}{|\{t \mid t[C] = c_i\}|}.$$

Without loss of generality we can assume that $\mu_1 \geq \mu_2 \geq \ldots \geq \mu_k$, otherwise we rename the categorical values appropriately to meet the above property. We then have the following theorem.

Theorem 4. *Among all the positive tests with subsets of categorical values, there exists a positive test of the form $C \in \{c_i \mid 1 \leq i \leq j\}$ that minimizes the entropy of the splitting.*

This theorem is due to Breiman et al.[2]. Thanks to this theorem, we only need to consider k tests of the form $C \in \{c_i \mid 1 \leq i \leq j\}$ to find the optimal test. We now prove the theorem by using techniques introduced in the previous section.

Proof of Theorem 4 We will prove the case of the minimization of the entropy. The case of maximization of the interclass variance can be shown similarly.

Since $\mu_1 \geq \mu_2 \geq \ldots \geq \mu_k$, for $1 \leq j \leq k$, $C \in \{c_i \mid 1 \leq i \leq j\}$ is a positive test. Assume that among positive tests with subsets of categorical values, there does not exist any $1 \leq j \leq k$ such that $C \in \{c_i \mid 1 \leq i \leq j\}$ minimizes the entropy, which we will contradict in what follows. Then, there exists $1 \leq h \leq k$ such that test $C \in \{c_i \mid 1 \leq i \leq h-1\} \cup V$, where V is a non-empty subset of $\{c_i \mid h < i \leq k\}$, is positive and minimizes the entropy. $\{c_i \mid 1 \leq i \leq h-1\} \cup V$ contains consecutive values from c_1 to c_{h-1} but lacks c_h. Let h be the minimum number that satisfies this property.

With each subset W of $\{c_1, c_2, \ldots, c_k\}$, we associate

$$p(W) = (\ |\{t \mid t[C] \in W\}|, \sum_{t[C] \in W} t[A]\ )$$

in the Euclidean plane. $Ent(p(W))$ is the entropy of the splitting by the test $C \in W$. The slope of the line between the origin and $p(W)$ gives the average value of the objective attribute A among records in $\{t \mid t[C] \in W\}$. We will denote the average by μ_W; namely,

$$\mu_W = \frac{\sum_{t[C] \in W} t[A]}{|\{t \mid t[C] \in W\}|}.$$

Let $\{c_1, \ldots, c_{h-1}\}$ denote $\{c_i \mid 1 \leq i \leq h-1\}$. Figure 2 shows $p(\{c_1, \ldots, c_{h-1}\})$, $p(\{c_1, \ldots, c_{h-1}\} \cup V)$, and $p(\{c_1, \ldots, c_{h-1}, c_h\} \cup V)$. Since $p(\{c_1, \ldots, c_{h-1}\} \cup V)$ is associated with a positive test, it lies in the upper side of the line between the origin and $p(\{c_1, c_2, \ldots, c_k\})$. $\mu_{\{c_1, \ldots, c_{h-1}\}} \geq \mu_{\{c_h\}} \geq \mu_V$, because $\mu_1 \geq \mu_2 \geq \ldots \geq \mu_k$, and V is a subset of $\{c_i \mid h < i \leq k\}$. It is easy to see that $p(\{c_1, \ldots, c_{h-1}\})$ and $p(\{c_1, \ldots, c_{h-1}, c_h\} \cup V)$ are also in the upper side of the line between the origin and $p(\{c_1, c_2, \ldots, c_k\})$.

Suppose that $p(\{c_1, \ldots, c_{h-1}, c_h\} \cup V)$ is in the upper side of the line passing through the origin and $p(\{c_1, \ldots, c_{h-1}\} \cup V)$. The left in Figure 3 shows this situation. Suppose that the line passing through $p(\{c_1, \ldots, c_{h-1}, c_h\} \cup V)$ and $p(\{c_1, \ldots, c_{h-1}\} \cup V)$ hits the line between the origin and $p(\{c_1, \ldots, c_k\})$ at Q. From the concavity of the entropy function

$$\min(Ent(p(\{c_1, \ldots, c_{h-1}, c_h\} \cup V)), Ent(Q)) \leq Ent(p(\{c_1, \ldots, c_{h-1}\} \cup V)).$$

Since $Ent(Q)$ is maximum,

$$Ent(p(\{c_1, \ldots, c_{h-1}, c_h\} \cup V)) \leq Ent(p(\{c_1, \ldots, c_{h-1}\} \cup V)),$$

which contradicts the choice of h.

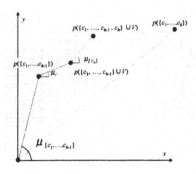

Fig. 2. $p(\{c_1,\ldots,c_{h-1}\})$, $p(\{c_1,\ldots,c_{h-1}\}\cup V)$, and $p(\{c_1,\ldots,c_h\}\cup V)$

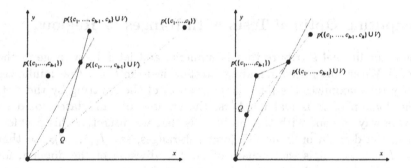

Fig. 3. The left (resp. right) figure illustrates the case when $p(\{c_1,\ldots,c_{h-1},c_h\}\cup V)$ is in the upper (lower) side of the line connecting the origin and $p(\{c_1,\ldots,c_{h-1}\}\cup V)$.

We now consider the opposite case when $p(\{c_1,\ldots,c_{h-1},c_h\}\cup V)$ is in the lower side of the line connecting the origin and $p(\{c_1,\ldots,c_{h-1}\}\cup V)$. See the right in Figure 3. Suppose that the line passing through $p(\{c_1,\ldots,c_{h-1},c_h\}\cup V)$ and $p(\{c_1,\ldots,c_{h-1}\}\cup V)$ hits the line between the origin and $p(\{c_1,\ldots,c_{h-1}\})$ at Q. If $Ent(p(\{c_1,\ldots,c_{h-1}\}\cup V)) < Ent(Q)$, from the concavity of the entropy function, we have

$$Ent(p(\{c_1,\ldots,c_{h-1},c_h\}\cup V)) \leq Ent(p(\{c_1,\ldots,c_{h-1}\}\cup V)),$$

which contradicts the choice of h. If $Ent(Q) \leq Ent(p(\{c_1,\ldots,c_{h-1}\}\cup V))$, we have

$$Ent(p(\{c_1,\ldots,c_{h-1}\})) \leq Ent(Q),$$

because $Ent(0,0)$ is maximum, and $Ent(x,y)$ is a concave function. Thus we have

$$Ent(p(\{c_1,\ldots,c_{h-1}\})) \leq Ent(p(\{c_1,\ldots,c_{h-1}\}\cup V)),$$

which again contradicts the choice of h.   □

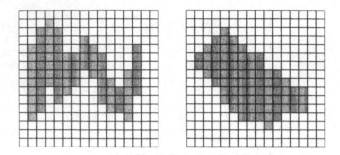

Fig. 4. An x-monotone region (left) and a rectilinear convex region (right)

4 Computing Optimal Tests with Ranges or Regions

Let B be a conditional attribute that is numeric, and let I be a range of the domain of B. We are interested in finding a test of the form $B \in I$ that minimizes the entropy (or maximizes the interclass variance) of the splitting by the test. When the domain of B is real numbers, the number of candidates could be infinite. One way to cope with this problem is that we discretize this problem by dividing the domain of B into disjoint sub-ranges, say $I_1, \ldots, I_N$, so that the union $I_1 \cup \ldots \cup I_N$ is the domain of B. The division of the domain, for instance, can be done by distributing the values of B in the given set of records into equal-sized sub-ranges. We then concatenate some successive sub-ranges, say $I_i, I_{i+1}, \ldots, I_j$, to create a range $I_i \cup I_{i+1} \cup \ldots \cup I_j$ that optimizes the criteria of interest.

It is natural to consider the two-dimensional version. Let B and C be numeric conditional attributes. We also simplify this problem by dividing the domain of B (resp. C) into N_B (N_C) equal-sized sub-ranges. We assume that $N_B = N_C = N$ without loss of generality as regards our algorithms. We then divide the Euclidean plane associated with B and C into $N \times N$ pixels. A *grid* region is a set of pixels, and let R be an instance. A record t satisfies test $(B, C) \in R$ if $(t[B], t[C])$ belongs to R. We can consider various types of grid regions for the purpose of splitting a relation in two. In the literature two classes of regions have been well studies [4, 3, 12, 8]. An *x-monotone* region is a connected grid region whose intersection with any vertical line is undivided. A *rectilinear convex* region is an x-monotone region whose intersection with any horizontal line is also undivided. Figure 4 shows an x-monotone region in the left and a rectilinear convex region in the right.

In the case of computing the optimal range by concatenating some consecutive sub-ranges of N sub-ranges, we may consider $O(N^2)$ sequences of successive sub-ranges, but to this end, Katoh [6] presents an $O(N \log N)$-time algorithm.

On the other hand, the number of x-monotone regions and the number of rectilinear convex regions is more than 2^N. It is non-trivial to efficiently find such a region R that minimizes the entropy (maximizes the interclass variance)

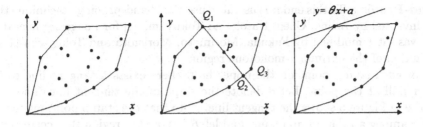

Fig. 5. The left figure presents the convex hull of stamp points. The middle illustrates P, Q_1, Q_2 and Q_3 in Proposition 1. The right shows the hand probing technique.

of the splitting by the test $(B, C) \in R$. Here we review some techniques for this purpose.

Convex Hull of Stamp Points Let $\mathcal{R}$ denote the family of x-monotone regions or the family of rectilinear convex regions. Let A be the objective attribute. When A is Boolean, we treat "true" and "false" as real numbers "1" and "0". With each region R in $\mathcal{R}$, we associate a *stamp point* (x, y) where $x = |\{t \mid t \text{ meets } (B, C) \in R\}|$ and $y = \sum_{\{t \mid t \text{ meets } (B,C) \in R\}} t[A]$. Since the number of regions in $\mathcal{R}$ is more than 2^N, we cannot afford to calculate all the point associated, and hence we simply assume their existence.

Let S denote the set of stamp points for a family of regions $\mathcal{R}$. A *convex polygon* of S has the property that any line connecting arbitrary two points of S must itself lies entirely inside the polygon. The *convex hull* of S is the smallest convex polygon of S. The left in Figure 5 illustrates the convex hull. The upper (lower) half of a convex hull is called the *upper* (*lower*) hull, in short.

Proposition 1. *Let $R \in \mathcal{R}$ be the region such that test $(B, C) \in R$ minimizes the entropy (or maximizes the interclass variance). The stamp point associated with R must be on the convex hull of S.*

Proof. Otherwise there exists such a point P inside the convex hull of S that minimizes the entropy. Select any point Q_1 on the convex hull, draw the line connecting P and Q_1, and let Q_2 be another point where the line between P and Q_1 crosses the convex hull. From the concavity of the entropy function, $\min(Ent(Q_1), Ent(Q_2)) \leq Ent(P)$, and there exists a point Q_3 on the convex hull such that $Ent(Q_3) \leq Ent(Q_2)$ (see Figure 5). Thus, $Ent(Q_3) \leq Ent(P)$, which is a contradiction. $\qquad\square$

If T is the positive (negative, resp.) test that minimizes the entropy among all the positive tests of the form $(B, C) \in R$, from Proposition 1 the stamp point associated with T must be on the upper (lower) hull. We then present how to scan the upper hull to search the stamp point that minimizes the entropy.

Hand-Probing To this end it is useful to use the "hand-probing" technique that was invented by Asano, Chen, Katoh and Tokuyama [1] for image segmentation and was later modified by Fukuda, Morimoto, Morishita and Tokuyama [4] for extraction of the optimal x-monotone region.

For each stamp point on the upper hull, there exists a tangent line to the upper hull at the point. Let θ denote the slope of the tangent line. The right picture in Figure 5 shows the tangent line. Note that the stamp point maximizes $y - \theta x$ among all the stamp points, and let R denote the region that corresponds to the stamp point. We now present a roadmap of how to construct R.

Let $p_{i,j} (1 \leq i, j \leq N)$ denote the (i,j)-th pixel in $N \times N$ pixels. A grid region is a union of pixels. Let $u_{i,j}$ be the number of records that meet $(B, C) \in p_{i,j}$, and let $v_{i,j}$ be the sum of the objective attribute values of all the records that satisfy $(B, C) \in p_{i,j}$, which is $\sum_{t \text{ meets } (B,C) \in p(i,j)} t[A]$. Using those notations, we can represent the stamp point associated with R by $(\sum_{p_{i,j} \subseteq R} u_{i,j}, \sum_{p_{i,j} \subseteq R} v_{i,j})$, which maximizes $y - \theta x$. Since

$$\sum_{p_{i,j} \subseteq R} v_{i,j} - \theta \sum_{p_{i,j} \subseteq R} u_{i,j} = \sum_{p_{i,j} \subseteq R} (v_{i,j} - \theta u_{i,j}),$$

R maximizes $\sum_{p_{i,j} \subseteq R} (v_{i,j} - \theta u_{i,j})$.

We call $v_{i,j} - \theta u_{i,j}$ the *gain* of the pixel $p_{i,j}$. The problem of computing the region that maximizes the sum of gains of pixels in the region has been studied. For a family of x-monotone regions, Fukuda, Morimoto, Morishita, and Tokuyama presents an $O(N^2)$-time algorithm [4]. For a family of rectilinear convex regions, Yoda, Fukuda, Morimoto, Morishita, and Tokuyama gives an $O(N^3)$-time algorithm [12]. Due to the space limitation, we do not introduce those algorithms. Those algorithms use the idea of dynamic programming, and they connect an interval in each column from lower index i to higher one to generate an x-monotone (or, rectilinear) region.

Since we have an efficient algorithm for generating the region associated with the stamp point on the convex hull at which the line with a slope θ touches, it remains to answer how many trials of hand-probing procedure are necessary to find the region that minimizes the entropy. If n is the number of given records, there could be at most n stamp points on the upper hull, and therefore we may have to do n trials of hand-probing by using n distinct slopes. Next we present a technique that is expected to reduce the number of trials to be $O(\log n)$ in practice.

Guided Branch-and-Bound Search Using a tangent line with the slope $\theta = 0$, we can touch the rightmost point on the convex hull. Let a be an aribitrary large real number such that we can touch the leftmost point on the convex hull by using the tangent line with slope a. Thus using slopes in $[0, a]$, we can scan all the points on the upper hull. We then perform the binary search on $[0, a]$ to scan the convex hull.

During the process we may dramatically reduce the search space. Figure 6 shows the case when we use two tangent lines to touch two points P and Q on

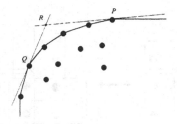

Fig. 6. Guided Branch-and-Bound Search

the convex hull, and R denotes the point of intersection of the two lines. Let X be an arbitrary point inside the triangle PQR. From the concavity of the entropy function, we immediately obtain

$$\min\{Ent(P), Ent(Q), Ent(R)\} \leq Ent(X).$$

If $\min\{Ent(P), Ent(Q)\} \leq Ent(R)$, we have $\min\{Ent(P), Ent(Q)\} \leq Ent(X)$, which implies that it is useless to check whether or not there exists a point between P and Q on the hull whose entropy is less than $\min\{Ent(P), Ent(Q)\}$. In practice, most of subintervals of slopes are expected to be pruned away during the binary search. This guided branch-and-bound search strategy has been experimentally evaluated [3, 8]. According to experimental tests the number of trials of hand-probing procedure is $O(\log n)$.

5   Computing Optimal Conjunctions and Disjunctions

Suppose that we are given a set S of tests on conditional attributes. We also assume that S contains the negation of an arbitrary test in S. We call a conjunction *positive* (*negative*, resp.) if it is a positive (negative) test. We will show that it is NP-hard to compute the positive conjunction (the positive disjunction, resp.) that minimizes the entropy among all positive conjunctions (positive disjunctions) of tests in S. Also, it is NP-hard to compute the positive conjunction (positive disjunction) that maximizes the interclass variance.

Let $T_1 \wedge \ldots \wedge T_k$ be a positive conjunction of tests in S. Observe that the entropy (the interclass variance, resp.) of the splitting by $T_1 \wedge \ldots \wedge T_k$ is equal to the entropy (the interclass variance) of the splitting by $\neg(T_1 \wedge \ldots \wedge T_k)$. $\neg(T_1 \wedge \ldots \wedge T_k)$ is equivalent to $\neg T_1 \vee \ldots \vee \neg T_k$, which is a negative disjunction of tests in S. Thus the negation of the optimal positive conjunction gives the optimal negative disjunction. As remarked in Subsection 2.5, computing a negative test can be done by using a way of computing a positive test, and therefore we will prove the intractability of computing the optimal disjunction.

Theorem 5. *Given a set S of tests on conditional attributes such that S contains the negation of any test in S, it is NP-hard to compute the positive dis-*

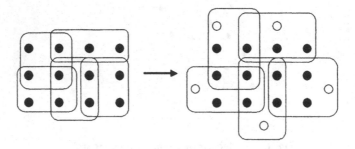

Fig. 7. Each subset is extended with a unique white element.

junction of tests in S that minimizes the entropy value among all positive disjunctions. It is also NP-hard to compute the positive disjunction that maximizes the interclass variance.

Proof. Here we present a proof for the case of the entropy. The case of the interclass variance can be proved in a similar manner. We reduce the difficulty of the problem to the NP-hardness of MINIMUM COVER [5]. Let V be a finite set, and let C be a collection of subsets of V. A sub-collection $C'(\subset C)$ is a cover of V if any element in V belongs to one of C'. Suppose that C_{min} is a cover that minimizes the number of subsets in it. It is NP-hard to compute C_{min}.

Suppose that V contains a elements, and C contains c subsets of V. We call elements in V *black*. Let b be a number greater than a and c, generate a set W of new b elements, and call them *white*. We then extend each subset in C by adding a unique white element that does not appear elsewhere. If $b > c$, $b - c$ white elements are not used for this extension. Figure 7 illustrates this operation. In the figure each hyperedge shows a subset in C. After this extension, C and C_{min} become collections of subsets of $V \cup W$.

In what follows, we treat elements in $V \cup W$ as records in a database. We assume that the objective attribute is true (false, resp.) for black (white) records in $V \cup W$. We then identify each subset in C with a test such that all elements in the subset meets the test, while none of elements outside the subset satisfy the test. We also identify a collection $C'(\subseteq C)$ with the disjunction of tests that correspond to subsets in C'. We then show that the disjunction corresponds to C_{min} minimizes the entropy, which means that finding the optimum disjunction is NP-hard.

With each sub-collection $C'(\subseteq C)$ such that the disjunction identified with C' is positive, we associate a point (x, y) in an Euclidean plane such that x is the number of records in C', and y is the number of black records in C'. See Figure 8. $Ent(x, y)$ gives the entropy of the disjunction identified with C'. Let k denote the number of subsets in the minimum cover C_{min}. $(a + k, a)$ is associated with C_{min}. We prove that all the points associated with collections of subsets of C fall in the gray region in Figure 8.

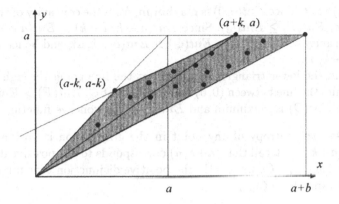

Fig. 8. Points Associated with Sub-collections

All the points lie in the upper side or on the line connecting the origin and $(a+b, a)$, because each point corresponds to a positive disjunction. We show that all the points lie under or on the line between $(a + k, a)$ and $(a - k, a - k)$. To this end, it is enough to prove that any $C' \subset C$ that contains $a - l (l \leq 1)$ black records must also have at least $k - l$ white records. The proof is an induction on l, and consider the case when $l = 1$. Suppose that the number of white records is less than $k - 1$. We can immediately construct a cover of V by adding to C' a subset X that contains the remaining black record. Note that the number of white records in $C' \cup \{X\}$ is less than k, which contradicts the choice of C_{min}. The argument carries over to the case when $l > 1$.

We then prove that $Ent(a - k, a - k) > Ent(a + k, a)$ for $k \geq 1$.

$$Ent(a - k, a - k) = \frac{a - k}{a + b} ent(\frac{a - k}{a - k}) + \frac{b + k}{a + b} ent(\frac{k}{b + k})$$

$$= \frac{1}{a + b}(-k \ln \frac{k}{b + k} - b \ln \frac{b}{b + k}) \quad \text{Because } ent(1) = 0$$

$$Ent(a + k, a) = \frac{a + k}{a + b} ent(\frac{a}{a + k}) + \frac{b - k}{a + b} ent(\frac{a - a}{b - k})$$

$$= \frac{1}{a + b}(-k \ln \frac{k}{a + k} - a \ln \frac{a}{a + k}) \quad \text{Because } ent(0) = 0$$

Let $f(x)$ denote $-k \ln \frac{k}{x + k} - x \ln \frac{x}{x + k}$. We then have $Ent(a - k, a - k) = \frac{1}{a+b} f(b)$ and $Ent(a + k, a) = \frac{1}{a+b} f(a)$. Since $f'(x) = \ln \frac{x+k}{x} > 0$ for $x > 0$. Because $b > a > 0$, we have $f(b) > f(a)$, and hence $Ent(a - k, a - k) > Ent(a + k, a)$.

From Theorem 1, $Ent(x, y)$ is maximum at any point (x, y) on the line between $(0, 0)$ and $(a+b, a)$, and $Ent(x, y)$ is a concave function on the gray region of Figure 8. Let P be an arbitrary point in the gray region.

– If P is in the upper triangle, draw the line that passes through the origin and P, and suppose that the line hits the line connecting $(a - k, a - k)$ and

$(a+k, a)$ at Q. Since $Ent(0,0)$ is maximum, from the concavity of the entropy function, $Ent(P) \geq Ent(Q)$. Since $Ent(a - k, a - k) > Ent(a + k, a)$, from the concavity of the entropy, $Ent(Q) > Ent(a + k, a)$, and hence $Ent(P) > Ent(a + k, a)$.

- If P is in the lower triangle, suppose that the line going through $(a + k, a)$ and P hits the line between $(0, 0)$ and $(a+b, a)$ at Q. $Ent(P) > Ent(a+k, a)$, because $Ent(Q)$ is maximum and $Ent(x, y)$ is a concave function.

In both cases, the entropy of any point in the gray region is no less than the entropy of $(a+k, a)$. Recall that $(a+k, a)$ corresponds to the positive disjunction associated with C_{min}. Consequently the positive disjunction that minimizes the entropy corresponds to C_{min}.   □

Design of Polynomial-Time Approximation Algorithm is Hard. We remark that it is hard to design a polynomial-time algorithm that approximates the optimal positive disjunction of tests. In the proof of Theorem 5, we reduce MINIMUM COVER to the problem of computing the optimal positive disjunction. Since it is NP-hard to find C_{min}, it is natural to further ask whether or not there exists an efficient way of approximating C_{min}. To be more precise, we wish to design a polynomial-time approximation algorithm that finds a cover of V such that the number l of white points in it is close to the number k of the white points in C_{min}; that is, the approximation ratio l/k is as small as possible. According to the result by Lund and Yannakakis [7], unless P=NP the best possible approximation ratio is $\Theta(\log n)$, where n is the number of points in a maximum-size subset.

Acknowledgements

This research is partly supported by Grant-in-Aid for Scientific Research on Priority Areas "Discovery Science" from the Ministry of Education, Science and Culture, Japan.

References

1. T. Asano, D. Chen, N. Katoh, and T. Tokuyama. Polynomial-time solutions to image segmentations. In *Proc. 7th ACM-SIAM Symposium on Discrete Algorithms*, pages 104–113, 1996.
2. L. Breiman, J. H. Friedman, R. A. Olshen, and C. J. Stone. *Classification and Regression Trees*. Wadsworth, 1984.
3. T. Fukuda, Y. Morimoto, S. Morishita, and T. Tokuyama. Constructing efficient decision trees by using optimized association rules. In *Proceedings of the 22nd VLDB Conference*, pages 146–155, Sept. 1996.
4. T. Fukuda, Y. Morimoto, S. Morishita, and T. Tokuyama. Data mining using two-dimensional optimized association rules: Scheme, algorithms, and visualization. In *Proceedings of the ACM SIGMOD Conference on Management of Data*, pages 13–23, June 1996.

5. M. R. Garey and D. S. Johnson. *Computer and Intractability. A Guide to NP-Completeness.* W. H. Freeman, 1979.
6. N. Katoh. Private communication, Jan. 1997.
7. C. Lund and M. Yannakakis. On the hardness of approximating minimization problems. *J.ACM*, 41(5):960–981, 1994.
8. Y. Morimoto, H. Ishii, and S. Morishita. Efficient construction of regression trees with range and region splitting. In *Proceedings of the 23rd VLDB Conference*, pages 166–175, Aug. 1997.
9. J. R. Quinlan. Induction of decision trees. *Machine Learning*, 1:81–106, 1986.
10. J. R. Quinlan. *C4.5: Programs for Machine Learning.* Morgan Kaufmann, 1993.
11. J. R. Quinlan and R. L. Rivest. Inferring decision trees using minimum description length principle. *Information and Computation*, 80:227–248, 1989.
12. K. Yoda, T. Fukuda, Y. Morimoto, S. Morishita, and T. Tokuyama. Computing optimized rectilinear regions for association rules. In *Proceedings of the Third International Conference on Knowledge Discovery and Data Mining*, pages 96–103, Aug. 1997.

Appendix

Proof of Theorem 1

We first prove the concavity of $Ent(x, y)$. For any $x>y>0$ and any real numbers δ_1 and δ_2, let V denote $\delta_1 x + \delta_2 y$. It suffices to prove $\partial^2 Ent(x, y)/\partial V^2 \leq 0$. Recall that

$$Ent(x, y) = \frac{x}{n}ent(\frac{y}{x}) + \frac{n - x}{n}ent(\frac{m - y}{n - x}).$$

Define $f(x, y) = \frac{x}{n}ent(\frac{y}{x})$. Then,

$$Ent(x, y) = f(x, y) + f(n - x, m - y) \tag{1}$$

To prove $\partial^2 Ent(x, y)/\partial V^2 \leq 0$, it is sufficient to show the following inequalities:

$$\partial^2 f(x, y)/\partial V^2 \leq 0 \qquad \partial^2 f(n - x, m - y)/\partial V^2 \leq 0.$$

Here we will prove the former inequality. The latter can be proved in a similar way. When $\delta_1, \delta_2 \neq 0$, we have:

$$f(x, y) = \frac{1}{n}(-y \log \frac{y}{x} - (x - y) \log(1 - \frac{y}{x}))$$

$$\frac{\partial f(x, y)}{\partial V}$$

$$= \frac{\partial f(x, y)}{\partial x}\frac{\partial x}{\partial V} + \frac{\partial f(x, y)}{\partial y}\frac{\partial y}{\partial V}$$

$$= \frac{1}{n}(-\log(x - y) + \log x)\frac{1}{\delta_1} + \frac{1}{n}(-\log y + \log(x - y))\frac{1}{\delta_2}$$

$$\frac{\partial^2 f(x, y)}{\partial V^2}$$

$$= \frac{1}{n}((-\frac{1}{x-y}+\frac{1}{x})\frac{1}{\delta_1}+\frac{1}{x-y}\frac{1}{\delta_2})\frac{1}{\delta_1}+\frac{1}{n}(\frac{1}{x-y}\frac{1}{\delta_1}+(-\frac{1}{y}-\frac{1}{x-y})\frac{1}{\delta_2})\frac{1}{\delta_2}$$

$$= -\frac{1}{n\delta_1^2\delta_2^2 x(x-y)y}(\delta_2 y - \delta_1 x)^2$$

$$\leq 0 \quad (x>y>0)$$

When $\delta_1 = 0, \delta_2 \neq 0$, $V = \delta_2 y$, and we have:

$$\frac{\partial f(x,y)}{\partial V} = \frac{1}{n}(-\log y + \log(x-y))\frac{1}{\delta_2}$$

$$\frac{\partial^2 f(x,y)}{\partial V^2} = \frac{1}{n}(-\frac{1}{y}-\frac{1}{x-y})\frac{1}{\delta_2^2} = \frac{1}{n}\frac{-x}{(x-y)y}\frac{1}{\delta_2^2} \leq 0 \quad (x>y>0)$$

The case when $\delta_1 \neq 0$ and $\delta_2 = 0$ can be handled in a similar way.

Next we prove that $Ent(x,y)$ is maximum when $y/x = m/n$. From Equation (1), observe that for any $0 \leq y \leq x$, $Ent(x,y) = Ent(n-x, m-y)$. According to the concavity of $Ent(x,y)$, we have

$$Ent(x,y) = \frac{1}{2}Ent(x,y) + \frac{1}{2}Ent(n-x, m-y)$$

$$\leq Ent(\frac{x+n-x}{2}, \frac{y+m-y}{2})$$

$$= Ent(\frac{n}{2}, \frac{m}{2}),$$

which means that $Ent(x,y)$ is maximum when $(x,y) = (\frac{n}{2}, \frac{m}{2})$. Finally we can prove that $Ent(x,y)$ is constant on $y = (m/n)x$, because

$$Ent(x,y) = \frac{x}{n}(-\frac{m}{n}\log\frac{m}{n} - (1-\frac{m}{n})\log(1-\frac{m}{n})) +$$

$$\frac{n-x}{n}(-\frac{m}{n}\log\frac{m}{n} - (1-\frac{m}{n})\log(1-\frac{m}{n}))$$

$$= -\frac{m}{n}\log\frac{m}{n} - (1-\frac{m}{n})\log(1-\frac{m}{n}).$$

Since $(\frac{n}{2}, \frac{m}{2})$ is on $y = (m/n)x$, $Ent(x,y)$ is maximum when $y/x = m/n$. □

Proof of Theorem 2

The interclass variance can be transformed as follows:

$$|R_1|(\mu(R_1) - \mu(R))^2 + |R_2|(\mu(R_2) - \mu(R))^2$$
$$= -|R|\mu(R)^2 + (|R_1|\mu(R_1)^2 + |R_2|\mu(R_2)^2),$$

because $|R| = |R_1| + |R_2|$ and $|R_1|\mu(R_1) + |R_2|\mu(R_2) = |R|\mu(R)$. Since $|R|$ and $\mu(R)$ are constants, the maximization of the interclass variance is equivalent to the maximization of $|R_1|\mu(R_1)^2 + |R_2|\mu(R_2)^2$.

On the other hand, the intraclass variance can be transformed as follows:

$$\frac{\sum_{t \in R_1}(t[A] - \mu(R_1))^2 + \sum_{t \in R_2}(t[A] - \mu(R_2))^2}{|R|}$$

$$= \frac{\sum_{t \in R} t[A]^2 - (|R_1|\mu(R_1)^2 + |R_2|\mu(R_2)^2)}{|R|},$$

because $\sum_{t \in R_1} t[A] = |R_1|\mu(R_1)$ and $\sum_{t \in R_2} t[A] = |R_2|\mu(R_2)$. Since R is fixed, $\sum_{t \in R} t[A]^2$ is a constant. Thus the minimization of the intraclass variance is equivalent to the maximization of $|R_1|\mu(R_1)^2 + |R_2|\mu(R_2)^2$ that is also equivalent to the maximization of the interclass variance.   □

Proof of Theorem 3

The proof is similar to the proof of Theorem 1. We first prove that $Var(x, y)$ is a convex function for $0 < x < n$. For any $0 < x < n$ and any δ_1 and δ_2, let V denote $\delta_1 x + \delta_2 y$. We will the case when $\delta_1, \delta_2 \neq 0$. It is sufficient to prove that $\partial^2 Var(x, y)/\partial V^2 \geq 0$. Recall that

$$Var(x, y) = x(\frac{y}{x} - \frac{m}{n})^2 + (n - x)(\frac{m - y}{n - x} - \frac{m}{n})^2.$$

Define $g(x, y) = x(\frac{y}{x} - \frac{m}{n})^2$. Then,

$$Var(x, y) = g(x, y) + g(n - x, m - y). \tag{2}$$

We prove $\partial^2 Var(x, y)/\partial V^2 \geq 0$ by showing the following two inequalities:

$$\partial^2 g(x, y)/\partial V^2 \geq 0 \qquad \partial^2 g(n - x, m - y)/\partial V^2 \geq 0.$$

We prove the former case. The latter can be shown in a similar manner.

$$\frac{\partial g(x, y)}{\partial V} = \frac{\partial g(x, y)}{\partial x}\frac{\partial x}{\partial V} + \frac{\partial g(x, y)}{\partial y}\frac{\partial y}{\partial V} = \frac{1}{\delta_1}\{(\frac{m}{n})^2 - (\frac{y}{x})^2\} + \frac{2}{\delta_2}(\frac{y}{x} - \frac{m}{n})$$

$$\frac{\partial^2 g(x, y)}{\partial V^2} = \frac{2}{x}(\frac{y}{\delta_1 x} - \frac{1}{\delta_2})^2 \geq 0$$

Next we prove that $Var(x, y)$ is minimum when $y/x = m/n$. From Equation (2), we have $Var(x, y) = Var(n - x, m - y)$. From the convexity of $Var(x, y)$,

$$Var(x, y) = \frac{1}{2}Var(x, y) + \frac{1}{2}Var(n - x, m - y)$$

$$\geq Var(\frac{x + n - x}{2}, \frac{y + m - y}{2})$$

$$= Var(\frac{n}{2}, \frac{m}{2}),$$

which implies that $Var(x, y)$ is minimum when $(x, y) = (\frac{n}{2}, \frac{m}{2})$. It is easy to see that $Var(x, y) = 0$ when $y/x = m/n$. Since $(\frac{n}{2}, \frac{m}{2})$ is on $y/x = m/n$, $Var(x, y)$ is minimum when $y/x = m/n$.   □

Knowledge Discovery in Biological and Chemical Domains

Stephen Muggleton

Department of Computer Science, University of York, Heslington, York, YO1 5DD,
United Kingdom.

1 Extended abstract

The pharmaceutical industry is increasingly overwhelmed by large-volume-data.
This is generated both internally as a side-effect of screening tests and combinatorial chemistry, as well as externally from sources such as the human genome
project. The industry is predominantly knowledge-driven. For instance, knowledge is required within computational chemistry for pharmacophore identification, as well as for determining biological function using sequence analysis.

From a computer science point of view, the knowledge requirements within
the industry give higher emphasis to "knowing that" (declarative or descriptive
knowledge) rather than "knowing how" (procedural or prescriptive knowledge).
Mathematical logic has always been the preferred representation for declarative
knowledge and thus knowledge discovery techniques are required which generate
logical formulae from data. Inductive Logic Programming (ILP) [6, 1] provides
such an approach.

This talk will review the results of the last few years' academic pilot studies
involving the application of ILP to the prediction of protein secondary structure [5, 8, 9], mutagenicity [4, 7], structure activity [3], pharmacophore discovery [2] and protein fold analysis [10]. While predictive accuracy is the central
performance measure of data analytical techniques which generate procedural
knowledge (neural nets, decision trees, etc.), the performance of an ILP system
is determined both by accuracy and degree of stereo-chemical insight provided.
ILP hypotheses can be easily stated in English and exemplified diagrammatically.
This allows cross-checking with the relevant biological and chemical literature.
Most importantly it allows for expert involvement in human background knowledge refinement and for final dissemination of discoveries to the wider scientific
community. In several of the comparative trials presented ILP systems provided
significant chemical and biological insights where other data analysis techniques
did not.

In his statement of the importance of this line of research to the Royal Society
[8] Sternberg emphasised the aspect of joint human-computer collaboration in
scientific discoveries. Science is an activity of human societies. It is our belief
that computer-based scientific discovery must support strong integration into
existing the social environment of human scientific communities. The discovered
knowledge must add to and build on existing science. The author believes that

the ability to incorporate background knowledge and re-use learned knowledge together with the comprehensibility of the hypotheses, have marked out ILP as a particularly effective approach for scientific knowledge discovery.

Acknowledgements

This work was supported partly by the Esprit Long Term Research Action ILP II (project 20237), EPSRC grant GR/K57985 on Experiments with Distribution-based Machine Learning and an EPSRC Advanced Research Fellowship held by the author. We would also like to thank both Pfizer UK and Smith-Kline Beecham for their generous support of some of this work.

References

1. I. Bratko and S. Muggleton. Applications of inductive logic programming. *Communications of the ACM*, 38(11):65–70, 1995.
2. P. Finn, S. Muggleton, D. Page, and A. Srinivasan. Pharmacophore discovery using the inductive logic programming system Progol. *Machine Learning*, 30:241–271, 1998.
3. R. King, S. Muggleton, R. Lewis, and M. Sternberg. Drug design by machine learning: The use of inductive logic programming to model the structure-activity relationships of trimethoprim analogues binding to dihydrofolate reductase. *Proceedings of the National Academy of Sciences*, 89(23):11322–11326, 1992.
4. R. King, S. Muggleton, A. Srinivasan, and M. Sternberg. Structure-activity relationships derived by machine learning: the use of atoms and their bond connectives to predict mutagenicity by inductive logic programming. *Proceedings of the National Academy of Sciences*, 93:438–442, 1996.
5. S. Muggleton, R. King, and M. Sternberg. Protein secondary structure prediction using logic-based machine learning. *Protein Engineering*, 5(7):647–657, 1992.
6. S. Muggleton and L. De Raedt. Inductive logic programming: Theory and methods. *Journal of Logic Programming*, 19,20:629–679, 1994.
7. A. Srinivasan, S. Muggleton, R. King, and M. Sternberg. Theories for mutagenicity: a study of first-order and feature based induction. *Artificial Intelligence*, 85(1,2):277–299, 1996.
8. M. Sternberg, R. King, R. Lewis, and S. Muggleton. Application of machine learning to structural molecular biology. *Philosophical Transactions of the Royal Society B*, 344:365–371, 1994.
9. M. Sternberg, R. Lewis, R. King, and S. Muggleton. Modelling the structure and function of enzymes by machine learning. *Proceedings of the Royal Society of Chemistry: Faraday Discussions*, 93:269–280, 1992.
10. M. Turcotte, S.H. Muggleton, and M.J.E. Sternberg. Protein fold recognition. In C.D. Page, editor, *Proc. of the 8th International Workshop on Inductive Logic Programming (ILP-98)*, LNAI 1446, pages 53–64, Berlin, 1998. Springer-Verlag.

Random Case Analysis of Inductive Learning Algorithms

Kuniaki Uehara

Research Center for Urban Safety and Security
Kobe University
Nada, Kobe, 657-8501 Japan

Abstract. In machine learning, it is important to reduce computational time to analyze learning algorithms. Some researchers have attempted to understand learning algorithms by experimenting them on a variety of domains. Others have presented theoretical methods of learning algorithm by using approximately mathematical model. The mathematical model has some deficiency that, if the model is too simplified, it may lose the essential behavior of the original algorithm. Furthermore, experimental analyses are based only on informal analyses of the learning task, whereas theoretical analyses address the worst case. Therefore, the results of theoretical analyses are quite different from empirical results. In our framework, called random case analysis, we adopt the idea of randomized algorithms. By using random case analysis, it can predict various aspects of learning algorithm's behavior, and require less computational time than the other theoretical analyses. Furthermore, we can easily apply our framework to practical learning algorithms.

1 Introduction

The main objective of this paper is to understand the learning behavior of inductive learning algorithms under various conditions. To this end, many researchers have studied systematic experimentation on a variety of problems in order to find empirical regularities. For example, UCI machine learning database [9] is widely used for experimental analysis. Since some domains may contain noise and other domains may include many irrelevant attributes, the algorithm's learning rate is affected by these invisible factors. Thus, experimental analysis leads to findings on the average case accuracy of an algorithm.

Others have carried out theoretical analyses based on the paradigm of probably approximately correct learning [4]. The PAC model has led to many important insights about the capabilities of machine learning algorithms, the PAC model can only deal with too simplified version of the actual learning algorithms. Furthermore, the PAC model produces overly-conservative estimates of error and does not take advantage of information available in actual training set [3].

Average-case analysis [10] was proposed to unify the theoretical and empirical approaches to analyze the behavior of learning algorithms under various conditions, along with information about the domain. Average-case analysis can explain empirical observations, predict average-case learning curves, and guide the development of new learning algorithms. The main drawback of average-case analysis is that it requires a detailed analysis of an individual algorithm to determine the condition under which the algorithm change the hypothesis. Therefore researchers have studied on the simple learning algorithms (i.e., wholist [10], one-level decision trees [5], Bayesian classifiers [7] and nearest neighbor algorithm [8]).

Pazzani [10] anticipated that the average-case analysis framework would scale to similar algorithms using more complex hypotheses, it would be possible, but computationally expensive, to model more complex learning algorithms with more expressive concepts. However, due to this complexity, only the limited number of training examples are considered by the average-case analysis.

In our framework, called random case analysis, we adopt the idea of randomized algorithms [6][13]. By randomized algorithms we mean algorithms which make random choices in the course of their execution. As a result, even for a fixed input, different runs of a randomized algorithm may give different results. By using random case analysis, we can predict various aspects of learning algorithm's behavior, requiring much less computational time than previous analyses. Furthermore, we can easily apply our framework to practical learning algorithms, such as ID3 [11], C4.5 [12], CBL [1] and AQ [15].

2 Random Case Analysis

2.1 Basic idea

The idea of randomized algorithms are applied in the fields of number theory, computational geometry, pattern matching and data structure maintenance. Many of the basic ideas of randomization were discovered and applied quite early in the field of computer science. For example, random sampling, random walk and randomized rounding can be used effectively in algorithm design.

We now consider the use of random sampling for the problem of random case analysis. Random sampling is based on the idea of drawing random samples from L (population) in order to determine the characteristics of L, thereby reducing the size of the problem. That is, random case analysis takes advantage of random sampling so as to evaluate specific learning algorithm under various conditions even with the large number of training sets, thereby reducing the computational cost for evaluation of learning algorithms.

Fig. 1 shows the simple framework of random case analysis. Note that N is the number of total trials. L is the original training set. K is a performance measure such as performance accuracy. In Fig. 1, independent trials of random case analysis are Bernoulli trials in the sense that successive trials are independent and at each trial the probability of appearance of a 'successful' classification remains constant. Then the distribution of success is given by the binomial distribution.

Now consider the fact that the number of possible concept descriptions grows exponentially with the number of training examples and the number of attributes. Thus, it would be interesting to obtain a reasonable number of trails.

1. $n \leftarrow 0, X \leftarrow 0$.
2. Draw from L a random sample $\hat{L}$ of size l.
3. Apply the learning algorithm to the training examples $\hat{L}$, and if
   the hypothesis can classify the test example correctly, $X \leftarrow X + 1$.
4. $n \leftarrow n + 1$.
5. If $n \geq N$ then goto (6), if $n < N$ then goto (2).
6. $K \leftarrow X/N$.

Fig. 1. A simple framework of random case analysis.

We are faced with a question in designing a random case analysis: how large must
N be so that the approximate performance accuracy achieves a higher level of
confidence. Tight answer to this question comes from a technique known as the
Chernoff bound [13].

Theorem 1 *Let X_i, $1 \leq i \leq n$ be independent Bernoulli trials such that* $\Pr(X_i = 1) = p_i$ *and* $\Pr(X_i = 0) = 1 - p_i$. *Let $X = \sum_{i=1}^{N} X_i$ and $\mu = \sum_{i=1}^{N} p_i > 0$. Then
for a real number $\delta \in (0, 1]$,*

$$\Pr(X < (1 - \delta)\mu) < \exp(-\mu\delta^2/2) \stackrel{\text{def}}{=} F^-(\mu, \delta) \tag{1}$$

$$\Pr(X > (1 + \delta)\mu) < \left[\frac{\exp(\delta)}{(1 + \delta)^{(1+\delta)}} \right]^{\mu} \stackrel{\text{def}}{=} F^+(\mu, \delta) \tag{2}$$

(1) indicates the probability that X is below $(1 - \delta)\mu$, whereas (2) shows the
probability that X exceeds $(1 + \delta)\mu$. These inequalities seek the bounds on the
"tail probabilities" of the binomial distribution.

Now consider random case analysis by using the Chernoff bound. Let X_i
be 1 if the hypothesis correctly classifies the test example at i-th trial and 0
otherwise. Let $X = \sum_{i=1}^{N} X_i$ be the number of correct classification out of N
trials. The performance accuracy is $K = X/N$ and its expectation is $\mu_K = \mu/N$.
Then we can get the following theorem.

Theorem 2 *Let $X_1, \cdots, X_N$ be independent Bernoulli trials with $\Pr(X_i = 1) = p_i$, $p_i \in (0, 1)$. Let $X = \sum_{i=1}^{N} X_i$ and $\mu = \sum_{i=1}^{N} p_i > 0$. Then for $\delta \in (0, 1]$,*

$$\Pr(K < (1 - \delta)\mu_K) < \exp(-\mu_K \delta^2 N/2) \stackrel{\text{def}}{=} F^-(\mu_K, \delta) \tag{3}$$

$$\Pr(K > (1 + \delta)\mu_K) < \left[\frac{\exp(\delta)}{(1 + \delta)^{(1+\delta)}} \right]^{\mu_K N} \stackrel{\text{def}}{=} F^+(\mu_K, \delta) \tag{4}$$

By using these inequalities (3) and (4), we will consider the minimal number of
trials for a given δ.

2.2   The minimal number of trials

We assume that $F^-(\mu_K, \delta)$ and $F^+(\mu_K, \delta)$ are constant. As N increases, we can derive the following inequalities from (3) and (4).

$$\Pr(K < (1 - \delta)\mu_K) < \exp(-\mu_K \delta^2 N/2) \stackrel{N \to \infty}{\longrightarrow} +0 \qquad (5)$$

$$\Pr(K > (1 + \delta)\mu_K) < \left[\frac{\exp(\delta)}{(1 + \delta)^{(1+\delta)}}\right]^{\mu_K N} \stackrel{N \to \infty}{\longrightarrow} +0 \qquad (6)$$

From (5), we can say that $K \geq (1 - \delta)\mu_K$ for any δ as N increases. Furthermore, as δ gets smaller, we can obtain the inequality $K \leq \mu_K$. In the same way, from (6), as N increases, $K \leq \mu_K$. From this observation, we can say that $K = \mu_K$ if N is a large number.

We now return to grappling with the simplification of (3) and (4) to obtain the minimal number of trials N_{min}^-. First of all, we will define the function $f^-(F^-, \delta)$ such that

$$f^-(F^-, \delta) \stackrel{\text{def}}{=} \frac{-2 \ln F^-(\mu_K, \delta)}{\delta^2} \qquad (7)$$

$f^-(F^-, \delta)$ determines the minimal number of trials where $K \geq (1 - \delta)\mu_K$. By using (3) and (7), we can obtain the minimal value N_{min}^-

$$N = \frac{-2 \ln F^-(\mu_K, \delta)}{\mu_K \delta^2} = \frac{f^-(F^-, \delta)}{\mu_K} \stackrel{\text{def}}{=} N_{min}^- \qquad (8)$$

In the same way, we will define the function $f^+(F^+, \delta)$

$$f^+(F^+, \delta) \stackrel{\text{def}}{=} \log_{\frac{\exp(\delta)}{(1+\delta)^{(1+\delta)}}} F^+(\mu_K, \delta) \qquad (9)$$

$f^+(F^+, \delta)$ determines the minimal number of trials where $K \geq (1 + \delta)\mu_K$. By using (4) and (9), we can obtain the minimal value N_{min}^+

$$N = \frac{\log_{\frac{\exp(\delta)}{(1+\delta)^{(1+\delta)}}} F^+}{\mu_K} = \frac{f^+(F^+, \delta)}{\mu_K} \stackrel{\text{def}}{=} N_{min}^+ \qquad (10)$$

By using (8) and (10), we will define the function

$$f(F^+, F^-, \delta) \stackrel{\text{def}}{=} \max\{f^+(F^+, \delta), f^-(F^-, \delta)\} \qquad (11)$$

This function determines the minimal number of trials where $(1 - \delta)\mu_K < K < (1+\delta)\mu_K$. The minimal value N_{min} which satisfies both (3) and (4) is as follows:

$$N_{min} = \max\{N_{min}^+, N_{min}^-\} = \frac{\max\{f^+(F^+, \delta), f^-(F^-, \delta)\}}{\mu_K}$$

$$= \frac{f(F^+, F^-, \delta)}{\mu_K} \qquad (12)$$

(12) means that if μ_K gets smaller the minimal number of trials exponentially increases. In order to avoid this problem, we will re-visit Theorem 2.

Now we will define the misclassification rate $K_{miss} = \bar{X}/N$ where $\bar{X} = \sum_{i=1}^{N} \bar{X}_i$. Let μ_{Kmiss} be the expectation of misclassification rate. Suppose that the misclassification rate K_{miss} achieves a high-level confidence, then we can obtain that $K = 1 - K_{miss}$. Given this, the probability that the misclassification rate K_{miss} is below $(1 - \delta)\mu_{Kmiss}$ is as follows:

$$\Pr(K_{miss} < (1 - \delta)\mu_{Kmiss}) < \exp(-(1 - \mu_K)\delta^2 N/2) \tag{13}$$

Furthermore, the probability that the misclassification rate exceeds $(1+\delta)\mu_{Kmiss}$ is derived from (4).

$$\Pr(K_{miss} > (1 + \delta)\mu_{Kmiss}) < \left[\frac{\exp(\delta)}{(1 + \delta)^{(1+\delta)}} \right]^{(1-\mu_K)N} \tag{14}$$

By using (7) and (9), we can transform (13) and (14) into (15) and (16) respectively.

$$N = \frac{-2 \ln F^-(\mu_K, \delta)}{(1 - \mu_K)\delta^2} = \frac{f^-(F^-, \delta)}{1 - \mu_K} \overset{\text{def}}{=} N_{miss\,min}^- \tag{15}$$

$$N = \frac{\log_{\frac{\exp(\delta)}{(1+\delta)^{(1+\delta)}}} F^+}{1 - \mu_K} = \frac{f^+(F^+, \delta)}{1 - \mu_K} \overset{\text{def}}{=} N_{miss\,min}^+ \tag{16}$$

Using (11), we arrive at

$$N_{missmin} = \max\{N_{missmin}^+, N_{missmin}^-\} = \frac{\max\{f^+(F^+, \delta), f^-(F^-, \delta)\}}{1 - \mu_K}$$

$$= \frac{f(F^+, F^-, \delta)}{1 - \mu_K} \tag{17}$$

From (17), we can obtain the minimal number of trials $N_{miss\,min}$ where the misclassification rate has a high-level confidence.

(12) and (17) specify that if we want to achieve a high-level confidence of random case analysis, we should repeat trials of random case analysis (Fig. 1) until either of the following conditions is satisfied.

$$N \geq N_{min} = \frac{f(F^+, F^-, \delta)}{\mu_K} \tag{18}$$

$$N \geq N_{miss\,min} = \frac{f(F^+, F^-, \delta)}{1 - \mu_K} \tag{19}$$

Given $\mu_K = X/N$, (18) and (19) are transformed into (20) and (21) respectively.

$$X \geq f(F^+, F^-, \delta) \tag{20}$$

$$N - X \geq f(F^+, F^-, \delta) \tag{21}$$

(20) and (21) are the conditions where the random case analysis algorithm can be terminated. In other words, the number of trials in random case analysis is less than $2 \times f(F^+, F^-, \delta)$ and greater than $f(F^+, F^-, \delta)$, thereby random case analysis can reduce the computational cost for evaluation of learning algorithms.

2.3 Improvement of random case analysis

Fig. 2 shows an improved version of random case analysis whose algorithm is extended by adding conditions (20) and (21). **EXAMPLE**(α) takes α as input and generates a case $(T_c, T_a) = (I_c, (a_1, \cdots, a_t))$ where I_c is the category of a case $(a_1, \cdots, a_t)$. α can be described as $(f, (p_1, p_2, \cdots, p_t))$, where f determines the category of a case and p_i is the set of probabilities of attribute values a_{ik} occurring in a case. Note that k is the number of attribute values of a_i. [1]. **LEARN**(L) learns a conceptual description β from a training set L. **LEARN**(L) is the inductive learning algorithm to be evaluated. **CLASSIFY**(β, T_a) determines the class of the case T_a according to the conceptual description β.

3 Implications for Learning Behavior

3.1 Comparison with PAC learning and average case analysis

First of all, we will compare the learning curves of the wholist algorithm based on the framework of random case analysis, PAC learning and average case analysis. The wholist algorithm [2] is a predecessor of the one-sided algorithm for pure conjunctive concepts. Although wholist is a relatively simple algorithm for random case analysis, the theoretical prediction has already been studied by using PAC model and average case model.

PAC model of wholist indicates that the minimal number of training examples for learning a monotone monomial concept is as follows:

$$m > \frac{1}{\epsilon}\left(n + \log_2 \frac{1}{\delta}\right) \tag{22}$$

where ϵ is an accuracy parameter and δ is a confidence parameter. n is the number of variables in a monotone monomial concept. Given this, we have

$$1 - \epsilon < 1 - \frac{1}{m}\left(n + \log_2 \frac{1}{\delta}\right) \tag{23}$$

---

[1] An important practical consideration that is ignored in Fig. 2 is the existence of noise in the source of examples. The noise can take on various forms and is modeled in two categories (i.e., class noise and attribute noise) in our research. We have extended **EXAMPLE**(α) so as to generate a case which contains the noise, but we will not mention the detail algorithm for the sake of simplicity.

RandomCaseAnalysis(α, l, F^+, F^-, δ)

α : concept description and the probabilities of attribute values
l : a size of training examples
F^+: upper bound
F^-: lower bound
δ : confidence parameter

begin
    $X \leftarrow 0$;
    $N \leftarrow 0$;
    while $(X < f(F^+, F^-, \delta)$ **and** $N - X < f(F^+, F^-, \delta))$ **do**
    begin
        $L \leftarrow \phi$;
        $N \leftarrow N + 1$;
        repeat l **do**
            $L \leftarrow L \cap \{\textbf{EXAMPLE}(\alpha)\}$;
        $\beta \leftarrow \textbf{LEARN}(L)$;
        $(T_c, T_a) \leftarrow \textbf{EXAMPLE}(\alpha)$;
        if $T_c = \textbf{CLASSIFY}(\beta, T_a)$ **then**
            $X \leftarrow X + 1$;
    end ;
    output X/N ;
end.

Fig. 2. Algorithm of random case analysis.

The left hand side shows the worst performance accuracy and the right hand side indicates its upper bound.

Fig. 3 indicates that the predicted accuracy of random case analysis is very similar to that of average case analysis. On the other hand, the predicted accuracy of PAC learning is over-conservative and its learning curve is quite different from the others. Note that the experimental condition is as follows: $\delta = 0.1$, $n = 3$ for PAC learning, $mono_{[001]}(x_1, x_2, x_3) = x_3$ for average case analysis, $mono_{[001]}(x_1, x_2, x_3) = x_3$, $F^+ = 0.05$, $F^- = 0.05$, $\delta = 0.005$ for random case analysis. Since the number of possible concept descriptions grows exponentially with the number of training examples and the number of attributes, average case analysis is limited to a small number of training examples (i.e., 10).

Fig. 4 shows the computational time required for random case analysis and average case analysis[2]. This figure shows that random case analysis can reduce the computational cost for evaluation of learning algorithms even with the large number of training examples, whereas computational time of average case analysis grows exponentially with the number of training examples.

---

[2] We used SGI Indy workstation with R4600 133 MHz CPU and 64MB memory under IRIX 5.3 OS. The program code is written in GNU Common Lisp 2.2.

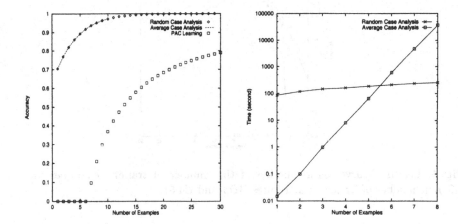

Fig. 3. Comparison of random case analysis with PAC learning and average case analysis.

Fig. 4. Computational time required for analyzing the wholist algorithm.

3.2 Behavioral implications of ID3 and C4.5

To model the induction of full decision trees is too difficult for both PAC learning and average case analysis, researchers [5] focused on a simple algorithm that constructs one-level decision trees. Recently, we have studied the average case analysis of N-level decision trees in the framework of average case analysis [14]. In this experiment, we will show the performance accuracies of ID3 and C4.5 by using random case analysis. We hold the number of classes constant at 3, and the number of attributes constant at 3. Each attribute has 3 values. We hold their probabilities of occurrences constant at $1/3$. Furthermore, we hold F^+ constant at 0.05, F^- constant at 0.05 and δ constant at 0.01. Under this condition, we generated a target concept from 24 examples.

Fig. 5 shows the learning curves of ID3 and C4.5 as a function of the number of training examples for two different number of irrelevant attributes (i.e., no irrelevant attribute and 2 irrelevant attributes) when other domain parameters are held constant. In this figure, we can say that ID3 and C4.5 are unaffected by the number of irrelevant attributes.

Fig. 6 shows the predicted effects of training examples and two levels of class noise (i.e., 0% and 10%) on performance accuracy and learning rate. The important point to observe is that, unlike the number of irrelevant attributes, the class noise mainly affects the overall rate of improvement. Another point is that C4.5 is somewhat robust with respect to class noise, whereas ID3 is not.

Fig. 7 presents similar results on the interaction between attribute noise and the number of training examples. This figure also indicates that attribute noise

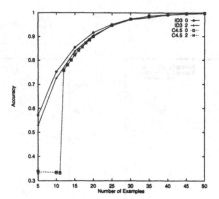

Fig. 5. Learning curves as a function of the number of training examples for two different number of irrelevant attributes (ID3 and C4.5)

affects the overall rate of improvement. That is, increasing the noise level flattens the learning curves of ID3 and C4.5. Here we should point out the fact that the performance accuracy of C4.5 is too bad if the number of training examples is small (i.e., the number of training example is less than 11). This means that C4.5 generates a tree with no leaf (i.e., a root) if the number of training examples is not enough for learning. This kind of learning behavior is hard to investigate without random case analysis.

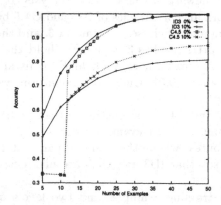

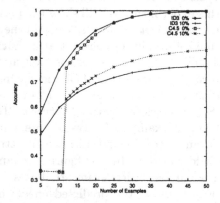

Fig. 6. Learning curves as a function of the number of training examples for two levels of class noise (ID3 and C4.5).

Fig. 7. Learning curves as a function of the number of training examples for two levels of attribute noise (ID3 and C4.5).

3.3 Behavioral implications of CBL

In recent years there has been growing interest in methods that store cases or instances in memory, and that apply these cases directly new situations. This method is called case-based learning [1]. The simplest and most widely studied class of techniques is nearest neighbor algorithm. Aha reported the experimental result (which they called CBL1). Langley [8] also studied the theoretical result of average case analysis. Although the basic nearest neighbor algorithm is simple enough to construct the average case model, slightly modified version of nearest neighbor algorithm (which they called CBL4) is difficult to construct the model. CBL4 is an extension of CBL1 in the sense that 1) it can reduce storage requirements, 2) tolerate noisy cases, and 3) learn attribute importance, represented as attribute weight settings.

Fig. 8 shows the interaction between two levels of irrelevant attributes and the number of training examples. This figure shows that CBL4 is unaffected by the number of irrelevant attributes whereas CBL1 is more sensitive to irrelevant attributes. This figure reveals that CBL4 tends to have slower learning rates because it must search for good attribute weight settings. These observations are consistent with Aha's report [1] on the sensitivity of CBL4 to the number of irrelevant attributes.

Fig. 9 shows the predicted effects of training examples and two levels of attribute noise. In [1], Aha applied CBL1 and CBL4 to only two noisy domains (i.e., LED-24 and Waveform-40) in UCI machine learning database and concluded that CBL4 outperforms CBL1. But our random case analysis reveals that Aha's experimental results does not prove that CBL4 will always outperforms CBL1 in more general domains.

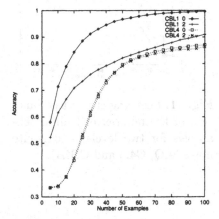 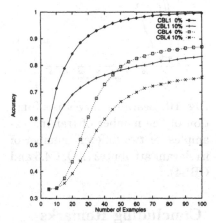

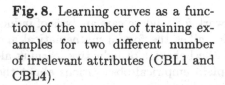

Fig. 8. Learning curves as a function of the number of training examples for two different number of irrelevant attributes (CBL1 and CBL4).

Fig. 9. Learning curves as a function of the number of training examples for two levels of attribute noise (CBL1 and CBL4).

Finally we shows the comparative results of AQ (a simplified version of AQ11), C4.5 and CBL4. Fig. 10 and Fig. 11 indicates that C4.5 outperforms both AQ and CBL4. Fig. 10 shows that AQ is affected by the number of irrelevant attributes, although C4.5 and CBL4 are unaffected by the number of irrelevant attributes. Fig. 11 shows that AQ and C4.5 are also dramatically affected by attribute noise, whereas CBL4 is robust compared with the other algorithms.

The second result to note is the peculiar 'S' shape of CBL4's learning curves, whereas the other learning curves begin to improve and then level off. Furthermore, increasing the noise level flattens or stretches the S shape, although increasing the number of irrelevant attributes shifts the learning curves somewhat to the right. Finally, quite surprising is that the performance accuracy of AQ decreases as the number of training examples increases. We have no idea to explain this phenomenon right now. In spite of this fact, the other theoretical approaches cannot reveal this phenomenon and such a learning behavior can only be found by random case analysis or experimental analysis with artificial data.

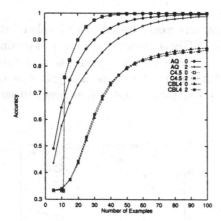

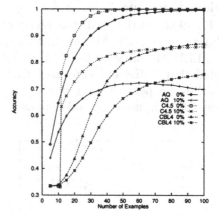

Fig. 10. Learning curves as a function of the number of training examples for two different number of irrelevant attributes (AQ, C4.5 and CBL4).

Fig. 11. Learning curves as a function of the number of training examples for two levels of attribute noise (AQ, C4.5 and CBL4).

4   Concluding Remarks

In this paper, we have presented a framework for random case analysis of inductive learning algorithms. Our framework unifies the formal mathematical and the empirical approaches to understand the behavior of inductive learning algorithms. Random case analysis can explain empirical observations about the behavior of various algorithms, make predictions, and guide the development of new learning algorithms. We have applied the framework to four different algorithms. We have verified through experimentation that the analysis accurately predict the expected performance accuracy.

The framework requires much more information about the training examples than both the average case model and the PAC model. Especially, although the average case analysis can obtain the precise theoretical value if the average case model is correct, our framework is randomized approximation algorithm for analyzing learning algorithms. That is, in our framework, we must carefully determine the values of F^+, F^- and δ so as to calculate the number of trails. Furthermore, the inequalities to calculate the number of trials, (12) and (17), are almost proportional to $1/\delta^2$, we must consider the tradeoff between the preciseness of analysis and the computational time to be executed.

References

1. Aha, D. W.: Case-Based Learning Algorithms, *Proc. of Case-Based Reasoning Workshop*, pp.147–158 (1991).
2. Bruner, J. S., Goodnow, J. J. and Austin, G. A.: *A Study of Thinking*, Wiley (1956).
3. Buntine, W.: A Critique of the Valiant Model, *Proc. of the 11th IJCAI*, pp.837–842 (1989).
4. Haussler, D.: Probably Approximately Correct Learning, *Proc. of the 8th AAAI*, pp.1101–1108 (1990).
5. Iba, W. and Langley, P.: Induction of One-Level Decision Trees, *Proc. of the 9th International Workshop on Machine Learning*, pp.233–240 (1992).
6. Karp, R. M.: An Introduction to Randomized Algorithms, *Discrete Applied Mathematics*, Vol.34, pp.165–201 (1991).
7. Langley, P., Iba, W. and Thompson, K.: An Analysis of Bayesian Classifiers, *Proc. of the 10th AAAI*, pp.223–228 (1992).
8. Langley, P. and Iba, W.: Average-Case Analysis of a Nearest Neighbor Algorithm, *Proc. of the 13th IJCAI*, pp.889–894 (1993).
9. Murphy, P. M. and Aha, D. W.: UCI Repository of Machine Learning Databases, Technical report, University of California, Department of Information and Computer Science, Irvine (1992).
10. Pazzani, M. J. and Sarett, W.: Average Case Analysis of Conjunctive Learning Algorithm, *Proc. of the 7th International Conference on Machine Learning*, pp.339–347 (1990).
11. Quinlan, J. R.: Induction of Decision Trees, *Machine Learning*, Vol.1, pp.81–106 (1986).
12. Quinlan, J. R.: C4.5: Programs for Machine Learning, Morgan Kaufmann (1993).
13. Raghaven, P.: Lecture Notes on Randomized Algorithms, Research report RC 15340, IBM (1990).
14. Sakamoto, T. and Uehara, K.: Induction of N-Level Decision Trees, *Trans. of IPSJ*, Vol.38, No.3, pp.419–428 (1997, in Japanese).
15. Shi, Z.: Principles of Machine Learning, International Academic Publishers (1992).

On Variants of Iterative Learning

Steffen Lange[1] and Gunter Grieser[2]

[1] Universität Leipzig, Institut für Informatik
Augustusplatz 10–11, 04109 Leipzig, Germany
slange@informatik.uni-leipzig.de

[2] Technische Universität Darmstadt, Fachbereich Informatik
Alexanderstraße 10, 64283 Darmstadt, Germany
gunter@intellektik.informatik.tu-darmstadt.de

Abstract. Within the present paper, we investigate the principal learning capabilities of iterative learners in some more details. The general scenario of iterative learning is as follows. An iterative learner successively takes as input one element of a text (an informant) of a target concept as well as its previously made hypothesis, and outputs a new hypothesis about the target concept. The sequence of hypotheses has to converge to a hypothesis correctly describing the target concept.

We study the following variants of this basic scenario. First, we consider the case that an iterative learner has to learn on redundant texts or informants, only. A text (an informant) is redundant, if it contains every data item infinitely many times. This approach guarantees that relevant information is, in principle, accessible at any time in the learning process. Second, we study a version of iterative learning, where an iterative learner is supposed to learn independent on the choice of the initial hypothesis. In contrast, in the basic scenario of iterative inference, it is assumed that the initial hypothesis is the same for every learning task which allows certain coding tricks.

We compare the learning capabilities of all models of iterative learning from text and informant, respectively, to one another as well as to finite inference, conservative identification, and learning in the limit from text and informant, respectively.

1 Introduction

Induction constitutes an important feature of learning. The corresponding theory is called inductive inference. Inductive inference may be characterized as the study of systems that map evidence on a target concept into hypotheses about it. The investigation of scenarios in which the sequence of hypotheses stabilizes to an accurate and finite description of the target concept is of some particular interest. The precise definitions of the notions evidence, stabilization, and accuracy go back to Gold [3] who introduced the model of learning in the limit.

The general situation investigated in Gold's model (cf. [3]) can be described as follows: Given more and more information concerning the concept to be learnt, the learning device has to produce hypotheses about the phenomenon to be inferred. The information sequence may contain only positive data, i.e., exactly all elements contained in the concept to be recognized, as well as both positive and negative data, i.e., all elements of the underlying learning domain which are classified with respect to their containment to the unknown concept. Those information sequences are called text and informant, respectively. The sequence of hypotheses has to converge to a hypothesis correctly describing the object to be learnt. Consequently, the inference process is an ongoing one.

However, Gold's model (cf. [3]) makes the unrealistic assumption that the learner has access to the whole initial segment of the information sequence provided so far. If huge data sets are around, no learning algorithm can use all the data or even large portions of it simultaneously for computing hypotheses about concepts represented by the data. Since each practical learning system has to deal with the limitations of space, variants of the general approach described above restricting the accessibility of input data have been discussed (cf., e.g., [14,9]). An intensively studied example is iterative learning. Here, the learning device (henceforth called iterative learner) is required to produce its actual hypothesis exclusively from its previous one and the next element in the information sequence.

Within the present paper, we investigate the principal learning capabilities of iterative learners in some more details. Thereby, we confine ourselves to study the learnability of indexable concept classes (cf. [1, 15]), only. The motivation of our study is based on the rather simple observation that there is no learning *per se*. Learning is embedded into scenarios of a more comprehensive usage. Such an environment is usually putting constraints on the way information is accessible, requirements hypotheses have to meet, and so on.

For illustration, consider the following scenario which is typical for several approaches to case-based reasoning (cf. [6]). A given case-based reasoning system is in use, i.e., some user is putting in repeatedly query cases and receives as the system's response proposals how to proceed with the query cases. If the proposals are satisfying, nothing has to be changed. If the outputs do not meet the users expectations or the environmental needs, (s)he is requested to provide data illustrating the misbehaviour. Based on this information, the system is supposed to change its state, and thereby to modify its behaviour appropriately. Thus, learning, in particular, some kind of iterative learning takes place. Learning succeeds, if the initial state is successfully transfered into a goal state (which meets all the users expectations) by processing only finitely many information.

In order to gain a better understanding of the principal learning capabilities of those case-based reasoning systems, it seems to be reasonable to consider them as a certain kind of iterative learners. Since the basic model of iterative learning does not reflect all their specifics very well, some modifications are in order.

First, in the learning scenario discussed above, it is highly desirable that every possible initial state of the system can be transformed into a goal state. The initial state of the system reflects a treasure of experiences that have proved their usefulness in the past; so it is justified to keep them if possible. Our notion of iterative learning with variable initial hypotheses (cf. Definition 4) reflects this intention. In contrast, in the basic model of iterative inference (cf. Definition 3), it is assumed that an iterative learner starts with an *a priori* fixed hypothesis. The initial hypothesis is the same for all learning tasks, so it does not carry any message. Note that this approach has some benefits, too. It gives the learner the freedom to code, up to a certain extent, information about the progress made in the actual learning task directly into intermediate hypotheses. In the modified model, such coding is meaningless, since one has to be aware that every intermediate hypothesis serves as initial hypothesis of different learning tasks, as well.

Second, in the basic model of iterative learning, a learner is supposed to learn on every possible information sequence. Thus, it may happen that relevant data items occur only once in the given information sequence. This may lead to situations, in which relevant data items will be overlooked, since they appear at the wrong time, and therefore learning may fail. However, this contradicts daily life experiences: if information is really important, it will not be presented only once. Redundant information sequences have the property that every data item appears infinitely many times, and therefore relevant data are, in principle, accessible at any time in the learning process. The corresponding learning model is called iterative learning from redundant texts and informants, respectively (cf. Definition 5).

As we will see, iterative learners that are supposed to learn from redundant information sequences, only, are much more powerful than those that have to be successful on every text and informant, respectively. When learning from positive data is concerned, iterative learning from redundant information sequences is exactly as powerful as conservative inference which itself is less powerful than learning in the limit (cf. Corollary 3 and Proposition 1). Interestingly, iterative learning from redundant informants is exactly as powerful as learning in the limit from positive and negative data (cf. Corollary 11). Consequently, if exclusively redundant information sequences have to be processed, it is justified to use iterative learners instead of unconstrained ones.

Surprisingly, even iterative learning with variable initial hypotheses from redundant informants turns out to be of the same learning power as learning in the limit (cf. Corollary 11). When learning from positive data is concerned, the situation changes. There are concept classes iteratively learnable from arbitrary texts that cannot be iteratively learnt with variable initial hypotheses even in case that exclusively redundant texts have to be processed (cf. Theorem 5).

As one may expect, the power of iterative learning with variable initial hypotheses from arbitrary texts and informants, respectively, is rather limited. In both cases, the corresponding learning model is incomparable to finite learning which itself is known to be very restrictive (cf. Theorems 8 and 15).

2 Preliminaries

$I\!N = \{0, 1, 2, ...\}$ is the set of all natural numbers. We set $I\!N^+ = I\!N \setminus \{0\}$. By $\langle ., . \rangle : I\!N \times I\!N \to I\!N$ we denote Cantor's pairing function. We write $A \# B$ to indicate that the sets A and B are incomparable, i.e., $A \setminus B \neq \emptyset$ and $B \setminus A \neq \emptyset$.

Any recursively enumerable set $\mathcal{X}$ is called a learning domain. By $\wp(\mathcal{X})$ we denote the power set of $\mathcal{X}$. Let $\mathcal{C} \subseteq \wp(\mathcal{X})$, and let $c \in \mathcal{C}$; then we refer to $\mathcal{C}$ and c as to a concept class and a concept, respectively. A concept class $\mathcal{C}$ is said to be inclusion-free iff $c \not\subset c'$ for all concepts $c, c' \in \mathcal{C}$. A concept class is said to be superfinite iff it contains all finite concepts over the learning domain $\mathcal{X}$ and at least one infinite concept.

In the sequel we deal with the learnability of indexable concept classes with uniformly decidable membership defined as follows (cf. [1]). A class of non-empty concepts $\mathcal{C}$ is said to be an indexable class with uniformly decidable membership provided there are an effective enumeration $(c_j)_{j \in I\!N}$ of all and only the concepts in $\mathcal{C}$ and a recursive function f such that for all $j \in I\!N$ and all $x \in \mathcal{X}$ we have:

$$f(j, x) = \begin{cases} 1, & \text{if } x \in c_j, \\ 0, & \text{otherwise.} \end{cases}$$

In the following we refer to indexable classes with uniformly decidable membership as to indexable classes, for short.

Next, we describe some well-known examples of indexable classes. Let Σ denote any fixed finite alphabet of symbols, and let Σ^* be the free monoid over Σ. Then $\mathcal{X} = \Sigma^*$ as well as $\mathcal{X} = \Sigma^+ = \Sigma^* \setminus \{\varepsilon\}$ (where ε is the empty string) serve as learning domains. As usual, we refer to subsets $L \subseteq \Sigma^*$ as to languages (instead of concepts). Then, the set of all context sensitive languages, context free languages, regular languages, and pattern languages, respectively, form indexable classes (cf. [1, 4]).

Let $X_n = \{0, 1\}^n$ be the set of all n-bit Boolean vectors. We consider $\mathcal{X} = \bigcup_{n \geq 1} X_n$ as learning domain. Then, the set of all concepts expressible as monomial, k-CNF, k-DNF, and k-decision list form indexable classes (cf. [12, 11]).

2.1 Gold-style learning from positive data

Let $\mathcal{X}$ be the underlying learning domain, let $c \subseteq \mathcal{X}$ be a concept, and let $t = (x_n)_{n \in I\!N}$ be an infinite sequence of elements from c such that $content(t) = \{x_n \mid n \in I\!N\} = c$. Then t is said to be a positive presentation or, synonymously, a text for c. By $text(c)$ we denote the set of all texts for c. A text t is said to be redundant provided that every element from c appears infinitely many times, i.e., for all $x \in c$, there are infinitely many $n \in I\!N$ with $x_n = x$. By $text^r(c)$ we denote the set of all redundant texts for c. Moreover, let t be a text, and let y be a number. Then, t_y denotes the initial segment of t of length $y + 1$, and $t_y^+ = \{x_n \mid n \leq y\}$. Furthermore, let $\sigma = x_0, \ldots, x_n$ be any finite sequence. Then we use $|\sigma|$ to denote the length of σ. Additionally, by $\sigma \diamond \tau$ we denote the concatenation of two finite sequences σ and τ.

As in [3] we define an inductive inference machine (abbr. IIM) to be an algorithmic device working as follows: The IIM takes as its input larger and larger initial segments of a positive presentation. After processing an initial segment, the IIM outputs a hypothesis, i.e., a number encoding a certain computer program. More formally, an IIM maps finite sequences of elements from $\mathcal{X}$ into numbers in $I\!N$.

The numbers output by an IIM are interpreted with respect to a suitably chosen hypothesis space $\mathcal{H}$. Since we exclusively deal with indexable classes $\mathcal{C}$, we always take as a hypothesis space an indexable class $\mathcal{H} = (h_j)_{j \in \mathbb{N}}$. The indices are regarded as suitable finite encodings of the concepts described by the hypotheses. When an IIM outputs a number j, we interpret it to mean that the machine is hypothesizing h_j. Clearly, $\mathcal{H}$ must be defined over the same learning domain $\mathcal{X}$ over which $\mathcal{C}$ is defined, and, moreover, $\mathcal{H}$ must comprise the target concept class $\mathcal{C}$.

Let t be a positive presentation, and let $y \in \mathbb{N}$. Then we use $M(t_y)$ to denote the hypothesis produced by M when fed the initial segment t_y. The sequence $(M(t_y))_{y \in \mathbb{N}}$ is said to converge to the number j iff all but finitely many terms of the sequence $(M(t_y))_{y \in \mathbb{N}}$ are equal to j.

Now, we define some models of learning. We start with learning in the limit.

Definition 1 ([3]). *Let $\mathcal{C}$ be an indexable class, let c be a concept, and let $\mathcal{H} = (h_j)_{j \in \mathbb{N}}$ be a hypothesis space. An IIM M $LimTxt_{\mathcal{H}}$–identifies c iff, for every $t \in text(c)$, there exists a $j \in \mathbb{N}$ such that the sequence $(M(t_y))_{y \in \mathbb{N}}$ converges to j and $c = h_j$.*

M $LimTxt_{\mathcal{H}}$–identifies $\mathcal{C}$ iff, for each $c \in \mathcal{C}$, M $LimTxt_{\mathcal{H}}$–identifies c.

Finally, let $LimTxt$ denote the collection of all indexable classes $\mathcal{C}$ for which there are an IIM M and a hypothesis space $\mathcal{H}$ such that M $LimTxt_{\mathcal{H}}$–identifies $\mathcal{C}$.

In the above definition Lim stands for "limit". Suppose, an IIM identifies some concept c. That means, after having seen only finitely many data of c the IIM reaches its (unknown) point of convergence and it computes a correct and finite description of the target concept. Hence, some form of learning must have taken place.

In general, it is not decidable whether or not an IIM M has already converged on a text t for the target concept c. Adding this requirement to Definition 1 results in finite learning (cf. [3]). The corresponding learning type is denoted by $FinTxt$.

Now, we define conservative IIMs. Intuitively, conservative IIMs maintain their actual hypothesis at least as long as they have received data that "provably misclassify" it.

Definition 2 ([1]). *Let $\mathcal{C}$ be an indexable class, let c be a concept, and let $\mathcal{H} = (h_j)_{j \in \mathbb{N}}$ be a hypothesis space. An IIM M $ConsvTxt_{\mathcal{H}}$–identifies c iff M $LimTxt_{\mathcal{H}}$–identifies c, and, for every $t \in text(c)$ and for all $y \in \mathbb{N}$, if $M(t_y) \neq M(t_{y+1})$ then $t_{y+1}^+ \nsubseteq h_{M(t_y)}$.*

M $ConsvTxt_{\mathcal{H}}$–identifies $\mathcal{C}$ iff, for each $c \in \mathcal{C}$, M $ConsvTxt_{\mathcal{H}}$–identifies c.

The learning type $ConsvTxt$ is analogously defined as above.

As it turned out, for proving some of the stated results, it is conceptually simpler to use the characterization of conservative learning equating it with set-driven inference (cf. [10]). Set-drivenness has been introduced in [13] and describes the requirement that the output of an IIM is only allowed to depend on the range of its input. More formally, an IIM M is said to be set-driven with respect to $\mathcal{C}$ iff $M(t_y) = M(\hat{t}_{y'})$ for all $y, y' \in \mathbb{N}$, and all texts $t, \hat{t}$ for concepts in $\mathcal{C}$ provided $t_y^+ = \hat{t}_{y'}^+$. By $s\text{-}LimTxt$ we denote the collection of all indexable classes for which there are a hypothesis space $\mathcal{H}$ and a set-driven IIM M that $LimTxt_{\mathcal{H}}$–identifies C.

2.2 Formalizing variants of iterative learning from positive data

Looking at the above definitions, we see that an IIM M has always access to the whole history of the learning process, i.e., in order to compute its actual guess M is fed all examples seen so far. In contrast to that, next we define iterative inductive inference machines. An iterative IIM is only allowed to use its last guess and the next element in the positive presentation of the target concept for computing its actual guess.

More formally, let $\mathcal{X}$ be the underlying learning domain. Then, an iterative IIM M is an algorithmic device that maps elements from $I\!N \times \mathcal{X}$ into $I\!N$. Let $t = (x_n)_{n \in I\!N}$ be any text for some concept $c \subseteq \mathcal{X}$, and let k be M's initial hypothesis. Then, we denote by $(M_n(k, t))_{n \in I\!N}$ the sequence of hypotheses generated by M when successively fed t, i.e., $M_0(k, t) = M(k, x_0)$ and, for all $n \in I\!N$, $M_{n+1}(k, t) = M(M_n(k, t), x_{n+1})$. Within the next definition we assume that M's initial hypothesis equals 0.

Definition 3 ([14]). *Let C be an indexable class, let c be a concept, and let $\mathcal{H} = (h_j)_{j \in I\!N}$ be a hypothesis space. An iterative IIM M $ItTxt_{\mathcal{H}}$–identifies c iff, for every $t \in text(c)$, there exists a $j \in I\!N$ such that the sequence $(M_n(0, t))_{n \in I\!N}$ converges to j and $c = h_j$.*

Finally, M $ItTxt_{\mathcal{H}}$–identifies C iff, for each $c \in C$, M $ItTxt_{\mathcal{H}}$–identifies c.

The resulting learning type $ItTxt$ is analogously defined as above.

Subsequently, we use the following convention. Let σ be any finite sequence of elements over the relevant learning domain. Then, we denote by $M_*(k, \sigma)$ the last hypothesis output by M when successively fed σ (as above, k denotes the initial hypothesis).

Within the following definition we consider a variant of iterative learning, where an iterative IIM has to learn successfully no matter which initial hypothesis has been selected.

Definition 4. *Let C be an indexable class, let c be a concept, and let $\mathcal{H} = (h_j)_{j \in I\!N}$ be a hypothesis space. An iterative IIM M $It^v Txt_{\mathcal{H}}$–identifies c iff, for every $t \in text(c)$ and every initial hypothesis $k \in I\!N$, there exists a $j \in I\!N$ such that the sequence $(M_n(k, t))_{n \in I\!N}$ converges to j and $c = h_j$.*

Finally, M $It^v Txt_{\mathcal{H}}$–identifies C iff, for each $c \in C$, M $It^v Txt_{\mathcal{H}}$–identifies c.

The resulting learning type $It^v Txt$ is analogously defined as above.

Finally, we define versions of the models of iterative learning introduced above, where for the iterative IIMs it is sufficient to learn from redundant texts, only. More formally:

Definition 5. *Let C be an indexable class, let c be a concept, and let $\mathcal{H} = (h_j)_{j \in I\!N}$ be a hypothesis space. An iterative IIM M $ItTxt_{\mathcal{H}}^r$ $[It^v Txt_{\mathcal{H}}^r]$–identifies c iff, for every redundant text $t \in text^r(c)$ [and every initial hypothesis $k \in I\!N$], there exists a $j \in I\!N$ such that the sequence $(M_n(0, t))_{n \in I\!N}$ $[(M_n(k, t))_{n \in I\!N}]$ converges to j and $c = h_j$.*

Finally, M $ItTxt^r_{\mathcal{H}}$ $[It^v Txt^r_{\mathcal{H}}]$–identifies C iff, for each $c \in C$, M $ItTxt^r_{\mathcal{H}}$ $[It^v Txt^r_{\mathcal{H}}]$–identifies c.

The learning types $ItTxt^r$ and $It^v Txt^r$ are analogously defined as above.

3   Iterative learning from positive data

In this section, we compare the learning capabilities of all models of iterative learning from positive data to one another as well as to finite inference, learning in the limit and conservative identification from text.

First, we summarize the previously known results (cf. [7–10]).

Proposition 1. *FinTxt $\subset$ ItTxt $\subset$ ConsvTxt $=$ s-LimTxt $\subset$ LimTxt.*

In case that it is guaranteed that the relevant information appears infinitely many times in a text, any conservative learner can be simulated by an iterative IIM that has the same learning power. More formally:

Theorem 1. *ConsvTxt $\subseteq$ ItTxt^r.*

Proof. Let $\mathcal{X}$ be the relevant learning domain over which C is defined. Assume $C \in ConsvTxt$. Applying the characterization of $ConsvTxt$ from [8], we know that there are a hypothesis space $\mathcal{H} = (h_j)_{j \in \mathbb{N}}$ and a computable function T that assigns a finite telltale set T_j to every hypothesis h_j. More formally, on every input $j \in \mathbb{N}$, T enumerates a finite set T_j and stops. Furthermore, T_j has the following features: (1) $T_j \subseteq h_j$ and (2), for all $k \in \mathbb{N}$, $T_j \subseteq h_k$ implies $h_k \not\subset h_j$.

Let $\mathcal{F} = (F_j)_{j \in \mathbb{N}}$ denote any repetition free enumeration of all finite subsets of $\mathcal{X}$, where $F_0 = \emptyset$. Furthermore, we assume an effective procedure computing, for every finite set $F \subseteq \mathcal{X}$, its uniquely determined index $\#(F)$ in $\mathcal{F}$. In order to show that $C \in ItTxt^r$, we select the following hypothesis space $\hat{\mathcal{H}} = (\hat{h}_{\langle j,k,l\rangle})_{j,k,l \in \mathbb{N}}$. For all $j, k, l \in \mathbb{N}$, let $\hat{h}_{\langle 0,k,l\rangle} = F_k$ and $\hat{h}_{\langle j+1,k,l\rangle} = h_j$.

The desired iterative IIM M is defined as follows. Suppose any $\langle j,k,l\rangle \in \mathbb{N}$ and any $x \in \mathcal{X}$. Note that, by definition of Cantor's pairing function, $\langle 0,0,0\rangle = 0$ which is, by definition, M's initial hypothesis.

IIM M: "On input $\langle j,k,l\rangle$ and x do the following:

Determine $F' = F_k \cup \{x\}$.

If $j = 0$ or $x \notin h_{j-1}$, check, for all $z = 0, \ldots, l+1$, whether or not $T_z \subseteq F' \subseteq h_z$. If there is a z passing this test, fix the least one, say $\hat{z}$, and output the hypothesis $\langle \hat{z}+1, \#(F'), l+1\rangle$. Otherwise, output the hypothesis $\langle 0, \#(F'), l+1\rangle$.

If $x \in h_{j-1}$, test, for all $z = 0, \ldots, l$, whether or not $x \in T_z$. If such a z is found, output the hypothesis $\langle j, \#(F'), l\rangle$. Otherwise, output the hypothesis $\langle j,k,l\rangle$."

Due to space limitations, a verification of M's correctness is omitted.   ∎

Interestingly, iterative IIMs cannot outperform conservative learners, even if the iterative IIM has to learn from redundant texts, only. Thus, the weakness of iterative learners (compared to the capabilities of unconstrained IIMs) cannot be compensated, even if the relevant data appear infinitely many times in the texts.

In order to elaborate the result mentioned above we heavily exploit the fact that conservative learners are exactly as powerful as set-driven IIMs (cf. Proposition 1). Thereby, we adapt an idea from [5] and [9].

Theorem 2. $ItTxt^r \subseteq s\text{-}LimTxt$.

Proof. Let $\mathcal{X}$ be the relevant learning domain over which $\mathcal{C}$ is defined, and assume $\mathcal{C} \in ItTxt^r$. Then there are an IIM M and a hypothesis space $\mathcal{H} = (h_j)_{j \in I\!N}$ such that M $ItTxt^r_{\mathcal{H}}$–identifies $\mathcal{C}$. For proving $\mathcal{C} \in s\text{-}LimTxt$, first we construct a suitable hypothesis space $\hat{\mathcal{H}} = (\hat{h}_j)_{j \in I\!N}$. Let $\mathcal{F} = (F_j)_{j \in I\!N}$ and $\#(F)$ be defined as in the demonstration of Theorem 1 above. Then, we define $\hat{h}_{2j} = h_j$ and $\hat{h}_{2j+1} = F_j$ for every $j \in I\!N$. For every non-empty finite set $T \subseteq \mathcal{X}$, we define $ref(T) = x_0, x_1, \ldots, x_{card(T)-1}$ to be the repetition free enumeration of all the elements of T in lexicographical order. Furthermore, we set $exh(T) = x_0 \diamond x_0, x_1 \diamond x_0, x_1, x_2 \diamond \cdots \diamond x_0, x_1, \ldots, x_{card(T)-1}$.

The desired set-driven IIM $\hat{M}$ takes as its inputs finite sets T, and is defined as follows:

IIM $\hat{M}$: "On input T do the following:
  Determine $exh(T)$. Check, for all $x \in T$, whether or not $M_*(0, exh(T)) = M_*(0, exh(T) \diamond x)$.
  If it is, output $2 \cdot M_*(0, exh(T))$, and request the next input.
  Otherwise, output $2 \cdot \#(T) + 1$, and request the next input."

By definition, $\hat{M}$ is set-driven. For showing that $\hat{M}$ $LimTxt_{\hat{\mathcal{H}}}$–infers $\mathcal{C}$, let $c \in \mathcal{C}$, and let $t \in text(c)$. We distinguish the following cases.

Case 1. c is finite.

Then, there exists an $n \in I\!N$ with $t_n^+ = c$. It suffices to show that $c = \hat{h}_{\hat{M}(c)}$. If $\hat{M}(c) = 2 \cdot \#(c) + 1$, we are done, by construction. Otherwise, for all $x \in c$, we have $M_*(0, exh(c)) = M_*(0, exh(c) \diamond x)$. Let $j = M_*(0, exh(c))$. Hence, M converges to j when fed the redundant text $exh(c) \diamond ref(c) \diamond ref(c) \diamond ref(c) \diamond \cdots$ from $text^r(c)$. Since M learns c, we are done.

Case 2. c is infinite.

Let $t^c = (x_j)_{j \in I\!N}$ be the lexicographically ordered positive presentation for c. Thus, $t^{exh} = x_0 \diamond x_0, x_1 \diamond x_0, x_1, x_2 \diamond \cdots$ is a redundant text from $text^r(c)$. Since M $ItTxt^r$–learns c from t^{exh}, there exists an $n_0 \in I\!N$ such that $M_*(0, t_{n_0}^{exh}) = M_*(0, t_n^{exh})$ for all $n \geq n_0$. By the choice of t^{exh}, we immediately obtain that $M_*(0, t_{n_0}^{exh} \diamond x) = M_*(0, t_{n_0}^{exh})$ for all $x \in c$. Since M learns c from t^{exh} we have $c = h_j = \hat{h}_{2j}$. Let $\sigma = t_{n_0+1}^{exh}$. Finally, since $t \in text(c)$, there exists an index m_0 such that $\sigma^+ \subseteq t_{m_0}^+$. Thus, σ is a prefix of $exh(t_{m_0}^+)$, and hence $\hat{M}(t_m^+) = 2j$ for all $m \geq m_0$. ∎

Furthermore, taking into consideration that $ItTxt \subset ConsvTxt$ (cf. [9]), we may easily conclude:

Corollary 3. $ItTxt \subset ItTxt^r = ConsvTxt$.

Next, we show that intermediate hypotheses have to be used to reflect the progress made in the learning process. Without this option, iterative learners are unable to exploit the whole extra information that is provided within redundant texts. In order to achieve the announced insight, we start with a theorem that illuminates the structural properties of $It^v\,Txt^r$–learnable concept classes.

Theorem 4. *Let C be an indexable class. $C \in It^v\,Txt^r$ iff C is inclusion-free.*

Proof. First, suppose that an inclusion-free indexable class $C = (c_j)_{j \in I\!N}$ is given. Select the hypothesis space $\mathcal{H} = (h_{\langle j,n \rangle})_{j,n \in I\!N}$ that meets, for all $j, n \in I\!N$, $h_{\langle j,n \rangle} = c_j$. Then, the following IIM M $It^v\,Txt^r_\mathcal{H}$–identifies C. For all $k \in I\!N$ and all possible input data x, let $M(k, x) = min\{j \mid j \geq k,\, x \in h_j\}$. We omit the details.

Next we show that $It^v\,Txt^r$-identifiable classes must be inclusion-free. To see this assume, for a moment, that there is an indexable class $C \in It^v\,Txt^r$ that is not inclusion-free. Hence, there are an iterative IIM M and a hypothesis space $\mathcal{H}$ such that M $It^v\,Txt^r_\mathcal{H}$–identifies C. Let $c, c' \in C$ with $c' \subset c$, let k be some final hypothesis[1] of M for c, and let t and t' be any redundant text for c and c', respectively. Since $c' \subset c$ and due to choice of the initial hypothesis k, we know that, for all $n \in I\!N$, $M_*(k, t_n) = M_*(k, t'_n) = k$, and therefore M fails to learn at least one of both concepts. ∎

Theorem 5. *$ItTxt \# It^v\,Txt^r$.*

Proof. Consider the class of all finite concepts C_{fin} over the learning domain $\mathcal{X}$. Clearly, $C_{fin} \in ItTxt$, but C_{fin} is not inclusion-free, and therefore $C_{fin} \notin It^v\,Txt^r$, by Theorem 4. On the other hand, let C be the class of all concepts c_j over the learning domain $\mathcal{X} = \{a\}^+$ with $c_j = \{a\}^+ \setminus \{a^{j+1}\}$. Clearly, C is inclusion-free, and, by Theorem 4, $C \in It^v\,Txt^r$. Since $C \notin ItTxt$ (cf. [9]), we are done. ∎

Furthermore, since $It^v\,Txt \subseteq It^v\,Txt^r$ and $It^v\,Txt \subseteq ItTxt$, we obtain:

Corollary 6.

(a) $It^v\,Txt \subset It^v\,Txt^r$.

(b) $It^v\,Txt \subset ItTxt$.

Our next result puts the weakness of $It^v\,Txt$ into the right perspective.

Theorem 7. *$FinTxt \setminus It^v\,Txt \neq \emptyset$.*

Proof. Let C be the indexable class that contains exactly all $c \subseteq \{a\}^+$ with $card(c) = 2$. Obviously, $C \in FinTxt$. On the other hand, one easily verifies that even the finite subclass C' that contains the concepts $\{a, a^2\}$, $\{a, a^3\}$, and $\{a^2, a^3\}$ does not belong to $It^v\,Txt$. To see this suppose that there are an iterative IIM M and a hypothesis space $\mathcal{H}$ such that M $It^v\,Txt$–identifies C'. Let k be some final hypothesis of M for $\{a^2, a^3\}$. Thus, $M(k, a^2) = M(k, a^3) = k$. Consequently, M, when starting with the initial hypothesis k, outputs exactly the same sequence of hypotheses when fed the texts $t = a^2, a, a, \ldots$ for $\{a, a^2\}$ and $t' = a^3, a, a, \ldots$ for $\{a, a^3\}$. Thus, M must fail to learn at least one of both concepts, a contradiction. ∎

However, $It^v\,Txt$ may outperform $FinTxt$, as well.

Theorem 8. *$FinTxt \# It^v\,Txt$.*

Proof. By Theorem 7, it remains to show that $It^v\,Txt \setminus FinTxt \neq \emptyset$. Let $(\varphi_j)_{j \in I\!N}$ denote any fixed programming system of all (and only all) partial recursive functions over $I\!N$, and let $(\phi_j)_{j \in I\!N}$ be any associated complexity measure (cf. [2]). Let C be the indexable class that contains all concepts $c_{2j} =$

---

[1] k is said to be a final hypothesis of M for c provided that $M(k, x) = k$ for all $x \in c$. Note that k must exist, since M, in particular, $It^v\,Txt^r_\mathcal{H}$–identifies c.

$\{a^j b, a^j b^{\phi_j(j)+100}\}$ and $c_{2j+1} = \{a^j b, a^j b^{\phi_j(j)+200}\}$. Note that $c_{2j} = c_{2j+1} = \{a^j b\}$ in case that $\varphi_j(j)$ is undefined. One easily verifies that $\mathcal{C} \in It^v Txt$. On the other hand, an IIM M that finitely learns $\mathcal{C}$ can be used to decide the halting problem. On input $j \in \mathbb{N}$, one has to check in parallel, for all $z = 0, 1, \ldots$, whether $\phi_j(j) \leq z$ or M, when processing the initial segment t_z of the text $t = a^j b, a^j b, \ldots$, outputs its final and correct hypothesis. One easily verifies that in the latter case $\varphi_j(j)$ must be undefined, since, otherwise, M cannot finitely learn both, c_{2j} and c_{2j+1}. ∎

It is quite obvious that $FinTxt$ contains exclusively indexable concept classes that are inclusion-free. Hence, we may conclude:

Corollary 9. $FinTxt \subset It^v Txt^r$.

The following picture displays the achieved separations and coincidences of the considered learning types. Each learning type is represented as a vertex in a directed graph. A directed edge (or path) from vertex A to vertex B indicates that A is a proper subset of B, and no edge (or path) between these vertices imply that A and B are incomparable.

$$
\begin{array}{c}
LimTxt \\
\uparrow \\
ItTxt^r = ConsvTxt = s\text{-}LimTxt \\
ItTxt \qquad It^v Txt^r \\
FinTxt \qquad It^v Txt
\end{array}
$$

4 Iterative learning from positive and negative data

Next, we study iterative learning from positive and negative data. Thus, we have to introduce some more notations and definitions.

Let $\mathcal{X}$ be the underlying learning domain, let $c \subseteq \mathcal{X}$ be a concept, and let $i = ((x_n, b_n))_{n \in \mathbb{N}}$ be any sequence of elements of $\mathcal{X} \times \{+, -\}$ such that $content(i) = \{x_n \mid n \in \mathbb{N}\} = \mathcal{X}$, $i^+ = \{x_n \mid n \in \mathbb{N}, b_n = +\} = c$ and $i^- = \{x_n \mid n \in \mathbb{N}, b_n = -\} = \mathcal{X} \setminus c = \bar{c}$. Then, we refer to i as an informant. By $info(c)$ we denote the set of all informants for c, and by $info^r(c)$ the set of all redundant informants for c, i.e., informants having the property that, for all $x \in \mathcal{X}$, there are infinitely many $n \in \mathbb{N}$ with $x_n = x$. We use i_y to denote the initial segment of i of length $y + 1$, and define $i_y^+ = \{x_n \mid n \leq y, b_n = +\}$ and $i_y^- = \{x_n \mid n \leq y, b_n = -\}$.

Furthermore, let $c \subseteq \mathcal{X}$, and let $(x, b) \in \mathcal{X} \times \{+, -\}$. Then, c is said to be consistent with (x, b) (abbr. $cons(c, (x, b))$) provided that $x \in c$, if $b = +$, and $x \notin c$, otherwise.

The learning models $LimInf$ and $FinInf$ are defined analogously as their text counterparts by replacing text by informant. Finally, we extend the definitions of all variants of iterative learning in the same way, and denote the resulting learning types by $ItInf$, $It^v Inf$, $ItInf^r$, and $It^v Inf^r$, respectively.

As in the previous section, we first summarize the known results (cf. [7]).

Proposition 2. $FinInf \subset ItInf \subset LimInf$.

In contrast to the text case, iterative learning from redundant positive and negative data is at least as powerful as learning in the limit from informant. This add-on in learning power can also be observed, if iterative learners have to be successful no matter which initial hypothesis has been selected.

Theorem 10. *Let C be an indexable class. Then, $C \in It^v Inf^r$.*

Proof. Let $C = (c_j)_{j \in I\!N}$ be an indexable concept class. Select the hypothesis space $\mathcal{H} = (h_{\langle j,n \rangle})_{j,n \in I\!N}$ that meets, for all $j, n \in I\!N$, $h_{\langle j,n \rangle} = c_j$. We claim that the following IIM M $It^v Inf^r_{\mathcal{H}}$–identifies C. For all $k \in I\!N$, and all data $(x,b) \in \mathcal{X} \times \{+,-\}$, we let $M(k,(x,b)) = min\{j \mid j \geq k, \, cons(h_j,(x,b))\}$.

Clearly, M implements the identification by enumeration principle (cf. [3]), and M, when fed any redundant informant for some $c \in C$, converges to the least $j \geq k$ that meets $h_j = c$, where k is M's initial hypothesis. ∎

Finally, since, by definition, $It^v Inf^r \subseteq ItInf^r$ and since every indexable concept class is $LimInf$–identifiable (cf. [3]), we can conclude:

Corollary 11. $It^v Inf^r = ItInf^r = LimInf$.

The picture changes drastically, if iterative learning from non-redundant informants is considered. However, in contrast to the text case, $It^v Inf$ contains relatively rich concept classes.

Observation 12. $C_{fin} \in It^v Inf$.

Proof. As in the proof of Theorem 1, let $\mathcal{F} = (F_j)_{j \in I\!N}$ denote any repetition free enumeration of all finite subsets of the learning domain $\mathcal{X}$ and assume any effective procedure computing, for every finite set $F \subseteq \mathcal{X}$, its uniquely determined index $\#(F)$ in $\mathcal{F}$. We choose $\mathcal{F}$ as hypothesis space and define the needed iterative learner M as follows. Let $k \in I\!N$ and (x,b) be given. Then, we let $M(k,(x,b)) = \#(F_k \cup \{x\})$, if $b = +$, and $M(k,(x,b)) = \#(F_k \setminus \{x\})$, if $b = -$. Further details are omitted. ∎

If the target concept class contains both finite and infinite concepts, it might be inevitable to select the initial hypothesis appropriately. To see this let C_s be the indexable class that contains the concept $c_0 = \{a\}^+$ and all singleton concepts $c_j = \{a^j\}$, $j \in I\!N^+$, over the learning domain $\mathcal{X} = \{a\}^+$.

Observation 13. $C_s \notin It^v Inf$.

Proof. Suppose to the contrary that there are an iterative learner M and a hypothesis space $\mathcal{H} = (h_j)_{j \in I\!N}$ such that M $It^v Inf_{\mathcal{H}}$–identifies C_s. Since M, in particular, learns c_0, there has to be some final hypothesis k of M for c_0, i.e., $M(k,(a^m,+)) = k$ for all $m \in I\!N^+$. Next, consider the hypotheses $M_*(k,i_n)$, $n \in I\!N$, generated by M when successively processing the canonical informant $i = (a,+),(a^2,-),(a^3,-),\ldots$ for the concept $c_1 \in C$. Since M has to infer c_1, there have to be $j \in I\!N$ and $z \in I\!N^+$ that satisfy $h_j = c_1$, $M_*(k,i_z) = j$, and $M(j,(a^m,-)) = j$ for all $m \geq z$. Finally, consider the informant $\hat{\imath} = (a^z,+),(a^2,-),\ldots,(a^{z-1},-),(a^{z+1},-),(a,-),(a^{z+2},-),(a^{z+3},-),\ldots$ and $\tilde{\imath} = (a^{z+1},+),(a^2,-),\ldots,(a^{z-1},-),(a^z,-),(a,-),(a^{z+2},-),(a^{z+3},-),\ldots$ for $c_z = \{a^z\}$ and $c_{z+1} = \{a^{z+1}\}$, respectively. One immediately sees that, for all $n \in I\!N$, $M_*(k,\hat{\imath}_n) = M_*(k,\tilde{\imath}_n)$, and thus M fails to infer at least one of both concepts, a contradiction. ∎

One easily verifies that the concept class C_s is a subclass of the well-known family of pattern languages (cf. [1]). Moreover, the proof idea presented above can easily be adapted to show that there is no superfinite class at all that belongs to $It^v Inf$. Hence, we have:

Corollary 14.

(a) *The set of all pattern languages C_{pat} is not $It^v Inf$–identifiable.*
(b) *Let C be any indexable class that is superfinite. Then, $C \notin It^v Inf$.*

Moreover, it is well-know that $C_{pat} \in FinInf$ and $C_{fin} \notin FinInf$ (cf., e.g., [15]). Hence, we may conclude:

Theorem 15. $FinInf \# It^v Inf$.

Since $FinInf \subset ItInf$ (cf. Proposition 2) and $It^v Inf \subseteq ItInf$, we obtain the missing part in the picture for the informant case.

Corollary 16. $It^v Inf \subset ItInf$.

The following figure summarizes the established relations of the considered learning types for the informant case. The semantics is analogous to that of the corresponding figure in Section 3.

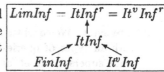

$$LimInf = ItInf^r = It^v Inf^r$$
$$\uparrow$$
$$ItInf$$
$$FinInf \qquad It^v Inf$$

References

1. Angluin, D., Inductive inference of formal languages from positive data, *Information and Control* **45**, 117–135, 1980.
2. Blum, M., A machine independent theory of the complexity of recursive functions, *Journal of the ACM* **14**, 322–336, 1967.
3. Gold, E.M., Language identification in the limit, *Information and Control* **10**, 447–474, 1967.
4. Hopcroft, J.E., and Ullman, J.D., "Formal Languages and their Relation to Automata," Addison-Wesley, 1969.
5. Kinber, E., and Stephan, F., Language learning from texts: Mind changes, limited memory and monotonicity, *in* "Proceedings 8th Annual ACM Conference on Computational Learning Theory," pp. 182–189, ACM Press, 1995.
6. Kolodner, J.K., An introduction to case-based reasoning. *Artificial Intelligence Review* **6**, 3–34, 1992.
7. Lange, S., and Zeugmann, T., Types of monotonic language learning and their characterization, *in* "Proceedings 5th Annual ACM Workshop on Computational Learning Theory," pp. 377–390, ACM Press, 1992.
8. Lange, S., and Zeugmann, T., Language learning in dependence on the space of hypotheses, *in* "Proceedings 6th Annual ACM Conference on Computational Learning Theory," pp. 127–136, ACM Press, 1993.
9. Lange, S., and Zeugmann, T., Incremental learning from positive data, *Journal of Computer and System Sciences* **53**, 88–103, 1996.
10. Lange, S., and Zeugmann, T., Set-driven and rearrangement-independent learning of recursive languages, *Mathematical Systems Theory* **29**, 599–634, 1996.
11. Rivest, R., Learning decision lists, *Machine Learning* **2**, 229–246, 1987.
12. Valiant, L.G., A theory of the learnable, *Communications of the ACM* **27**, 1134–1142, 1984.
13. Wexler, K., and Culicover, P., "Formal Principles of Language Acquisition," MIT Press, 1980.
14. Wiehagen, R., Limes–Erkennung rekursiver Funktionen durch spezielle Strategien, *Journal of Information Processing and Cybernetics (EIK)* **12** , 93–99, 1976.
15. Zeugmann, T., and Lange, S., A guided tour across the boundaries of learning recursive languages, *in* "Algorithmic Learning for Knowledge-Based Systems," Lecture Notes in Artificial Intelligence, Vol. 961, pp. 193–262, Springer-Verlag, 1995.

Uniform Characterizations of Polynomial-Query Learnabilities

Yosuke Hayashi[1], Satoshi Matsumoto[2],
Ayumi Shinohara[1], and Masayuki Takeda[1]

[1] Department of Informatics, Kyushu University 33, Fukuoka 812-8581, JAPAN
[2] Faculty of Science, Tokai University, Kanagawa 259-1291, JAPAN
{yosuke,ayumi,takeda}@i.kyushu-u.ac.jp, matumoto@ss.u-tokai.ac.jp

Abstract. We consider the exact learning in the query model. We deal with all types of queries introduced by Angluin: membership, equivalence, superset, subset, disjointness and exhaustiveness queries, and their weak (or restricted) versions where no counterexample is returned. For each of all possible combinations of these queries, we uniformly give complete characterizations of boolean concept classes that are learnable using a polynomial number of polynomial sized queries. Our characterizations show the equivalence between the learnability of a concept class C using queries and the existence of a good query for any subset H of C which is guaranteed to reject a certain fraction of candidate concepts in H regardless of the answer. As a special case for equivalence queries alone, our characterizations directly correspond to the lack of the approximate fingerprint property, which is known to be a sufficient and necessary condition for the learnability using equivalence queries.

1 Introduction

With the remarkable advances in computer and network technology, a large quantity of data obtained from scientific experiments is available. It is an urgent and very important problem to establish methods to discover some rules which explain such a large quantity of data. Because the data is so large, it is expected to use computers to analyze the data. One approach is to apply a machine learning system which learns concepts from examples, in order to discover rules automatically. Moreover, a successful learning algorithm using queries would give us a good strategy to make experiments within a reasonable amount of time, in order to identify underlying rules. For these purpose, we have to clarify the possibilities and limitations for computers to learn concepts from examples.

The exact learning model due to Angluin [1] is one of the most popular models in the field of learning theory. In this model, a learner is required to identify a target concept exactly using queries which give partial information about the target concept to the learner. Angluin introduced six kinds of queries, membership, equivalence, superset, subset, disjointness, and exhaustiveness. In some cases, *weak* (or *restricted*) versions of queries are often used, where no counterexample is provided to a learner.

Among these queries, membership and equivalence queries have been focused on especially, and there have been some individual approaches to show combinatorial properties in order to characterize the learnability using each of three combinations of membership and equivalence queries. For equivalence queries alone, Angluin [2] introduced a notion of *approximate fingerprint property* as a tool for proving non-learnability. Gavaldà [4] showed that the property (with a slight modification) can be used to prove the converse: if a concept class does not have an approximate fingerprint property, then the concept class is exactly learnable using a polynomial number of polynomial sized equivalence queries. (See also [3]). For membership queries alone, Goldman and Kearns [5] showed that the teaching dimension gives a lower bound for the number of membership queries required to learn. Hegedüs [6] generalized it so that the generalized teaching dimension of a concept class is polynomial if and only if the concept class is learnable using a polynomial number of membership queries alone. For the combination of membership and equivalence queries, Hellerstein et al. [7] and Hegedüs [6] independently gave an elegant combinatorial property called *polynomial certificates* as a necessary and sufficient condition for polynomial-query learning.

In this paper, we give combinatorial properties which uniformly characterize the learnability for *each of any possible combinations* of all queries introduced above. We will give the characterizations in an abstract form so that we can easily generalize it for another kind of queries, not specific to these six queries.

Our characterizations are based on the following two intuitions: The first one is that if a learner can ask a *good* question to a teacher about the unknown target concept, then the concept is easy to learn. Otherwise, it might be hard to learn. Here, we regard that a question is *good* if at least a certain fraction of concepts will be rejected, no matter how the answer is returned by the teacher. If there always exists a good question for any subset of a concept class, then the learner can use it to reduce the hypothesis space efficiently. Otherwise, that is, if there is no good question for some subset of a concept class, adversary teacher can answer maliciously so that little information will be given to the learner to identify the target concept. In fact, this was a key idea to prove that the lack of approximate fingerprint property is a necessary and sufficient condition for the learnability using equivalence queries alone [2, 4].

The second intuition is that a learner can identify any target concept *exactly* if and only if the learner can confirm that the hypothesis is absolutely correct by using queries. We introduce a notion of *specifying queries* in order to capture the intuition. When equivalence queries are available, it is a trivial task, since the learner can directly confirm whether the hypothesis is correct or not.

We apply these intuitions for each type of queries, and capture the essence of query complexity of exact learning using *each of any possible combinations* of all these queries. The technicalities of the proofs may not be quite new since they are rather straightforward extensions appeared in the literature [2, 4, 6]. However, our characterizations will be applied for any kind of queries, not restricted to the ones mentioned above. Since a query corresponds to an experiment in scientific discovery, we hope that our characterizations will lead us an efficient strategy to choose and perform experiments among a large number of possible experiments.

2 Preliminaries

We adopt the terminology from [7,8]. Let Σ be an alphabet. Then Σ^* denotes the set of all finite length strings over Σ, and Σ^n denotes the set of all strings over Σ of length exactly n. For a string $w \in \Sigma^*$, $\|w\|$ denotes the length of w, for a set S, $|S|$ denotes the cardinality of S.

A *Representation of concepts* $\mathcal{R} = \langle \Sigma, \Delta, R, \mu \rangle$ is a 4-tuple where Σ and Δ are finite alphabets, R is a subset of Δ^*, and μ is a map from R to subsets of Σ^*, called *concepts*. R is a set of *representations*, and μ is the map that specifies which concept is represented by a given representation. For any concept c, χ_c denotes the characteristic function of c. For any string w, $\chi_c(w) = 1$ if w is in c and $\chi_c(w) = 0$ otherwise. The *size* of a concept c is $\min\{\|r\| : \mu(r) = c\}$.

Throughout this paper, we assume that for any representation class $\mathcal{R}$, the following problems are computable:

1. For a given string $r \in \Delta^*$, decide if $r \in R$.
2. For a given string $w \in \Sigma^*$ and $r \in \Delta^*$, decide if $w \in \mu(r)$.

The *concept class* $\mathcal{C}$ by $\mathcal{R}$ is a set of concepts that have representations in R. For any positive integer m, $\mathcal{C}_m = \{\mu(r) : r \in R, \|r\| \leq m\}$, and $\mathcal{C} = \bigcup_{m \geq 1} \mathcal{C}_m$.

In this paper, we deal with *boolean concept classes* only. Thus let us assume that $\Sigma = \{0, 1\}$. A boolean concept c is a subset of Σ^n for any positive integer n. When it causes no confusion, we will use c itself to denote χ_c. If $\mathcal{R}$ is a *boolean representation class*, each $r \in R$ will represent a boolean formula over n variables, and the concept is a set of assignments to the variables that satisfies the function. For any positive integers m and n, let $\mathcal{C}_{m,n} = \{\mu(r) : \|r\| \leq m \text{ and } \mu(r) \subseteq \Sigma^n\}$, and $\mathcal{C}_n = \bigcup_{m \geq 1} \mathcal{C}_{m,n}$. In the sequel, we identify a concept $c \in \mathcal{C}$ with its representation $r \in \mathcal{R}$ with $\mu(r) = c$ when it is clear from the context.

We assume several oracles which give some information about a target concept c^* to a learner. We may regard them as experiments to identify the target concept. In the literatures, six oracles have been introduced as follows. For each string $v \in \Sigma^n$, the *membership oracle* MEM returns "Yes" if $c^*(v) = 1$ and "No" otherwise. Moreover, for each concept $h \in \mathcal{C}$, we define *Equivalence* (EQU), *Superset* (SUP), *Subset* (SUB), *Disjointness* (DIS), *Exhaustiveness* (EXH) oracles and their *weak* versions (wEQU, wSUP, wSUB, wDIS, wEXH) as in Table 1. However, in this paper, we do not have to restrict the queries to those ones, since our characterizations would not be specific to these oracles. For a query σ and a concept c, we denote by $c[\sigma]$ the set of possible answers for c when asking σ. We denote by $\|\sigma\|$ the length of a query σ. For example, for a membership query σ_1, $c[\sigma_1]$ is $\{$"Yes"$\}$ or $\{$"No"$\}$, and for an equivalence query σ_2, $c[\sigma_2]$ is $\{$"Yes"$\}$ or the set of all counterexamples.

Table 1. The definitions of oracles. The first row represents the types of oracles. The second row represents conditions when each oracle returns "Yes", and the third row "No". The last row shows the condition which a counterexample should satisfy. For instance, the weak equivalence oracle wEQU answers "Yes" if $h = c^*$, and answers "No" if $h \neq c^*$. The equivalence oracle EQU answers "Yes" if $h = c^*$, and returns a counterexample w with $w \in (h \cup c^*) - (h \cap c^*)$ if $h \neq c^*$.

	Yes	No	w
MEM	$c^*(v) = 1$	$c^*(v) = 0$	
EQU, wEQU	$h = c^*$	$h \neq c^*$	$w \in (h \cup c^*) - (h \cap c^*)$
SUP, wSUP	$h \supseteq c^*$	$h \not\supseteq c^*$	$w \in c^* - h$
SUB, wSUB	$h \subseteq c^*$	$h \not\subseteq c^*$	$w \in h - c^*$
DIS, wDIS	$h \cap c^* = \emptyset$	$h \cap c^* \neq \emptyset$	$w \in h \cap c^*$
EXH, wEXH	$h \cup c^* = \Sigma^n$	$h \cup c^* \neq \Sigma^n$	$w \in \Sigma^n - (h \cup c^*)$

The *query complexity* of learning algorithm $\mathcal{A}$ is the sum of the lengths of queries and counterexamples returned by oracles. Note that the length of a counterexample is always n, since we consider only boolean concepts.

Definition 1. *Let $\mathcal{Q}$ be a set of queries. A concept class $\mathcal{C}$ is polynomial-query learnable using $\mathcal{Q}$ if there exists an algorithm $\mathcal{A}$ and a polynomial $p(\cdot, \cdot)$ such that, for any positive integers m, n and an unknown target concept $c^* \in \mathcal{C}_{m,n}$:*

1. *$\mathcal{A}$ gets n as input.*
2. *$\mathcal{A}$ may ask queries in $\mathcal{Q}$.*
3. *$\mathcal{A}$ eventually halts and outputs $r \in R$ with $\mu(r) = c^*$.*
4. *The total query complexity of $\mathcal{A}$ is at most $p(m, n)$.*

In Section 3, we consider the case where the size m of a target concept is additionally given to a learner.

3 Good Queries

We introduce a notion of good queries in order to characterize polynomial-query learnability where the size of a target concept is known to a learner. Intuitively, a query is good for a set $\mathcal{T}$ of concepts, if a certain fraction of $\mathcal{T}$ are eliminated by the query no matter how the answer is returned.

Definition 2. *For a concept class $\mathcal{T}$, a query σ and its answer α, we define $Cons(\mathcal{T}, \sigma, \alpha)$ be the set of concept in $\mathcal{T}$ that is consistent with σ and α. That is,*

$$Cons(\mathcal{T}, \sigma, \alpha) = \{h \in \mathcal{T} \mid \alpha \in h[\sigma]\}.$$

Definition 3. *A query σ is δ-good for a concept class $\mathcal{T}$ if for any answer α,*

$$|Cons(\mathcal{T}, \sigma, \alpha)| \leq (1 - \delta)|\mathcal{T}|.$$

```
Algorithm LEARNER1(m, n : positive integers)
Given Q : available queries
begin
    H := C_{m,n};
    while |H| ≥ 2 do
        Find a query σ ∈ Q that is 1/q(m,n)-good for H;        (*)
        Let α is the answer to the query σ;
        H := Cons(H, σ, α)
    endwhile;
    if |H| = 1 then output the unique hypothesis h in H
    else output "Target concept is not in C_{m,n}"
end.
```

Fig. 1. Algorithm LEARNER1, where the size m of a target concept is known.

Theorem 1. *Assume the size m of the target concept is known to a learner. A concept class C is polynomial-query learnable using Q if and only if there exist polynomials $q(\cdot, \cdot)$ and $p(\cdot, \cdot)$ such that for any positive integers m, n and any $T \subseteq C_{m,n}$ with $|T| \geq 2$, there exists a query σ in Q with $\|\sigma\| \leq p(m,n)$ that is $1/q(m,n)$-good for T.*

Proof. (if part)   Let $p(\cdot, \cdot)$ and $q(\cdot, \cdot)$ be polynomials such that for any positive integers m, n and any $T \subseteq C_{m,n}$ with $|T| \geq 2$, there exists a query σ that is $1/q(m,n)$-good for T. We show a learning algorithm using queries in Q in Figure 1. It is not hard to verify that all procedures in the algorithm, such as *Cons*, are computable, since we only deal with boolean concepts.

First we show the correctness of the algorithm. Since H is initialized as $C_{m,n}$, and the target concept c^* is assumed to be in $C_{m,n}$, H contains c^* before the first stage. Since c^* is consistent with any answer returned by the oracles in Q, and at any stage H is updated so that only inconsistent concepts are eliminated from H, c^* is never eliminated. Moreover, whenever $|H| \geq 2$, we can find a query that is $1/q(m,n)$-good for H in Q. Thus the output of the algorithm is guaranteed to be exactly equal to the target concept c^*.

We now show that the total number of queries is $O(m \cdot p(m,n))$. We denote the set H at i-th stage of the algorithm by H_i, and l be the number of the stages. We can show that for any stage $i = 1, 2, ..., l - 1$,

$$|H_i| \leq \left(1 - \frac{1}{p(m,n)}\right) \cdot |H_{i-1}|,$$

regardless of the answer from an oracle in Q. Since H_0 is initialized as $C_{m,n}$, we have

$$|H_i| \leq \left(1 - \frac{1}{p(m,n)}\right)^i \cdot |C_{m,n}|$$

for any i. We can show that the right part becomes at most one if $i > p(m,n) \cdot \ln|\mathcal{C}_{m,n}|$ by simple calculations, which ensures the termination of the algorithm. Recall that $|\mathcal{C}_{m,n}| \leq (|\Delta|+1)^{m+1}$ for any m and n, since any concept in $\mathcal{C}_{m,n}$ is represented by a string over Δ of length at most m. Since at each stage, exactly one query is asked to an oracle, the total number of query is $O(m \cdot p(m,n))$. Since the length of each query is at most $p(m,n)$ and the length of each counterexample returned by oracles is n, the query complexity of the algorithm is $O(m \cdot p(m,n)(p(m,n)+n))$, which is a polynomial with respect to m and n.

(only if part)   Assume that for any polynomials $p(\cdot,\cdot)$ and $q(\cdot,\cdot)$, there exist positive integers m, n and a set $\mathcal{T} \subseteq \mathcal{C}_{m,n}$ with $|\mathcal{T}| \geq 2$ such that there exists no query σ that is $1/q(m,n)$-good for $\mathcal{T}$. Suppose to the contrary that there exists a learning algorithm $\mathcal{A}$ that exactly identifies any target concept using queries in $\mathcal{Q}$, whose query complexity is bounded by a polynomial $p'(m,n)$ for any m and n. Let $p(m,n) = p'(m,n)$ and $q(m,n) = 2p'(m,n)$.

We construct an adversary teacher who answers for each query σ in $\mathcal{Q}$ as follows: If $||\sigma|| > p(m,n)$, the teacher may answer arbitrarily, say "Yes". (Since the query complexity of $\mathcal{A}$ is bounded by $p'(m,n) = p(m,n)$, actually $\mathcal{A}$ can never ask such a query.) If $||\sigma|| \leq p(m,n)$, the teacher answers α such that $|Cons(\mathcal{T},\sigma,\alpha)| > \left(1 - \frac{1}{q(m,n)}\right)|\mathcal{T}|$. By the assumption, there always exists such a *malicious* answer. The important point is that for any query σ, its answer α returned by the teacher contradicts less than $1/q(m,n)$ fraction of concepts in $\mathcal{T}$. That is,

$$|\mathcal{T}| - |Cons(\mathcal{T},\sigma,\alpha)| < \frac{1}{q(m,n)}|\mathcal{T}|.$$

Since the query complexity of $\mathcal{A}$ is $p'(m,n)$, at most $p'(m,n)$ queries can be asked to the teacher.

Thus the learner can eliminate less than $(p'(m,n)/q(m,n))|\mathcal{T}|$ concepts after $p'(m,n)$ queries. Since $q(m,n) = 2p'(m,n)$, more than $(1/2)|\mathcal{T}|$ concepts in $\mathcal{T}$ are consistent with all the answers. Moreover, since $|\mathcal{T}| \geq 2$, at least two distinct concepts from $\mathcal{T}$ are consistent with all the answers so far. Since $\mathcal{A}$ is deterministic, the output of $\mathcal{A}$ will be incorrect for at least one concept in $\mathcal{T}$, which is a contradiction. □

4   Specifying Queries

This section deals with the case where the size m of a target concept is unknown to a learner. The standard trick to overcome this problem is to guess m incrementally and try to learn: initially let $m = 1$, and if there is no concept in $\mathcal{C}_{m,n}$ that is consistent with the answers given by oracles, we double m and repeat. For some cases, such that the equivalence query is available, or both subset and superset queries are available, we can apply the trick correctly, since the learner can confirm the hypothesis is correct or not by asking these queries. The next definition is an abstraction of the notion.

> **Algorithm** LEARNER2 (n : positive integer)
> **Given** Q : available queries
> **begin**
> $m = 1$;
>   **repeat**
>   simulate LEARNER1(m, n) using Q;
>   **if** LEARNER1 outputs a hypothesis h **then**
>   Let Q be a set of specifying queries for h in C_n;
>   **if** h is consistent with the answers for all queries in Q **then**
>   **output** h and **terminate**
> $m = m * 2$
>   **forever**
> **end.**

Fig. 2. Algorithm LEARNER2, where the target size m is unknown

Definition 4. *A set Q of queries is called* specifying queries *for a concept c in $\mathcal{T}$ if the set of consistent concept in $\mathcal{T}$ is a singleton of c for any answer. That is,*

$$\{h \in \mathcal{T} \mid h[\sigma] = c[\sigma] \text{ for all } \sigma \in Q\} = \{c\}.$$

For instance, if the equivalence oracle is available, the set $\{\text{EQU}(c)\}$ is a trivial specifying queries for any c in C. Moreover, if both the superset and subset oracles are available, the $\{\text{WSUP}(c), \text{WSUB}(c)\}$ is also specifying queries for any c in C. If the only membership oracle is available, our notion corresponds to the notion of *specifying set* [6].

Theorem 2. *A concept class C is polynomial-query learnable using Q if and only if there exist polynomials $q(\cdot, \cdot)$, $p(\cdot, \cdot)$ and $r(\cdot, \cdot)$ such that for any positive integers m and n, the following two conditions hold:*

(1) for any $\mathcal{T} \subseteq C_{m,n}$ with $|\mathcal{T}| \geq 2$, there exists a query σ in Q with $\|\sigma\| \leq p(m, n)$ that is $1/q(m, n)$-good for $\mathcal{T}$.

(2) for any concept $c \in C_n$, there exist specifying queries $Q \subseteq Q$ for c in C_n such that $\|Q\| \leq r(m, n)$.

Proof. (if part)   We show a learning algorithm LEARNER2 in Fig. 2, assuming that the two conditions hold. The condition (2) guarantees that the output of LEARNER2 is exactly equal to the target concept, while the condition (1) assures that LEARNER1 will return a correct hypothesis as soon as m becomes greater than or equal to the size of the target concept.

(only if part)   Assume that the concept class C is polynomial-query learnable by a learning algorithm $\mathcal{A}$ using queries in Q. We have only to show the condition (2), since Theorem 1 implies the condition (1). Let $n > 0$ and $c \in C_n$ be arbitrarily fixed. Let Q be the set of queries asked by $\mathcal{A}$ when the target concept is c. We can verify that Q is specifying queries, since the output of $\mathcal{A}$ is always equal to the target concept c. Since the size of Q is bounded by a polynomial, the condition holds. □

Let us notice that the above theorem uniformly gives complete characterizations of boolean concept classes that are polynomial-query learnable for each of all possible combinations of the queries such as membership, equivalence, superset, subset, disjointness and exhaustiveness queries, and their weak versions.

Moreover, as a special case, we get the characterization of learning using equivalence queries alone in terms of the approximate fingerprint property. We say that a concept class $\mathcal{C}$ has an *approximate fingerprint property* if for any polynomials $p(\cdot, \cdot)$ and $q(\cdot, \cdot)$, there exist positive integers m, n and a set $\mathcal{T} \subseteq \mathcal{C}_{m,n}$ with $|\mathcal{T}| \geq 2$ such that for any concept $h \in \mathcal{C}_{p(m,n),n}$, we have $|\{c \in \mathcal{T} \mid h(w) = c(w)\}| < \frac{1}{q(m,n)}|\mathcal{T}|$ for some $w \in \Sigma^n$. Since equivalence queries contain a trivial single specifying query for each concept, we get the following result.

Corollary 1 ([2, 4]). *A concept class $\mathcal{C}$ is polynomial-query learnable using equivalence queries if and only if $\mathcal{C}$ does not have an approximate fingerprint property.*

5 Conclusion

We have shown uniform characterizations of the polynomial-query learnabilities using each of any combinations of all queries, such as membership, equivalence, superset queries, etc. Our results reveal that the polynomial-query learnability using a set of oracles is equivalent to the existence of a good query to the oracles which eliminate a certain fraction of any hypothesis space. This is quite intuitive.

In this paper, we only dealt with boolean concepts. We will generalize our results to treat general concepts in future works. Moreover, it is also interesting to investigate the computational complexity of the learning task for honest concept classes with polynomial query-complexity, in the similar way as shown by Köbler and Lindner [8], where they showed that Σ_2^p oracles are sufficient for the learning using each of three possible combinations of membership and equivalence queries.

Acknowledgements

This research is partly supported by Grant-in-Aid for Scientific Research on Priority Areas "Discovery Science" from the Ministry of Education, Science and Culture, Japan.

References

1. D. Angluin. Queries and concept learning. *Machine Learning*, 2:319–342, 1988.

2. D. Angluin. Negative results for equivarence queries. *Machine Learning*, 5:121–150, 1990.
3. N. Bshouty, R. Cleve, R. Gavalda, S. Kannan, and C. Tamon. Oracles and queries that sufficient for exact learning. *Journal of Computer and System Sciences*, 52:421–433, 1996.
4. R. Gavaldà. On the power of equivarence queries. In *Proceedings of the 1st European Conference on Computational Learning Theory*, pages 193–203, 1993.
5. S. Goldman and M. Kearns. On the complexity of teaching. *Journal of Computer and System Sciences*, 50:20–31, 1995.
6. T. Hegedüs. Generalized teaching dimensions and the query complexity of learning. In *Proceedings of the 8th Annual Conference on Computational Learning Theory*, pages 108–117, 1995.
7. L. Hellerstein, K. Pillaipakkamnatt, V. Raghavan, and D. Wilkins. How many queries are needed to learn? *Journal of the ACM*, 43(5), 1996.
8. J. Köbler and W. Lindner. Oracles in Σ_2^p are sufficient for exact learning. In *Proceedings of 8th Workshop on Algorithmic Learning Theory*, pages 277–290. Springer-Verlag, 1997.

Inferring a Rewriting System from Examples*

Yasuhito Mukouchi, Ikuyo Yamaue and Masako Sato

Department of Mathematics and Information Sciences
College of Integrated Arts and Sciences
Osaka Prefecture University, Sakai, Osaka 599-8531, Japan
e-mail: {mukouchi, sato}@mi.cias.osakafu-u.ac.jp

Abstract. In their previous paper, Mukouchi and Arikawa discussed both refutability and inferability of a hypothesis space from examples. If a target language is a member of the hypothesis space, then an inference machine should identify it in the limit, otherwise it should refute the hypothesis space itself in a finite time. They pointed out the necessity of refutability of a hypothesis space from a view point of machine discovery.

Recently, Mukouchi focused sequences of examples successively generated by a certain kind of system. He call such a sequence an *observation* with time passage, and a sequence extended as long as possible a *complete observation*. Then the set of all possible complete observations is called a *phenomenon* of the system.

In this paper, we introduce phenomena generated by rewriting systems known as $0L$ systems and pure grammars, and investigate their inferability in the limit from positive examples as well as refutable inferability from complete examples.

First, we show that any phenomenon class generated by $0L$ systems is inferable in the limit from positive examples. We also show that the phenomenon class generated by pure grammars such that left hand side of each production is not longer than a fixed length is inferable in the limit from positive examples, while the phenomenon class of unrestricted pure grammars is shown not to be inferable. We also obtain the result that the phenomenon class of pure grammars such that the number of productions and that of axioms are not greater than a fixed number is inferable in the limit from positive examples as well as refutably inferable from complete examples.

1 Introduction

Inductive inference is a process of hypothesizing a general rule from examples. As a correct inference criterion for inductive inference of formal languages and models of logic programming, we have mainly used Gold's *identification in the limit*[4]. An inference machine M is said to identify a language L in the limit, if the sequence of guesses from M, which is successively fed a sequence of examples of L, converges to a correct expression of L.

* Supported in part by Grant-in-Aid for Scientific Research on Priority Areas No. 10143104 from the Ministry of Education, Science and Culture, Japan.

In this criterion, a target language, whose examples are fed to an inference machine, is assumed to belong to a hypothesis space which is given in advance. However, this assumption is not appropriate, if we want an inference machine to infer or to discover an unknown rule which explains examples or data obtained from scientific experiments. That is, the behavior of an inference machine is not specified, in case we feed examples of a target language not in a hypothesis space in question. By noting this point, many types of inference machines have been proposed (cf. Blum and Blum[2], Sakurai[20], Mukouchi and Arikawa[13, 15], Lange and Watson[10], Mukouchi[14], Kobayashi and Yokomori[9], and so on).

In their previous paper, Mukouchi and Arikawa[13, 15] discussed both refutability and inferability of a hypothesis space from examples of a target language. If a target language is a member of the hypothesis space, then an inference machine should identify it in the limit, otherwise it should refute the hypothesis space itself in a finite time. They showed that there are some rich hypothesis spaces that are refutable and inferable from complete examples (i.e. positive and negative examples), but refutable and inferable classes from only positive examples (i.e. text) are very small.

Recently, Mukouchi[16] focused sequences of examples successively generated by a certain kind of system. He call such a sequence an *observation* with time passage, and a sequence extended as long as possible a *complete observation*. Then the set of all possible complete observations is called a *phenomenon* of the system. For example, we consider positions of a free fall with a distinct initial value. Then the sequence $(0, -1, -4, -9, -16, \cdots)$ of positions is a complete observation, and the sequence $(-1, -2, -5, -10, -17, \cdots)$ is another complete observation generated by a physical system. Then we consider the set of such sequences a phenomenon generated by the system. We can regard a language as a phenomenon each of which complete observation consists of each word in the language. Furthermore, we can also regard a total function $f : N \to N$ as a phenomenon whose complete observation consists of a unique sequence $(f(0), f(1), f(2), \cdots)$ over N, where N is the set of all natural numbers. Thus, by thinking learnability of phenomena, we can deal with learnability of both languages and total functions uniformly. On learnability of phenomena, Mukouchi[16] has presented some characterization theorems.

In the present paper, we discuss inferability of rewriting systems known as L *systems* and *pure grammars*. L systems are rewriting systems introduced by Lindenmayer[6] in 1968 to model the growth of filamentous organism in developmental biology. There are two types of L systems: One is an IL system, in which we assume there are interactions, and the other is a $0L$ system, in which no interactions. In this paper, we consider only $0L$ systems. A $0L$ system and a pure grammar are both rewriting systems that have no distinction between terminals and nonterminals. In the world of a $0L$ system, at every step in a derivation every symbol is rewritten in parallel, while in a pure grammar just one word is rewritten at each step. Many properties concerning languages generated by L systems and pure grammars are extensively studied in Lindenmayer[6, 7], Gabrielian[3], Maurer, Salomaa and Wood[8] and so on. Inductive inferability in the limit of languages generated by L systems and pure grammars from positive examples are studied in Yokomori[22], Tanida and Yokomori[21] and so on.

In developmental systems with cell lineages, the observed sequences of branching (tree) structures are represented in terms of sequences of strings (cf. Jürgensen and Lindenmayer[5]). In this paper, we define phenomena by all possible sequences of strings whose initial segments, i.e., observations are derivations generated by $0L$ systems or pure grammars, and investigate their inferability in the limit from positive examples as well as refutable inferability from complete examples. First, we show that any phenomenon class generated by $0L$ systems is inferable in the limit from positive examples. We also show that the phenomenon class generated by pure grammars such that left hand side of each production is not longer than a fixed length is inferable in the limit from positive examples, while the phenomenon class of unrestricted pure grammars is shown not to be inferable. We also obtain the result that the phenomenon class of pure grammars such that the number of productions and that of axioms are not greater than a fixed number is inferable in the limit from positive examples as well as refutably inferable from complete examples. These results may be useful to construct a machine learning/discovery system in the field of developmental biology.

2 Rewriting Systems

Let Σ be a fixed finite alphabet. Each element of Σ is called a constant symbol. Let Σ^+ be the set of all nonnull constant strings over Σ and let $\Sigma^* = \Sigma^+ \cup \{\lambda\}$, where λ is the null string.

Let $N = \{0, 1, 2, \cdots\}$ be the set of all natural numbers.

For a string $w \in \Sigma^*$, the length of w is denoted by $|w|$.

Let $\mu = (w_0, w_1, \cdots)$ be a (possibly finite) sequence over Σ^* and let $n \in N$ be a nonnegative integer. The length of μ is denoted by $|\mu|$. In case μ is an infinite sequence, we regard $|\mu| = \infty$. By $\mu[n]$, we denote the $(n+1)$-st element w_n of μ, and by $\mu^{(n)}$ we also denote the initial segment $(w_0, w_1, \cdots, w_{n-1})$ of length n, if $n \geq |\mu|$, otherwise it represents μ itself.

Let S be a set of sequences over Σ^*. We put $S^{(n)} = \{\mu^{(n)} \mid \mu \in S\}$ and $\mathcal{O}(S) = \{\mu \mid \mu$ is a finite initial segment of some sequence in $S\}$. A pair (T, F) of sets of finite sequences over Σ^* is said to be consistent with S, if $T \subseteq \mathcal{O}(S)$ and $F \cap \mathcal{O}(S) = \phi$ hold.

2.1 L Systems and Pure Grammars

A $0L$ system was introduced by Lindenmayer[6] in 1968 and a pure grammar was introduced by Gabrielian[3] in 1981. We recall their languages, and then introduce phenomena generated by them.

Definition 1. A *production* is an expression of the form $\alpha \rightarrow \beta$, where $\alpha \in \Sigma^+$ is a nonnull constant string and $\beta \in \Sigma^*$ is a constant string.

A 0L *system* is a triple $G = (\Sigma, P, w)$, where Σ is a finite alphabet, $w \in \Sigma^+$ is a nonnull constant string which we call an *axiom* and P is a finite set of productions which satisfies the following two conditions:

(1) For any production $(\alpha \to \beta) \in P$, $|\alpha| = 1$, i.e., $\alpha \in \Sigma$ holds.

(2) For any $a \in \Sigma$, there exists at least one production $(\alpha \to \beta) \in P$ such that $\alpha = a$ holds.

A 0L system $G = (\Sigma, P, w)$ is *deterministic*, or a D0L system for short, if for any $a \in \Sigma$, there exists just one production $(\alpha \to \beta) \in P$ such that $\alpha = a$ holds.

A 0L system $G = (\Sigma, P, w)$ is *propagating*, or a P0L system for short, if for any $(\alpha \to \beta) \in P$, $|\beta| \geq 1$ holds.

A 0L system is a PD0L *system*, if it is a P0L system as well as a D0L system.

We denote by $0\mathcal{L}$, $\mathcal{D}0\mathcal{L}$, $\mathcal{P}0\mathcal{L}$ or $\mathcal{PD}0\mathcal{L}$ the class of all 0L systems, that of all D0L systems, that of all P0L systems or that of all PD0L systems, respectively.

Definition 2. A *pure grammar* is a triple $G = (\Sigma, P, S)$, where Σ is a finite alphabet, P is a finite set of productions and $S \subseteq \Sigma^+$ is a finite set of constant strings each of which we call an *axiom*.

Let $n \geq 1$ be a positive integer. A pure grammar $G = (\Sigma, P, S)$ is a Pure$_{\leq n}$ *grammar*, if for any production $(\alpha \to \beta) \in P$, $1 \leq |\alpha| \leq n$ holds.

A pure grammar $G = (\Sigma, P, S)$ is *context-free*, or a PCF grammar for short, if G is a Pure$_{\leq 1}$ grammar.

We denote by $\mathcal{P}ure$, $\mathcal{P}ure_{\leq n}$ or $\mathcal{PCF}$ the class of all pure grammars, that of all Pure$_{\leq n}$ grammars or that of all PCF grammars, respectively.

Definition 3. Let $G = (\Sigma, P, w)$ be a 0L system. We define a binary relation $\Rightarrow_G$ over Σ^* as follows: For any $\alpha \in \Sigma^+$ and any $\beta \in \Sigma^*$, $\alpha \Rightarrow_G \beta$ if and only if there exist productions $(a_i \to \beta_i) \in P$ $(1 \leq i \leq n)$ such that $\alpha = a_1 a_2 \cdots a_n$ and $\beta = \beta_1 \beta_2 \cdots \beta_n$ hold, where $n = |\alpha|$.

Let $G = (\Sigma, P, S)$ be a pure grammar. We define a binary relation $\Rightarrow_G$ over Σ^* as follows: For any $\alpha \in \Sigma^+$ and any $\beta \in \Sigma^*$, $\alpha \Rightarrow_G \beta$ if and only if there exists a production $(\alpha' \to \beta') \in P$ and constant strings $u, v \in \Sigma^*$ such that $\alpha = u\alpha'v$ and $\beta = u\beta'v$ hold.

Let G be a 0L system or a pure grammar.

When no ambiguity occurs, by omitting the explicit reference to G, we simply write $\alpha \Rightarrow \beta$ to denote $\alpha \Rightarrow_G \beta$. Furthermore, by $\Rightarrow_G^*$ we represent the reflexive and transitive closure of $\Rightarrow_G$.

Definition 4. We define the *language* $L(G)$ *generated by a* 0L *system* $G = (\Sigma, P, w)$ as follows:

$$L(G) = \{u \mid w \Rightarrow_G^* u\}.$$

We define the *language* $L(G)$ *generated by a pure grammar* $G = (\Sigma, P, S)$ as follows:

$$L(G) = \{u \mid \exists w \in S \text{ s.t. } w \Rightarrow_G^* u\}.$$

Definition 5. Let $G = (\Sigma, P, w)$ be a 0L system. A *complete observation gener-ated by* G is a (possibly finite) sequence $\mu = (w_0, w_1, \cdots)$ over Σ^* which satisfies the following three conditions:

(1) $w_0 = w$.

(2) For any i with $0 \le i < |\mu|$, $w_i \Rightarrow_G w_{i+1}$ holds.

(3) If μ is a finite sequence which ends with w_n, then there is no $u \in \Sigma^*$ such that $w_n \Rightarrow_G u$ holds.

Let $G = (\Sigma, P, S)$ be a pure grammar. A *complete observation generated by* G is a (possibly finite) sequence $\mu = (w_0, w_1, \cdots)$ over Σ^* which satisfies the following three conditions:

(1) $w_0 \in S$.

(2) For any i with $0 \le i < |\mu|$, $w_i \Rightarrow_G w_{i+1}$ holds.

(3) If μ is a finite sequence which ends with w_n, then there is no $u \in \Sigma^*$ such that $w_n \Rightarrow_G u$ holds.

Let G be a 0L system or a pure grammar. A *phenomenon* $\mathcal{P}(G)$ *generated by* G is the set of all complete observations generated by G.

We call any finite initial segment of a complete observation generated by G an *observation*.

Now we present examples of languages and phenomena generated by a 0L system and a pure grammar.

Example 1. Let $G = (\{a\}, \{a \to a, \ a \to a^2\}, a)$ be a 0L system. Then the language $L(G)$ and the phenomenon $\mathcal{P}(G)$ generated by G are as follows:

$$L(G) = \{a^n \mid n \ge 1\},$$
$$\mathcal{P}(G) = \left\{ \begin{array}{l} (a, a, a, \cdots), \quad (a, a, a^2, \cdots), \\ (a, a^2, a^2, \cdots), (a, a^2, a^3, \cdots), (a, a^2, a^4, \cdots), \\ \cdots \end{array} \right\}$$
$$= \{(a^{n_0}, a^{n_1}, a^{n_2}, \cdots) \mid n_0 = 1, \ n_{i-1} \le n_i \le 2n_{i-1} \ (i \ge 1)\}.$$

Let $G = (\{a, b\}, \{a \to ab, \ b \to ba\}, \{ab\})$ be a PCF grammar. Then the language $L(G)$ and the phenomenon $\mathcal{P}(G)$ generated by G are as follows:

$$L(G) = \{ab, abb, aba, abaa, abab, abba, abbb, \cdots\},$$
$$\mathcal{P}(G) = \left\{ \begin{array}{l} (ab, abb, abbb, \cdots), \ (ab, abb, abab, \cdots), \ (ab, abb, abba, \cdots), \\ (ab, aba, abba, \cdots), \ (ab, aba, abaa, \cdots), \ (ab, aba, abab, \cdots), \\ \cdots \end{array} \right\}.$$

2.2   Some Properties

Definition 6. For two 0L systems $G = (\Sigma, P, w)$ and $G' = (\Sigma, P', w')$, we write $G \le G'$, if $P \subseteq P'$ and $w = w'$ hold.

For two pure grammars $G = (\Sigma, P, S)$ and $G' = (\Sigma, P', S')$, we write $G \le G'$, if $P \subseteq P'$ and $S \subseteq S'$ hold, and we also write $G \lneq G'$, if $G \le G'$ and $G \ne G'$ hold.

Let T be a set of sequences over Σ^* and let G be a $0L$ system or a pure grammar. Then we say G is *reduced with respect to* T, if (i) $T \subseteq \mathcal{O}(\mathcal{P}(G))$ holds and (ii) for any $G' \leq G$, $T \not\subseteq \mathcal{O}(\mathcal{P}(G'))$ holds. Moreover, we simply say G is *reduced*, if G is reduced with respect to $\mathcal{O}(\mathcal{P}(G))$.

Proposition 7. *Let $n \geq 1$ be a positive integer and let $G = (\Sigma, P, S)$ be a reduced* Pure$_{\leq n}$ *grammar.*

Then every production in P is used in the derivations in $\mathcal{P}(G)^{(k_1)}$, where $k_1 = 1 + k_0 + k_0^2 + \cdots + k_0^n$ with $k_0 = |\Sigma|$.

Proposition 8. *Let $G = (\Sigma, P, w)$ be a reduced $0L$ system.*

Then every production in P is used in the derivations in $\mathcal{P}(G)^{(k_1)}$, where $k_1 = 1 + |\Sigma|$.

Definition 9. For constant strings $u, v \in \Sigma^*$, we define sets of productions as follows ($n \geq 1$):

$$\text{Prod}_*(u,v) = \{\alpha \to \beta \mid \alpha \in \Sigma^+, \ \beta \in \Sigma^*, \ \exists s, t \in \Sigma^* \text{ s.t. } u = s\alpha t, \ v = s\beta t\},$$

$$\text{Prod}_{\leq n}(u,v) = \left\{\alpha \to \beta \ \middle| \ \begin{array}{l} \alpha \in \Sigma^+, \ \beta \in \Sigma^*, \ |\alpha| \leq n, \\ \exists s, t \in \Sigma^* \text{ s.t. } u = s\alpha t, \ v = s\beta t \end{array}\right\},$$

$$\text{Prod}_{0L}(u,v) = \left\{ a_1 \to \beta_1, \ a_2 \to \beta_2, \ \cdots, a_n \to \beta_n \ \middle| \ \begin{array}{l} n = |u|, \ a_i \in \Sigma, \ \beta_i \in \Sigma^*, \\ u = a_1 a_2 \cdots a_n, \\ v = \beta_1 \beta_2 \cdots \beta_n \end{array}\right\}.$$

For a set T of sequences over Σ^*, we put

$$\text{Prod}_*(T) = \bigcup_{\mu \in T} \ \bigcup_{0 \leq i < |\mu|} \text{Prod}_*(\mu[i], \mu[i+1]).$$

We also define $\text{Prod}_{\leq n}(T)$ and $\text{Prod}_{0L}(T)$, similarly.

It is easy to see that for any finite set T of finite sequences over Σ^*, $\text{Prod}_*(T)$, $\text{Prod}_{\leq n}(T)$ ($n \geq 1$) and $\text{Prod}_{0L}(T)$ are finite sets.

A finite-set-valued function $F(u, v)$ is said to be *recursively generable*, if there is an effective procedure that on inputs u and v enumerates all elements in $F(u, v)$ and then stops (cf. Lange and Zeugmann[11]).

Proposition 10. *For constant strings $u, v \in \Sigma^*$, the sets $\text{Prod}_*(u, v)$, $\text{Prod}_{\leq n}(u, v)$ ($n \geq 1$) and $\text{Prod}_{0L}(u, v)$ are recursively generable.*

3 Inferability of Rewriting Systems

Let $\mathcal{G}$ be a class of $0L$ systems or pure grammars. We denote by $L(\mathcal{G})$ (resp., $\mathcal{P}(\mathcal{G})$) the class of languages (resp., phenomena) generated by $0L$ systems or pure grammars in $\mathcal{G}$. For example, $L(0\mathcal{L})$ or $\mathcal{P}(0\mathcal{L})$ represents the class of languages or that of phenomena generated by $0L$ systems, respectively.

Due to the space limitation, we omit the detailed definitions of learnability. For definitions and properties of language learning, please refer to Angluin[1], Gold[4], Osherson, Stob and Weinstein[17], Lange and Zeugmann[12], and so on. For those of refutable learning and phenomenon learning, please refer to Mukouchi and Arikawa[13, 15], and Mukouchi[16].

3.1   Languages Classes Generated by Rewriting Systems

We summarize results concerning inferability of language classes generated by $0L$ systems or pure grammars.

Theorem 11 (Yokomori[22]). *(1) $L(0\mathcal{L})$ and $L(\mathcal{P}0\mathcal{L})$ are not inferable in the limit from positive examples.*
(2) $L(\mathcal{P}\mathcal{D}0\mathcal{L})$ is inferable in the limit from positive examples.

Theorem 12 (Tanida and Yokomori[21]). *$L(\mathcal{P}\mathcal{C}\mathcal{F})$ is not inferable in the limit from positive examples.*

We know from the latter theorem that $L(\mathcal{P}ure_{\leq n})$ and $L(\mathcal{P}ure)$ are not inferable in the limit from positive examples.

On refutable inferability from complete examples, the following results are valid:

Theorem 13. *(1) $L(0\mathcal{L})$ and $L(\mathcal{P}0\mathcal{L})$ are not refutably inferable from complete examples.*
(2) $L(\mathcal{P}\mathcal{D}0\mathcal{L})$ is refutably inferable from complete examples.

Theorem 14. $L(\mathcal{P}\mathcal{C}\mathcal{F})$, $L(\mathcal{P}ure_{\leq n})$, $L(\mathcal{P}ure)$ *are not refutably inferable from complete examples.*

Proof. We shall show only the case of $L(\mathcal{P}\mathcal{C}\mathcal{F})$. The other cases are shown similarly.

By Corollary 6 in Mukouchi and Arikawa[15], it is sufficient for us to show that $L(\mathcal{P}\mathcal{C}\mathcal{F})$ contains every nonempty finite language over Σ.

Let $L \subseteq \Sigma^*$ be a nonempty finite language. Put $G = (\Sigma, P, L)$, where $P = \{a \to a \mid a \in \Sigma\}$. Then G is a PCF grammar, and $L(G) = L$ holds. Therefore we have $L \in L(\mathcal{P}\mathcal{C}\mathcal{F})$.

Thus $L(\mathcal{P}\mathcal{C}\mathcal{F})$ contains every nonempty finite language.   □

3.2   Phenomenon Classes Generated by $0L$ Systems

Theorem 15. $\mathcal{P}(0\mathcal{L})$, $\mathcal{P}(\mathcal{P}0\mathcal{L})$, $\mathcal{P}(\mathcal{D}0\mathcal{L})$ and $\mathcal{P}(\mathcal{P}\mathcal{D}0\mathcal{L})$ *are inferable in the limit from positive examples.*

Proof. It is sufficient for us to show that $\mathcal{P}(0\mathcal{L})$ is inferable in the limit from positive examples, because each of other classes is a subclass of $\mathcal{P}(0\mathcal{L})$.

First, we consider the following algorithm:

Algorithm SubIIM($\mathcal{G}$, δ)
Input :   a class $\mathcal{G}$ of finitely many $0L$ systems,
            a finite sequence δ of observations;
Output : a $0L$ system G;
begin

let $m = |\mathcal{G}|$ and $\mathcal{G} = \{G_1, G_2, \cdots, G_m\}$;
let n be the length of δ and $\delta = \mu_1, \mu_2, \cdots, \mu_n$;
$T := \{\mu_1, \mu_2, \cdots, \mu_n\}$;
search for the least index i $(1 \le i \le m)$ such that

- $T \subseteq \mathcal{O}(\mathcal{P}(G_i))$, and

- $\forall j$ $(1 \le j \le m), [T \subseteq \mathcal{O}(\mathcal{P}(G_j)) \Rightarrow \mathcal{O}(\mathcal{P}(G_j)^{(n)}) \not\subseteq \mathcal{O}(\mathcal{P}(G_i)^{(n)})]$;

if such an index i is found **then** return G_i **else** return G_1;
end.

Similarly to the proof of Theorem 19 in Mukouchi[14], we can show the following claim:

Claim: Let $\mathcal{G}$ be a class of finitely many $0L$ systems, let $\mathcal{P} \in \mathcal{P}(\mathcal{G})$ be a phenomenon and let σ be a positive presentation of $\mathcal{P}$.

Then there exists an $n \in N$ and a $G \in \mathcal{G}$ such that for any $m \ge n$, SubIIM$(\mathcal{G}, \sigma[m]) = G$ and $\mathcal{P} = \mathcal{P}(G)$ hold, where $\sigma[m]$ represents the finite initial segment of σ of length m.

Now we consider an IIM which infers $\mathcal{P}(0\mathcal{L})$ in the limit from positive examples:

Algorithm IIM
Input :   positive examples;
Output : a $0L$ system G;
begin
    let δ be the null sequence;
    $T_0 := \phi$; $k_1 := 1 + |\Sigma|$; $i := 1$;
    repeat
        read the next example μ_i;
        $T_i := T_{i-1} \cup \{\mu_i\}$; $\delta_i := \delta_{i-1}, \mu_i$;
        $w_0 := \mu_i[0]$; $P_i := \mathrm{Prod}_{0L}(T_i^{(k_1)})$;   (1)
        $\mathcal{G}_i := \{(\Sigma, P, w_0) \mid P \subseteq P_i\}$;   (2)
        $G_i := \mathrm{SubIIM}(\mathcal{G}_i, \delta_i)$;
        output G_i;
        $i := i + 1$;
    forever;
end.

In this algorithm, we note that for any $i \ge 1$, P_i at (1) above is a finite set of productions, and thus $\mathcal{G}_i$ at (2) is a class of finitely many $0L$ systems.

Let $\mathcal{P} \in \mathcal{P}(0\mathcal{L})$ be a phenomenon, let σ be a positive presentation of $\mathcal{P}$ and let $k_1 = 1 + |\Sigma|$. Since $\mathcal{P}^{(k_1)}$ is a finite set of finite sequences, when we feed examples of σ successively to the algorithm above on its input requests, there exists an $i_0 \in N$ such that for any $i \ge i_0$, $T_i^{(k_1)} = \mathcal{P}^{(k_1)}$, and thus $P_i = P_{i_0}$ and $\mathcal{G}_i = \mathcal{G}_{i_0}$ hold. We note that w_0 at (1) does not change, because $0L$ system has a unique axiom.

Let $G = (\Sigma, P, w)$ be a reduced $0L$ system w.r.t. $\mathcal{P}$. Then, by Proposition 8, we know that $P \subseteq P_{i_0}$ holds. Furthermore, clearly $w = w_0$ holds. Therefore, $G \in \mathcal{G}_{i_0}$ holds, and thus we have $\mathcal{P} \in \mathcal{P}(\mathcal{G}_{i_0})$. Thus, by the claim above, we conclude that the algorithm above converges to G' with $\mathcal{P} = \mathcal{P}(G')$ for σ.   □

Lemma 16 (Mukouchi[16]). *If a class $\mathcal{C}$ of phenomena is refutably inferable from complete examples, then for any phenomenon $\mathcal{P}_0 \notin \mathcal{C}$, there exists a pair (T, F) of finite sets of finite sequences such that (i) (T, F) is consistent with $\mathcal{P}_0$ and (ii) (T, F) is inconsistent with every $\mathcal{P} \in \mathcal{C}$.*

Theorem 17. *$\mathcal{P}(0\mathcal{L})$ and $\mathcal{P}(\mathcal{P}0\mathcal{L})$ are not refutably inferable from complete examples.*

Proof. We shall show only the case of $\mathcal{P}(0\mathcal{L})$. The case of $\mathcal{P}(\mathcal{P}0\mathcal{L})$ is shown similarly.

Let $\mathcal{P}_0 = \{(a, b, b, b, \cdots), (a, bb, bb, bb, \cdots), \cdots, (a, b^i, b^i, b^i, \cdots), \cdots\} = \{(a, b^i, b^i, b^i, \cdots) \mid i \geq 1\}$ be a phenomenon. It is easy to see that $\mathcal{P}_0 \notin \mathcal{P}(0\mathcal{L})$ holds.

Let (T, F) be a pair of finite sets of finite sequences such that (T, F) is consistent with $\mathcal{P}_0$. Put $G = (\Sigma, P, a)$, where $P = \{a \to b^i \mid (a, b^i, b^i, \cdots, b^i) \in T\} \cup \{b \to b\}$. Then G is a $0L$ system, and $T \subseteq \mathcal{O}(\mathcal{P}(G))$ holds. Furthermore, we see that $\mathcal{O}(\mathcal{P}(G)) \subseteq \mathcal{O}(\mathcal{P}_0)$ holds, and thus (T, F) is consistent with $\mathcal{P}(G)$.

Therefore, by Lemma 16, it follows that $\mathcal{P}(0\mathcal{L})$ is not refutably inferable from complete examples.   □

Theorem 18. *$\mathcal{P}(\mathcal{D}0\mathcal{L})$ and $\mathcal{P}(\mathcal{P}\mathcal{D}0\mathcal{L})$ are refutably inferable from complete examples.*

3.3   Phenomenon Classes Generated by Pure Grammars

Theorem 19. *$\mathcal{P}(\mathcal{P}ure)$ is not inferable in the limit from positive examples.*

Proof. By Corollary 6 in Mukouchi[16], it is sufficient for us to show that there exists an infinite sequence of phenomena $\mathcal{P}_0, \mathcal{P}_1, \mathcal{P}_2, \cdots \in \mathcal{P}(\mathcal{P}ure)$ such that

$$\mathcal{O}(\mathcal{P}_1) \subsetneqq \mathcal{O}(\mathcal{P}_2) \subsetneqq \cdots \subsetneqq \mathcal{O}(\mathcal{P}_0) \quad \text{and} \quad \mathcal{O}(\mathcal{P}_0) = \bigcup_{i \geq 1} \mathcal{O}(\mathcal{P}_i).$$

Let $\Sigma = \{a, b, c\}$ be an alphabet. We define finite sets of productions as follows:

$$P_0 = \{a \to bcb, c \to c^2\}, \qquad P_i = \{a \to bcb, bcb \to bc^2b, \cdots, bc^{i-1}b \to bc^ib\}$$
$$(i \geq 1).$$

Let $G_i = (\Sigma, P_i, \{a\})$ be a pure grammar, and put $\mathcal{P}_i = \mathcal{P}(G_i)$ $(i \in N)$. Then

$$\mathcal{P}_0 = \{(a, bcb, bc^2b, \cdots)\}, \qquad \mathcal{P}_i = \{(a, bcb, bc^2b, \cdots, bc^ib)\} \quad (i \geq 1).$$

It is easy to see that these phenomena satisfy the condition above, and thus $\mathcal{P}(\mathcal{P}ure)$ is not inferable in the limit from positive examples.   □

Similarly to the proof of Theorem 15, we can show the following theorem:

Theorem 20. $P(\mathcal{PCF})$ and $P(\mathcal{Pure}_{\leq n})$ are inferable in the limit from positive examples.

Theorem 21. $P(\mathcal{PCF})$, $P(\mathcal{Pure}_{\leq n})$ and $P(\mathcal{Pure})$ are not refutably inferable from complete examples.

Proof. We shall show only the case of $P(\mathcal{PCF})$. The other cases are shown similarly.

Let $\mathcal{P}_0 = \{(a,b),\ (a,bb),\ \cdots,\ (a,b^i),\ \cdots\} = \{(a,b^i) \mid i \geq 1\}$ be a phenomenon. It is easy to see that $\mathcal{P}_0 \notin P(\mathcal{PCF})$ holds.

Let (T, F) be a pair of finite sets of finite sequences such that (T, F) is consistent with $\mathcal{P}_0$. Put $G = (\Sigma, P, \{a\})$, where $P = \{a \to b^i \mid (a, b^i) \in T\}$. Then G is a PCF grammar, and $T \subseteq \mathcal{O}(P(G))$ holds. Furthermore, we see that $\mathcal{O}(P(G)) \subseteq \mathcal{O}(\mathcal{P}_0)$ holds, and thus (T, F) is consistent with $P(G)$.

Therefore, by Lemma 16, it follows that $P(\mathcal{PCF})$ is not refutably inferable from complete examples. □

On the other hand, as shown below, when we restrict the number of productions and that of axioms by a fixed number, the phenomenon class is shown to be inferable in the limit from positive examples as well as refutably inferable from complete examples.

Definition 22. Let $n \geq 1$ be a positive integer. A pure grammar $G = (\Sigma, P, S)$ is a $\mathrm{Pure}^{\leq n}$ grammar, if $|P| \leq n$ and $|S| \leq n$ hold.
We denote by $\mathcal{Pure}^{\leq n}$ the class of all $\mathrm{Pure}^{\leq n}$ grammars.

Theorem 23. Let $n \geq 1$ be a positive integer.
(1) $P(\mathcal{Pure}^{\leq n})$ is inferable in the limit from positive examples.
(2) $P(\mathcal{Pure}^{\leq n})$ is refutably inferable from complete examples.

4   Concluding Remarks

In the present paper, we have introduced phenomena generated by $0L$ systems and pure grammars, and discussed inferability in the limit from positive examples as well as refutable inferability from complete examples. We can summarize as in Table 1 the results obtained so far, where for a class $\mathcal{C}$, $\mathcal{C} \in$ LIM-TXT means that $\mathcal{C}$ is inferable in the limit from positive examples and $\mathcal{C} \in$ REF-INF means that $\mathcal{C}$ is refutably inferable from complete examples.

We see by this table that the learning power is increased in general when we receive as examples observations of a phenomenon rather than words of a language.

As related works, Sakakibara[18] and Sakamoto[19] studied language learning of a class of context-free languages using structural informations. We note that phenomenon learning of classes generated by rewriting systems uses informations of derivations but never assumes their structural informations.

Table 1. Inferability of classes generated by rewriting systems

	Language Class		Phenomenon Class
$0L$ systems	$L(0\mathcal{L}) \notin$ LIM-TXT $L(0\mathcal{L}) \notin$ REF-INF	*1	$\mathcal{P}(0\mathcal{L}) \in$ LIM-TXT $\mathcal{P}(0\mathcal{L}) \notin$ REF-INF
$P0L$ systems	$L(\mathcal{P}0\mathcal{L}) \notin$ LIM-TXT $L(\mathcal{P}0\mathcal{L}) \notin$ REF-INF	*1	$\mathcal{P}(\mathcal{P}0\mathcal{L}) \in$ LIM-TXT $\mathcal{P}(\mathcal{P}0\mathcal{L}) \notin$ REF-INF
$D0L$ systems			$\mathcal{P}(D0\mathcal{L}) \in$ LIM-TXT $\mathcal{P}(D0\mathcal{L}) \in$ REF-INF
$PD0L$ systems	$L(\mathcal{P}D0\mathcal{L}) \in$ LIM-TXT $L(\mathcal{P}D0\mathcal{L}) \in$ REF-INF	*1	$\mathcal{P}(\mathcal{P}D0\mathcal{L}) \in$ LIM-TXT $\mathcal{P}(\mathcal{P}D0\mathcal{L}) \in$ REF-INF
Pure grammars	$L(\mathcal{P}ure) \notin$ LIM-TXT $L(\mathcal{P}ure) \notin$ REF-INF	*2	$\mathcal{P}(\mathcal{P}ure) \notin$ LIM-TXT $\mathcal{P}(\mathcal{P}ure) \notin$ REF-INF
Pure$_{\leq n}$ grammars	$L(\mathcal{P}ure_{\leq n}) \notin$ LIM-TXT $L(\mathcal{P}ure_{\leq n}) \notin$ REF-INF		$\mathcal{P}(\mathcal{P}ure_{\leq n}) \in$ LIM-TXT $\mathcal{P}(\mathcal{P}ure_{\leq n}) \notin$ REF-INF
PCF grammars	$L(\mathcal{P}\mathcal{C}\mathcal{F}) \notin$ LIM-TXT $L(\mathcal{P}\mathcal{C}\mathcal{F}) \notin$ REF-INF	*2	$\mathcal{P}(\mathcal{P}\mathcal{C}\mathcal{F}) \in$ LIM-TXT $\mathcal{P}(\mathcal{P}\mathcal{C}\mathcal{F}) \notin$ REF-INF
Pure$^{\leq n}$ grammars			$\mathcal{P}(\mathcal{P}ure^{\leq n}) \in$ LIM-TXT $\mathcal{P}(\mathcal{P}ure^{\leq n}) \in$ REF-INF

Note: The results marked as *1 are obtained in Yokomori[22] and those marked as *2 are obtained in Tanida and Yokomori[21].

References

1. D. Angluin: *Inductive Inference of Formal Languages from Positive Data*, Information and Control **45** (1980) 117–135.
2. L. Blum and M. Blum: *Toward a Mathematical Theory of Inductive Inference*, Information and Control **28** (1975) 125–155.
3. A. Gabrielian: *Pure Grammars and Pure Languages*, International Journal of Computer Mathematics **9** (1981) 3–16.
4. E.M. Gold: *Language Identification in the Limit*, Information and Control **10** (1967) 447–474.
5. H. Jürgensen and A. Lindenmayer: *Inference Algorithms for Developmental Systems with Cell Lineages*, Bulletin of Mathematical Biology **49** (1987) 93–123.
6. A. Lindenmayer: *Mathematical Models for Cellular Interactions in Development. Parts I, II*, Journal of Theoretical Biology **18** (1968) 280–299, 300–315.
7. A. Lindenmayer: *Developmental Systems without Cellular Interactions, their Languages and Grammars*, Journal of Theoretical Biology **21** (1971) 455–484.
8. H.A. Maurer, A. Salomaa and D. Wood: *Pure Grammars*, Information and Control **44** (1980) 47–72.
9. S. Kobayashi and T. Yokomori: *On Approximately Identifying Concept Classes in the Limit*, in Proceedings of the Sixth International Workshop on Algorithmic Learning Theory, Lecture Notes in Artificial Intelligence **997** (1995) 298–312.
10. S. Lange and P. Watson: *Machine Discovery in the Presence of Incomplete or Ambiguous Data*, in Proceedings of the Fifth International Workshop on Algorithmic Learning Theory, Lecture Notes in Artificial Intelligence **872** (1994) 438–452.

11. S. Lange and T. Zeugmann: *Types of Monotonic Language Learning and Their Characterization*, in Proceedings of the Fifth Annual ACM Workshop on Computational Learning Theory (1992) 377–390.
12. S. Lange and T. Zeugmann: *A Guided Tour across the Boundaries of Learning Recursive Languages*, Lecture Notes in Artificial Intelligence **961** (1995) 190–258.
13. Y. Mukouchi and S. Arikawa: *Inductive Inference Machines That Can Refute Hypothesis Spaces*, in Proceedings of the Fourth International Workshop on Algorithmic Learning Theory, Lecture Notes in Artificial Intelligence **744** (1993) 123–136.
14. Y. Mukouchi: *Inductive Inference of an Approximate Concept from Positive Data*, in Proceedings of the Fifth International Workshop on Algorithmic Learning Theory, Lecture Notes in Artificial Intelligence **872** (1994) 484–499.
15. Y. Mukouchi and S. Arikawa: *Towards a Mathematical Theory of Machine Discovery from Facts*, Theoretical Computer Science **137** (1995) 53–84.
16. Y. Mukouchi: *Inferring a System from Examples with Time Passage*, in Proceedings of the Eighth International Workshop on Algorithmic Learning Theory, Lecture Notes in Artificial Intelligence **1316** (1997) 197–211.
17. D. Osherson, M. Stob and S. Weinstein: "Systems That Learn: An Introduction to Learning Theory for Cognitive and Computer Scientists," MIT-Press, 1986.
18. Y. Sakakibara: *Learning Context-Free Grammars from Structural Data in Polynomial Time*, Theoretical Computer Science **76** (1990) 223–242.
19. H. Sakamoto: *Language Learning from Membership Queries and Characteristic Examples*, in Proceedings of the Sixth International Workshop on Algorithmic Learning Theory, Lecture Notes in Artificial Intelligence **997** (1995) 55–65.
20. A. Sakurai: *Inductive Inference of Formal Languages from Positive Data Enumerated Primitive-Recursively*, in Proceedings of the Second Workshop on Algorithmic Learning Theory (1991) 73–83.
21. N. Tanida and T. Yokomori: *Inductive Inference of Monogenic Pure Context-Free Languages*, IEICE Transactions on Information and Systems **E79-D(11)** (1996) 1503–1510.
22. T. Yokomori: *Inductive Inference of 0L Languages*, Lindenmayer Systems (Rozenberg and Salomaa, Eds.), Springer-Verlag (1992) 115–132.

Toward Genomic Hypothesis Creator:
View Designer for Discovery

Osamu Maruyama[1], Tomoyuki Uchida[2], Takayoshi Shoudai[3], and
Satoru Miyano[1]

[1] Human Genome Center, Institute of Medical Science, University of Tokyo
4-6-1 Shirokanedai, Minato-ku, Tokyo 108-8639, Japan
{maruyama,miyano}@ims.u-tokyo.ac.jp
[2] Faculty of Information Sciences, Hiroshima City University
3-4-1, Ozuka-Higashi, Asa-Minami-Ku, Hiroshima 731-3194, Japan
uchida@cs.hiroshima-cu.ac.jp
[3] Department of Informatics, Kyushu University 39
6-1 Kasuga-Kouen, Kasuga 816-8580, Japan
shoudai@i.kyushu-u.ac.jp

Abstract. Software tools for genomic researches like homology search
are very useful and have contributed on the progress of the genomic
researches. However, these tools are not designed directly toward scientific discovery and more discovery-oriented software tools are strongly
expected to assist scientific discovery in genomic researches. We have
designed and developed a multistrategic and discovery-oriented system
Genomic Hypothesis Creator by introducing two notions: view on data
and view space on data. With these newly defined notions, we describe a
View Designer, a component of Genomic Hypothesis Creator, which dy-
namically creates new views on data and searches a view space for more
appropriate views. A good view obtained from Genomic Hypothesis Creator makes it possible for us to understand the data and eventually attain
to the goal of discovery. Genomic Hypothesis Creator can be extended
by adding user's own views on data and hypothesis generators into the
system with plug-in interfaces. Therefore it would be feasible to apply
this system to other problems than genomic researches.

1 Introduction

Currently, genome sequencing projects have been organized for about 70 organisms, and sequence and structural databases have accumulated into gigabytes in
size. Some projects have already finished sequencing and complete genomes are
open to public; *Escherichia coli*, 4,639,221bp, (1997); *Haemophilus influenzae*,
1,830,135bp, (1995); *Bacillus subtilis*, 4,214,814bp, (1997); *Saccharomyces cerevisiae*, 12,069,313bp, (1996). In 1998, the sequencing project of Caenorhabditis
elegans, a multicellular organism, will be finished with the complete genome
of size 100Mbp. Recently, it has been announced that the human genome sequencing shall be finished until 2001 [16]. The human haploid genome consists
of 3,000Mbp of DNA and encodes 75,000~100,000 genes. Till the end of 1997,

more than 50,000 genes have been identified but only about 5,000 of them have known functions. Facing with a coming huge production of such genomic sequences, the issue of assisting scientific discovery by using data compiled in databases is a matter of utmost concern in Genome Science.

Machine learning and data mining technology have been used for knowledge discovery and prediction in many fields [10] and successful results for scientific data have been surveyed in [6]. In recent years, especially, the applications of machine learning techniques to a variety of "real-world" problems have been provided [9]. For genomic data which is one of "real-world" data, it is also very promising that the knowledge discovery using machine learning techniques will play an important role in the process of scientific discovery.

Many institutes (NCBI, EMBL/EBI, MIPS, DDBJ, TIGR, GenomeNet, etc.) provide services with the databases GenBank, EMBL (typical databases of DNA sequences), SWISS-PROT, PIR (typical databases of amino acid sequences of proteins), and many other databases [20] whose data are text files in specific record formats. The homology search and conventional information retrieval with keywords are the most common services. Although these services have made considerable contributions in assisting scientific discoveries in genomic researches, more discovery-oriented services would change the style of research and speed up the process of scientific discovery. There have been some researches on discovery of specific topics, e.g., [4, 15] and these contributions should be highly appreciated. However, it is a challenge to design and develop a general system which can strongly assist the process of scientific discovery with the above databases for Genome Science.

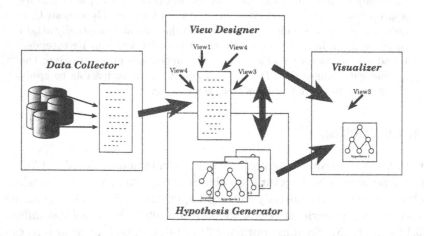

Fig. 1. The design concept of Genomic Hypothesis Creator.

In the history of science, it has been repeatedly witnessed that an invention of a new view on data is a key to scientific discovery. Aiming at discovery-oriented service for genome science with the motivation above, we have developed a system Genomic Hypothesis Creator which is designed to discover a new view on data in text databases. The process of knowledge discovery from databases starts with data collection and ends with knowledge as is overviewed in [7]. The components of Genomic Hypothesis Creator are shown in Fig. 1 which basically follows the KDD process in [7]. Contributions of this paper are notions of *view over sequences* and *view space over sequences* with which we have designed View Designer. Section 2 gives formal definitions of a view and a view space and their relation to visualization of hypotheses that is reflected to the design of Visualizer. View Designer is aiming at a tool for users to discover manually/automatically a new view on data which yields a better understanding of data. Distinctive features can be also found in Data Collector and Hypothesis Generator shown in section 3 and 4, respectively. For Data Collector, we employed and modified the text database management system SIGMA [1] which has been used for service at the Computer Center of Kyushu University since Genomic Hypothesis Creator assumes text data. As is indicated in [17], the main bottleneck for scientific knowledge discovery applications is not the lack of techniques for data analysis. The problem is to exploit and combine existing algorithms effectively. From this point, Genomic Hypothesis Creator employs the multi strategy principle [12] and the plug-in architecture used in Kepler [18]. By linking a view created with View Designer to a hypothesis generator selected from the pool of hypothesis generators, Genomic Hypothesis Creator provides a diversity of knowledge discovery tools.

2 View Designer for Discovery from Genomic Sequences

2.1 View and Discovery

Informally, a view on data provides terms with which we understand and explain the data. A discovery or definition of a new view on data is a key to scientific discovery. Discovering a view on data has been usually done by experts of the field since it is the most important part of discovery process. However, when we have to deal with data on which we cannot assume any experts, this process turns to be the most difficult obstacle in discovery.

In the design of Genomic Hypothesis Creator, we focus our attention on this matter. We define the notion of a view in a rather abstract way. With this definition, we can separate the process of discovery into the process of view design/discovery and the process of hypothesis generation.

Definition 1. Let Σ be a finite alphabet called a *data alphabet*. We call a string s in Σ^* a Σ-*sequence*. Let Γ be a finite alphabet called a *view representation alphabet*. A *view* over Σ-sequences is a pair $M = (V, L)$ of an algorithm V with two input parameters and a set $L \subseteq \Gamma^*$ satisfying the following conditions:

1. V takes two strings $s \in \Sigma^*$ and $\pi \in \Gamma^*$ as input. If $\pi \in L$, then V on (s, π) outputs a value $V(s, \pi)$ in a set W. Otherwise, V on (s, π) outputs "undefined".

2. V on (s, π) runs in polynomial time with respect to $|s|$ and $|\pi|$.

We call the algorithm V the *view interpreter*, an element $\pi \in L$ a *view element* and the set W the *view element value set*. For a set $S \subseteq \Sigma^*$, we call the $S \times L$ matrix S^M defined by $S^M(s, \pi) = V(s, \pi)$ for $s \in S$ and $\pi \in L$ the *data matrix* of S under the view M.

In the above definition, we have not exactly specified what kind of values are considered as view element values. They can be boolean values, real numbers, strings, etc. The first reason why we defined a view as an algorithmic process is that we can deal with various views flexibly. The second reason is that the time-series process of V on (s, π) gives the interpretation of the sequence s by the view element π that provides a bridge between view design and visualization of hypotheses.

Given a collection $S_0, \ldots, S_{m-1}$ of sets of sequences over Σ, the data matrices $S_0^M, \ldots, S_{m-1}^M$ are used by a hypothesis generator for producing a hypothesis for $S_0, \ldots, S_{m-1}$ which is represented in terms of the view elements of $M = (V, L)$. For example, when we want to discover an explanation discriminating two sets S_0 and S_1 of positive and negative examples, respectively, the data matrices S_0^M and S_1^M will be constructed for a hypothesis generator to create a hypothesis described with view elements in L.

View 1 *Alphabet Indexing and Regular Patterns.* We presented a system BON-SAI [14], which is designed for discovering classifications of amino acid residues automatically from positive and negative examples. BONSAI has a parameter which specifies the number k of categories into which the symbols describing the original sequences are classified. In the case of amino acid sequences of proteins, we set Σ to be the alphabet consisting of 20 amino acid residues. Let $\Sigma_k = \{0, 1, \ldots, k-1\}$ when the parameter k is chosen. An *alphabet indexing* is a mapping $\psi_k : \Sigma \to \Sigma_k$ which classifies the symbols in Σ into k categories. Given two sets S_0 and S_1 of strings over Σ, BONSAI discovers an alphabet indexing ψ_k and a small decision tree T whose internal nodes are labeled with patterns of the form $\pi = xwy$ $(w \in \Sigma_k^*)$ and leaves are labeled with 0 or 1. We denote by $L(T)$ the set of strings in Σ_k^* classified as 0 by T. The accuracy of the hypothesis (T, ψ) is defined by $\sqrt{\frac{|L(T) \cap \psi_k^*(S_1)|}{|\psi_k^*(S_1)|} \cdot \frac{|L(T) \cap \psi_k^*(S_0)|}{|\psi_k^*(S_0)|}}$, where $\psi_k^*(a_1 \cdots a_n) = \psi_k(a_1) \cdots \psi_k(a_n)$ for $a_i \in \Sigma$. Then a view employed by BONSAI is defined as follows: Let V_{ψ_k} be a polynomial-time algorithm which decides if for given $s \in \Sigma^*$ and a pattern $\pi = x_1 w x_2$ with $w \in \Sigma_k^*$, w occurs on $\psi_k^*(s)$, and shows all occurrences of w in $\psi_k^*(s)$ on the corresponding locations in the string s. Let $L_k = \{x_1 w x_2 \mid w \in \Sigma_k^*\}$. A view $B_{\psi_k} = (V_{\psi_k}, L_k)$ over Σ-sequences is employed by BONSAI. BONSAI could discover a biologically meaningful knowledge from 689 transmembrane domain data and 19256 non-transmembrane data [14].

View 2 *Approximate String Matching.* For an integer $k \geq 0$, let $X_k = \{(w, k) \mid w \in \Sigma^*\}$. Let A be a polynomial-time algorithm which decides if, for a string

$s \in \Sigma^*$ and (w, k) with $w \in \Sigma^*$ and $k \geq 0$, s contains a substring whose edit distance from w is at most k and shows all such substrings. Then $APPROX^k = (A, X_k)$ is a view over Σ-sequences. We call $APPROX^k$ the k-*mismatch view*. We can also consider a range on $|w|$, e.g., $p \leq |w| \leq q$, to define a new view.

View 3 *PROSITE View.* Let E be the set of all prosite patterns in PROSITE [3] and let R be a prosite pattern matching algorithm. Then $PROSITE= (P, E)$ provides a view over amino acid sequences. We call this view the *prosite view*.

View 4 *Range Restrictions.* Let $1 \leq i < j$ be positive integers. For a string $s = s[1]s[2] \cdots s[n]$ in Σ^* with $s[i] \in \Sigma$ for $1 \leq i \leq n$, let $s[i, j] = s[i^*] \cdots s[j^*]$ and $s[-j, -i] = s[n - j^* + 1] \cdots s[n - i^* + 1]$, where $i^* = \min\{i, n\}$ and $j^* = \min\{j, n\}$. $APPROX^k[i, j] = (A[i, j], X_k)$ $(APPROX^k[-j, -i] = (A[-j, -i], X_k))$ is a view given by the view interpreter $A[i, j]$ $(A[-j, -i])$ which runs A with $s[i, j]$ $(s[-j, -i])$ instead of s. Similarly, $PROSITE[i, j]$ and $PROSITE[-j, -i]$ are given by restricting the range of search to $s[i, j]$ and $s[-j, -i]$, respectively.

2.2   Searching View Space

A view gives us a method of interpreting sequences. In the process of discovery, however, one of the most important aspects is a discovery of a new view on data since a choice of a better view may lead to a better understanding of data. Considering this aspect of discovery, we define a view space and shows some examples which are implemented in Genomic Hypothesis Creator.

We use a simple abstraction of search strategies. For a set N called a search space, a *search strategy* σ over N is a procedure that specifies the start element ν_0 in N and determines the next element $\sigma^\varphi(\nu)$ in N to visit for any $\nu \in N$ when an arbitrary score function $\varphi : N \to \mathbf{R}$ is given to σ in the sense that σ can use the value $\varphi(\mu)$ for any $\mu \in N$, where $\mathbf{R}$ is the set of real numbers.

Definition 2. A *view space* over Σ-sequences is a pair $(\mathcal{M}, \sigma)$, where $\mathcal{M} = \{M_\nu\}_{\nu \in N}$ is a collection of views over Σ-sequences indexed by N and σ is a search strategy for N. When σ is clear from the context, we simply denote the view space by $\mathcal{M}$.

Let $\mathcal{M} = \{M_\nu\}_{\nu \in N}$ be a view space over Σ-sequences with a search strategy σ. For a view M_ν in $\mathcal{M}$, data sets $S_0, \ldots, S_{m-1}$ of Σ-sequences are transformed to data matrices $S_0^{M_\nu}, \ldots, S_{m-1}^{M_\nu}$ for a hypothesis generator $\mathcal{G}$ to generate a hypothesis h which explains $S_0, \ldots, S_{m-1}$. We assume a score function φ that measures the "goodness" of the hypothesis. On the other hand, we can regard this score function as a score function given to the view M_ν for the data $S_0, \ldots, S_{m-1}$. In this way we make a connection between a view space and a hypothesis generator. Since the definition of the score function for $\mathcal{G}$ is a matter which is dependent on $\mathcal{H}$ itself, we shall not discuss this matter further.

The following view spaces are realized in Genomic Hypothesis Creator.

ViewSpace 1 *Alphabet Indexing and Regular Patterns with Local Search Strategy.* BONSAI uses the view space $\mathcal{B}^{(k)} = \{B_{\psi_k}\}_{\psi_k \in (\Sigma_k)^\Sigma}$, where $(\Sigma_k)^\Sigma$ is the set of mappings $\psi : \Sigma \to \Sigma_k$. For a fixed view $B_{\psi_k} = (V_{\psi_k}, L_k)$, BONSAI with S_0 of positive examples and S_1 of negative examples generates a hypothesis by using $S_0^{B_{\psi_k}}$ and $S_1^{B_{\psi_k}}$. A hypothesis is represented as a small decision tree. By employing the accuracy of the decision tree as the score function, it starts with a randomly generated view and changes the view by local search and produces a hypothesis which attains a local maximum. In this way, BONSAI searches the view space $\mathcal{B}^{(k)}$ for a fixed parameter k which is specified by the user.

ViewSpace 2 *Approximate Matching with Exhaustive Search.* The view space $\mathcal{AX}^k[*, *] = \{APPROX^k[i, j] \mid 1 \leq i \leq j\}$ has numerical parameters i and j. When we can assume some rational properties on the score function φ of the hypothesis generator $\mathcal{G}$, some efficient algorithms are known for solving this optimization problem [8]. As long as we cannon assume any specific properties on the score function, we employ the exhaustive search strategy for finding an optimal interval $I = [i, j]$ by bounding $j - i$ by some constant. Similarly, we define view spaces $\mathcal{AX}^k[i, *] = \{APPROX^k[i, j] \mid 1 \leq i \leq j\}$ and $\mathcal{AX}^k[*, j] = \{APPROX^k[i, j] \mid 1 \leq i \leq j\}$. We also consider the view space $\mathcal{AX}^* = \{APPROX^k \mid k \geq 0\}$ over Σ-sequences with an exhaustive search strategy.

2.3 Operations on Views

In view design, it is convenient to combine several views to define a new view. View Designer of Genomic Hypothesis Creator is designed to have an ability to combine several views.

Definition 3. Let $M_i = (V_i, L_i)$ be a view over Σ-sequences with a view element value set W_i for $1 \leq i \leq l$. We assume $L_i \cap L_j = \emptyset$ for $i \neq j$. Let V^+ be an algorithm such that V^+ on (s, π) simulates V_i on (s, π) if π belongs to L_i. Then we define $M_1 + \cdots + M_l = (V^+, L^+)$, where $L^+ = L_1 \cup \cdots \cup L_l$. Furthermore, we define a view $M_1 \times \cdots \times M_l = (V^\times, L_1 \times \cdots \times L_l)$, where $V^\times$ be an algorithm such that $V^\times$ on $(s, (\pi_1, \ldots, \pi_l))$ runs V_i on (s, π_i) for all $1 \leq i \leq l$ and outputs $(V_1(s, \pi_1), \ldots, V_l(s, \pi_l))$.

Let $\mathcal{M} = \{M_\nu\}_{\nu \in N}$ and $\mathcal{M}' = \{M'_{\nu'}\}_{\nu' \in N'}$ be two view spaces with search strategies σ and σ', respectively. It is also convenient to create a new view space by combining view spaces $(\mathcal{M}, \sigma)$ and $(\mathcal{M}', \sigma')$. This view space shall consist of $\mathcal{M} + \mathcal{M}' = \{M_\nu + M_{\nu'}\}_{(\nu, \nu') \in N \times N'}$ (or $\mathcal{M} \times \mathcal{M}' = \{M_\nu \times M_{\nu'}\}_{(\nu, \nu') \in N \times N'}$) and a new search strategy σ'' for $N \times N'$. However, it is not possible to define σ'' in a uniform way.

Genomic Hypothesis Creator employs two kinds of search strategies; local search and exhaustive search shown in ViewSpace 1 and ViewSpace 2, respectively. Fortunately, we can define a search strategy in a natural way for any combination of exhaustive search and local search. In this way, Genomic Hypothesis Creator is designed to allow combinations of views and view spaces.

2.4   Visualization of Views and Hypotheses

Visualization of hypotheses is also an important problem in the discovery process for understanding hypotheses.

We defined a view over sequences as a pair $M = (V, L)$ of the view interpreter V and the set L of view elements. Then a hypothesis generator produces a hypothesis which is represented by using view elements in L. For example, if a hypothesis is described as a binary decision diagram with view elements as decision rules, then, in addition to the visualization of the diagram itself, we need to visualize the view elements on the nodes for better understanding of the hypothesis.

In Genomic Hypothesis Creator, the view interpreter is used for visualizing view elements in hypotheses. Given (s, π) in $\Sigma^* \times L$, the view interpreter on (s, π) exactly shows as an algorithmic process how the string s is viewed by the view element π. The views pre-installed in View Designer equip with view interpreters which animate this algorithmic process.

A view provided by a user can be also used for visualization if the user uses a view interpreter provided by View Designer or provides a view interpreter with a capability of visualization in a specified format.

In this way, a strong association of views and hypothesis visualization is realized in Genomic Hypothesis Creator.

2.5   View Designer of Genomic Hypothesis Creator

Genomic Hypothesis Creator deals with DNA sequences of genes of various organisms. We shall now describe views and view spaces which are implemented in Genomic Hypothesis Creator.

Organisms are roughly classified into two classes, *prokaryotes* and *eukaryotes*. *E. coli* is a prokaryotic organism which has a single circular chromosome and its DNA is described as a single DNA sequence of size about 4.6Mbp. In a prokaryotic DNA sequence, functionally related genes are coded in a series called an *operon* and will be transcripted at one time. Intuitively, a region which is concerned with one transcription process has a structure shown in Fig. 2. A prokaryotic transcription unit contains one or more coding regions, each of which encodes one protein. We assume that locations of coding regions are specified on the sequence. Eukaryotes have a different structure of genes. A gene consists of *exons* and *introns* and encodes one protein (Fig. 3). In the process of splicing, introns are removed and only exons are concatenated to produce a sequence which encodes a protein. A single cell eukaryotic organism *S. cerevisiae* has about 6,000 genes. Most of the genes of *S. cerevisiae* do not have any introns. However, the genes of *Homo sapiens* have a large variety in the number of exons (introns) and their lengths.

The locations of genes on DNA sequences are either tested by experiments or predicted by some softwares [19]. Databases [20] are constructed based on such information.

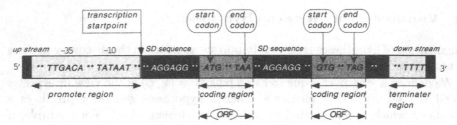

Fig. 2. Structure of prokaryotic genes. Patterns on the sequence are not unique.

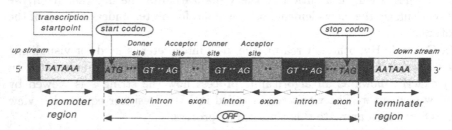

Fig. 3. Structure of a eukaryotic gene.

As a template, View Designer assumes on a DNA sequence the specific regions and locations, such as the transcription startsite, the start point of ORF, exon, intron, ORF, etc, which are shown in Fig. 2 and Fig. 3. Further, the translation of ORF to an amino acid sequence is also attached to the region of an ORF on the DNA sequence.

The 5' region of the transcription startsite is called the *upstream* while the 3' region is called the *downstream*. How to set the length of the upstream and the length of the terminator region becomes a matter of view design.

The View Designer of Genomic Hypothesis Creator is developed with the following principle so that it can cope with a diversity of requirements.

(V1) User can place views $APPROX^k$, $APPROX^k[i,j]$, $APPROX^k[-j,-i]$ on any regions by specifying parameters k, i, and j.

(V2) User can place views $PROSITE$, $PROSITE[i,j]$, $PROSITE[-j,-i]$ on the amino acid sequence located on the region of ORF by specifying the parameters.

(V3) User can place view spaces $\mathcal{AX}^k[*,*]$ and $\mathcal{B}^{(k)}$ any k specified by the user. We call these view spaces the *atomic view spaces*.

(V4) User can use user's own view by plugging-in the view $M = (V, L)$ in a specified format. We assume the set L of view elements is finite. User can also employ the view interpreters provided by View Designer to design user's view by defining a finite collection L of view elements.

(V5) User can combine arbitrary views to create a new view.

(V6) User can combine a view and a view space to create a new view space.

(V7) User can combine two view spaces to create a new view spaces under the condition that at most two atomic view spaces are included.

3   Data Collection from Text Databases

Genomic Hypothesis Creator employs SIGMA[1] for collecting data from databases. SIGMA is a general purpose text database management system that realizes very precise and fast one-way sequential processing of text files. The pattern matching algorithm [2] implemented in SIGMA can handle ten thousands of keywords simultaneously in one search both time and space efficiently. In addition to conventional keyword search, SIGMA realizes very fine searches and replaces which cannot be done with usual information retrieval systems which use inverted files. In SIGMA, we can define records on a text in almost arbitrary way by specifying some strings as record delimiters. This allows us to handle various kinds of text data.

For example, SIGMA can collect records which contain a sequence with segments cgatgacc, tagatt, taatgagttgg occurring in this order by a single search. It is also possible to collect records such that a given keyword repeats more than four times. For example, records with tandem repeats of a specific segment are collectable. For a given file I of keywords, it is possible to collect all records which contain at least one keyword in I through one text search. Furthermore, these searches can be combined into one search over the text. Collected records can be refiled into another format in a very flexible manner so that the refiled data can be used for another analysis. The text search principle realized in SIGMA makes such fine operations feasible. Genomic Hypothesis Creator employs SIGMA for searching and refiling. With this system, by using keywords in annotations, sequence segments, etc, user can easily collect segments of DNA sequences of target genes in a user specified format. The collection of data shall be transferred to a hypothesis generator through View Designer.

Various genome databases [20] also provide very useful information retrieval tools based on inverted files. However, they do not include such an ability of SIGMA that is most suited for our purpose.

4   Hypothesis Generator

The idea of multistrategy principle has been extensively discussed as an important aspect of knowledge discovery system [11, 12, 17, 18]. Genomic Hypothesis Creator is designed to allow user to select a hypothesis generator from the pool $\mathcal{H}$ of hypothesis generators, and enables us to plug-in an external tool in a specific format into $\mathcal{H}$ without redevelopment of the system core. Such a flexibility is also realized in a data mining system Kepler [18], which produces decision trees, neural networks, etc. In contrast with Kepler, Genomic Hypothesis Creator automatically repeats the generation of hypotheses according to views determined by a search strategy of a view space.

We consider a hypothesis generator as an algorithm $\mathcal{G}$ such that, given the data matrices $S_0^M, \ldots, S_{m-1}^M$ of $S_0, \ldots, S_{m-1} \subseteq \Sigma^*$ under a view M, it generates an expression representing a hypothesis h on $S_0, \ldots, S_{m-1}$ in terms of the view elements of M. A score function φ is associated with the hypothesis generator in order to evaluate hypotheses generated by $\mathcal{G}$.

The following hypothesis generators are implemented in the current prototype system of Genomic Hypothesis Creator.

Decision Tree The hypothesis generated in the BONSAI system is a small decision tree as is mentioned in ViewSpace 1. A hypothesis generator $\mathcal{G}_B$ of BONSAI executes the following: Given sets S_0 and S_1 of positive and negative examples, $\mathcal{G}_B$ chooses randomly small sets P and N of positive and negative examples from S_0 and S_1, respectively. Then BONSAI generates a decision tree T by using the algorithm of ID3 [13] which discriminates P and N perfectly. The score function φ_B is given as the accuracy function given in ViewSpace 1. The size $|P|$ (resp. $|N|$) is called the *window size* of positive (resp. negative) training examples. In BONSAI, the window size is set from 5 to 20 in order to keep T small. The size of decision tree can be controlled by the window sizes. This hypothesis generator has a special sense in knowledge discovery since the sets P and N used for constructing a decision tree can be regarded as representative examples of the hypothesis.

Binary Decision Diagram *Binary decision diagrams* (BDDs) are a useful representation of Boolean functions, which are regarded as directed acyclic graphs [5].

We have constructed a hypothesis generator $\mathcal{G}_{BDD}$ for binary decision diagrams in the following heuristic way: Given data matrices S_0^M and S_1^M, $\mathcal{G}_{BDD}$ makes a decision tree T by in the same way as BONSAI except that in this case, instead of randomly chosen small sets P and N, *all* examples in S_0 and S_1 are used for generation of a decision tree. This produces, in general, a large tree which classifies S_0 and S_1 as optimally as possible. Then $\mathcal{G}_{BDD}$ transforms the resulting tree T to an equivalent BDD by employing the method in [5]. This reduces the number of nodes to some extent if the same view elements occur in the tree. The score function φ_{BDD} is defined by the number of nodes of the BDD. The resulting BDD itself may provide an interesting knowledge if view elements are interpreted visually. Further, when a view space is placed on data, if we could discover a view under which the score of the resulting BDD is locally optimum, the view can provide a good understanding of the data S_0 and S_1 and can be regarded as a kind of discovery. It is not hard to extend this hypothesis generator for a k-decision diagram with $k \geq 3$.

Conclusion

Our motivation of this paper is to contribute to scientific discovery in genomic researches by formulating a method of scientific discovery and realizing it as Genomic Hypothesis Creator. We have defined new notions of view over sequences and view space over sequences that constitute the foundations of the discussion in this paper. With them, View Designer is designed for automatically creating new views on data, which may give user a key to discovery. User can add user's own views to the system through a plug-in interface. Currently the view space $\mathcal{B}^{(k)}$ used in the BONSAI system and the view space $\mathcal{AX}^k[*,*]$ in ViewSpace 2 are available.

In Genomic Hypothesis Creator users can take multistrategies for discovery by selecting hypothesis generators from the pool of hypothesis generators. The system also supports to plug-in external tools of hypothesis generators into the pool, which extend the capabilities of Genomic Hypothesis Creator in a specific way. In the process of searching a view space $(\mathcal{M}, \sigma)$, a view of $\mathcal{M}$ selected by σ is linked to a hypothesis generator which is selected from the pool of hypothesis generators. In this way Genomic Hypothesis Creator offers a very wide range of methods for scientific discovery from text databases.

It has been scheduled to carry out experiments, we will report experimental results with further discussions on theoretical aspects.

References

1. ARIKAWA, S., HARAGUCHI, M., INOUE, H., KAWASAKI, Y., MIYAHARA, T., MIYANO, S., OSHIMA, K., SAKAI, H., SHINOHARA, T., SHIRAISHI, S., TAKEDA, M., TAKEYA, S., YAMAMOTO, A., AND YUASA, H. The text database management system SIGMA: An improvement of the main engine. In *Proc. Berliner Informatik-Tage* (1989), pp. 72–81.
2. ARIKAWA, S., AND SHINOHARA, T. A run-time efficient realization of Aho-Corasick pattern matching machines. *New Generation Computing 2* (1984), 171–186.
3. BAIROCH, A. PROSITE: a dictionary of sites and patterns in proteins. *Nucleic Acids Res. 19* (1991), 2241–2245.
4. BRAZMA, A., VILO, J., UKKONEN, E., AND VALTONEN, K. Data mining for regulatory elements in yeast genome. In *Proc. 5th Int. Conf. Intelligent Systems for Molecular Biology (ISMB-97)* (1997), AAAI Press, pp. 65–74.
5. BRYANT, R. Graph-based algorithms for Boolean function manipulation. *IEEE Transactions on Computers 35* (1986), 677–691.
6. FAYYAD, U., HAUSSLER, D., AND STOLORZ, P. Mining scientific data. *Commun. ACM 39*, 11 (1996), 51–57.
7. FAYYAD, U., PIATETSKY-SHAPIRO, G., AND SMYTH, P. The KDD process for extracting useful knowledge from volumes of data. *Commun. ACM 39*, 11 (1996), 27–34.
8. FUKUDA, T., MORIMOTO, Y., MORISHITA, S., AND TOKUYAMA, T. Interval finding and its application to data mining. In *Proc. 7th International Symposium on Algorithms and Computation (ISAAC '96)* (1996), Lecture Notes in Computer Science 1178, Springer-Verlag, pp. 55–64.
9. KOHAVI, R., AND (EDS.), F. P. *Machine Learning.* Kluwer Academic Publishers, 1998.
10. MICHALSKI, R. S., BRATKO, I., AND KUBAT, M. *Machine Learning and Data Mining: Methods and Applications.* John Wiley & Sons, Ltd., 1998.
11. MICHALSKI, R. S., AND (EDS.), J. W. *Machine Learning.* Kluwer Academic Publishers, 1997.
12. MICHALSKI, R. S., KERSCHBERG, L., KAUFMAN, K., AND RIBEIRO, J. Mining for knowledge in databases: The INLEN architechure, initail implementation and first results. *J. Intelligent Information System: Integrating AI and Database Technologies 1* (1992).

13. QUINLAN, J. R. Induction of decision trees. *Machine Learning 1* (1986), 81–106.
14. SHIMOZONO, S., SHINOHARA, A., SHINOHARA, T., MIYANO, S., KUHARA, S., AND ARIKAWA, S. Knowledge acquisition from amino acid sequences by machine learning system BONSAI. *Trans. Information Processing Society of Japan 35* (1994), 2009–2018.
15. SRINIVASAN, A., AND KING, R. D. Feature construction with inductive logic programming: a study of quantitative predictions of biological activity by structural attributes. In *Proc. 6th International Workshop on Inductive Logic Programming (ILP-96)* (1997), Springer-Verlag, pp. 89–104.
16. WADMAN, M. Company aims to beat NIH human genome efforts. *NATURE 393* (1998), 101.
17. WIRTH, R., SHEARER, C., GRIMMER, U., REINARTZ, T., SCHLOSSER, J., BREITNER, C., ENGELS, R., AND LINDNER, G. Towards process-oriented tool support for knowledge discovery in databases. In *Proc. First European Symposium on Principles of Data Mining and Knowledge Discovery (PKDD '97)* (1997), Springer-Verlag.
18. WROBEL, S., WETTSCHERECK, D., SOMMER, E., AND EMDE, W. Extensibility in data mining systems. In *Proc. of the 2nd International Conference On Knowledge Discovery and Data Mining (KDD-96)* (1996), pp. 214–219.
19. XU, Y., EINSTEIN, J. R., MURAL, R. J., SHAH, M., AND UBERBACHER, E. C. An improved system for exon recognition and gene modeling in human DNA sequences. In *Proc. the Second Internatinal Conference on Intelligent Systems for Molecular Biology* (1994), AAAI Press, pp. 376–383.
20. http://genome-www.stanford.edu/,
    ftp://ncbi.nlm.nih.gov/genbank/genomes/,
    http://www.genetics.wisc.edu/,
    http://www.mbl.edu/html/riley/monica.html,
    http://www.genome.ad.jp/.

Visualization of Community Knowledge Interaction Using Associative Representation

Takashi Hirata[1], Harumi Maeda[2], and Toyoaki Nishida[1]

[1] Graduate School of Information Science, Nara Institute of Science and Technology
8916-5 Takayama, Ikoma, Nara 630-01, Japan
Phone : +81-743-72-5265, Fax : +81-743-72-5269
E-mail : {takash-h, nishida}@is.aist-nara.ac.jp
[2] Media Center, Osaka City University
3-3-138 Sugimoto, Sumiyoshi, Osaka 558-8585, Japan
Phone : +81-6-605-3375, Fax : +81-6-690-2736
E-mail : harumi@media.osaka-cu.ac.jp

Abstract. *This paper describes a method for supporting knowledge evolution and facilitating awareness in a community at the same time. We propose two ideas. One is associative representation for facilitating externalization of both personal and community information. The associative representation links heterogeneous information without defining the semantics strictly. We leave the interpretation of the semantics to human background knowledge. The other is visualization of information interaction in a community using talking-alter-egos metaphor. Taking-alter-egos metaphor mimics a salon in which alter-ego representing each community member interact with each others, thereby the community member can see how their own or others' knowledge interact.*
We have developed a called CoMeMo-Community that pursue collaborative story generation based on the talking-alter-egos metaphor. We investigated how far people can exchange ideas with associative representation and how people react the talking-alter-egos metaphor.

1 Introduction

Thought all people are autonomous information processing, they create their knowledge not by themselves but by interacting others, such as persons, articles and so forth. It will be necessary to interact other persons by communication in order to create more organized knowledge especially. People may make and join a group, when they collaborate with each other to solve common problems or fulfill common interests. A community is a group of persons with common interests.

The first step of spontaneous communication in a community is what one is aware of other's state, for example, who knows what, what are common interests and so forth. It is called awareness [1]. Communication has not only function of conversation but also role of knowledge interaction. The more awareness of a community is enhanced, the more knowledge interaction is active. But personal information tends to be heterogeneous and its structures are ambiguous. These will prevent community members from communicating and collaborating with each other. Such heterogeneous information should be exchanged and shared more freely.

When people consider and summarize their own ideas, they usually prefer to be alone. But people want to contact with others in order to find a clue, when their thoughts or ideas come to a deadlock once. The human knowledge creation is practiced by studying by themselves and exchanging opinions with each other. People evolve their knowledge with combining personal and others knowledge in turn.

Our research aims to support for evolution of both personal and community knowledge. We proposed two ideas. One is associative representation for facilitating externalization of both personal and community information. The other is visualization of information interaction using talking-alter-egos metaphor. We developed a system called CoMeMo-Community based on these two ideas.

In the next section, we describe overview of CoMeMo-Community. In section 3, we propose associative representation that facilitates externalization of human knowledge and describe how people generate and understand it. In section 4, we describe alter-ego metaphor that visualize community knowledge interaction and report how people react it.

2   CoMeMo-Community : a system for supporting community knowledge evolution

2.1   Overview of CoMeMo-Community

CoMeMo-Community is a system designed to support community knowledge evolution by enhancing community awareness. This system based on the following two ideas.

Associative representation : to facilitate externalization of the both personal and community knowledge

Talking-alter-ego metaphor : to visualize community members' knowledge interaction using human alter-ego

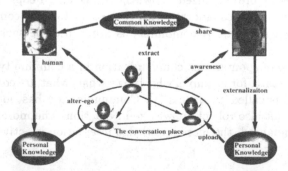

Fig. 1. Overview of CoMeMo-Community

Figure 1 is the overview of CoMeMo-Community. Community member externalize their own knowledge using associative representation and store it as personal knowledge. Uploading each stored personal knowledge, alter-egos representing each community member interact with each others on the conversation place. This knowledge interaction is visualized by talking-alter-ego metaphor and generates new knowledge. The generated knowledge is extracted as community knowledge and shared in a community. By setting in motion and observing virtual conversation among alter-egos of herself/himself or/and others, the user is facilitated awareness of community.

2.2 Cycle of community knowledge evolution

This system provides the user with two phases. One is personal phase for maintaining personal knowledge with associative representation. The other is community phase for observing community knowledge interaction using taking-alter-egos metaphor. With coming and going between personal and community phase, the user evolve knowledge. The cycle of community knowledge evolution is shown in Figure 2.

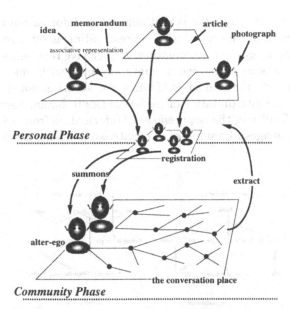

Fig. 2. Cycle of community knowledge evolution

When users join a virtual community for the first time, they must register alter-ego who retains their knowledge in order to be free to summon it. Thus one may run the virtual conversation as many times as he/she wants at any time, and observe what happens. The users update their own personal knowledge using associative representation in everyday life , for example, when the ideas occur to them. Because personal knowledge is updated day by day, one may see different conversation at each time even if the same set of keywords given. The users take useful information on the conversation place back to personal knowledge and reuse it. Thus the users evolve their knowledge with combining personal and others knowledge in turn.

3   Maintaining personal knowledge with associative representation

CoMeMo-Community supports maintaining personal knowledge by providing the user with facilitates of aggregating, browsing, editing, refining associative representation. The semantics of associative representation itself is left open. Associative representation permits raw information materials to be accumulated with minimal overhead. In return, interpretation of such information heavily relies on background knowledge.

3.1   Associative representation

In the following, we set a hypothesis that connecting information without defining the semantics using the associative representation is effective to handle a large amount of heterogeneous information. Associative representation is many-to-many hyper-link associating one or more key unit with one or more value unit. In our approach, the semantics of the associations is not defined strictly. Instead, we leave the interpretation of the semantics to human tacit background knowledge. This facilitates the acquirement of information from a variety of data (e.g. ideas, texts, images). Example of associative representation is illustrated in Figure 3.

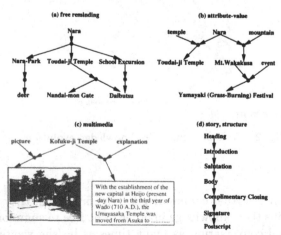

Fig. 3. Example of associative representation

– Free Reminding
 Figure 3(a) denotes that from given concepts "Nara",one may be reminded of "Nara Park", "Todai-ji Temple", and "School Excursion". Ten persons have ten different free reminding.
– Attribute-Value
 Figure 3(b) denotes that "Todai-ji Temple" is reminded when "Nara" and "Temple" are given as keys. Association in this kind of collections are value-attribute representation.
– Multimedia
 Reminded things are not only concepts but also images or texts. Figure 3(c) denotes that from given concepts "Kofuku-ji Temple", one may be reminded of "Picture of Kofuku-ji Temple" and "Explanation of Kofuku-ji Temple".
– Story, Structure
 Associations is able to show work-flow or story line too. Figure 3(d) denotes form of letter.

We call a set of associations collected from a particular point of view workspace. Any data items are represented by icons called unit on a workspace, e.g., concepts, texts, image files and so on. workspaces can be nested. A workspace, presented as icon, can be placed in another workspace.

User externalize own knowledge using associative representation and store as personal knowledge on the workspace. Example of personal knowledge about fishing is shown in Figure 4. User is able to add, delete, and edit units and reorganize its associations optionally by mouse handling. In human sight user access own personal knowledge through workspace.

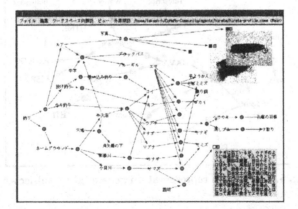

Fig. 4. Example of personal knowledge on workspace

3.2 Experiment - Generating and Understanding associative representation -

In this section, we report how people generate associative representations, and how people understand their semantics.

Method

Apparatus
CoMeMo[1]

Subjects
One Ph.D. students, four 2nd-year M.Eng. students, and one staff in our laboratory.

Procedure
(1) The subjects were trained how to generate associations (students only).
(2) They generated associations reminded by a keyword "agent" on the CoMeMo freely (students only).
(3) They were shown associations generated by other subjects and answered the following questions:
  - "Do you understand what are written ?"
  - "Do you identify who wrote this associative representation ?"
  - "If you identify who, why ?"
  - "Say anything you felt in this experiment"
  (students only)
(4) The same as (3). (the staff subject only)

Results and Discussion An example of associative representation generated by the Subject A is shown in Figure 5[2]

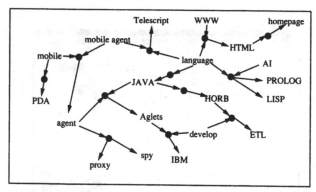

Fig. 5. Associative representation generated by subject-A

All subjects generated associations within 30 minutes. We analyze that adults who have computing skills can generate associations without difficulty.

All subjects understood the meaning of associations generated by others. Concerning an associative representation generated by the subject C, all other

---

[1] CoMeMo is a system for integrating heterogeneous information using associative representation [2].
[2] Concepts were originally written in Japanese.

subjects identified that it was made by him. We analyze that ideas can be transmitted using associations among people who share knowledge. All subjects except for the subject C laughed when they saw screens wrote by the subject C. 80% (4 out of 5 student subjects) said that they had some fun during the experiment. The staff subject said that "I can assume the subject's knowledge level concerning research topics", "I may want to ask a report for this subject", "I want to talk to this subject, because he/she may be interesting", and so on. We think that transmitting ideas using associations between groups leads to know people and therefore facilitates for human communication.

4 Visualizing community knowledge interaction using taking-alter-egos metaphor

CoMeMo-Community provides users with visualizing community knowledge interaction using taking-alter-egos metaphor. Alter-egos behaves on behalf of the user on the conversation place. By observing this behavior, the user learn lots of things about the community.

4.1 Talking-alter-egos metaphor

Talking-alter-egos metaphor consists of two components. One is an alter-ego that keeps the externalized knowledge of person. The other is a conversation place where alter-egos make utterances in turn.

The example of utterance mechanism is shown in Figure 6. In the beginning the user must put one or more keyword as topic of utterance on the conversation place and make them active. Now "Nara" and "Temple" are given as keywords and activated (Figure 6(a)). Each alter-ego monitors keyword on the conversation place whether they have information related to it in their knowledge. If there are something information to relate, the alter-ego links the information as new keyword by copying it from her/his knowledge. The new keyword is add to the right of the original keyword using associative representation. Alter-ego-A find out new keyword "Todaiji-temple" and linked it with "Nara" and "Temple" (Figure 6(b)). Then, the original keywords "Nara" and "Temple" are deactivated, the new keyword "Todaiji-temple" is activate. Each agents put to the information related to "Todaiji-temple" in a row (Figure 6(c)). Alter-ego-A put the new keywords "Nandai-mon Gate" and "Daibutsu" that are structures of Todaiji-temple. Alter-ego-B put the new keyword "Shuni-e Ceremony" that is event at Todaiji-temple. Such activation cycle will continue. Thus, alter-egos collaborate with each other to generate a story by alternately reproducing information fragments from its knowledge.

4.2 Story generation by talking-alter-egos metaphor

Example of actual story generation on CoMeMo-Community shows Figure 7.

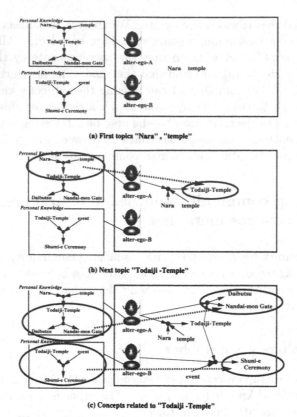

(a) First topics "Nara", "temple"

(b) Next topic "Todaiji -Temple"

(c) Concepts related to "Todaiji -Temple"

Fig. 6. Mechanism of taking-alter-egos metaphor

When user enter one virtual community, all registered alter-egos in it appear on the conversation place firstly. Each alter-ego is represented by each community members' image. There are two ways of choosing alter-egos. One is selection from a list of registered alter-egos and/or pointing to the alter-egos' image directly by user (Figure 7(a)). The other is the way that alter-egos who have information related to given keyword are chose by system.

In this case, three alter-egos are chose and put to the left of the conversation place. User put a keyword as first topic. Each alter-ego react by throwing back its own knowledge for a given keyword. As interaction among alter-egos continue, the story grows gradually. The original keyword was "Alcohol Drinks" (Figure 7(b)), which reminds "Wine" (Figure 7(c)), which in turn reminds "Pasta", "Italian Food", resulting in "Word-of-Mouth information" (Figure 7(d)).

4.3   Analysis of the conversation among alter-egos

Virtual conversation by alter-egos generates various stories. Story generation is influenced by a variety of factors such as choosing alter-egos, given keywords, the maximum keyword that an alter-ego can put on the conversation place and so on.

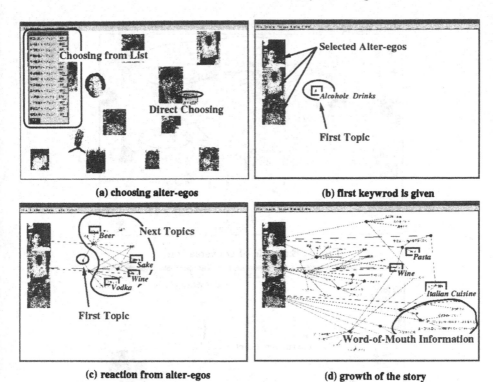

(a) choosing alter-egos

(b) first keywrod is given

(c) reaction from alter-egos

(d) growth of the story

Fig. 7. Example of story generation on CoMeMo-Community

Although mechanism of this virtual conversation is very simple, lots of interesting phenomenon are observed. Such conversation among alter-egos looks like conversation by humans.

We show interesting conversation among alter-egos in the following examples.

A typical example is shown in Figure 8. Figure 8 represents that alter-egos representing a staff and three students make conversation about laboratory project. The alter-ego of staff talks about main theme and research of project. Each alter-ego of students talk about their system in detail. This conversation is typical type of knowledge sharing .

Figure 9 shows the example of topic shift during conversation. At first, each alter-ego talk about "international conference" that is given as first topic. But one of them begins to talk about "ruins" at the topic "Mexico" as turning point.

Figure 10 shows the example of topic shift by a homonym. At the beginning of conversation, lower two alter-egos talk about "fishing" that is given as first topic. Each alter-ego begin to talk about "baseball field" and/or "fishing spot" respectively as soon as keyword "Home Ground" is put on the conversation place. Such phenomenon happens in real conversation.

4.4   Experiment - Reaction to talking-alter-egos metaphor -

In this section, we report how people react to the talking-alter-egos metaphor.

126   Takashi Hirata et al.

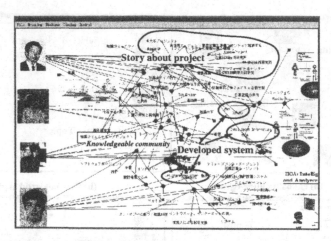

Fig. 8. Example of conversation

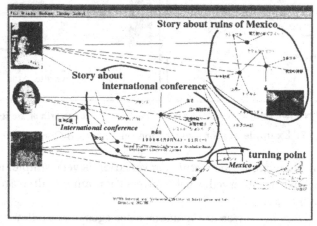

Fig. 9. Example of topic shift

Method

Apparatus
CoMeMo-Community

Subjects
Four 2nd-year M.Eng. students, and one 1st-year M.Eng. students in our laboratory.

Preparation
We created 14 alter-egos that represent a staff and students in our laboratory on CoMeMo-Community.

Procedure
(1) The subjects chose interesting alter-egos and put keywords on the conversation place.

(2) They were shown demonstration of CoMeMo-Community using choosing alter-egos and keywords by them and answered the following questions :

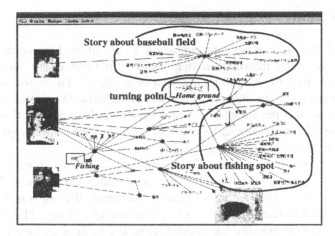

Fig. 10. Example of topic shift by a homonym

- "Could you understand the information displayed on the conversation place ?"[3]
- "What do you find out from this demonstration ?"
- "What do you think about that your information is opened to public with your image ?"
- "Please feel free to comment about this system."

Results and Discussion The reaction was all favorable.

All subjects were able to understand what is meant by associative representation. But the more associative representation increase, the harder most of them can understand it, because they are confused by overlapped units and links. We analyzed that adults who don't know about associative representation at first can mostly understand it. We also think that it is necessary to share background knowledge for the purpose of more comprehend associative representation.

All subjects said that it is easy to find out common topics and interests in this system because information interaction is shown visually and objectively. This suggests that this system is very effective for discovering helpful information in order to make contact with others in a community.

One subject said that "Why he has such information ? I'd like to talk with him personally.". Thus this system enable to facilitate communication not only on virtual community but also in real world.

Subjects all said that they would publish their own knowledge if it contributes to their community and the use of the facial image is effective in demonstration, in particular when the observer is acquainted with the person herself/himself each alter-ego represents.

As a result of investigation, we realized that our system enables community members to facilitate communication by sharing each member's information and enhancing awareness of a community.

---

[3] All subjects aren't knowledgeable about associative representation

5 Conclusion

We proposed the associative representation to integrate heterogeneous information such as static information (e.g. local sites information) and dynamic information created in word-of-mouth communication. The associative representation connect various information without defining the semantics strictly. We investigated how people generate and understand the associative representation. We found that ideas can be transmitted using associative representation among people who share knowledge.

We developed a system called CoMeMo-Community which support knowledge evolution in a community. This system visualizes community knowledge interaction and facilitates community awareness using alter-egos metaphor. We investigated the effectiveness of this system in our laboratory. The results suggest that the system enables community members to facilitate communication by sharing each members' information.

As a future research, We plan to apply this system to more large community over a network practically and evaluate the usefulness of the system.

References

1. Dourish, P., and Bly, S. "Portholes : Supporting Awareness in a Distributed Work Group", Proc. CHI'92, pp.541-548, 1992.
2. Harumi Maeda, Takashi Hirata, and Toyoaki Nishida. "CoMeMo : Constructing and Sharing Everyday Memory", Proceedings of the Ninth International Conference on Tools with Artificial Intelligence (ICTAI'97), pp.23-30, 1997.
3. Harumi Maeda and Toyoaki Nishida. "Generating and Understanding Weak Information Structures by Humans", IASTED International Conference "Artificial Intelligence and Soft Computing (ASC'98)", pp.74-78, 1998.
4. Takashi Hirata, Harumi Maeda, and Toyoaki Nishida. "Facilitating Community Awareness with Associative Representation", Proceedings of the Second International Conference on Knowledge-Based Intelligent Electronic Systems (KES'98), pp.411-416, 1998.

Discovering Characteristic Patterns
from Collections of Classical Japanese Poems

Mayumi Yamasaki[1], Masayuki Takeda[1],
Tomoko Fukuda[2], and Ichirō Nanri[3]

[1] Department of Informatics, Kyushu University 33, Fukuoka 812-8581, Japan
[2] Fukuoka Jo Gakuin College, Ogōri 838-0141, Japan
[3] Junshin Women's Junior College, Fukuoka 815-0036, Japan
{yamasaki, takeda}@i.kyushu-u.ac.jp

Abstract. WAKA is a form of traditional Japanese poetry with a 1300-year history. In this paper, we attempt to discover characteristics common to a collection of WAKA poems. As a formalism for characteristics, we use regular patterns where the constant parts are limited to sequences of auxiliary verbs and postpositional particles. We call such patterns FUSHI. The problem is to find automatically significant FUSHI patterns that characterize the poems.

Solving this problem requires a reliable significance measure for the patterns. Brāzma et al. (1996) proposed such a measure according to the MDL principle. Using this method, we report successful results in finding patterns from five anthologies. Some of the results are quite stimulating, and we hope that they will lead to new discoveries. Based on our experience, we also propose a pattern-based text data mining system. Further research into WAKA poetry is now proceeding using this system.

1 Introduction

WAKA is a form of traditional Japanese poetry with a 1300-year history. Most WAKA poems are in the form of TANKA, namely, they have five lines and thirty-one syllables, arranged thus: 5-7-5-7-7. This poetry was usually composed in momentary flashes of inspiration. Most frequently, it was used as a subtle means of communication between lovers and friends, and was therefore an important part of daily life in ancient Japan. WAKA poetry has been central to the history of Japanese literature, and has been studied extensively by many scholars. Recently, since an accumulation of about 450,000 WAKA poems became available in a machine-readable form, it is expected that computers will play an important role in research into WAKA poetry.

In this paper, we focus on the problem of discovering characteristics common to a collection of WAKA poems. As a formalism for characteristics, we use the class of *pattern languages* introduced by D. Angluin [2]. One of the most important subclasses is the class of *regular* pattern languages, in which each variable symbol appears only once. This subclass is sufficiently rich from a practical viewpoint [9]. To characterize WAKA poems, we limit the constant parts of regular patterns to sequences of *adjuncts*, i.e., auxiliary verbs and postpositional particles. For example, we consider patterns such as *BA*ZARAMASHIO *, where BA is a postpositional particle, and ZARAMASHIO is a chain of two auxiliary verbs ZARA and MASHI and a postpositional particle O. This pattern corresponds to the subjunctive mood. We call such patterns FUSHI, and call their constant parts *adjunct sequences*. This limitation of the constant parts to adjunct sequences is essential in our characterization. It should be noted that a Japanese sentence is a sequence of *segments*, each of which consists of a word and its subsequent adjuncts. The FUSHI pattern is a reliable model for techniques used in composing WAKA poems, and is closely related to their structure and rhythm.

The goal of this paper is to discover, more or less automatically, significant FUSHI patterns that characterize a given set of WAKA poems, and then report some features that may be due to the times or to the poets' personalities. The difficulties are summarized as follows.

- To identify adjuncts appearing in WAKA poems.
- To give an appropriate definition of the *significance* of FUSHI patterns.

Since we have no delimiters between segments in the Japanese language, the first item requires morphological analysis of WAKA poems. However, many ambiguities, which are not easy to resolve, will arise during this analysis. In this paper, we assume that any substring identical to an adjunct is the adjunct, and therefore we have only to perform pattern matching. This assumption simplifies the discussion: Let Σ be an alphabet, and let $* \notin \Sigma$ be the gap symbol. A *pattern* is a nonempty string over $\Sigma \cup \{*\}$. We say that a pattern p *matches* a string w in Σ^+ if w is obtained by substituting strings in Σ^* for occurrences of $*$ in p, respectively. A set Π of patterns is said to be a *covering* of a set S of strings in Σ^+ if, for any string in S, at least one pattern in Π exists that matches it.

Let $C \subseteq \Sigma^+$ be a set of *adjunct sequences*. A FUSHI pattern is a pattern of the form $*\alpha_1*\alpha_2*\cdots*\alpha_h*$, where $h > 1$ and $\alpha_1, \alpha_2, \ldots, \alpha_h \in C$. Our problem is then defined as follows.

Given a finite set S of strings in Σ^+, find the most *significant* covering Π of S that consists of FUSHI patterns.

Note that, in general, S has infinitely many coverings. If we have an appropriate definition of the significance of such coverings, then we can determine the most significant covering according to the definition. However, the problem remains of defining the significance appropriately. One such definition was given by Arimura et al. [3]. A *k-minimal multiple generalization* (abbreviated to *k-mmg*) of a set S of strings is defined as a minimally general set that is a covering of S containing at most k patterns. They showed in [3] that k-mmg is optimal from the viewpoint of inductive inference from positive data based on identification in the limit [7]. However, we are faced with the following difficulties: (a) we must give an integer k as an upper bound of the size of coverings in advance; (b) a polynomial-time algorithm exists for finding a k-mmg, but it is impractical in the sense that the time complexity will be very large for a relatively large value of k; (c) since a set S of strings has more than one k-mmg, we need a criterion to choose an appropriate one.

Another definition was proposed by Brāzma et al. [5]. The most significant covering of a set S is defined as the most probable collection of patterns likely to be present in the strings in S, assuming some simple, probabilistic model. This criterion is equivalent to Rissanen's Minimum Description Length (MDL) principle [8]. According to the MDL principle, the most significant covering is the one that minimizes the sum of the length (in bits) of the patterns and the length (in bits) of the strings when encoded with the help of the patterns. Since finding the optimal solution is NP-hard, a polynomial-time algorithm for approximating the optimal solution within a logarithmic factor is presented.

In this paper, we use the MDL principle to define the significance of FUSHI patterns, and apply the method developed by Brāzma et al. to the problem of finding significant patterns from a set of WAKA poems. The main contributions of this paper are summarized as follows.

1. A new schema is presented in which the constant parts of regular patterns are restricted to strings in a set C. Allowing C to be the set of adjunct sequences yields a reliable model for characterizing WAKA poems.
2. A new grammatical scheme of Japanese language is given as the basis of the above characterization. This scheme is far from standard, but constitutes a simple and effective tool.
3. Successful results of our experiment of finding patterns from five anthologies are reported, some of which are very suggestive. We hope that they will lead to new areas of research. The significance measure for patterns based on the MDL principle is proved to be useful.
4. A text data mining system is proposed, which consists of a *pattern matching part* and a *pattern discovery part*. Using this system, further research into WAKA poetry is now proceeding.

It should be emphasized that the FUSHI pattern of a WAKA poem is not conclusively established even when determined by non-computer-based efforts. The determined pattern may vary according to the particular interests of scholars, and to the other poems used for comparison. Our purpose is to develop a method for finding a set of significant patterns, some of which may give a scholar clues for further investigation. Similar settings can be found in the field of data mining. In other words, our goal is to develop a text data mining system to support research into WAKA poetry.

Unlike data mining in relational databases, data mining in texts that are written in natural language requires preprocessing based on natural language processing techniques with some domain knowledge, and such techniques and knowledge are the key to success [1, 6]. In our case, no such techniques are needed; the knowledge used here is merely the set of adjunct sequences, which are allowed to appear in FUSHI patterns as constant parts. It may be relevant to mention that the third author is a WAKA researcher, the fourth author is a linguist in Japanese language, and the first and the second authors are researchers in computer science.

2 Method

This section shows why the FUSHI pattern can be a reliable model for characterizing WAKA poems, and presents our grammatical scheme on which the characterization is based.

2.1 FUSHI pattern as a model of characteristics

Consider the problem of finding characteristics from a collection of WAKA poems. Most studies on this problem have been undertaken mainly from the viewpoint of preference for KAGO words. By KAGO, we mean the nouns, verbs and adjectives used in WAKA poems[1]. It might be considered that WAKA poems commonly containing some KAGO words are on the same subject matter: in other words, such characterization corresponds to similarity of subject matter. For example, Ki no Tsurayuki (ca. 872–945), one of the greatest of the early court poets, composed many WAKA poems on the cherry blossom. It is, however, simplistic thinking to conclude that the poet had a preference for cherry blossoms. The reason for this is that court poets in those days were frequently given a theme when composing a poem. We must assume that the poets could not freely use favored words.

What then should be considered as characteristics of WAKA poems? The three poems shown in Fig. 1 are very famous poems from the imperial anthology SHINKOKINSHŪ. These poems were arranged by the compilers in one section of the anthology, and are known as *the three autumn evening poems* (SANSEKI NO UTA). All the poems express a scene of autumn evening. However, the reason why the poems are regarded as outstanding poems on autumn evening is that they all used the following techniques:

1. Each poem has two parts: the first three lines and the remaining two lines.
2. The first part ends with the auxiliary verb KERI.
3. The second part ends with a noun.

Such techniques are basically modeled by FUSHI patterns, regular patterns in which the constant parts are limited to adjunct sequences.

WAKA poetry can be compared to IKEBANA, the traditional Japanese flower arrangement. The art of IKEBANA relies on the choice and combination of both *materials* and *containers*. Limitation on the choice of materials, that is, KAGO words, probably forced the poets to concentrate on the choice of containers, FUSHI patterns. The FUSHI pattern is thus a reliable model for characterizing WAKA poems.

---

[1] The term KAGO consists of two morphemes KA and GO, which mean 'poem' and 'word', respectively; therefore, it means words used in poetry rather than in prose. However, here, we mean the nouns, verbs and adjectives used in WAKA poems.

#361 (Priest Jakuren)

SABISHISA WA	*One cannot ask loneliness*
SONO IRO TO SHI MO	*How or where it starts.*
NAKARI KERI	*On the cypress-mountain,*
MAKI TATSU YAMA NO	*autumn evening.*
AKI NO YŪGURE.	

#362 (Priest Saigyō)

KOKORO NAKI	*A man without feelings,*
MI NI MO AWARE WA	*Even, would know sadness*
SHIRA RE KERI	*When snipe start from the marshes*
SHIGI TATSU SAWA NO	*On an autumn evening.*
AKI NO YŪGURE.	

#363 (Fujiwara no Teika)

MIWATASE BA	*As far as the eye can see,*
HANA MO MOMIJI MO	*No cherry blossom,*
NAKARI KERI	*No crimson leaf:*
URA NO TOMAYA NO	*A thatched hut by a lagoon,*
AKI NO YŪGURE.	*This autumn evening.*

Fig. 1. The three autumn evening poems from SHINKOKINSHŪ; blank symbols are placed between the words for readability. English translations are from [4].

2.2 Our grammatical scheme

In the standard framework of Japanese grammar, words are divided into two categories: independent words (or simply, *words*) and dependent words (or *adjuncts*). The former is a category of nouns, verbs, adjectives, adverbs, conjunctions and interjections, while the latter are auxiliary verbs and postpositional particles. A Japanese sentence is a sequence of segments, and each segment consists of a word and its subsequent adjuncts. Verbs, adjectives, and auxiliary verbs can be conjugated. It should be noted that most of the conjugated suffixes of verbs and adjectives are identical to some auxiliary verbs or to their conjugated suffixes. If we regard the conjugated suffixes of words as adjuncts, a segment can be viewed as a word stem and its subsequent adjuncts. This stem-adjunct scheme is far from being standard grammar, but it does constitute a simple and effective tool for our purposes.

We can see that the FUSHI pattern *REBA*KOSO*KERE* is common in the poems in Figure 2. The occurrences of BA and KOSO in these poems are postpositional particles. However the occurrences of RE and KERE have more than one grammatical category in the standard grammar. In fact, the occurrence of RE in each of the first two poems is a conjugated suffix of a verb, while RE in the last poem is a conjugated suffix of an auxiliary verb. The occurrence of KERE in each of the first two poems is a conjugated suffix of an adjective, while KERE in the last poem is an auxiliary verb. Although the occurrences of the FUSHI pattern in the three poems are thus different, we intend to treat them as if they were the same. Finding FUSHI patterns requires a new grammatical scheme that is different from the standard; our stem-adjunct scheme is appropriate. As stated previously, the strings appearing as the constant parts of FUSHI patterns are called adjunct sequences. Although an adjunct sequence consists of one or more adjuncts, we treat it as an indivisible unit.

3   Optimal covering based on MDL principle

This section presents the definition of optimal covering based on the MDL principle proposed by Brāzma et al. [5], and then shows an algorithm for approximating the optimal covering.

3.1   Definition

Let us denote by $L(\pi)$ the set of strings a pattern π matches. Consider a pattern

$$\pi = *\beta_1 * \cdots *\beta_h *   (\beta_1, \ldots, \beta_h \in \Sigma^+)$$

and a set $B = \{\alpha_1, \ldots, \alpha_n\}$ of strings such that $B \subseteq L(\pi)$. The set B can be described by the pattern π and the strings

$$\gamma_{1,0}\ \gamma_{1,1}\ \cdots\ \gamma_{1,h}$$
$$\gamma_{2,0}\ \gamma_{2,1}\ \cdots\ \gamma_{2,h}$$
$$\vdots$$
$$\gamma_{n,0}\ \gamma_{n,1}\ \cdots\ \gamma_{n,h}$$

such that $\alpha_i = \gamma_{i,0}\beta_1\gamma_{i,1}\cdots\gamma_{i,h-1}\beta_h\gamma_{i,h}$ for $i = 1, \ldots, n$. Such description of B is called *the encoding by pattern* π. We denote by $\|\alpha\|$ the description length of a string α in some encoding. For simplicity, we ignore the delimiters between strings. The description length of B is

$$\|\pi\| + \sum_{i=1}^{n} \sum_{j=0}^{h} \|\gamma_{i,j}\|.$$

KOKINSHŪ #193 (Ōe no Chisato)
TSUKI MIRE BA / CHIJI NI MONO KOSO / KANASHIKERE /
WAGA-MI HITOTSU NO / AKI NI WA ARA NE DO.

GOSENSHŪ #739 (Daughter of Kanemochi no asom)
YŪSARE BA / WAGA-MI NOMI KOSO / KANASHIKERE /
IZURE NO KATA NI / MAKURA SADAME M.

SHŪISHŪ   #271 (Minamoto no Shitagō)
OI NURE BA / ONAJI KOTO KOSO / SE RARE KERE /
KIMI WA CHIYO MASE / KIMI WA CHIYO MASE.

Fig. 2. WAKA poems containing pattern *REBA*KOSO*KERE*.

Let us denote by $c(\pi)$ the string obtained from π by removing all $*$'s. Assuming some symbolwise encoding, we have

$$\|\alpha_i\| = \sum_{j=1}^{h} \|\gamma_{i,j}\| + \|c(\pi)\|.$$

The description length of B is then

$$\|\pi\| + \sum_{i=1}^{n}(\|\alpha_i\| - \|c(\pi)\|) = \sum_{i=1}^{n} \|\alpha_i\| - \Big(\|c(\pi)\| \cdot |B| - \|\pi\|\Big).$$

Let A be a finite set of strings. A finite set $\Omega = \{(\pi_1, B_1), \ldots, (\pi_k, B_k)\}$ of pairs of a pattern π_i and a subset B_i of A is also said to be a *covering* of A if:

- $B_i \subseteq L(\pi_i)$ $(i = 1, \ldots, k)$.
- $A = B_1 \cup \cdots \cup B_k$.
- $B_1, \ldots, B_k$ are disjoint.

When the set B_i is encoded by π_i for each $i = 1, \ldots, k$, the description length of A is

$$M(\Omega) = \sum_{i=1}^{n} \|\alpha_i\| - C(\Omega),$$

where $C(\Omega)$ is given by

$$C(\Omega) = \sum_{j=1}^{k} \Big(\|c(\pi_j)\| \cdot |B_j| - \|\pi_j\|\Big).$$

Now, the *optimal covering* of the set A is defined to be the covering Ω minimizing $M(\Omega)$, or to be the set of patterns in it. Minimizing $M(\Omega)$ is equivalent to maximizing $C(\Omega)$.

3.2   Detail of encoding

In the above definition, the optimal covering varies depending on the encoding method. In the coding scheme in [5], the patterns and the strings for substitutions are coded together with delimiter symbols in some optimal symbolwise coding with respect to a probability distribution. Therefore, the formula of $C(\Omega)$ contains parameters that are the occurring probabilities of the delimiter symbols and the gap symbol $*$.

However, in our case the strings to be coded are of length less than $m = 32$ because we deal with only the poems consisting of thirty-one syllables. So, we can choose a simple way. We shall describe a string $w \in \Sigma^*$ as the pair of the length of w and the bit-string representing w in some optimal coding with respect to a probability distribution P over Σ. In practice, we can take $P(a)$ proportional to the relative frequency of a symbol $a \in \Sigma$ in a database. We denote by $\ell_P(w)$

the length of the bit-string representing w. We also denote by $n_*(\pi)$ the number of occurrences of $*$ in a pattern π. Assuming a positive number m such that $|w| < m$, we have

$$C(\Omega) = \sum_{j=1}^{k} \Big(u(\pi_j) \cdot |B_j| - v(\pi_j) \Big),$$

where

$$u(\pi) = \ell_P(c(\pi)) - n_*(\pi) \log_2 m + \log_2 m,$$
$$v(\pi) = \ell_P(c(\pi)) + n_*(\pi) \log_2 m.$$

3.3   Approximation algorithm

Since the problem of finding the optimal covering of a set of strings contains as a special case the set covering problem, it is NP-hard. Brāzma et al. [5] modified the problem as below:

> Given a finite set A of strings and a finite set Δ of patterns, find a covering Ω of A in which patterns are chosen from Δ that minimizes $M(\Omega)$.

They presented a greedy algorithm that approximates the optimal solution. It computes the values of

$$u(\pi) - \frac{v(\pi)}{|L(\pi) \cap U|}$$

for all possible patterns π at each iteration of a loop, and selects the pattern maximizing it to the covering. Here U is the set of strings in A not covered by any pattern that has already been selected. The value of $M(\Omega)$ for an approximate solution Ω obtained by this algorithm is at most $\log_2 |A|$ times with respect to the optimal one. The time complexity of the algorithm is $O(|\Delta| \cdot |A| \cdot \log_2 |A|)$ when excluding the computation of $\{(\pi, L(\pi) \cap A) \mid \pi \in \Delta\}$, which requires $O\Big(\sum_{\pi \in \Delta} |\pi| + |\Delta| \cdot \sum_{\alpha \in A} |\alpha|\Big)$ time to perform the pattern matching between the patterns in Δ and the strings in A.

4   Finding FUSHI patterns from WAKA poems

This section describes our experiment of seeking FUSHI patterns in a collection of WAKA poems. To apply the algorithm of Brāzma et al. to our problem, we need a way of identifying adjuncts with less misdetections and a formal definition of adjunct sequences, which are presented in Sections 4.1 and 4.2. Successful results of the experiment are then shown in Section 4.3.

4.1 How to avoid misdetection of adjuncts

Since we identify a string that matches an adjunct with the adjunct, there can be many misdetections. To avoid such misdetections, we adopted the following techniques.

– We restricted ourselves to the adjunct sequences appearing at the end of lines of WAKA poems. Obviously, an adjunct sequence can appear in the middle of a line when the line contains more than one segment. However, most of the important adjunct sequences related to FUSHI patterns appear at the end of lines.

– A WAKA poem was written in a mixture of Chinese characters and KANA characters. The former are ideograph characters whereas the latter are syllabic characters. An equivalent written in only KANA characters is attached to every WAKA poem in our database. Suppose that a line of one poem is equivalent to a line of another poem. If we use the one having a shorter KANA string at the end of line, then misdetection of an adjunct will be decreased because adjuncts are written in KANA characters. Based on this idea, we replaced each line of the poem by its 'canonical' form.

Although we cannot avoid all misdetections of adjuncts, our system is adequate for finding FUSHI patterns.

4.2 Definition of adjunct sequences

In our setting, the class of FUSHI patterns is defined by giving a set C of adjunct sequences. We therefore need a formal definition of adjunct sequences. For the definition, we give grammatical rules about concatenation of adjuncts. An adjunct sequence can be divided into three parts: first, a conjugated suffix of verb, adjective, or auxiliary verbs; second, a sequence of auxiliary verbs; third, a sequence of postpositional particles. Let us denote a conjugated suffix, an auxiliary verb and a postpositional particle by *Suf*, *AX*, and *PP*, respectively. There are syntactic and semantic constraints in concatenation of *Suf*, *AX*, and *PP*. The syntactic constraint is relatively simple, and is easy to describe. It depends on the combination of a word itself and the conjugated form of the preceding word. On the other hand, the semantic constraint is not so easy to describe completely. Here, we consider only the constraint between *AX* and *AX*, and between *PP* and *PP*. We classified the category of *AX* into five subcategories, and developed rules according to the classification. We also classified *PP* into six subcategories, and applied rules in a similar way. The set C of adjunct sequences was thus defined.

4.3 Experimental results for five anthologies

We applied the algorithm to five anthologies : KOKINSHŪ, SHINKOKINSHŪ, MINI-SHŪ, SHŪIGUSŌ, and SANKASHŪ. See Table 1. The first two are imperial anthologies, i.e., anthologies compiled by imperial command, the first completed in 922,

Table 1. Five collections of WAKA poems.

Anthology	Explanation	# poems
KOKINSHŪ	Imperial anthology compiled in 922	1,111
SHINKOKINSHŪ	Imperial anthology compiled in 1205	2,005
MINISHŪ	Private anthology by Fujiwara no Ietaka (1158–1237)	3,201
SHŪIGUSŌ	Private anthology by Fujiwara no Teika (1162–1241)	2,985
SANKASHŪ	Private anthology by Priest Saigyō (1118–1190)	1,552

and the second in 1205. The differences between the two anthologies, if any exist, may be due to the time difference in compilation. On the other hand, the others are private anthologies of poems composed by the three contemporaries: Fujiwara no Ietaka (1158–1237), Fujiwara no Teika (1162–1241), and the priest Saigyō (1118–1190). Their differences probably depend on the poets' personalities.

Table 2 shows the results of the experiments. A great number of patterns occur in each anthology, and therefore it is impossible to examine all of them manually. In the second column, the values in parentheses are the numbers of patterns occurring more than once. In the experiment, we used these sets of patterns as Δ, the sets of candidate patterns. The size of coverings is shown in the third column. For example, 191 of 8,265 patterns were extracted from KOKINSHŪ. The coverings are relatively small in order to examine all the patterns within them.

Table 2. Coverings of five anthologies.

Anthology	# occurring patterns		Size of covering
KOKINSHŪ	164,978	(8,265)	191
SHINKOKINSHŪ	233,187	(12,449)	270
MINISHŪ	187,014	(16,425)	369
SHŪIGUSŌ	214,940	(14,365)	335
SANKASHŪ	279,904	(12,963)	232

Table 3 shows the first five patterns emitted by the algorithm from KOKINSHŪ. The first pattern *KEREBA*BERANARI* contains the auxiliary verb BERANARI, which is known to be used mainly in the period of KOKINSHŪ. The fourth pattern *RISEBA*RAMASHI* corresponds to the subjunctive mood. The last pattern *WA*NARIKERI* corresponds to the expression "I have become aware of the fact that · · ·". The remaining two patterns are different correlative word expressions, called KAKARI-MUSUBI. The obtained patterns are thus closely related to techniques used in composing poems.

Next, we shall compare pattern occurrences in the five anthologies. Table 4 shows the first five patterns for each anthology, where each numeral denotes the occurring frequency of the pattern in the anthology. The following facts, for example, can be read from Table 4:

Table 3. FUSHI patterns from KOKINSHŪ.

Pattern	Annotation
*KEREBA*BERANARI*	use of auxiliary verb BERANARI
*ZO*SHIKARIKERU*	correlative word expression (KAKARI-MUSUBI)
*KOSO*RIKERE*	correlative word expression (KAKARI-MUSUBI)
*RISEBA*RAMASHI*	the subjunctive mood
*WA*NARIKERI*	the expression of awareness

1. Pattern *BAKARI*RAM* does not occur in either KOKINSHŪ or SHINKOKINSHŪ.
2. Pattern *WA*NARIKERI* occurs in each of the anthologies. In particular, it occurs frequently in SANKASHŪ.
3. Pattern *MASHI*NARISEBA* occurs frequently in SANKASHŪ.
4. Pattern *KOSO*RIKERE* does not occur in SHŪIGUSŌ.
5. Pattern *YA*RURAM* occur in each anthology except KOKINSHŪ.

It is possible that the above facts are important characteristics that may be due to the times or the poets' personalities. For example, (2) and (3) may reflect Priest Saigyō's preferences, and (5) may imply that the pattern *YA*RURAM* was not preferred in the period of KOKINSHŪ. Comparisons of the obtained patterns and their frequencies thus provide a WAKA researcher clues for further investigation.

Table 4. FUSHI patterns from five anthologies with frequencies, where A, B, C, D and E denote KOKINSHŪ, SHINKOKINSHŪ, MINISHŪ, SHŪIGUSŌ, and SANKASHŪ, respectively.

	Patterns	A	B	C	D	E		Patterns	A	B	C	D	E
	*KEREBA*BERANARI*	5	0	0	0	0		*BAKARI*RAM*[1]	0	0	11	8	3
	*ZO*SHIKARIKERU*	8	1	0	0	3		*NO*NARIKERI*	19	30	39	19	49
A	*KOSO*RIKERE*[4]	11	8	8	0	13	D	*RAZARIKI*NO*	0	0	1	6	1
	*RISEBA*RAMASHI*	5	2	0	0	4		*YA*RURAM*[5]	0	8	40	24	23
	*WA*NARIKERI*[2]	20	26	26	11	52		*NI*NARURAM*	0	2	8	8	7
	*KARISEBA*MASHI*	3	6	0	0	1		*MASHI*NARISEBA*[3]	0	2	1	0	10
	*NO*NIKERUKANA*	4	11	2	1	4		*KOSO*KARIKERE*	4	4	1	0	8
B	*WA*NARIKERI*[2]	20	26	26	11	52	E	*NARABA*RAMASHI*	1	0	0	0	8
	*KOSO*RIKERE*[4]	11	8	8	0	13		*O*UNARIKERI*	1	0	0	0	7
	*MO*KARIKERI*	4	11	8	5	7		*NO*RUNARIKERI*	4	3	4	0	10
	*BAKARI*RURAM*	0	0	6	0	3							
	*KOSO*NARIKERE*	4	0	5	0	5							
C	*YA*NARURAM*	0	2	16	4	7							
	*WA*NARIKERI*[2]	20	26	26	11	52							
	*NO*NARIKERI*	19	30	39	19	49							

5 A pattern-based text data mining system

Based on the experience of finding patterns from anthologies described in the previous section, we propose a text data mining system to support research into

WAKA poetry. The system consists of two parts: the pattern discovery part and the pattern matching part. The pattern discovery part emits a set of patterns, some of which stimulate the user to form hypotheses. To verify the hypotheses, the user retrieves a set of poems containing the patterns by using the pattern matching part, examines the retrieved poems, and then updates the hypotheses. The updated hypotheses are then verified again. Repeating this process will provide results that are worthwhile to the user.

In practice, a slightly modified pattern is often better than the original emitted by the pattern discovery part, that is, a slightly more general/specific pattern may be preferred. Say the user wants to browse the 'neighbors' of a pattern. The proposed system has as GUI a pattern browser for traversing the Hasse diagram of the partial-order on the set of patterns. We have implemented a prototype of this system, and a new style of research into WAKA poetry utilizing the prototype system is now proceeding.

Acknowledgments

The authors would like to thank Ayumi Shinohara and Hiroki Arimura for valuable discussions concerning this work. We also thank Setsuo Arikawa, Yuichiro Imanishi, and Masakatsu Murakami for their valuable comments.

References

1. H. Ahonen, O. Heinonen, M. Klementtinen, and A.I. Verkamo: Mining in the phrasal frontier. In *Proc. 1st European Symposium on Principles of Data Mining and Knowledge Discovery* (PKDD'97), 343–350, 1997.
2. D. Angluin: Finding patterns common to a set of strings. In *Proc. 11th Annual Symposium on Theory of Computing*, 130–141, 1979.
3. H. Arimura, T. Shinohara, and S. Otsuki: Finding minimal generalizations for unions of pattern languages and its application to inductive inference from positive data. In *Proc. 11th Annual Symposium on Theoretical Aspects of Computer Science* (STACS'94), 649–660, 1994.
4. G. Bownas and A. Thwaite: The Penguin Book of Japanese verse. Penguin Books Ltd., 1964.
5. A. Brāzma, E. Ukkonen, and J. Vilo: Discovering unbounded unions of regular pattern languages from positive examples. In *Proc. 7th International Symposium on Algorithms and Computation* (ISAAC'96), 95–104, 1996.
6. R. Feldman and I. Dagan: Knowledge discovery in textual databases (KDT). In *Proc. 1st International Conference on Knowledge Discovery and Data Mining* (KDD'95), 112-117, 1995.
7. E. M. Gold: Language identification in the limit. *Information and Control*, 10: 447–474, 1967.
8. J. Rissanen: Modeling by the shortest data description. *Automatica*, 14: 465–471, 1978.
9. T. Shinohara: Polynomial-time inference of pattern languages and its applications. In *Proc. 7th IBM Symposium on Mathematical Foundations of Computer Science*, 191–209, 1982.

Approximate Retrieval of High-Dimensional Data by Spatial Indexing

Takeshi Shinohara, Jiyuan An and Hiroki Ishizaka

Department of Artificial Intelligence
Kyushu Institute of Technology
Iizuka, 820-8502 Japan
shino@ai.kyutech.ac.jp

Abstract. High-dimensional data, such as documents, digital images, and audio clips, can be considered as spatial objects, which induce a metric space where the metric can be used to measure dissimilarities between objects. We propose a method for retrieving objects within some distance from a given object by utilizing a spatial indexing/access method R-tree. Since R-tree usually assumes a Euclidean metric, we have to embed objects into a Euclidean space. However, some of naturally defined distance measures, such as L_1 distance (or Manhattan distance), cannot be embedded into any Euclidean space. First, we prove that objects in discrete L_1 metric space can be embedded into vertices of a unit hyper-cube when the square root of L_1 distance is used as the distance. To take fully advantage of R-tree spatial indexing, we have to project objects into space of relatively lower dimension. We adopt FastMap by Faloutsos and Lin to reduce the dimension of object space. The range corresponding to a query (Q, h) for retrieving objects within distance h from a object Q is naturally considered as a hyper-sphere even after FastMap projection, which is an orthogonal projection in Euclidean space. However, it is turned out that the query range is contracted into a smaller hyper-box than the hyper-sphere by applying FastMap to objects embedded in the above mentioned way. Finally, we give a brief summary of experiments in applying our method to Japanese chess boards.

1 Introduction

Let S be a finite set of objects $\{O_1, O_2, \ldots, O_m\}$ and $D : S \times S \to N$ be a function which gives the distance between objects. We call (S, D) an *object space*. A *query* is given as a pair (Q, h) of an object $Q \in S$ and a natural number h. The *answer* $Ans(Q, h)$ to a query (Q, h) is the set of objects within distance h from Q, that is,

$$Ans(Q, h) = \{O_i \in S \mid D(Q, O_i) \le h\}.$$

The above setting of approximate retrieval as $Ans(Q, h)$ is very natural and general. When (S, D) is a Euclidean space, most spatial indexing structures are almost directly used to realize approximate retrieval. In many cases, however, unless objects are inherently geometrical like map information, object space is not Euclidean.

In this paper, we assume (S, D) can be considered as a metric space based on discrete L_1 (or, Manhattan) distance, that is,

$$S \subseteq N^n \text{ and } D(O_i, O_j) = \sum_{k=1}^{n} |O_i^{(k)} - O_j^{(k)}|,$$

where $O_i^{(k)}$ and $O_j^{(k)}$ are the k-th coordinates of objects O_i and O_j, respectively. Most of the difference measures might be captured as a discrete L_1 distance. For example, a natural definition of distance between objects consisting of several attribute values may be the sum of the symmetric differences between each attribute values. This definition can be applied to many sort of objects, such as, documents, digital images, and game boards.

We adopt R-tree [1, 5] as a spatial indexing/access method. As Otterman pointed out [7], R-tree can efficiently be used only for relatively low-dimensional objects. Therefore, we have to map high-dimensional objects into a subspace of lower dimension. We can use the FastMap method by Faloutsos and Lin [4] to project objects in Euclidean space into its subspace. Since FastMap is based on orthogonal projection in Euclidean space, we have to embed objects into a Euclidean space. However, L_1 distance cannot be embedded into any Euclidean space, in general. As we will see in Section 2, if we take the square root of L_1 distance as the distance, the objects can be embedded into a Euclidean space. In other words, if we define

$$D^{\frac{1}{2}}(X, Y) = \sqrt{D(X, Y)},$$

$(S, D^{\frac{1}{2}})$ can be embedded into a Euclidean space. If we appropriately map objects to vertices of unit n_0-cube, then the Euclidean distance between vertices coincides with the square root of the L_1 distance between objects.

Here, we briefly explain the FastMap method. Consider a set of objects $\{O_1, O_2, \ldots, O_m\}$ in a Euclidean space, where $d(O_i, O_j)$ gives the Euclidean distance between objects O_i and O_j. Let take arbitrarily a pair (O_a, O_b) of objects, which is called a *pivot*. The first coordinate X_i of an object O_i is given by

$$X_i = \overline{O_a E} = \frac{(d(O_a, O_i))^2 + (d(O_a, O_b))^2 - (d(O_b, O_j))^2}{2d(O_a, O_b)}$$

where E is the image of O_i by the orthogonal projection to the straight line $O_a O_b$ (Figure 1). Here, we should note that distances between objects are enough to calculate the coordinate X_i and any coordinates of objects are not necessary. Let O_i' be the image of O_i by the orthogonal projection to the hyper-plane that is orthogonal to the straight line $O_a O_b$. The distance between O_i' and O_j' is given by

$$(d(O_i', O_j'))^2 = (d(O_i, O_j))^2 - (X_i - X_j)^2.$$

Thus, we can repeatedly apply the above projection to get the second and other coordinates of objects. One of the most important issues in applying FastMap may be how to select pivots. Intuitively, the better pivot should provide the more selectivity in retrieval. Details are discussed in [4].

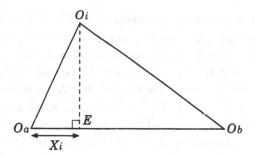

Fig. 1. Orthogonal projection to pivot line

Let p be an orthogonal projection to $\boldsymbol{R}^{k_0}$ which is obtained by FastMap, where (S, D) is the original object space and $D^{\frac{1}{2}}$ is used as the distance function in applying FastMap. We call $\boldsymbol{R}^{k_0}$ the *index space*. Since p is an orthogonal projection, the distance between images of objects in the index space is not larger than the square root of the distance between objects, that is, $d(p(O_i), p(O_j)) \leq D^{\frac{1}{2}}(O_i, O_j)$. For a query (Q, h), we have

$$\{O_i \in S \mid d(p(Q), p(O_i)) \leq \sqrt{h}\} \supseteq Ans(Q, h).$$

Therefore, we can retrieve all the necessary objects even after reducing dimension by FastMap. Such a retrieval is easily realized by using spatial access method like R-tree. The result from the method may include irrelevant objects to the query, which is caused by FastMap projection. To get exact answer, screening might be needed.

From the experiments of our method, we observed that the image of the query range in the index space $\boldsymbol{R}^{k_0}$, which is naturally considered as a k_0-sphere with radius $h^{\frac{1}{2}}$, is too large to get all the necessary objects. Precisely, we can prove

$$\{O_i \in S \mid |p^{(k)}(Q) - p^{(k)}(O_i)| \leq \lambda_k h \text{ for all } k = 1, \ldots, k_0\} \supseteq Ans(Q, h),$$

where $p^{(k)}(O)$ is the k-th coordinate of the image of O in the index space and λ_k is a constant which is usually much smaller than 1. Thus, the query range of k_0-box, which is smaller than the k_0-sphere, is enough to retrieve the correct answer. This phenomenon, which is derived from the combination of our object embedding into unit n_0-cube and FastMap, will be theoretically explained as the contraction of query range by FastMap in Section 3.

2   Embedding L_1 distance into Euclidean space

Theorem 1. *For any object space (S, D), $(S, D^{\frac{1}{2}})$ can be embedded into Euclidean space.*

Proof. Without loss of generality, we assume that $S = \{O_1, \ldots, O_m\} \subseteq \mathbf{N}^n$. For each $k = 1, \ldots, n$, we use a bit vector of length b_k where $b_k = \max\{O_i^{(k)} \mid i = 1, \ldots, m\}$ and map the k-th coordinate value v to $u_{b_k}(v) = 1^v 0^{b_k - v}$, which is a bit vector such that the first v bits are 1 and other bits are 0. Here we identify bit vectors and bit strings. For each object O_i, we map O_i to a bit vector $u(O_i) = u_{b_1}(O_i^{(1)}) u_{b_2}(O_i^{(2)}) \cdots u_{b_n}(O_i^{(n)})$. Clearly, for any $k \in \{0, \ldots, n\}$ and any $v, v' \in \{0, \ldots, b_k\}$, $(d(u_{b_k}(v), u_{b_k}(v')))^2 = |v - v'|$. Therefore, $d(u(O_i), u(O_j)) = D^{\frac{1}{2}}(O_i, O_j)$. Thus, we can embed (S, D) into unit n_0-cube, where $n_0 = b_1 + \cdots + b_n$. $\square$

For example, consider points $O(0,0)$, $A(0,1)$, $B(0,2)$, and $C(1,1)$ in x-y plane as in Figure 2. Distances between these points based on L_1 metric are $D(O, A) = D(A, B) = D(A, C) = 1$ and $D(O, B) = D(O, C) = D(B, C) = 2$. As long as using D as the metric, A should be on the straight line OB and $O, B,$ and C should make a regular triangle no matter what embedding is used. The height of regular triangle OBC is the square root of 3, which contradicts to $D(A, C) = 1$. Thus, there is no Euclidean space where these four points are embedded keeping the metric as it is. The maximum values of x and y coordinates are 1 and 2, respectively. Therefore, we map each point to a bit vector of length $n_0 = 1 + 2 = 3$. We can regard the first bit as representing if the x coordinate is equal to 1, the second as if the y coordinate is greater than or equal to 1, and the third as if the y coordinate is equal to 2. Clearly Euclidean distances between bit vectors are equal to respective L_1 distances between points in x-y plane.

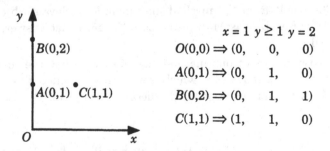

Fig. 2. Embedding L_1 distance into unit n_0-cube

From Theorem 1, we can apply FastMap to $(S, D^{\frac{1}{2}})$, which is embedded in Euclidean space of n_0-dimension. Here we should note that only distances between objects are sufficient for applying FastMap and actual values of coordinate in the Euclidean space are not necessary. Thus, the dimension n_0 of the Euclidean space, which may be quite larger than that of the original object space, does not the matter when we use FastMap.

3   Contraction of Query Range by FastMap

From Theorem 1, we can assume that every object is a vertex of a unit n_0-cube and the value of each coordinate is 0 or 1. Let P_0 be the vector between two objects used as the first pivot for FastMap. Let e be a unit vector of any coordinate in $\boldsymbol{R}^{n_o}$. Then the length of the image of e by orthogonal projection to the first pivot is given by

$$\frac{|e \cdot P_0|}{|P_0|}.$$

Since every component of P_0 is either -1, 0, or 1, the inner product $e \cdot P_0$ is also -1, 0, or 1. Therefore,

$$\frac{|e \cdot P_0|}{|P_0|} \leq \frac{1}{|P_0|}.$$

Let define λ_1 as the right side of the above inequation. Consider two objects O_1 and O_2 such that $D(O_1, O_2) = h$ and a vector v between O_1 and O_2. Clearly, exactly h components of v are -1 or 1 and all the other components are 0. Therefore, the length of the image of v is less than or equal to $h\lambda_1$. Since $|P_0|$ is usually larger than 1, λ_1 is relatively small. For the second and other projections by FastMap, similar phenomena can be derived. In what follows, to avoid a little bit complicated discussion, we give only the result.

Let P_k be the $(k+1)$-th pivot. Let $\Pi_0 = P_0$, and Π_k be the image of P_k by orthogonal projection to the hyper-plane H_{k-1}, which is orthogonal to $\Pi_0, \Pi_1, \ldots$, and Π_{k-1}. Define $\beta(k,l)$, and $\gamma(k,l)$ for each $k > l \geq 0$ by

$$\beta(k,l) = \frac{P_k \cdot P_l}{|P_l|},$$

$$\gamma(k,l) = \beta(k,l) - \sum_{i=l+1}^{k-1} \beta(k,i)\gamma(i,l).$$

Then, for each $k > 0$, Π_k is given as

$$\Pi_k = P_k - \sum_{l=0}^{k-1} \gamma(k,l)P_l.$$

Finally, as for the length of the image of a unit vector e by the k-th orthogonal projection of FastMap, we have the following upper bound λ_k.

$$\frac{|e \cdot \Pi_k|}{|\Pi_k|} \leq \frac{1 + \sum_{i=0}^{n-1} |\gamma(k,i)|}{|\Pi_k|} = \lambda_k$$

Theorem 2. *For any query (Q, h) and any FastMap p,*

$$\{O_i \in S \mid |p^{(k)}(Q) - p^{(k)}(O_i)| \leq \lambda_k h \text{ for all } k = 1, \ldots, k_0\} \supseteq Ans(Q, h).$$

The answer $Ans(Q, h)$ to a query (Q, h) is given as a set of objects within a n_0-sphere whose center and radius are Q and $h^{\frac{1}{2}}$, respectively, where $D^{\frac{1}{2}}$ is used as a Euclidean distance, which we call a *query range*. Since a mapping p obtained by FastMap is an orthogonal projection into Euclidean space, the image of a query range by p is a k_0-sphere of the same radius $h^{\frac{1}{2}}$ in the index space. On the other hand, Theorem 2 says that all the objects in $Ans(Q, h)$ are projected by p into a k_0-box whose center and radius of the k-th coordinate are $p(Q)$ and λ_k, respectively. Here we should note that the constant λ_k is usually much smaller than 1, and therefore, the k_0-box has a smaller volume than the k_0-sphere for relatively small h. Let $\lambda_0 = \max\{\lambda_k \mid 1 \leq k \leq k_0\}$. Then, the volume V_B of the k_0-box is less than or equal to $(2\lambda_0 h)^{k_0}$. On the other hand, the volume V_S of the k_0-sphere is $C_{k_0} h^{\frac{1}{2}k_0}$, where C_r is a constant determined by r and $C_r > 1$ for any $r \leq 12$. Therefore, $V_B < V_S$ whenever $2\lambda h \leq 1$ and $k_0 \leq 12$. Although this estimation is very rough, in many cases we may expect the contraction of query range by FastMap, which is illustrated in Figure 3. Since the square root of h is used as the radius of k_0-sphere query range while $\lambda_k h$ is used for k_0-box, as low as possible dimension should be selected to get much effect of contraction of query range by FastMap.

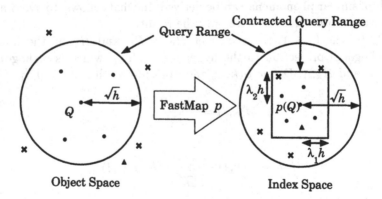

Fig. 3. Contraction of Query Range by FastMap

4   Experimental Results – Japanese Chess Boards

In this section, we give a brief summary of the experiments in applying our indexing method to retrieval of Japanese chess (shogi) boards analogous to given one from 40,412 boards drawn by 500 play records.

Shogi uses 40 pieces of 8 sorts and reverse side of 6 sorts of pieces. A shogi board consists of $9 \times 9 = 81$ positions, each of which may be possessed by one of $(8 + 6) \times 2 = 28$ sorts of pieces, and two sets of captured pieces, which is a subset of 38 (all but 2 Kings) pieces.

4.1   Distance between boards and embedding into unit hyper-cube

For each position, we define the difference between two boards O_i and O_j depending on what pieces are each on the position. When two positions are the same, that is, they have the same piece or both of them have no piece, the difference is 0. When one has a piece and the other has no, the difference is 1. Otherwise, they have different pieces and the difference is defined as 2. For captured pieces, the difference is the sum of the symmetric difference of the numbers of pieces of each sort. We define the distance $D(O_i, O_j)$ as the sum of differences for all positions and captured pieces. Note that the largest possible distance between boards is 80 because all 40 pieces should be put on some position or included in captured pieces.

By using 28 bits for each of 81 positions and 38 bits for each of two sets of captured pieces, we can put shogi boards in a unit hyper-cube of $28 \times 81 + 38 \times 2 = 2,344$ dimensions, where the distance $D(O_i, O_j)$ is given by L_1 metric. Thus, we can regard shogi pieces as object space.

4.2   FastMap projection and R-tree spatial indexing

FastMap projection was applied to 40,412 shogi boards to reduce the dimension of boards and efficiently utilize R-tree spatial indexing. Selection of pivot for each step of FastMap was done by randomly choosing 500 candidates and selecting one that maximizes the variance of coordinate values. As for the dimension k_0 of the index space, we adopted 5, 7, and 10. We used off-line packed R-trees [6] based on Hilbert space filling curves [2, 3].

4.3   Effect of contraction of query range by FastMap

After projecting boards by FastMap for each $k_0 = 5$, 7, or 10, we measured the maximum lengths of the image of a unit vector in unit hyper-cube on each coordinate in index space, which are between 0.12 and 0.21. On the other hand, Theoretical bounds derived from Theorem 2 are between 0.12 and 0.25. The gap between actual measurements and theoretical bounds seems to suggest that there are no worst combinations within current shogi boards. Since the maximum possible distance between boards is 80, the lower bound of λ_k is given by

$$\lambda_k \geq \frac{1}{\sqrt{80}} = 0.112,$$

for each $1 \leq k \leq k_0$. From this, we observe that query ranges are contracted into relatively small k_0-box by FastMap projection.

4.4 Approximate retrieval of boards

Finally, we made experiments of retrieval of boards analogous to given one by using R-tree index. For boards given as the centers of queries, we randomly selected 700 boards from 40,412 boards. For the radius of queries we gave $h = 0$, 2, 4, 6, 8. Retrievals in case $h = 0$ are exact ones. The averages of elapsed time for retrievals are summarized in Table 1, where the column "none" represents the average time of retrieval without indexing. From Table 1, our indexing are efficient for small radius. Within 5, 7, and 10, the best for the dimension k_0 of index space is 7 for all radiuses.

Table 1. Elapsed time of retrieval

k_0	$h = 0$	$h = 2$	$h = 4$	$h = 6$	$h = 8$
5	0.0098	0.0501	0.2351	0.8642	1.8316
7	0.0097	0.0452	0.2326	0.8384	1.8104
10	0.0134	0.0682	0.2549	0.8720	1.8983
none	1.9803	2.3000	2.4301	2.4517	2.5608

As mentioned in Section 4, the lower dimension is desired for contraction of query range. For example, let $\lambda_0 = \max\{\lambda_k \mid 1 \le k \le k_0\} = 0.25$, which is consistent with our experiments. The volume $V_B(k_0)$ of k_0-cube with radius $\lambda_0 h$ and the volume $V_S(k_0)$ of k_0-sphere with radius $h^{\frac{1}{2}}$ are given by $V_B(k_0) = (2\lambda_0 h)^{k_0}$ and $V_S(k_0) = C_{k_0} h^{\frac{1}{2}k_0}$, respectively, where $C_5 = \frac{8}{15}\pi^2 = 5.26$, $C_7 = \frac{16}{105}\pi^3 = 4.72$, and $C_{10} = \frac{1}{5!}\pi^5 = 2.55$. These volumes are summarized in Table 2. From this, V_B is larger than V_S for all cases of $h = 8$ and a case of $h = 6$ and $k_0 = 10$, which may in part explain the tradeoff between the dimension of index space and elapsed time of retrieval.

Table 2. Hyper-box vs. Hyper-sphere as Query Range

k_0	V_B				V_S			
	$h = 2$	$h = 4$	$h = 6$	$h = 8$	$h = 2$	$h = 4$	$h = 6$	$h = 8$
5	1	32	243	1024	29.8	168	464	952
7	1	128	2187	16384	53.4	604	2497	6835
10	1	1024	59049	1048576	81.6	2611	19829	83558

The tradeoff also may be explained by a nature of R-trees. In other words, higher dimension of index space gives more precise image but more difficulty in spatial indexing by R-tree.

5 Concluding Remarks

In this paper, we have proposed a method for approximate retrieval by using spatial indexing/access method like R-tree, where dissimilarities between objects are measured by L_1 distance. As Theorem 1, we proved that objects with L_1 distance can be embedded into a Euclidean space preserving the square root of L_1 distance as distance. In Theorem 2, we pointed out that contraction of query range by FastMap can be expected when our embedding is used. Although the experiments on approximate retrieval of Japanese chess boards seem to suggest that our method can be successfully applied to many other cases, we should run experiments in other natural applications of our method to analyze its applicability.

As for future work, we should enlarge the applicability of our method. Although we assumed that distance between objects is measured by discrete L_1 distance, proposed embedding of taking the square root of distances might be applicable to other distances, such as edit distance between strings. As for strings with edit distance, we made experiments that reported no inconsistency in embedding and FastMap projection. The restriction of discreteness might be relaxed.

References

1. N. Beckmann, H.P. Kriegal, R. Schneider, and B. Seeger. The R*-tree: An Efficient and Robust Access Method for Points and Rectangles. In *Proc. ACM SIGMOD International Conference on Management of Data*, 19(2):322–331, 1990.
2. T. Bially. Space-filling Curves: Their Generation and Their Application to Bandwidth Reduction. *IEEE Trans. on Information Theory*, IT-15(6):658–664, 1969.
3. C. Faloutsos and S. Roseman. Fractals for Secondary Key Retrieval. In *Proc. 8th ACM SIGACT-SIGMOD-SIGART Symposium on Principles of Database Systems*, pp. 247–252, 1989.
4. C. Faloutsos and K.I. Lin. FastMap: A Fast Algorithm for Indexing, Data-Mining and Visualization of Traditional and Multimedia Datasets. In *Proc. ACM SIGMOD International Conference on Management of Data*, 24(2):163–174, 1995.
5. A. Guttman. R-tree: A Dynamic Index Structure for Spatial Searching. In *Proc. ACM SIGMOD*, pp. 47–57, 1984.
6. I. Kamel and C. Faloutsos. On Packing R-trees. In *Proc. 2nd. International Conference on Information and Knowledge Management*, pp. 490–499, 1993.
7. M. Otterman. *Approximate Matching with High Dimensionality R-trees*. M. Sc. Scholarly paper, Dept. of Computer Science, Univ. of Maryland, 1992.

Practical Algorithms for On-line Sampling

Carlos Domingo[1]*, Ricard Gavaldà[1]**, and Osamu Watanabe[2]***

[1] Department de LSI, Universitat Politècnica de Catalunya
Campus Nord, Mòdul C5, 08034-Barcelona, Spain
{carlos, gavalda}@lsi.upc.es
[2] Dept. of Mathematical and Computing Sciences
Tokyo Institute of Technology, Tokyo 152-8552, Japan
watanabe@is.titech.ac.jp

Abstract. One of the core applications of machine learning to knowledge discovery is building a hypothesis (such as a decision tree or neural network) from a given amount of data, so that we can later use it to predict new instances of the data. In this paper, we focus on a particular situation where we assume that the hypothesis we want to use for prediction is a very simple one so the hypotheses class is of feasible size. We study the problem of how to determine which of the hypotheses in the class is almost the best one. We present two on-line sampling algorithms for selecting a hypothesis, give theoretical bounds on the number of examples needed, and analyze them experimentally. We compare them with the simple batch sampling approach commonly used and show that in most of the situations our algorithms use a much smlaler number of examples.

1 Introduction and Motivation

The ubiquity of computers in business and commerce has lead to generation of huge quantities of stored data. A simple commercial transaction, phone call, or use of a credit card is usually stored in a computer. Today's databases are growing in size and therefore there is a clear need for automatic tools for analyzing and understanding these data. The field known as knowledge discovery and data mining aims at understanding and developing all the issues involved in extraction of patterns from vast amount of data. Some of the techniques used are basically machine learning techniques. However, due to the restriction that the data available is very large, many machine learning techniques do not always scale well and simply cannot be applied.

One of the core applications of machine learning to knowledge discovery consists of building a function from a given amount data (for instance a decision

* Partially supported by the ESPRIT Working Group NeuroCOLT2 (No.27150), the ESPRIT project ALCOM-IT (No.20244), and CICYT TIC97-1475-CE.
** Partially supported by the ESPRIT Working Group NeuroCOLT2 (No.27150), DGES project KOALA (PB95-0787), and CIRIT SGR 1997SGR-00366.
*** Partially supported by the Minsitry of Education, Science, Sports and Culture, Grant-in-Aid for Scientific Research on Priority Areas (Discovery Science) 1998.

tree or a neural network) such that we can later use it to predict the behavior of new instances of the data. This is commonly know as concept learning or supervised learning.

Most of the previous research in machine learning has focused on developing efficient techniques for obtaining *highly* accurate predictors. For achieving high accuracy, it is better that learning algorithms can handle complicated predictors, and developing efficient algorithms for complicated predictors has been studied intensively in machine learning.

On the other hand, for knowledge discovery, there are some other aspects of concept learning that should be considered. We discuss one of them in this paper. We study concept learning (or, more simply, hypothesis selection) for a particular situation that we describe in the following. Our assumption is that we have a class $\mathcal{H}$ of very simple hypotheses, and we want to select one of the reasonably accurate hypothesis from it, by using a given set of data, i.e., labeled examples. Since hypotheses we deal with are very simple, we cannot hope, in general, to find highly accurate hypotheses in $\mathcal{H}$. On the other hand, the size of hypothesis space $\mathcal{H}$ is relatively small and feasible. We also assume that the amount of data available is huge, and thus, we do not want to use the whole dataset to find the accurate hypothesis. Simple hypotheses have been studied before by several researchers and it has been reported that in some cases they can achieve surprisingly high accuracy (see, e.g., [9, 6, 1]). Moreover, with the discovery of new voting methods like boosting [4] or bagging [2], several of these hypotheses can be combined in a way that the overall accuracy becomes extremely high.

The obvious approach for solving this problem, which is commonly used in computational learning theory, is to first choose randomly a certain number m of examples from the dataset, and then select the hypothesis that performs best on the selected sample. We will call this simple approach *Batch Selection* (BS) in this paper. The number m is calculated so that the best hypothesis in the selected sample is close to the really best one with high probability; such m can be calculated by using uniform convergence bounds like the Chernoff or the Hoeffding bound (see, e.g., [8] for some examples of this approach). However, if we want to apply this method in a real setting we will encounter two problems. First, the theoretical bounds are usually too pessimistic and thus the bounds obtained are not practical. Second, to obtain this bounds we need to have certain knowledge about the accuracy of hypotheses in a given hypothesis space. What is usually assumed is that we know a lower bound on the accuracy of the best hypothesis. Again, this lower bound might be far from the real accuracy of the best hypothesis and thus the theoretical bound becomes too pessimistic. Or even worse, in many applications we just do not know anything about the accuracy of the hypotheses.

In this paper we propose two algorithms for solving this problem, obtain theoretical bounds of their performance, and evaluate them experimentally. Our goal is to obtain algorithms that are useful in practice but also have certain theoretical guarantees about their performance. The first distinct characteristic is that we obtain the examples in an on-line manner rather than in batch. The second is that the number of examples has less dependency on the lower bound of the accuracy than the above obvious Batch Selection. More specifically, if γ_0 is the accuracy of the best hypothesis, and γ is the lower bound for γ_0 that we would use, then the sample size m for Batch Selection given by the theoretical bound is $\mathcal{O}(1/\gamma^2)$ (ignoring dependencies on other parameters). On the other hand, the sample size of our first algorithm is $\mathcal{O}(1/\gamma\gamma_0)$, and that of the second one is $\mathcal{O}(1/\gamma_0^2)$.

Our algorithms can be regarded as *agnostic PAC learning* algorithms. Recall that in the agnostic PAC model [5, 7, 1] one tries to find the almost best hypothesis in a fixed class with no assumption about an underlying target concept. Given a finite and "small" class of hypothesis, one can easily derive an agnostic PAC learning algorithm just by exhaustively finding the best fitting hypothesis with a set of examples. The goal of our algorithms is to carry out this task using as few examples as possible.

In this extended abstract, we omit some of the proofs due to the space limit, which can be found in [3].

2   On-line Selection Algorithms and Their Analysis

Here we present our two on-line selection algorithms and investigate their reliability and efficiency theoretically. Before doing this, we will give some definitions.

Throughout this paper, we use $\mathcal{H}$ and n to denote the set of hypotheses and its cardinality, and use $\mathcal{D}$ to denote a distribution on instances. For any $h \in \mathcal{H}$, let $\mathrm{prc}_{\mathcal{D}}(h)$ denote the accuracy of h, that is, the probability that h gives a correct prediction to x for a randomly given x under the distribution $\mathcal{D}$. Let h_0 denote the best hypothesis in $\mathcal{H}$ (w.r.t. $\mathcal{D}$); that is, $\mathrm{prc}_{\mathcal{D}}(h_0) = \max\{\mathrm{prc}_{\mathcal{D}}(h) | h \in \mathcal{H}\}$. Let γ_0 denote $\mathrm{prc}_{\mathcal{D}}(h_0) - 1/2$; that is, $\mathrm{prc}_{\mathcal{D}}(h_0) = 1/2 + \gamma_0$. We will denote by $\#_t(h)$ the number of examples for which hypothesis h succeeds within t steps. In our analysis we count each-while-iteration as one step; thus, the number of steps is equal to the number of examples needed in the algorithm. It will be also useful for our analysis to partition the hypothesis space in two sets depending on the precision of each hypothesis. Thus, let $\mathcal{H}_{\mathrm{good}}$ (resp., $\mathcal{H}_{\mathrm{bad}}$) denote the set of hypotheses h such that $\mathrm{prc}_{\mathcal{D}}(h) \geq 1/2 + \gamma_0/2$ (resp., $\mathrm{prc}_{\mathcal{D}}(h) < 1/2 + \gamma_0/2$). This partition is done in an arbitrary way. The complexity of our algorithms depends on it but can be easily adapted to a more restrictive condition (e.g. $h \in \mathcal{H}_{\mathrm{good}}$ if $\mathrm{prc}_{\mathcal{D}}(h) \geq 1/2 + 3\gamma_0/4$) if it is needed for a particular application. Obviously, the more demanding is the definition of $\mathcal{H}_{\mathrm{good}}$ the greater is the complexity of our algorithms.

Our main tool will be the following bound on tail probabilities (see e.g. [8]).

Theorem 1. *(Hoeffding bound)   For any $t \geq 1$ and p, $0 \leq p \leq 1$, consider t independent random variables $X_1, \ldots, X_t$ each of which takes values 0 and 1 with probabilities $1 - p$ and p. Then there is a universal constant c_{H} such that* *for any $\varepsilon > 0$, we have*

$$\Pr\{\sum_{i=1}^{t} X_i > pt + \varepsilon t\} < \exp(-c_H \varepsilon^2 t), \quad \Pr\{\sum_{i=1}^{t} X_i < pt - \varepsilon t\} < \exp(-c_H \varepsilon^2 t).$$

Constant c_H is equal to 2 in this general bound, but we leave it undetermined since, as we will see later, this value is too pessimistic for many combinations of t, p, and ε.

By using this bound, we can estimate the sufficient number of examples to guarantee that Batch Selection, the simple hypothesis selection algorithm, yields a hypothesis of reasonable accuracy with high probability. (In the following, we use $BS(\delta, \gamma, m)$ to denote the execution of Batch Selection for parameters δ, γ and m, the sample size. Recall that the hypotheses space, its size, and the accuracy of best hypothesis is fixed, throughout this paper, to $\mathcal{H}$, n, and $1/2 + \gamma_0$.)

Theorem 2. *For any γ and δ, $0 < \gamma, \delta < 1$, if $m = 16\ln(2n/\delta)/(c_H \gamma^2)$ and $\gamma \leq \gamma_0$, then with probability more than $1 - \delta$, $BS(\gamma, \delta, m)$ yields some hypothesis h with $prc_{\mathcal{D}}(h) \geq 1/2 + \gamma_0/2$.*

2.1   Constrained Selection Algorithm

We begin by introducing a function that is used to determine an important parameter of our algorithm. For a given n, δ, and γ, define $b_{CS}(n, \delta, \gamma)$ by

$$b_{CS}(n, \delta, \gamma) = \frac{16}{c_H \gamma^2} \cdot \ln\left(\left(\frac{2n}{\delta}\right)\left(\frac{16e}{c_H(e-1)\gamma^2}\right)\right) = \frac{16}{c_H \gamma^2} \cdot \ln\left(\frac{32en}{c_H(e-1)\delta\gamma^2}\right).$$

Our first algorithm, denoted by CS from constrained selection, is stated as follows.

Algorithm $CS(\delta, \gamma)$
　　$B \leftarrow 3\gamma b_{CS}(n, \delta, \gamma)/4$;
　　set $w(h) \leftarrow 0$ for all $h \in \mathcal{H}$;
　　while $\forall h \in \mathcal{H} \, [w(h) < B]$ **do**
　　　$(x, b) \leftarrow EX_{\mathcal{D}}()$;
　　　$\mathcal{H}' \leftarrow \{h \in \mathcal{H} : h(x) = b\}$; $n' \leftarrow |\mathcal{H}'|$;
　　　for each $h \in \mathcal{H}$ **do**
　　　　if $h \in \mathcal{H}'$ **then** $w(h) \leftarrow w(h) + 1 - n'/n$;
　　　　　　　else $w(h) \leftarrow w(h) - n'/n$;
　　　output $h \in \mathcal{H}$ with the largest $w(h)$;

Note that the number n' of successful hypotheses may vary at each step, which makes our analysis difficult. For avoiding this difficulty, we approximate n' as $n/2$; that is, we assume that half of the hypotheses in $\mathcal{H}$ always succeeds on a given example.

Assumption. After t steps, for each $h \in \mathcal{H}$ we have $w(h) = \#_t(h) - t/2$.

Remark 1. In fact, we can modify CS to the one satisfying this assumption; that is, use a fixed, i.e., $1/2$, decrement term instead of n'/n. As our experiments show, both algorithms seem to have almost the same reliability, while the modified algorithm has more stable complexity. We believe, however, that the original algorithm is more efficient in many practical applications. (See the next section for our experiments and discussion.)

First we investigate the reliability of this algorithm.

Theorem 3. *For any γ and δ, $0 < \gamma, \delta < 1$, if $\gamma \le \gamma_0$, then with probability more than $1 - \delta$, $CS(\gamma, \delta)$ yields some hypothesis $h \in \mathcal{H}_{\text{good}}$.*

Proof. We estimate the error probability P_{err}, i.e., the probability that CS chooses some hypothesis with $\text{prc}_D(h) < 1/2 + \gamma_0/2$, and show that it is less than δ, in the following way.

$$
\begin{aligned}
P_{\text{err}} &= \Pr_{\text{CS}}\{ \bigcup_{t \ge 1} [\,\text{CS terminates at the } t\text{th step and yields some } h \in \mathcal{H}_{\text{bad}}\,] \,\} \\
&\le \Pr_{\text{CS}}\{ \bigcup_{t \ge 1} [\exists h \in \mathcal{H}_{\text{bad}}[\,w(h) \ge B \text{ at the } t\text{th step for the first time}\,] \\
&\qquad\qquad \wedge\ \forall h \in \mathcal{H}_{\text{good}}[\,w(h) < B \text{ within } t-1 \text{ steps}\,]] \,\} \\
&\le \sum_{h \in \mathcal{H}_{\text{bad}}} \Pr_{\text{CS}}\{ \bigcup_{t \ge 1} [[\,w(h) \ge B \text{ at the } t\text{th step for the first time}\,] \\
&\qquad\qquad \wedge\ [\,w(h_0) < B \text{ within } t-1 \text{ steps}\,]]\}.
\end{aligned}
$$

Let $\tilde{t}_0 = b_{\text{CS}}(n, \delta, \gamma)$ and $t_0 = (\gamma/\gamma_0)\tilde{t}_0$. (Note that $t_0 \le \tilde{t}_0$.) We estimate the above probability considering two cases, $t \le t_0$ and $t \ge t_0 + 1$, as follows.

$$
P_1(h) = \Pr_{\text{CS}}\{ \bigcup_{t \le t_0} [[\,w(h) \ge B \text{ at the } t\text{th step for the first time}\,] \wedge [\,w(h_0) < B \text{ within } t-1 \text{ steps}\,]] \,\}, \quad \text{and}
$$

$$
P_2(h) = \Pr_{\text{CS}}\{ \bigcup_{t_0+1 \le t} [[\,w(h) \ge B \text{ at the } t\text{th step for the first time}\,] \wedge [\,w(h_0) < B \text{ within } t-1 \text{ steps}\,]] \,\}.
$$

The following can be shown.

Lemma 1. *For any $h \in \mathcal{H}_{\text{bad}}$, we have $P_1(h) \le \delta/2n$.*

Lemma 2. *For any $h \in \mathcal{H}_{\text{bad}}$, we have $P_2(h) \le \delta/2n$.*

Therefore we have:

$$
P_{\text{err}} \le \sum_{h \in \mathcal{H}_{\text{bad}}} P_1(h) + P_2(h) \le n\left(\frac{\delta}{2n} + \frac{\delta}{2n}\right) = \delta.
$$

$\square$

Though valid, our estimation is not tight, and it may not give us a useful bound B for practical applications. We can derive a much better formula for computing B under the following assumption (i.e. independence of the hypothesis) for any $h \in \mathcal{H}_{\text{bad}}$.

$$
\begin{aligned}
&\Pr\{[\,w(h) \ge B \text{ in } t \text{ steps}\,] \wedge [\,w(h_0) < B \text{ in } t-1 \text{ steps}\,]\} \\
&= \Pr\{[\,w(h) \ge B \text{ in } t \text{ steps}\,]\} \times \Pr\{[\,w(h_0) < B \text{ in } t-1 \text{ steps}\,]\}
\end{aligned}
$$

Theorem 4. *Consider a modification of CS where $b_{CS}(n, \delta, \gamma) = \frac{16 \ln(2n/\delta)}{c_H \gamma^2}$, and assume that the above condition holds. Then we can show the same reliability for CS as Theorem 3 for the modified algorithm.*

It may be unlikely that h_0 is independent from *all* hypotheses in $\mathcal{H}_{bad}$. We may reasonably assume, however, that for any $h \in \mathcal{H}_{bad}$, there exists some $h' \in \mathcal{H}_{good}$ such that h and h' are (approximately) independent, and our proof above works similarly for such an assumption. Thus, in most cases, we may safely use the simplified version of b_{CS}.

We now discuss the complexity (number of examples) of our algorithm CS on some $\delta > 0$ and $\gamma \le \gamma_0$. It is easy to see that, after t steps, the weight of h_0 becomes $\gamma_0 t$ on average. Thus, on average, the weight reaches B in B/γ_0 steps. Therefore, the average complexity of $CS(\delta, \gamma)$ is $t_{CS}(n, \delta, \gamma, \gamma_0) = B/\gamma_0 = 12 \ln(2n/\delta)/c_H \gamma \gamma_0$.

2.2 Adaptive Selection Algorithm

In this section we give a different algorithm that does not use any knowledge on the accuracy of the best hypothesis in the class (recall that algorithm CS used the knowledge of a lower bound on γ_0.). To achieve this goal, we modify the condition of the while-loop so it is changing adaptively according to the number of examples we are collecting. We call the algorithm AS from adaptive selection. The algorithm is stated as follows.

Algorithm AS(δ)
   $S \leftarrow \emptyset$; $t \leftarrow 0$; $\epsilon \leftarrow 1/5$;
   while $\forall h \in \mathcal{H}$ $[\#_t(h) \le t/2 + 5t\varepsilon/2]$ **do**
      $(x, b) \leftarrow EX_{\mathcal{D}}()$;
      $S \leftarrow S \cup \{(x, b)\}$; $t \leftarrow t + 1$;
      $\varepsilon \leftarrow \sqrt{4 \ln(3n/\delta)/(c_H t)}$;
   output $h \in \mathcal{H}$ with the largest $\#_t(h)$;

The condition of the while-loop is trivially satisfied while the algorithm has collected less than $4 \ln(3n/\delta)/(c_H (1/5)^2)$ examples, so in practice we collect first these many examples in batch.

Again we begin by investigating the reliability of this algorithm.

Theorem 5. *For any δ, $0 < \delta < 1$, with probability more than $1 - \delta$, AS(δ) yields some hypothesis $h \in \mathcal{H}_{good}$.*

Proof. Our goal is to show that when the algorithm stops it yields $h \in \mathcal{H}_{good}$ with high probability. Thus, we want to show that the following is larger than $1 - \delta$.

$$P_{crct} = \Pr_{AS}\{ \bigcup_{t \ge 1} [\text{AS stops at the } t\text{th step and yields some } h \in \mathcal{H}_{good}] \}$$

$$= \sum_{t \ge 1} \Pr_{AS}\{ \text{AS yields some } h \in \mathcal{H}_{good} \mid \text{AS stops at the } t\text{th step} \}$$
$$\times \Pr_{AS}\{ \text{AS stops at the } t\text{th step} \}.$$

Consider any $t \geq 1$, and assume in the following that the algorithm stops at the tth step. (Thus, we discuss here probability under the condition that AS stops at the tth step.) Let ε_t and S_t be the value of ε and S at the tth step. Also let h be the hypothesis that AS yields; that is, $\#_t(h)$ is the largest at the tth step.

By our choice of ε_t, we know that $t = 4\ln(3n/\delta)/(c_H\varepsilon_t^2)$. Thus, we can apply the following lemma (whose proof is omitted in this version):

Lemma 3. *For a given ε, $0 < \varepsilon \leq 1$, consider the point in the execution of the algorithm just after the tth step with $t = 4\ln(3n/\delta)/(c_H\varepsilon^2)$. Then for any $h \in \mathcal{H}$ such that $\#_t(h) \geq \#_t(h_0)$ we have*

$$\Pr_{AS}\{\,[\,prc_{\mathcal{D}}(h_0) \leq prc_{\mathcal{D}}(h) + \varepsilon\,] \wedge [\,|prc_{\mathcal{D}}(h) - \#_t(h)/t| \leq \varepsilon/2\,]\} > 1 - \delta.$$

Then the following inequalities hold with probability $> 1 - \delta$.

$$\mathrm{prc}_{\mathcal{D}}(h_0) \leq \mathrm{prc}_{\mathcal{D}}(h) + \varepsilon_t, \quad \text{and} \quad |\mathrm{prc}_{\mathcal{D}}(h) - \#_t(h)/t| \leq \varepsilon_t/2.$$

From the second inequality, we have $\#_t(h) \leq t(\varepsilon_t/2 + \mathrm{prc}_{\mathcal{D}}(h))$, and since we know that $1/2 + \gamma_0 = \mathrm{prc}_{\mathcal{D}}(h_0) \geq \mathrm{prc}_{\mathcal{D}}(h)$, we get $\#_t(h) \leq t/2 + t\gamma_0 + t\varepsilon_t/2$. Moreover, the algorithm stopped, so the condition of the loop is falsified and thus, the following holds.

$$t/2 + 5t\varepsilon_t/2 \leq \#_t(h) \leq t/2 + t\gamma_0 + t\varepsilon_t/2$$

This implies that $\varepsilon_t \leq \gamma_0/2$. With this fact together with the first inequality above (i.e., $\mathrm{prc}_{\mathcal{D}}(h_0) \leq \mathrm{prc}_{\mathcal{D}}(h) + \varepsilon_t$), we can conclude that $1/2 + \gamma_0/2 \leq \mathrm{prc}_{\mathcal{D}}(h)$.

Therefore, for any $t \geq 1$, we have $\Pr_{AS}\{$ AS yields some $h \in \mathcal{H}_{good} \mid$ AS stops at the tth step $\} > 1 - \delta$ and since $\sum_{t \geq 1} \Pr_{AS}\{$ AS stops at the tth step $\} = 1$ we are done. $\qquad\square$

Next we discuss the complexity of the algorithm.

Theorem 6. *For any δ, $0 < \delta < 1$, with probability more than $1 - \delta$, AS(δ) terminates within $64\ln(3n/\delta)/c_H\gamma_0^2$ steps.*

Proof. Notice first that while we are in the while-loop, the value of ε is always strictly decreasing. Suppose that at some step t, ε_t has become small enough so that $4\varepsilon_t < \gamma_0$. Then from Lemma 3 (the condition of the lemma always holds due to our choice of ε), with probability $> 1 - \delta$, we have that $t(\mathrm{prc}_{\mathcal{D}}(h) - \varepsilon_t/2) \leq \#_t(h)$, and $\mathrm{prc}_{\mathcal{D}}(h_0) - \varepsilon_t \leq \mathrm{prc}_{\mathcal{D}}(h)$. Putting these two inequalities together, we obtain that $t/2 + t\gamma_0 - t\varepsilon_t - t\varepsilon_t/2 \leq \#_t(h)$ (since $\mathrm{prc}_{\mathcal{D}}(h_0) = 1/2 + \gamma_0$). Since we assumed that $4\varepsilon_t < \gamma_0$, we can conclude that $t/2 + 5t\varepsilon_t/2 < \#_t(h)$, and thus, the condition of the loop is falsified. That is, the algorithm terminates (at least) after the tth while-iteration.

Recall that ε_t is defined to be $\sqrt{4\ln(3|\mathcal{H}|/\delta)/(c_H t)}$ at any step. Thus, when we reach to the tth step with $t = 64\ln(3|\mathcal{H}|/\delta)/(c_H\gamma_0^2)$ it mush hold that $\varepsilon_t < \gamma/4$ and, by the above argument, the algorithm terminates with probability larger than $1 - \delta$. $\qquad\square$

3   Experimental Evaluation of the Algorithms

We first summarize the three selection algorithms considered, and state the functions that bound the sufficient number of examples to guarantee, in theory, that the algorithm selects a hypothesis h such that $prc_\mathcal{D}(h) \geq 1/2 + \gamma_0/2$.

Batch Selection: $BS(n, \delta, \gamma)$, the standard selection algorithm.
    Bound: $t_{BS}(n, \delta, \gamma) = 16 \ln(2n/\delta)/(c_H \gamma^2)$ worst case. Condition: $\gamma \leq \gamma_0$.
Constrained Selection: $CS(n, \delta, \gamma)$
    Bound: $t_{CS}(n, \delta, \gamma) = 12 \ln(2n/\delta)/(c_H \gamma \gamma_0)$ average. Condition: $\gamma \leq \gamma_0$.
Adaptive Selection: $AS(n, \delta)$
    Bound: $t_{AS}(n, \delta) = 64 \ln(3n/\delta)/(c_H \gamma_0^2)$ worst case. Condition: None.

First we describe the setup used in our experiments. We have decided to use synthetic data instead of real datasets so that we can investigate our algorithms in a wider range of parameter values. In future work we are planning to evaluate also them with real data.

The common fixed parameters involved in our experiments are δ, the confidence parameter, and n, the number of hypotheses in $\mathcal{H}$. Note that these two parameters are inside a logarithm in the bounds; thus, results are not really affected by modifying them.

In fact, we verified this experimentally, and based on those results we set them to 18 for n, and 0.01 for δ; that is, we require confidence of 99%. The other parameter is the accuracy of the best hypothesis, which is specified by γ_0. In our experiments the value of γ_0 ranges from 0.04 to 0.3 with a increment of 0.004 (that is, the accuracy of the best hypothesis ranges from 54% to 80% with a increment of 0.4%) and we have a total of 65 different values. For each γ_0, we distributed the 18 hypotheses in 9 groups of 2 hypotheses, where the accuracy of the hypotheses in each group is set $1/2 - \gamma_0$, $1/2 - 3\gamma_0/4$, ..., $1/2 + 3\gamma_0/4$, $1/2 + \gamma_0$. The choice of the distribution of hypotheses accuracy does not affect the performance of neither BS nor AS (because their performance depends only on the accuracy of the best hypothesis). On the other hand, it affects the performance of CS. For this reason, we also tried other distributions of the hypotheses accuracy for CS.

For each set of parameters, we generated a *success pattern* for each hypothesis h. A *success pattern* is a 0/1 string of 1000 bits that are used to determine whether the hypothesis h predicts correctly for a given example. That is, to simulate the behavior of h on examples from $EX_\mathcal{D}()$, we just draw a random number i between 1 and 1000, and decide h predicts correctly/wrongly on the current example if the ith bit of the success pattern is 1/0. Finally, for every fixed setting of all the parameters, we run these experiments 30 times, i.e., run each algorithm 30 times, and averaged the results.

1. The Tightness of Theoretical Bounds: From the bounds summarized first, one may think that, e.g., CS is more efficient than BS. It turned out, however, it is not the case. Our experiment shows that the number of required examples is similar among three algorithms (when γ_0 is known, see next subsection for a

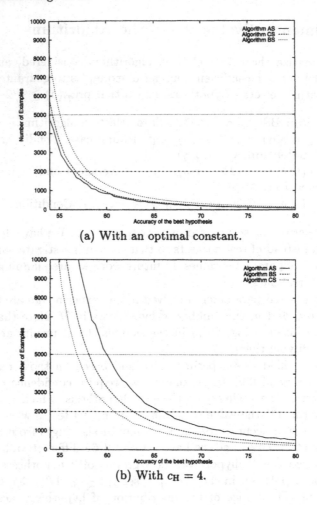

(a) With an optimal constant.

(b) With $c_H = 4$.

Fig. 1. Number of examples used by BS, CS, and AS.

discussion on this point), and the difference is the tightness of our theoretical bounds.

For fixed n and δ, we checked that the number of examples "necessary and sufficient" is proportional to $1/\gamma_0^2$ and we changed the parameter c_H to get the tightest bounds. For each algorithm, we obtained the smallest c_H for which there is no mistake in 30 runs. Graph (a) of Figure 1 shows the number of examples needed by three algorithms with optimal constants. There is not so much difference, in particular, between CS and AS (of course, for the case when γ_0 is known as we mentioned before).

It is, however, impossible in real applications to estimate the optimal constant and get the tightest bound. Nevertheless, we can still get a better bound by simple calculation. Recall that the Hoeffding bound is a general bound for tail probabilities of Bernoulli trials. While it may be hard to improve the constant

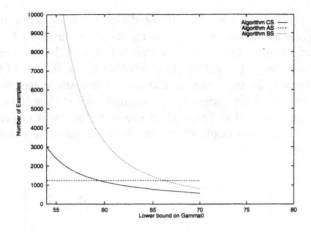

Fig. 2. Behavior of the algorithms changing γ (while $\gamma_0 = 0.2$)

c_H in general, we can numerically calculate it for a given set of parameters. For instance, for our experiments, we can safely use $c_H = 4$ instead of $c_H = 2$, and the difference is around half; e.g., $t_{BS}(18, 0.01, 0.1)$ (the sample size when the best hypothesis has 60% of accuracy) is 6550 with $c_H = 2$ but 3275 with $c_H = 4$. The graph (b) of Figure 1 shows the number of examples needed by three algorithms with $c_H = 4$. Thus, when using these algorithms, it is recommended to estimate first an appropriate constant c_H, and use it in the algorithms. For such usage, CS is the most efficient for the set of parameters we have used.

2.Comparison of the Three Algorithms: Graph (b) in Figure 1 indicates that CS is the best (at least for this range of parameters) but only if γ_0 or a good approximation of it is *known*. For example, if $\gamma_0 = 20\%$ but it is underestimated as 5%, then BS and CS needed 13101 and 2308 examples respectively, while AS needed only 1237 examples, so in that case AS is the most efficient. This phenomenon is shown in the graph in Figure 2 where we have fixed γ_0 to be 0.2 (so the accuracy of the best hypothesis is 70%), $c_H = 4$ and we have changed the value of the lower bound γ from 0.04 to 0.2. Algorithm AS is not affected by the value of γ and thus it uses the same number of examples (the horizontal line). With this graph we can see that, for instance, when γ ranges from 0.04 to 0.058 algorithm AS is the most efficient while from 0.058 to 0.2 algorithm CS becomes the best. We can also see that BS is by far the worst and that, if γ_0 is very much underestimated, the number of examples required by BS can be huge.

3. CS: Constant *dec* vs. Variable *dec*: For simplifying our theoretical analysis, we assumed that $dec = n'/n$ is constant $1/2$. In fact, we have two choices: either (i) to use constant *dec*, or (ii) to use variable *dec*. We investigated whether it affects the performance of the algorithm CS and verified that it does not affect at all the reliability of CS. On the other hand, it affected the efficiency of CS, i.e., the number of examples needed by CS.

Intuitively the following is clear: If the distribution of hypotheses accuracy is symmetric (like in the above experiment), then the number of successful hypotheses, at each step, is about $n/2$; thus, $dec \approx 1/2$, and the number of examples does not change between (i) and (ii). On the other hand, if most of the hypotheses are better than $1/2$ (resp., most of the hypotheses are worse than $1/2$), then the number of examples gets larger (resp., smaller) in (ii) than in (i). We verified this intuition experimentally. Thus, when the distribution is negatively biased, which is the case in many applications, we recommend to use the original CS with variable dec.

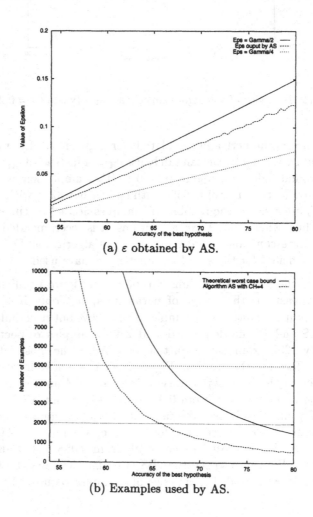

(a) ε obtained by AS.

(b) Examples used by AS.

Fig. 3. Results on algorithm AS.

4. **AS: ε vs. γ_0:** From the theoretical analysis of Theorem 6 we obtained that the algorithm stops with high probability when ε becomes smaller than $\gamma_0/4$. On the

other hand, to guarantee the correctness of our algorithm (Theorem 5) we just need to conclude that ε is smaller than $\gamma_0/2$. This difference gets reflected in the bound on the number of examples since the constant there is $64/c_H$ but in our experiments we verified that the number of examples is in fact much smaller. The reason is that, experimentally, as soon as ε becomes slightly smaller than $\gamma_0/2$ (in our experiments it became $\gamma_0/2.38$) the condition of the loop is falsified and the algorithm finishes. Graph (a) in Figure 3 shows the ε obtained by algorithm AS (using $c_H = 4$) against $\gamma_0/2$ and $\gamma_0/4$ and graph (b) shows the examples used and the theoretical worst case upper bound (again using $c_H = 4$).

References

1. Peter Auer, Robert C. Holte, and Wolfgang Mass. Theory and applications of agnostic PAC-learning with small decision trees. Proceedings of the 12th International Conference on Machine Learning, 21-29, 1995.
2. Leo Breiman. Bagging predictors. *Machine Learning*, 26(2):123-140, 1996.
3. Carlos Domingo, Ricard Gavaldà, and Osamu Watanabe, Practical Algorithms for On-line Sampling, Research Report C-123, Dept. of Math. and Comput. Sci, Tokyo Inst. of Tech. (1998), http://www.is.titech.ac.jp/research/technical-report/index.html.
4. Yoav Freund and Robert E. Schapire. A decision-theoretic generalization of on-line learning and an application to boosting. *Journal of Computer and System Science*, 55:1, 119–139, 1997.
5. D. Haussler. Decision theoretic generalization of the PAC-model for neural nets and other learning applications. *Information and Computation*, 100:78–150, 1992.
6. Robert C. Holte. Very simple classification rules perform well on most common datasets. *Machine Learning*, 11:63–91, 1993.
7. M.J. Kearns, R.E. Schapire, and L.M. Sellie. Towards efficient agnostic learning. *Proc. 5th ACM Workshop on Computational Learning Theory*, 341–352, 1992.
8. M.J. Kearns and U.V. Vazirani. *An Introduction to Computational Learning Theory*. Cambridge University Press, 1994.
9. S.M. Weiss, R.S. Galen, and P.V. Tadepalli. Maximizing the predictive value of production rules. *Artificial Intelligence*, 45, 47–71, 1990.

Discovering Conceptual Differences among People from Cases

Tetsuya Yoshida[1] and Teruyuki Kondo[1,2]

[1] Dept. of Systems and Human Science, Osaka Univ., Osaka 560-8531, Japan
[2] Presently with Mitsubishi Heavy Industries, LTD.

Abstract. We propose a method for discovering conceptual differences (CD) among people from cases. In general different people seem to have different ways of conception and thus can have different concepts even on the same thing.

Removing CD seems especially important when people with different backgrounds and knowledge carry out collaborative works as a group; otherwise they cannot communicate ideas and establish mutual understanding even on the same thing. In our approach knowledge of users is structured into decision trees so that differences in concepts can be discovered as the differences in the structure of trees. Based on the candidates suggested by the system with our discovering algorithms, the users then discuss each other on differences in their concepts and modify them to reduce the differences. CD is gradually removed by repeating the interaction between the system and users. Experiments were carried out on the cases for motor diagnosis with artificially encoded CD. Admittedly our approach is simple, however, the result shows that our approach is effective to some extent as the first step toward dealing with the issue of CD among people.

1 Introduction

It is required to support collaborative works with the participation of various people by extending conventional information processing technologies in accordance with the need for dealing with large-scale accummulated cases. In addition, the importance of facilitating interdisciplinary collaboration among people with different backgrounds has been recognized these days. As for supporting collaborative works among people, various researches have been carried out in the field of CSCW (Computer Supported Cooperative Work) [9, 1]. Most researches tend to put emphasis on providing media to *enable* collaboration among people, however, the issues of *what kind of information should be dealt with* to facilitate collaboration and *how it should be represented* have not been focused on.

We aim at supporting mutual understanding among people when they collaboratively work as a group. Generally different people seem to have different ways of conception and thus can have different concepts even on the same thing. Since Conceptual Difference (CD) due to different conceptions or viewpoints can hinder effective collaboration, we try to remove CD toward facilitating smooth collaboration. Removing CD seems especially important when people with different backgrounds and knowledge carry out collaborative works as a group; otherwise they cannot communicate ideas and establish mutual understanding even on the same thing.

We propose a method for discovering CD among people from cases. Although there are various levels of concepts in general, we focus on CD at the *symbol* level in this paper and deal with the differences in the usage of symbols for concepts. When users specify their knowledge as cases, which consist of attributes, values and classes, they are structured into decision trees so that the trees reflect conceptual structures of users. Then the candidates for CD are suggested based on the structural characteristics of decision trees. By seeing the suggested candidates, users then discuss each other on differences in their concepts and modify them to reduce the differences. By repeating the above processes it is possible to interactively reduce CDs among users toward facilitating their collaboration, which also contributes to clarifying their conceptions or viewpoints as well.

Our approach is new in the sense that concepts of people are *concretely structured* into decision trees to clarify their differences [5]. The framework of representing concepts of people with hierarchical or network structures is also utilized in researches on semantic net [2] and decision making support systems [4, 7], however, few have dealt with the issue of *differences in concepts* among people via the *differences in structures* as in our approach. Our system can be utilized as a module in group knowledge acquisition systems [10] by acquiring knowledge from members in a group interactively. In addition, seeing concepts on things by others helps to notice different viewpoints, which will useful for coming up new ideas from other viewpoints. Thus, our system can also be used as a module in creative thinking support systems [3].

We first describe the framework and definition of Conceptual Difference (CD) in our approach, followed by the description of the system architecture for discovering CD. Then the algorithms for discovering 6 kinds of CD is explained in detail. Experiments to evaluate our system are then described, which show the effectiveness of our approach.

2   Framework of Discovering Conceptual Difference

2.1   Conceptual Difference

In general when people recognize events in the world, they first perceive external events and form internal concepts on them, albeit the exact information contained in concepts might still be nebulous or ambiguous. When it is necessary to communicate internal concepts, they need to be expressed externally by being denoted with symbols or terms. Admittedly symbols might not be sufficient for reflecting all the information or ideas which is intuitively conceived by the notion "concepts". However, it is necessary to use symbols to communicate internal concepts in any case and they still reflect information included in concepts to some extent. Thus, as the first step it is still significant and important to deal with symbols to tackle the issue of concepts.

In this paper we focus on dealing with "Conceptual Difference (CD)" at the symbol level. Usually different symbols are used to denote different concepts, however, the same symbol can be used to denote different concepts depending on the viewpoint of people and the context in which the symbol is used. In contrast, different symbols can be used to represent the same concept. These can occur especially among people with different backgrounds and knowledge. CD dealt with in this paper are defined as follows:

- **Type 1: different symbols are used to denote the same concept.**
- **Type 2: the same symbol is used to denote different concepts.**

Suppose there are an expert in electric engineering called Adam and an expert in mechanical engineering called Bob. When they carry out the diagnosis of motor failure, Adam might point out "anomaly in voltage frequency" and Bob might point out "anomaly in revolutions" for the same symptom. The above two symbols or terms represent different concepts in general, however, they can be considered as denoting the same concept when they are used in the context of the diagnosis of motor failure.

As another example, it is hardly the case that the ranges of temperature in which Adam feels "hot" and Bob feels "hot" are the same and usually there are some differences between the ranges. Thus, there can be some range in which Adam feels "hot" and Bob feels "warm". This means that there are differences in the concept which is denoted with the symbol "hot" between Adam and Bob and that there is some overlap between Adam's concept "hot" and Bob's concept "warm".

2.2 Discovering Conceptual Difference

The kind of problem which can be dealt with in our approach is classification such as diagnosis and the *class* of cases is determined based on the *attributes* and *values*, which characterizes the cases [6]. The system tries to construct the decision tree for cases which is most effective for their classification based on the information theory. A node in decision trees holds the attribute to characterize cases. Each link below a node holds the value for the attribute at the node and cases are divided gradually by following links. The class of cases is determined at the leaf which is reached as the result of link following. We utilize ID3 algorithm [8] to construct decision trees since it is fast and thus is suitable for interactive systems.

The system architecture which incorporates the descovering method in this paper is shown in Fig 1. Currently the system requires that two users represent their knowledge as cases with their respective symbols for them. By accepting the cases as input the system constructs decision trees for them and tries to detect CD in attributes, values and classes based on the structural differences in trees. Since there are 2 types of CD for 3 entities, the system tries to discover 6 kinds of CD and shows the candidates for them in the descending order of the possibility to users. The system also displays the decision trees for cases.

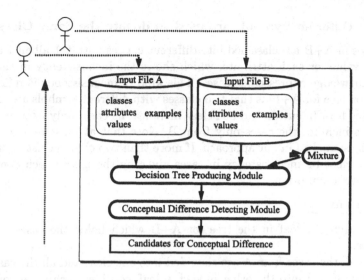

Fig. 1. System Architecture

Visualizing their knowledge as concrete decision trees is expected to help them modify their knowledge and to stimulate further conception. The above processes are repeated interactively to remove CD gradually. In future we plan to extend the system so that it be applicable to more than two users.

Cases specified by users and the synthesized cases from these cases are given to the module for constructing decision trees as input. All classes and attributes in cases from users are utilized in the representation of synthesized cases. In the synthesis of cases it is assumed that the attribute represented with the same symbol have the same value in the original cases. Cases from users are simply merged to construct the synthesized cases, however, it can occur that some attributes are used only by one user and it is impossible to determine their values for the cases from another user. When such an attribute exists, new cases are created for each value of the attribute with the assumption that the attribute can have each value with equal probability.

3   Algorithm

This section describes the algorithms for discovering CD in Fig 1. The algorithms are basically heuristics and designed based on the structural characteristics of decision trees when there exists CD in the cases from users. Hereafter the set of cases from each user and that of synthesized cases are called as A, B, A+B, respectively. The algorithms are based on the following assumptions:

- There is no inconsistency in the classification of each set of cases which are given by the user.
- The same attribute has the same set of values for the attribute.

3.1   C1: Different Symbols are used to denote the Same Class

Some cases in A+B are classified into different classes with **C1** albeit they have the same value for each attribute, which shows the inconsistency in the classification knowledge expressed as cases. This kind of inconsistency is reflected on the decision tree for A+B as that two classes with different symbols are classified at the same leaf. In such a leaf one class includes the cases only from A and the other class includes the cases only from B. Note that at most two classes can exist at the same leaf in our approach. If more than two classes exist at the same leaf, there must be inconsistency in cases given by the user, which contradicts to the above assumption.

Algorithm for C1

Step.1 Search the leaf in the tree for A+B which holds the cases with two classes.

Step.2 If all the cases from A are classified into one class and all the cases from B are classified into the other in such a leaf, consider such a pair of classes as a candidate for **C1**.

Step.3 Repeat **Step.1** and **Step.2** and count the number of detection for each candidate.

Step.4 Sort the candidates according to the number of detection and show them in descending order.

3.2   C2 : the Same Symbol is used to denote Different Classes

Although some cases from A and B are classified into the same class in A+B, their attributes and values differ with **C2** and thus suggest inconsistency. In the decision tree for A+B, the class with **C2** resides in the leaf which holds only the cases from A and also in that which holds only the cases from B.

Algorithm for C2

Step.1 Search the same symbol for class in A and B.

Step.2 If such a symbol is found, search a leaf with it in A+B which holds only the cases from A.

Step.3 If such a leaf is found, search the leaf with the symbol in A+B which holds only the cases from B.

Step.4 Keep the symbol for such a pair of leaves as a candidate for **C2**.

Step.5 Repeat **Step.1** ~ **Step.4** and count the number of detection for each candidate.

Step.6 Sort the candidates according to the number of detection and show them in descending order.

3.3   A1 : Different Symbols are used to denote the Same Attribute

In the decision trees for A and B the distribution of cases below the node for the attribute with **A1** would be similar. The degree of difference in the distribution of cases are calculated as follows:

Algorithm for the Degree of Difference in the Distribution of Cases

Suppose there are n number of classes in A+B. Receive nodes from the decision trees for A and B, respectively, as the input. Choose the child nodes of the nodes for A and also those of the nodes for B with the *same* value. Suppose there are $A_1 \sim A_n$ number of cases for each class below the child nodes in the decision tree for A and $B_1 \sim B_n$ number of cases below the child nodes in the decision tree for B as well.

Step.1 Construct the following n dimension vectors $\overrightarrow{OA}$ and $\overrightarrow{OB}$ as:

$$\overrightarrow{OA} = (A_1, A_2, \cdots, A_n), \overrightarrow{OB} = (B_1, B_2, \cdots, B_n)$$

Step.2 Normalize each vector as: $\overrightarrow{OA'}$ and $\overrightarrow{OB'}$ as:

$$\overrightarrow{OA'} = \frac{\overrightarrow{OA}}{\|\overrightarrow{OA}\|}, \overrightarrow{OB'} = \frac{\overrightarrow{OB}}{\|\overrightarrow{OB}\|}$$

$\|\ \|$: norm of vector, which is defined as the sum of its elements. Calculate their difference as:

$$\overrightarrow{AB'} = \overrightarrow{OB'} - \overrightarrow{OA'}$$

Step.3 Treat the norm $\|\overrightarrow{AB'}\|$ as the degree of difference in the distribution of cases for the value.

Step.4 Repeat **Step.1** $\sim$ **Step.3** for all values in the pair of attributes and calculate their mean. Treat the mean as the degree of difference in the distribution of cases for the pair of attributes.

The possibility that the pair of attributes are the same gets higher as the degree of difference gets smaller. In addition to utilizing the number of detection in sorting the candidates, it is reasonable to utilize the degree of difference in the distribution of cases as well. Thus, the algorithm for **A1** is as follows:

Algorithm for A1

Step.1 Check whether there are attributes with the same set of values in A and B. If found, construct a pair of attributes.

Step.2 Search the nodes with the attributes in **Step.1** from the decision trees for A and B.

Step.3 Calculate the degree of difference in the distribution of cases for the nodes with the above algorithm.

Step.4 Repeat **Step.1** $\sim$ **Step.3**.

Step.5 Choose the pair of attributes with the least degree of differences. Treat the xth degree of difference in the pairs of attributes as the threshold and remove the pairs of attributes with larger difference from the candidates.

Step.6 Count the number of detection for each candidate.

Step.7 Sort the candidates according to the number of detection and show them in descending order.

x in **Step.5** is the minimum number of detection for each pair of attributes and is set by the user. When x is small, candidates are ordered by emphasizing the degree of difference. In contrast, the candidates are ordered by emphasizing the number of detection with large x.

3.4   A2: the Same Symbol is used to denote Different Attributes

In the decision trees for A and B the distribution of cases below the node for the attribute with **A2** would be different. Thus, the algorithm for **A1** is modified so as to choose the pair of nodes with *larger* degree of difference in the distribution of cases.

Algorithm for A2

Step.1 Check whether there is an attribute with the same symbol in A and B.

Step.2 If such an attribute is found, search the nodes with the attribute from the decision trees for A and B.

Step.3 Calculate the degree of difference in the distribution of cases for the nodes.

Step.4 Repeat **Step.1** ~ **Step.3**.

Step.5 Choose the pair of attributes with the largest degree of differences. Treat the *x*th degree of difference in the pairs of attributes as the threshold and remove the pairs of attributes with smaller difference from the candidates.

Step.6 Count the number of detection for each candidate.

Step.7 Sort the candidates according to the number of detection and show them in descending order.

3.5   V1: Different Symbols are used to denote the Same Value

CD for values means that the ranges covered by values differ and thus it is necessary to consider *to what extent* the ranges overlap. This is different from CD for class and attribute in the sense that it is not sufficient to deal with it by considering whether CD simply exist or not exist. Fig 2 illustrates the relationship between the ranges of value "Normal/inc/INC" for the attribute "Temperature" for this kind of CD.

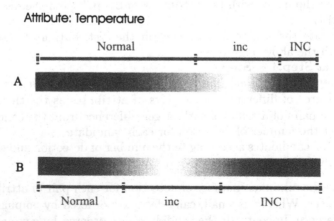

Fig. 2. An example of conceptual difference for value.

Although some cases in A and B have different values, they are classified into the same class in the synthesized cases with **V1**. Suppose cases are classified with respect to an attribute with **V1** in the decision tree for A+B. Then ideally cases are classified so that the child node for one value has only the cases from A and the other child node for the other value has only the cases from B. However, such a clear separation of cases with respect to values rarely occurs, especially at the upper nodes in decision trees since classification of cases usually requires several nodes to follow.

We pay attention to how cases are classified *before* and *after* the test for one attribute depending on the value for it. For instance, suppose 10 cases from A and 20 cases from B are going to be tested for one attribute. Also suppose 10 cases from A and 2 cases from B have the same value for it. Then the difference in the ratio of change of cases is calculated as 100 % - 10 % = 90 %. Likewise, suppose 18 cases from B and 0 cases from A have the other same value for the attribute. This time the difference in the ratio of change of cases is calculated as 90 % - 0 % = 90 %. When both differences are larger than the threshold specified by the user, the system points out that these values can be the same even though different symbols are used to denote. The above process is summarized as follows:

Algorithm for V1

Step.1 Check whether there is an attribute with the same symbol in A and B.

Step.2 If such an attribute is found, construct the pairs of values for the attribute.

Step.3 Search the nodes with the attribute in the decision tree for A+B.

Step.4 As for the pairs in **Step.2**, calculate the differences in the ratio of change of cases before and after the test for the attribute as described above.

Step.5 Keep the pair of values as the candidate for **V1** when both differences are larger than the specified threshold t.

Step.6 Repeat **Step.1** $\sim$ **Step.5** and count the number of detection for each candidate.

Step.7 Sort the candidates according to the number of detection and show them in descending order.

t in **Step.5** can take any percentage between 0% and 100%. When t is small, more pairs of values tend to be considered as the candidates. In contrast, when t is large, the algorithm tends to pick up the pairs with clear separation of cases. This can result in the failure of detection.

3.6 V2: the Same Symbol is used to denote Different Values

Some cases from A and B are classified into different classes even with the same value with **V2**. This means that when cases in the same class are tested for an attribute, the difference in the ratio of change of cases gets larger for such a value. Thus, it is possible to detect the value by calculating the difference in the ratio of change of cases for each value, not the pair of values as in the algorithm for **V1**.

Algorithm for V2

Step.1 Check whether there is an attribute with the same symbol in A and B.

Step.2 If such an attribute is found, search the nodes with the attribute.

Step.3 For each value of the attribute in **Step.3**, calculate the differences in the ratio of change of cases before and after the test for the attribute.

Step.4 Keep the value as the candidate when the difference is larger than the specified threshold t.

Step.5 Repeat **Step.1** ~ **Step.4** and count the number of detection for each candidate.

Step.6 Sort the candidates according to the number of detection and show them in descending order.

4   Experiments and Evaluations

A prototype system has been implemented on the UNIX workstation and each module of the system in Fig 1 is written in C language. The experiments on motor diagnosis cases were carried out to evaluate our approach. In experiments two persons specified their knowledge in the form of thirty cases (as shown in Fig 3), which were composed of six attributes, two or three values and five classes, respectively. CD is artificially encoded into the cases in B by modifying the original cases. Some of CD encoded are:

Input Data: 30 **case for A, B**

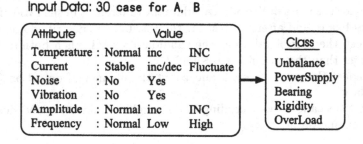

Fig. 3. Input data in the experiment.

Examples of Conceptual Differences

C1 Class "PowerSupply" in A and B
   → modify "PowerSupply" into "Insulation" in B

C2 Class "Unbalance" in A and B
   → modify the attributes or values for "Unbalance" in B

A1 Attribute "Noise"in A and B
   → modify "Noise"into "Stench" in B

A2 Attribute "Vibration" in A and B
   → modify the values for "Vibration" in B

V1 Value "inc" for attribute "Amplitude" in A and B
  → modify "inc" into "Normal" in B
V2 Value "Normal" in A and value "inc" in B for attribute "Amplitude"
  → modify "inc" into "Normal" in B.

The first experiment was carried out for the set of cases in which each kind of CD existed separately. The sets of cases in the trials were constructed by modifying the symbols in cases or the selection of cases randomly. The system enumerated CDs which were detected as up to the third candidate in trials. The result is shown in Table 1. The result shows that all CD for class and attribute could be detected as up to the third candidate. As for the CDs for value, the system could detect them as the first candidate in 4 trials out of 6. Admittedly not all CDs could be detected by the system as the first candidate with the current algorithms. However, since the role of the system is to point out the possibility and to suggest the candidates to the user, the result shows that it has sufficient capability for discovery when each kind of CD exists separately.

Table 1. The result of experiment in which each conceptual difference exists separately.

	number of trials	first candidate	second candidate	third candidate
C1	5	5	0	0
C2	5	5	0	0
A1	6	6	0	0
A2	6	2	2	2
V1	6	4	0	0
V2	6	4	0	0

x is set to 3 with respect to attributes.
t is set to 50% with respect to values.

The second experiment was carried out for the set of cases in which multiple kinds of CD existed. Table 2 shows the candidates for each kind of CD which were suggested by the system. Candidates in italic are the actual CD encoded into the cases. Candidates are ordered according to the number of detection in () and the number in [] represents their ordinal. For this set of cases all the actual CD could be detected as up to the second candidate by the system. With the result in Table 2 the users modified CDs for **C1**, **C2**, **A1**, **V1**, **V2** which had been detected as the first candidate in the cases so that the same symbols were used for them respectively. Then the system could detect CD for A2 as the first candidate with the modified cases, as shown in Table 3. The experiments show that the algorithms are sufficient for discovering 6 kinds of CD. Especially, actual CDs tend to be discovered with higher probability with the repetition of the suggestion by the system and the modification of cases by the users. Thus, the system is effective to reduce CD gradually through the interaction with users.

Table 2. Candidates for conceptual differences (CD) which were suggested by the system when multiple kinds of CD exist in cases.

C1 *[1] A:PowerSupply B:Insulation (2)*   **C2** *[1] Unbalance (21)*
    [1] A:Unbalance   B:Rigidity (2)   [1] Rigidity (12)
    [3] A:Unbalance   B:Bearing (1)   [3] Bearing (2)
A1 [1] A:Noise   B:Vibration (3)   **A2** *[1] Vibration (4)*
    [2] A:Noise   B:Stench (2)   [2] Current (3)
    [3] A:Vibration B:Stench (1)   [3] Frequency (1)
                                                                          V2 *[1] attr<Amplitude> Normal (3)*
                                                                              [1] attr<Amplitude>inc (3)
V1 *[1] attr<Amplitude> A:inc   B:Normal* (3) [3] attr<Vibration> No (2)
    [2] attr<Frequency> A:Normal B:Low (2)   [3] attr<Vibration> Yes (2)
    [2] attr<Vibration> A:No   B:Yes (2)   [3] attr<Frequency> Normal (2)
    [4] attr<Current>   A:inc/dec B:Stable (1)   [3] attr<Frequency> Low (2)
    [4] attr<Frequency> A:Normal B:High (1)   [3] attr<Frequency> High (1)
                                                     [7] attr<Current>   inc/dec (1)
                                                     [7] attr<Current>   Stable (1)

Table 3. Candidates after correcting the first condidates suggested by the system.

A1 *[2] A:Noise   B:Stench (3)*
    [2] A:Vibration B:Stench (1)
    [3] A:Noise   B:Vibration (1)

5   Conclusion

We have proposed a method for discovering conceptual differences (CD) among people from cases. Removing CD seems especially important when people with different backgrounds and knowledge carry out collaborative works as a group; otherwise they cannot communicate ideas and establish mutual understanding even on the same thing. In our approach knowledge of users is structured into decision trees and candidates for CD are suggested by the system which are based on the structural characteristics of decision trees. Experiments were carried out on the cases for motor diagnosis with artificially encoded CD. Admittedly there are various levels of concepts, and our approach is simple and remains at the symbol level. However, the result shows that our approach is effective to some extent as the first step toward dealing with the issue of CD among people. We plan to extend the algorithms based on further experiments with more cases. Currently it is assumed that there is no inconsistency in the knowledge provided by each user, however, we plan to extend our approach to detect such inconsistency.

6   Acknowledgments

This research is partially supported by Grant in Aid for Scientific Research from the Japanese Ministry of Education, Science and Culture (10875080, 09450159, 10143211).

References

1. R. Baecker, editor. *Readings in Groupware and Computer-Supported Cooperative Work – Assisting Human-Human Collaboration.* Morgan Kaufmann, San Mateo, 1993.
2. Philip J. Hayes. On semantic nets, frames and associations. In *Proceedings of the Fifth International Joint Conference on Artificial Intelligence (IJCAI-77)*, pages 99–107, 1977.
3. Koichi Hori. A system for aiding creative concept formation. *IEEE Transactions on Systems, Man, and Cybernetics*, 24(6):882–894, 1994.
4. E. J. Horvitz, J. S. Breese, and M. Henrion. Decision theory in expert systems and artificial intelligence. *International Journal of Approximate Reasoning*, 2:247–302, 1988.
5. T. Kondo, T. Yoshida, and S. Nishida. Design of the Interfaces to Detect Conceptual Difference among Different People. *Information Processing Society of Japan*, 39(5):1195 – 1202, 1997. in Japanese.
6. D. Michie, D. J. Spiegelhalter, and C. C. Taylor, editors. *Machine Learning, Neural and Statistical Classification.* Ellis Horwood, Chichester, England, 1994.
7. R. M. Oliver and J. Q. Smith, editors. *Influence Diagrams, Belief Nets and Decision Analysis.* Wiley, New York, 1990.
8. J. R. Quinlan. Induction of decision trees. *Machine Learning*, 1:81–106, 1986.
9. M. Stefik, G. Foster, D.G. Bobrow, K. Kahn, S. Lanning, and L. Suchman. Beyond the Chalkboard: Computer-Support for Collaboration and Problem Solving in Meeting. *Communications of the ACM*, 30(1):32–47, 1987.
10. K. Tsujino, G. Vlad Davija, and S. Nishida. Interactive Improvement of Decision Trees through Flaw Analysis and Interpretation. *International Journal on Human-Computer Studies*, (45):449–526, 1996.

Discovery of Unknown Causes from Unexpected Co-occurrence of Inferred Known Causes

Yukio Ohsawa[1]* and Masahiko Yachida[1]

Dept. of Systems and Human Science, Graduate School of Engineering Science, Osaka University, Toyonaka, Osaka 560-8531, JAPAN

Abstract. We or computers cannot have perfect knowledge about real events. Even in a very simple case, it is very difficult to have knowledge including every event which may have caused observed events. This makes it difficult to have the computer infer significant causes of observed events. However, the computer inference may supply good hints for humans, because unexpected relations detected between known factors by a computer suggest unknown events to humans, being combined with the vast human knowledge acquired by rich experience. In this paper, we aim at making the computer express "unknown" hidden causes, occurring at the observed time but not included in the given knowledge. This is for discovering temporary events, not acquiring permanent knowledge which generalizes observed data. Our method is to have the computer infer known causes of time-series of observed events, for detecting unexpectedly strong co-occurrences among known events. Then the detected relations are expressed to humans, which make significant unknown causal events easily understood. The Cost-based Cooperation of Multi-Abducers (CCMA), which we employ for the inference of causes, contributes for our purpose of discovery because CCMA can infer well about non-modelable time-series involving unknown causes.

1 Introduction

We cannot have perfect knowledge about real events, because human-made knowledge is biased by human experience. Also, knowledge automatically learned by a computer is biased by the presented data and the prepared representation language. In the artificial intelligence area this may be taken as the frame problem, i.e., knowledge describes only a partial world in the vast real world.

For example, concerning electronic circuit failures, a computer may know constraints on voltages of terminals under the normal connection state of the circuit, and of breaks in conduction lines (for description examples see Eq.(4) to Eq.(6)). With all such knowledge, however, in some real cases the computer can not understand occurring events. That is, when some terminals strongly tend to have an equal voltage, the case cannot be diagnosed because the knowledge does not include information about shorts between terminals. In fact, having knowledge about all possible shorts in a circuit is quite memory-consuming because a

---

* e-mail:osawa@sys.es.osaka-u.ac.jp

short may occur among multiple terminals or conduction lines. Thus, even in a simple case we see it is difficult to have knowledge of every event which may have caused the observed event. This is essential for understanding observed events [1], making it difficult for a computer to infer real causes directly.

However, the inference of computer may supply good hints for the discoveries by a human. In the recent challenges of Knowledge Discovery in Databases (KDD), patterns of data are obtained by computation and presented to serve understandable information to humans [2].

In this paper, we aim at making the computer express "unknown" hidden causal events. Here, by unknown events we do not mean "hidden" (not observed) events, but events occurring at the time of observation but not included in the given knowledge-base. It should be noted that this is the discovery of temporarily occurring significant events, rather than acquiring knowledge which generalizes observed data. For example, discovering under-ground active faults rather than earth quake mechanism from observed quakes, or discovering virus before studying how they affect human bodies from observed symptoms is our target.

The remainder of this paper is as follows. In section 2, the approach and the discovery system outline is presented. Then section 3 reviews the inference applied to the system, and empirically evaluates how the inference suits our purpose of discovery. Section 4 shows how the *discovered* events are presented after the inference, with empirical evaluations.

2   The Outline of the Hypotheses Discovery System

Our goal is a system to supply hints to people for discovering unknown hidden causes of observed events. In existing KDD approaches, such hints have been obtained from data employing regression, machine learning, and other computation methods as surveyed in [2]. Compared to these strategies, our approach is related to the typical source of discovery – the gap between observation and given knowledge [3]. More generally speaking, we focused our attention to unexpected relations among known events as a key to discovery. Specifically, our key is the unexpected co-occurrence of known events, considered to enhance human attentions and lead to discovery (e.g., *spots on a shirt* in [3]). In fact, unexpected co-occurrences may imply unknown common causes of known events [4].

Our system is designed to output groups of known events (events given in the hypothetical knowledge), which co-occur much more often than are expected from given knowledge. For making the output good hints, the system is desired to present *qualitative* information about the events-groups which makes people understand the output without vagueness as exemplified in [5]. For this purpose, a logical approach is taken for detecting co-occurring events, i.e., we employ abductive inference exploited in the artificial intelligence area. By abductive inference, we mean inferring the causal events which are not observed directly.

Simply putting our approach above, we aim at a system to supply good hints for suggesting unknown causes which are not observed and are unknown.

Leaving the details to be stated in the following sections, let us outline the system process here. First the system infers which of known events caused the observed events, by a cooperative abductive inference method called the Cost-based Cooperation of Multi- Abducers (CCMA). By CCMA, the most probable causal events are obtained for the time-series observation. Based on these inference results, events likely to simultaneously occur are grouped. These groups are presented to people looking at the system output. According to the later mentioned empirical results, the presented groups supplies good hints or expressions of unknown significant causes of the observations.

3   The Inference Module by Cost based Cooperation of Multi-Abducers (CCMA)

First the system infers events-sets which are likely to occur simultaneously, by Cost-based Cooperation of Multi-Abducers (CCMA) [6]. CCMA is an inference method for understanding time-series of events. The remarkable feature of CCMA is that it is good at inferring about *non-modelable* changes, i.e. changes in unpredictable speed caused by unknown factors.

Compared with CCMA, inference using *transition probabilities* representing the frequencies (speed, in other words) of state-changes are being established for understanding time-series events [7]. However, in some real world problems recently coming up, there are cases where the changing speed in the environment can not be easily measured. For example, the user interest of a document search engine may change for various reasons, so even the speed may change with time while being measured [6]. Even worse, the errors in transition probabilities are accumulated in the posterior probability of one whole time-series.

More concretely, variance in the changing speed becomes large if the involved physical principles are *non-modelable*, i.e. if there are many or significant (heavily affecting the observation) events the computer does not know, causing unpredictable changes in the observation. This non-modelability becomes severe when we think of discovering unknown events, which increase the variance of changing speed and make a time-series non-modelable.

For the reasons above, we employ CCMA for the inference in discovering unknown events. Because CCMA does not use transition probabilities in sequential events, it has no risk of being deceived by compulsorily given values of transition probabilities when the environmental situation changes non-modelably. In fact, search-engine users' unpredictably changing interests were finely understood by CCMA in the Index Navigator [6], and CCMA worked better than inference with transitional probabilities (e.g. dynamic Bayesian networks [7]). In this section, we review what problem CCMA solves (3.1) and how CCMA works (3.2).

3.1   The Problem for CCMA

CCMA explains time-series by *solutions*, which are the hypotheses-sets $h_j (j = 1, 2, ...N)$ satisfying the constraints in Eq.(1)-(2). Here, N is the length of the time-series, i.e., the number of events observed in series. Σ and H denote the background knowledge (knowledge given as true) and hypothetical knowledge (a set of hypotheses each of which may be false or true and may contradict others), respectively. G_j is the observation (goal) at the j-th moment. A *solution* $h_j (\subset H)$ must support G_j if combined with Σ (as in Eq.(1)), without contradiction between events at a moment (as in Eq.(2), with the empty clause ϕ).

$$h_j \cup \Sigma \vdash G_j, \tag{1}$$

$$h_j \cup \Sigma \not\vdash \phi. \tag{2}$$

Σ in propositional Horn clauses are dealt in CCMA. A Horn clause is as in Eq.(3), meaning that y is true if $x_1, x_2, ...x_n$ are all true, where $x_1, x_2, ...x_n$ and y are atoms, each denoting an event occurrence.

$$y :- x_1, x_2, ..., x_n. \tag{3}$$

If y is the empty clause ϕ, then Eq.(3) means that $x_1, x_2, ...x_n$ cannot be true simultaneously and $x_1, x_2, ...x_n$ are said to be inconsistent. Also, we can write a probabilistic rule "y is true if $x_1, x_2, ...x_{n-1}$ are all true, by conditional probability $p(x_n)$," by letting x_n in Eq.(3) denote an atom to be true by $p(x_n)$.

For example, output z_2 of a normal AND gate A with inputs u_2 and v_2 satisfies Eq.(4), where x_{ON} (x_{OFF}) denotes the state where voltage x is ON, i.e., of the higher (lower) voltage. Also, two terminals P and R connected by line PR can be constrained as in Eq.(5), Eq.(6) and so on, where P_{ON} means the voltage at terminal P is ON, and PR_{broken} (PR_{OK}) means a break- (normal-) state in the line PR, respectively.

$$z_{2ON} :- u_{2ON}, v_{2ON}, \tag{4}$$

$$P_{ON} :- R_{ON}, PR_{broken} \tag{5}$$

$$P_{OFF} :- R_{ON}, PR_{OK}, \tag{6}$$

An inconsistency constraint in Eq.(2) forbids distinct states of one component, e.g. $\phi :- PR_{OK}, PR_{broken}$ to occur simultaneously.

Further, we should consider *coherence*, i.e. h_i and h_{i+1} should be identical if the real situations do not change from the i th to the $i+1$ th moment, and different if the situation really changed. It is here that we should consider that transition probabilities may not be available between moments.

Without transition probabilities, CCMA compromises to believe in a hypothesis *extremely* strongly, if it is believed at adjacent moments. To be specific, CCMA obtains h_j, supporting G_j, of the minimal C_{1j} where

$$C_{1j} = \Sigma_{\eta_i, \, s.t., \, \eta_i \in h_j \wedge \eta_i \notin h_{j+1} \cup h_{j-1}} w_{ij}. \tag{7}$$

Here, η_i is the i th hypothesis in H and has weight w_{ij} in explaining G_j.

Intuitively Eq.(7) means that, in the solution h_j at the j th moment, hypotheses *not believed* at adjacent moments (i.e. not in $h_{j-1} \cup h_{j+1}$) are excluded as much as possible. The semantics of Eq.(7) is supported by the probabilistic semantics of *Cost-based Abduction (CBA)* [8]. That, defining weight w_{ij} as

$$w_{ij} = -log(p(\eta_i)),   (8)$$

where $p(\eta_i)$ is the prior probability of η_i to be true, h_j of the least C_{1j} becomes the maximum a-posteriori (MAP) explanation of G_j, assuming all hypotheses in $h_{j-1} \cup h_{j+1}$ to be believed at the j th moment. $p(\eta)$ in Eq.(8) is computed in manners relying on the problem in hand. For example, if η stands for the break in a terminal P in an electronic circuit, $p(\eta)$ is computed as

$$p(\eta) = \frac{times\ of\ past\ observed\ breaks\ in\ P}{all\ the\ moments\ when\ P\ was\ monitored}.   (9)$$

Reader interested in more semantics of Eq.(7) is referred to [6], where the embodied computation of $p(\eta)$ is also presented for an example of document retrieval.

Because we aim at discovering *temporary* events, the problem is formalized here as of reasoning rather than learning i.e., acquiring permanent knowledge as in prevalent frameworks in inductive logic programming, etc.

3.2   The Mechanism of CCMA

In CCMA, C_{1j} in Eq.(7) is minimized by cooperative reasoning. For each i, w_{ij} is given initially as w_i^0 common to each j, and revised in the following process.

The j th *abducer* α_j solves its assigned CBA, i.e., infers to obtain the solution h_j of the minimal *cost* in Eq.(10) for its local observation G_j.

$$cost = \Sigma_{\eta_i \in h_j} w_{ij}.   (10)$$

On obtaining h_j, α_j sends/receives messages to/from adjacent abducers, i.e. α_{j-1} and α_{j+1}. Here, α_j's reaction to a *message* is to reduce the weight w_{ij} of hypothesis η_i in α_j to 0, if η_i is included in $h_{j-1} \cup h_{j+1}$ (i.e., believed to be true by α_{j-1} or α_{j+1}). Solving CBA and then sending and receiving messages, in the sense above, forms one *cycle* of an abducer. Note that a message is passed to both adjacent abducers, i.e. for the past and the future. Reducing w_{ij} to 0 intends to set $p(\eta_i)$ to 1 at the j th moment, according to Eq.(8). Intuitively, this means making α_j believe in η_i strongly if α_{j-1} or α_{j+1} also believe in η_j. Formally, solving CBA for α_j under "$w_{ij} = 0\ \forall \eta_i \in h_{j-1} \cup h_{j+1}$" minimizes C_{1j} in Eq.(7). Cycles run until solutions are no longer revised. In summary, each abducer explains its goal for the assigned moment, revising its own belief according to adjacent abducers' solutions.

For example in Fig. 1-b, nodes at the top (G_1, G_2 and G_3) denote goals, each being the observed voltages (x, y, z) at a moment, in the OR circuit of Fig. 1-a, valued 1 for ON and 0 for OFF. That is, (1,0,0) at time $t - dt$ (G_1), (0,1,0) at t (G_2) and (1,1,0) at $t + dt$ (G_3). The leaves and edges, respectively, are the parts of hypotheses in H and knowledge Σ supporting each goal. For example, G_2 can be supported by two rules, first by "$z = 0$ for $(x, y) = (0, 1)$ if η_1 is true

(R is broken) and η_5 is true," and second by "$z = 0$ for $(x, y) = (0, 1)$ if η_3 is true." η_4, η_5 and η_6 denote the absence of moisture which may short terminal R, at time $t - dt$, t and $t + dt$ respectively, with probability of less than but nearly 1. η_7 is some event not stated here (e.g. "short between P and Q"), but will be referred to later in Fig.2-b and c. Here let us ignore fault states not in Fig.1.

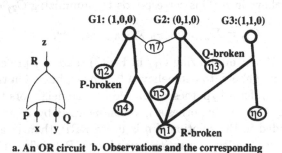

a. An OR circuit b. Observations and the corresponding
knowledge-base

Fig. 1. A knowledge-base for an OR circuit.

For explaining this time-series, the simplest version of CCMA runs as in Fig. 2-a. In Fig. 2-a,b, and c, each line for the respective value of k depicts abducers' behavior in one cycle. The tree beneath each goal (G_1, G_2, or G_3), shown in each frame for $\alpha_i (i = 1, 2, 3)$ in the figure-top, is the relevant (to explaining each goal) part of the knowledge in Fig.1.

The values depicted beside the nodes in Fig. 2 are the weights of hypotheses. In Cycle 1, abducers infer the fault sources independently and obtain h_1: $\{\eta_2\}$, h_2: $\{\eta_3\}$ and h_3: $\{\eta_1, \eta_6\}$, i.e., the least-cost explanations for G_1, G_2 and G_3 respectively. Messages depicted by the thick dashed arrows in Fig. 2-a change the weights of the hypotheses at the arrowheads, e.g. in Cycle 1 the solution of α_3 in η_1 enhances the belief of α_2 in η_1. In Cycle 3, the process stops because the solutions do not change further. The obtained explanation is that R is broken throughout the sequence. As stated in [6], the messages by weight-changing enables the inference about unpredictable changes in the time-series.

Also, relative and absolute *refractory periods* are used for reducing the risk of *infinite cycles*. That is, if a message is sent as simply as in Fig. 2-a from abducer α to its adjacent abducer β, the message may be sent back to α in the next cycle, which may result in infinite cycles. For example, in Cycle 2 of Fig. 2-b, α_1 sends its belief in η_7 to α_2, and α_2 sends its belief in η_1 to α_1. Then these messages are sent back to their senders in Cycle 3. The states in Cycle 2 and 4 become identical, hence infinite cycles.

For reducing the risk of infinite cycles, abducer α must wait without receiving messages, until its adjacent abducer β's belief somehow converges by communications with abducers other than α. However, it is meaningless to make all the abducers wait for an equal number of cycles, because in that manner abducers α and β after the waiting cycles return to the same states as in the cycle before the message was sent from α to β. Therefore, α should wait longer than β, if α receives weaker messages from its adjacent abducers than β does.

For realizing this, *relative refractory periods* (briefly RRP) are introduced, during which an abducer α_j receives messages on hypothesis η iff *both* adjacent abducers of α_j believe in η. This corresponds to minimizing C_{2j} defined as

$$C_{2j} = \Sigma_{\eta_i, \ s.t., \ \eta_i \in h_j \wedge \eta_i \notin h_{j+1} \cap h_{j-1}} w_{ij}. \tag{11}$$

In the simplest CCMA, minimizing C_{1j} in Eq.(7) made α_j prefer a hypothesis believed by *some* neighbors, to one believed by *no* neighbor. On the other hand, Eq.(11) makes α_j prefer a hypothesis believed by *two* neighbors to one believed by *one* neighbor. That is, hypothesis η recommended by a stronger message tends to be included in the solution if η is in an RRP, than in a period of the fundamental process (called a *reactive period*, or RP briefly) in which Eq.(7) rules the selection of hypotheses. By mixing RP and RRP, α_j comes to wait longer than its adjacent abducer β if α_j receives weaker messages than β does.

Also, for making abducers forget messages sent many cycles before, *absolute refractory periods* (briefly ARP) are used. In an ARP of hypothesis η_i in α_j, w_{ij} is returned to its initial value w_i^0 iff neither α_{j-1} nor α_{j+1} believes in η_i.

Consequently, the process of CCMA is described as follows. In the following description, variables are defined as, k : the count of cycles, t_{ij} : the truth value of η_i in the j th abducer defined as 1 for true, 0 for false. m_{ij} : the strength of the message on η_i to α_j, i.e., $t_{i(j-1)} + t_{i(j+1)}$. w_i^0, as stated above, is the value of w_{ij} given when k was 0. $X \leftarrow Y$ means substituting X with Y.

The procedure of CCMA (begin with $k = 0$)
1. CBA phase:
  1.1 $k \leftarrow k+1$. Obtain the least cost solution h_j for G_j by the j th abducer, for $j = 1, 2, ...N$ where N is the length of the time-series, i.e., the number of events observed in series.
  1.2 For $j = 1, 2, ...N$, for all $\eta_i \in H$ in the j th abducer, if $t_{ij} = 1$ put η_i in ARP.
  Otherwise:
    If t_{ij} changed from 1 to 0 in **1.1**, put η_i in RRP.
    Otherwise, put η_i in RP.
2. Message sending phase: For $j = 1, 2, ...N$, for all $\eta_i \in H$ in the j th abducer, change w_{ij} as:
  For η_i in an RP,
    $w_{ij} \leftarrow w_{ij}$ if $m_{ij} = 0$. $w_{ij} \leftarrow 0$ if $m_{ij} \geq 1$,.
  For η_i in an RRP,
    $w_{ij} \leftarrow w_{ij}$ if $m_{ij} \leq 1$. $w_{ij} \leftarrow 0$ if $m_{ij} = 2$.
  For η_i in an ARP,
    $w_{ij} \leftarrow w_i^0$ if $m_{ij} = 0$. $w_{ij} \leftarrow w_{ij}$ if $m_{ij} \geq 1$.
3. Halting condition: Stop if no weight changed in **2** or $k = 100$. Otherwise go to **1**.

See Fig. 2-c, for an example of thus improved CCMA process.

We fixed the length of each ARP or RRP to one cycle, because the inference time increases with the length of refractory periods. Increasing this length does not improve the ability of avoiding infinite cycles, because merely long refractory periods make all the abducers wait equally long, and the contents of messages to be sent after a refractory period do not depend on how long abducers wait.

The notion of refractory periods are taken from neuroscience. Neurons go through similar periods to hypotheses in CCMA. That is, on firing with stimuli from adjacent neurons in a *reactive period*, a neuron gets into an *absolute refractory period*, and then a *relative refractory period* [12]. It is of an inter-field interest that refractory periods work for reducing infinite cycles in CCMA, because the role of refractory periods has not been clarified in neuroscience.

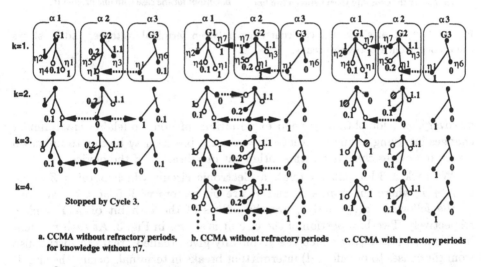

a. CCMA without refractory periods,   b. CCMA without refractory periods   c. CCMA with refractory periods
for knowledge without η7.

● : An atom inferred or given as truth (in an absolute refractory period if it is a hypothesis, in c)
⊘ : A hypothesis in a relative refractory period
○ : One of other atoms (in a reactive period if it is a hypothesis, in c)

Fig. 2. The CCMA process. **a** and **b** are the simplest version of CCMA running for a simple and a rather complex case, respectively. The infinite cycles in **b** are avoided by refractory periods in **c**.

3.3   Tests of CCMA performance for non-modelable changes

Among fault diagnosis of electronic circuits as popular example problems for probabilistic or abductive reasoning e.g. [9], we take intermittent faults with non-modelable changes in the fault states. Our aim in selecting this example is to test whether CCMA is a good tool or not for the inference in hypotheses

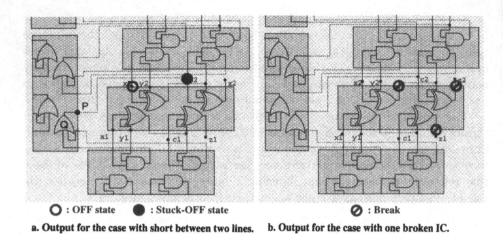

○ : OFF state   ● : Stuck-OFF state   ⊘ : Break

a. Output for the case with short between two lines.   **b. Output for the case with one broken IC.**

Fig. 3. Output **a.** shows the co-occurrence between terminal voltages, and **b.** shows the co-occurrence between breaks.

discovery, i.e., for obtaining good explanations of non-modelable environment changes involving unknown events. In a circuit, breaks may be due to unknown cracks in devices changing with subtle heat or vibration of the device.

We took a 3-bit adder circuit, a electronic circuit which obtains Z equal to $X + Y$ where the inputs X and Y are bit-vectors with 0 or 1 in each bit. In the following, x_i, y_i and z_i are the values of the i th bit of X, Y and Z respectively. Two-bits portion of the circuit is shown in Fig. 3. At each moment in a time-series, during which the artificially given (called "real" to distinguish from the causes to be inferred) intermittent breaks in terminals occur, the circuit computes. Values of X, Y and computed Z are observed at each moment, and CCMA explains them.

Knowledge Σ given here is formed by gathering Horn clauses as in Eq.(4) - (6), or as depicted in Fig. 1. The only faults considered are breaks, by having hypotheses in H represent possible states of connections, e.g., $\{P_{OK}, P_{broken}, ...\}$. Each hypothesis η was assigned the initial weight of $-log(p(\eta))$, $p(\eta)$ denoting the probability of η to be true (see Eq.(8)), given artificially. Goals are the observed voltages, e.g., $\{x_1 = 1, x_2 = 0, ..., z_2 = 0\}$.

First let us show an example result. CCMA explained $series_1$, the observations in Eq.(12), as that the connection at P in Fig. 3-a was broken for G_2 and G_4, but was closed at other moments. By exchanging the observations of G_1 and G_3 (call this $series_2$), P was explained to be broken for G_2, G_3 and G_4.

For comparison, we constructed a dynamic Bayesian network where the probabilities of transitions between normal and fault states of a terminal were set to given values. By this Bayesian network both $series_1$ and $series_2$ were explained to be with broken P only for G_2, G_4 and G_6, i.e., as changing constantly at the given speed.

$$G_1 : 011 + 100 = 111, \ G_4 : 101 + 011 = 110,$$
$$G_2 : 011 + 011 = 100, \ G_5 : 010 + 101 = 111,$$
$$G_3 : 001 + 101 = 110, \ (in \ G_j : x_1 x_2 x_3 + y_1 y_2 y_3 = z_1 z_2 z_3) \quad (12)$$

Next we controlled u, a factor concerning the modelability of changes. u is defined by the number of faults not included in Σ, changing twice as fast as other faults in terminal connections. That is, u represents the number of transient events not acquired into Σ, e.g., intermittent stuck-OFF of the output of gate Q in Fig.3 by transient breaks due to hidden cracks in the IC.

"Real" time-series each made of 50 moments were given and tested. That is, causal states were given as connection states and the values of X and Y. The results are shown in Fig. 4. Each dot depicts the *precision* of the five inference methods tested in [6], **1**: CCMA, **2**: dynamic Bayesian networks [7], **3**: the FA/C distributed inference [10], **4**: Explanatory Coherence [11], and **5**: abduction without constraints between different moments. Here, *precision* is defined as C/N, where C and N denote the number of connection nodes at which the voltages in the solutions corresponded to the "real" ones, and the total number of nodes.

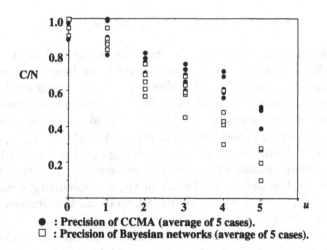

● : Precision of CCMA (average of 5 cases).
□ : Precision of Bayesian networks (average of 5 cases).

Fig. 4. Precision of each method for the diagnosis of an adder circuit.

In Fig.4, only methods **1** and **2** are recorded because methods **3** to **5** did not obtain better precision than these two in any single case. Here we observe that CCMA did well for large u. For larger u, Bayesian networks tended to explain with solutions changing radically because the posterior probability of a coherent solutions-set was low for a quickly changing time-series. On the other hand, CCMA explained by solutions with less changes, as if knowing that unknown faults may enlarge the variance of changing speed and cause unexpectedly quick changes, without appearing to the solutions.

These results show that CCMA well explains non-modelable time-series with many unknown events. This also supports the ability of Index Navigator to understand search engine user interest [6], because many undefined emotions affect human interests.

4   Hypotheses discovery based on event co-occurrences

In the hypotheses discovery system, we use the inference results of CCMA as the basis of discovering new hypotheses. Here, relations between hypotheses are computed and shown to a human.

Among other relations among known hypotheses, co-occurrence are considered in the system (see section 1). The system obtains groups of hypotheses which co-occur, i.e. tend to be in the solution at the same moment. A hypotheses-set H_g of the maximum co in Eq.(13) is taken as a co-occurrence group. Here, co means the degree of co-occurrence of hypotheses in group H_g, compared to when the events represented by these hypotheses independently occur.

$$co = p(\eta \in h, \ \forall \eta \ s.t., \ \eta \in H_g)/\Pi_{\eta \in H_g} p(\eta \in h). \tag{13}$$

For the circuit example, the output is shown as in Fig.3. Here, x_i, y_i, z_i are identically defined as in 3.3. The adder considered is one hand-made using ready-made Integrated Circuits (ICs, the darker region), not constructed in one IC. Because breaks and shorts are much more likely at hand-made connections than in ICs, we gave only knowledge about breaks at hand-made connections.

When a co-occurring hypotheses group H_g is detected (e.g., H_g : $\{x_2 = 0, p = 1, P_{broken}\}$), H_g is divided into subgroups each (e.g., g : $\{x_2 = 0, p = 1, P_{broken}\}$) of which is the only possible explanation of an event (e : $\{c_2 \ stuck \ OFF\}$, for g) existing in Σ. The output is the set of these co-occurring events (like e) obtained from H_g, meaning the most fundamental and the simplest expression of co-occurring events.

For example, Fig. 3-a is the output of terminals marked with circles, meaning "$x_2 = 0$ and the stuck-OFF state of c_2 co-occurs." This figure was obtained when we assumed an intermittent short between the x_2 and the line between c_2 and P, in the "real" causes. The system did not mean "two lines touched intermittently" directly, but the figure indicates this for a human. On the other hand, Fig. 3-b shows terminals where simultaneous breaks are inferred by CCMA. It is very apparent from the figure that the breaks were caused by some accident in one of the machine-made ICs, e.g. overheat on the way of the time-series.

The 144 cases tested included 132 time-series with faults by causes unknown to the system, but known and given by authors. For all these cases, we recollected the given unknown causes and found the system looked like meaning to directly represent the *discovered* events.

5   Conclusions

We showed that co-occurring known events inferred by CCMA and presented in appropriate manners supply good *hints* serving almost as *answers* for a human engaged in discovery of unknown causes.

For a human who knows that the events to be discovered cause the events known to the computer, as in our experiments, this is the discovery of *unnoticed* events. On the other hand, for one without such knowledge, this system guides in revealing unknown events and unknown causal rules in the real world. In both cases, the human can give knowledge about the discovered events and rules to the computer, so this is both discovery and knowledge acquisition for the computer.

The discovering effect of this system comes from combining the computer output with the rich human experience. Our next challenge will be to verify the effect in large scale real problems, and realize a discovery system which is automatic in a stronger sense by combining heterogeneous knowledge-bases, as we combined human and computer knowledge in this paper.

References

1. Hanson, N.R., *Patterns of Discovery*, Cambridge University Press (1958)
2. Fayyad, U., Shapiro, G.P. and Smyth, P., From Data Mining to Knowledge Discovery in Databases, *AI magazine*, Vol.17, No.3, 37–54 (1996)
3. Langley, P., Simon, H.A., Bradshaw, G.L., and Zytkow, J.M., *Scientific Discovery – Computational Explorations of the Creative Processes*, MIT Press (1992)
4. Sunayama,W.,Ohsawa,Y. and Yachida,M., Validating Questionnaire Data Analysis Systems by Shrinking Iterative Questionnaires, *Workshop Notes on Validation, Verification and Refinement of AI*, IJCAI'97 (1997)
5. Glymour, C., Available Technology for Discovering Causal Models, Building Bayes Nets, and Selecting Predictors, The TETRAD 2 Program, *Proc. KDD'95*, 130 – 135 (1995)
6. Ohsawa, Y. and Yachida, M., An Index Navigator for Understanding and Expressing User's Coherent Interest, *Proc. IJCAI'97*, 722 – 729 (1997)
7. Binder, J., Murphy, K. and Russel, S., Space-efficient inference in dynamic probabilistic networks, *Proc. IJCAI'97*, 1292–1296 (1997)
8. Charniak, E. and Shimony, S.E.: Probabilistic Semantics for Cost Based Abduction, *Proc. AAAI'90*, 106–111 (1990)
9. Poole, D: Probabilistic Horn Abduction and Bayesian Networks, *Artificial Intelligence*, 81–130 (1993)
10. Carver, N., Lesser, V., The DRESUN Tested for Research in FA/C Distributed Situation Assessment, *Proc. Int'l Conf. Multiagent Systems*, 33-40 (1995)
11. Ng, H. and Mooney, R., On the Role of Coherence in Abductive Explanation, *Proc. AAAI'90*, 337–342 (1990)
12. Koester, J., Voltage-Gated channels and the Generation of the Action Potential, *Principles of Neural Science*, 2nd ed.(Kandel, E.R. et al eds.), Elsevier, Part 2, 75–86 (1985)

Refining Search Expression by Discovering Hidden User's Interests

Wataru Sunayama[1]*, Yuji Nomura[1], Yukio Ohsawa[1] and Masahiko Yachida[1]

Dept. of Systems and Human Science, Graduate School of Engineering Science, Osaka University, Toyonaka, Osaka 560-8531, Japan

Abstract. When an Internet user wants to know about a certain topic, the user uses a search engine to find pages related to that topic. However, there are so many pages in World Wide Web that the user cannot access to the pages directly, because most search engines output URLs matching the exact words entered by the user. This paper presents a Interest Hypothesis Discovering System which discovers hypotheses impling user's hidden interests. As an application of the system, we show a method to support user's expression of interests when the user uses a search engine by offering words related to the hidden interest. The user refines the search condition with words presented to get URLs.

This work includes a new modelling of user's interest expressions. That is, our discovering system employs a Hypothesis Creation System which creates new hypotheses implying relationships among input words for search. New hypotheses are interpreted as a user's hidden interests, which enable the discovering system to present effective new words for the user to search.

1 Introduction:WWW and Information Searching Technique

In recent years, with the growth of the Internet, the quantity of available information being dealt with has been increasing. When an Internet user searches for information, there comes to be a need of technique to search. Nowadays, the form of information most frequently used are World Wide Web (WWW) documents, called Web **pages**, and the service most well known to search those information are search engines.

A search engine has database of URL (Universal Resource Locator) and supplies URLs matching a user input search condition. Though there are search engines databases where URLs are classified into items [13] or tagged by keywords, in this paper a **database** of URLs are uncared one; we assume full text search engines. Because full text search engines has so many pages are in the database. With a full text search engine, it is difficult for a user to get pages he or she wants at once. This is because

---

* e-mail:sunayama@yachi-lab.sys.es.osaka-u.ac.jp

1. Words, user input search conditions, may be used in different meanings from what the user intended.
2. The pages a user wants to get are displayed by different words from what the user input.
3. The words are insufficient to specify the user's interests.

The problems of 1 and 2 are called disagreement of vocabulary [5]. Problem 1 decreases the Precision value, the rate of proper pages in output, and problem 2 decreases Recall value, the rate of extracted proper pages. (Precision and Recall are defined in Chapter 5.) Problem 3 is caused by the lack of knowledge about the target, or because the user can't hit upon a proper words immediately. To prevent these problems and to improve full text search, intelligent search techniques were appeared. For example, a search engine expands a user's **search expression** (combination of keywords to search) by using a synonym dictionary or by using a correlation model of words learned from sample documents. In other search engines, words prepared in advance are only used to search and if a user gave a word not included in prepared set, the search engine chooses one from prepared set by relationships between a word the user input and words in prepared set [3]. However, these intelligent search techniques cannot solve the above problems clearly. Because,

1. The words made up by a search engine are not always effective to search.
2. The words made up by a search engine do not always agreed with user's interest.

The reason why above two problems are still is; those intelligent search techniques have been ignored user's viewpoints. The words made up are not depending on each user and not depending on the Web pages at the time of search. Therefore we propose a support system where a user can refine his or her own search expression by keywords related to the user's interests. That is, our system discovers hypotheses representing the user's hidden interests and supplies keywords related to inferred interests. A hidden interest isn't exposed positively in search expression.

[7] shows a method to learn proper words for search by inductive learning from a history of user's action. However, these methods have a hard problem in that the system must learn user's temporary interests from a brief time period of observation because user's interest and information on WWW changes time by time. Then, our support system infers user's interests from only one input of search expression. [1, 10] show an approach where the system supplies a user with words representing the user's interests. But our keywords for expressing user's intersests are extracted from **real** Web pages not depending on a knowledge base the system has. Our system for discovery is based on KDD [4] to discover new knowledge.

In this paper, in chapter 2 we explain the interest hypotheses discovering system which discovers hypotheses related with user's interests by the input of search expressions. Chapter 3 describes the support system for searching Web pages which supplies keywords related to the words used in search expression and in chapter 4, we evaluate the system with some experiments. Chapter 5 gives a way of refining of search expressions, and chapter 6 concludes this paper.

2   Discovery:Hypotheses representing user's hidden Interests

In this chapter, we describe the Interest Hypotheses Discovering System(IHDS) Fig. 1 which discovers hypotheses representing user's hidden interests, **interest hypotheses**, from a user's search expression and Web pages.

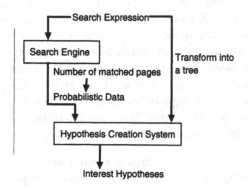

Fig. 1. Interest Hypotheses Discovering System

2.1   Search expression

A search expression is given as boolean. **A Search expression** is the combination of keywords using the logical operations "AND" and "OR". For example, a search expression "Tokugawa AND Ieyasu" matches pages which include both words, and the expression "Toyotomi OR Hideyoshi [1] " matches pages which include either word.

2.2   Search expression tree

A search expression is transformed into a tree structure, that is a directed acyclic graph including "AND" and "OR", as input to the Hypotheses Creation System (Section 2.3).

The method of transformation is that leaf-nodes denote the keywords included in search expression, and middle nodes denote partial search expressions. Then, the direction is established from leaf to root. This means user's interest leads to some keywords and keywords lead to expression of user's interest (Fig.2).

---

[1] Ieyasu Tokugawa and Hideyoshi Toyotomi are names of shogun

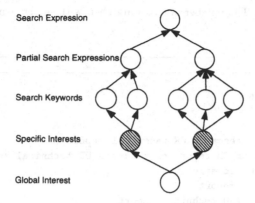

Fig. 2. Interest causes Expression

For example, Fig.3 shows the search expression for "induction AND (paper OR report of research OR technical report)". If there is a parent node which has more than or equal to 3 children nodes, AND-parent nodes are added above every pair of children. This is the way of partially examining precise relationship. This network will be the input to the Hypotheses Creation System.

2.3   Hypotheses Creation System

To discover hypotheses implying user's hidden interests, we use the Hypotheses Creation System (HCS)[12]. The Hypotheses Creation System is a system that; when a network expressing logical relationship among events and each event's probability are given respectively, outputs new hypotheses representing unknown fundamental causes. To put it concretely, by estimating the difference between each event's observed probability and combined probability under the assumption that events are independent, relationships among events are calculated, (some events are likely to occur simultaneously as inclusion, some events aren't likely to occur simultaneously as exclusion) and unknown events which can explain those dependencies are output. A new hypothesis will be created as a common cause under the events which are (not) likely to occur simultaneously. Discovery from simultaneity of events is also seen in [8].

To discover an interest hypothesis, we input the search expression and probabilities to HCS by regarding each keyword included in a search expression as event and the search expression as relationship rules among events. The probabilities are also given as rates of pages matched with search expression in all Web pages the search engine dealing with. Thus, HCS calculates dependencies among the words and output new hypotheses impling some words are commonly appeared. For example, if a hypothesis which implies "bat" and "glove" are likely to be used simultaneously is created, we can interpret the new hypothesis as "Because there is a baseball". Namely, if a user searched using some words which are likely to be used simultaneously, we think of those words as representing interest and the user is searching based on those interests. Those interests are generally

Table 1. The number of pages matched with search expression

Search expression	no. of pages
paper	92407
report of research	11871
technical report	937
induction	1548
paper OR report of research OR technical report	100109
induction AND (paper OR report of research OR technical report)	537
paper AND report of research	4506
paper AND technical report	541
report of research AND technical report	265

hidden because a user uses the words unconciously, or because the user doesn't use (doesn't know) proper words corresponding to the correlated words.

The user's original search expression is an expression of the user's global interest, what the user wants. However we take **interest** as a specific part, diagonal lined nodes in Fig.2, of global interest expressed in the search expression. Namely, a global interest is consist of some specific interests when a user includes some thoughts in the search expression, so we think an interest of a user is once broken and recomposed as the search expression (Fig.2). A new created hypothesis representing specific user's interest is called an **interest hypotheses** in the IHDS. In other words, IHDS discovers interest hypotheses which represent user's specific interests.

2.4   An example action of IHDS

In this section, we show an example action of IHDS. We used goo [6] as a search engine and search expression was given by boolean. The objects to be searched are 5500000 documents which were administrated by the search engine goo in December, 1997.

A user created the search expression

"induction AND (paper OR report of research OR technical report)"

to survey previous works about inductive learning and inductive reasoning and gave it to the IHDS. First, the system transformed the search expression into a tree structure like diagonal nodes and links among them in Fig.3. Then, since the nodes "paper", "report of research" and "technical report" were parallel, parent nodes expressing AND conjunctions were added for each pair (inside the right framework in Fig.3). Once the tree was constructed, probabilities for each nodes were calculated. The probability is defined as the number of pages matched with the search expression (table.1) divided by 5500000, and those probabilities were given to each node.

IHDS created a new node representing the user's hidden interest shown in Fig.3 in the lower part. This new node implies "paper" and "report of research"

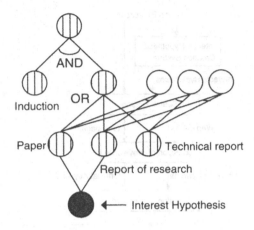

Fig. 3. An example action of IHDS

are commonly appear in Web pages and both are related to report of study, therefore we can interpret one of user's intentions to search is there.

The reason why we use HCS to estimate user's interests is the data needed is small in quantity [12]. The data needed in HCS are the probabilities of each node in the tree, and each node needs only one probability. Compared with Bayesian learning method, a Bayesian network needs large transaction data of frequency. If we intend to count simultaneity of each word in each page, we have to continue to search for exponential times of the words, so the time would be too long to wait. The search engine goo returns the number of pages matched with a search expression, thus the probability of each word is easily to obtain.

Factor analysis also discovers unknown factor, but factor analysis doesn't have a framework to discover relationship between knowledge and unknown factors.

3 Supply:Keywords for refining search expression

In this chapter, we explain the **Support System for Interests Expression**, SSIE in short, which can support user's interests expression by suppling **keywords for search** which are related to the user's interests inferred from IHDS. Fig.4 shows the flow of the system.

3.1 Acquisition of Web pages from a interest hypothesis

The purpose of SSIE is to find some keywords related to user's interests and supplies the keywords to the user. The system gets Web pages related to an interest hypothesis by making a search expression for the interest hypothesis. The search expression is given as AND conjunction of parent nodes of an interest hypothesis, because we want to get other keywords from Web pages appearing

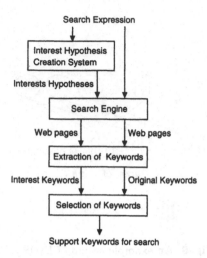

Fig. 4. Support System for User's Interests Expression

with original keywords simultaneously. For example, if an interest hypothesis is created as a child node of "paper" and "report of research" as in Fig.3, a search expression "paper AND report of research" is made and this expression is input to a search engine so that this system can get Web pages.

3.2   Keyword extraction from Web pages

In this step, the system extracts keywords from Web pages obtained by a search engine. The Web pages for keyword extraction are gotten from search expressions, original one and one about user's interest constructed in 3.1. The maximum number of pages is 50 pages ordered by the search engine.

To put this into practice, first we extract nouns from the documents (Web pages). Web pages and Keywords we used were only in Japanese through all our experiments. We extracted nouns by morpheme analysis but some inproper words like "home", "page", used frequently but no making a sense, are excluded in advance. Morpheme analysis is an algorithm that divides a sentence into words. Our system uses ChaSen [2], a Japanese morpheme analysis system. Extracted nouns, keyword candidates, are evaluated by the value given by the equation (1). In the SSIE, $value(A)$, the value of word A, is given as follows.

$$value(A) = \sum_d \log(1 + tf(A, d)) \log(1 + length(A)) \qquad (1)$$

$tf(A, d)$ is the frequency of word A in document d and $length(A)$ is the length of the word A. This value will be large if a word frequently appeares in a set of all documents obtained by a search expression. This evalation function is defined refering to tf×idf ([11]), but different from ordinary function to extract keywords. Since we want keywords which can be used for search expressions, keywords commonly used are better than keywords which are used a few times. The details of the function were decided by some experiments to output good results.

3.3   Selection of keywords to support searching

As output of SSIE, the keywords are selected from two kind of keywords as follows;

1. Original keywords: Keywords extracted from the original search expression.
2. Interest keywords: Keywords extracted from the search expression related to user's interest.

Original and interest keywords are extracted (each maximum is fixed at 16). As output of the SSEU for search support, keywords are supplied to a user from two types of keywords above (maximum is fixed at 16). At the time of output, keywords included both types of keywords are surely selected because those keywords are concerned with both original search expression and user's interests, and the others are selected by the value of equation (1).

4   Experiments:Discovering interests and supplying keywords

In this chapter, we evaluate our SSEU by some experiments. We experimented with a SUN SPARC Station 10GX(64MB) using the programming language Perl and goo [6] as the search engine. The keywords were all in Japanese.

4.1   Experiment 1:Search expression about Heian literature)

A user prepared the search expression,
"Murasakishikibu AND Seishonagon AND (literature OR novelist)"
to investigate literature in the period of Murasakishikibu and Seishonagon, novelists in the Heian period(794-1192) of Japan. Once the user input this search expression into the system, two interest hypotheses were created to imply the user's interests between "literature" and "novelist" and between "Murasakishikibu" and "Seishonagon". In other words, the user's global interest consists of two specific interests above. Keywords underlined in Fig.5 were supplied to the user.

With the keyword output, a user could see new keywords like "Hyakunin Isshu [2]" and "Izumishikibu [3]", these keywords weren't extracted by the original search expression, but should not be excluded when considering the literature of the Heian period.

---

[2] Hyakunin Isshu is a set of tanka, a Japanese poem of thirty-one syllables, by 100 people of the Heian period.
[3] Izumishikibu is a famous novelist in the Heian period.

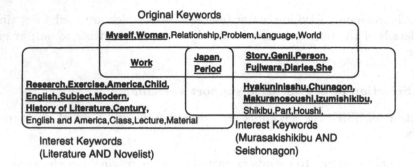

Fig. 5. Result of experiment about Heian literature

4.2   Experiment 2:Search expression about video games

A user prepared the search expression

"(Simulation OR Action OR Role playing) AND game software AND introduction"

by using above words, the user instantly hit upon, to search introduction articles of game software [4]. As a result of our system, one interest hypothesis implying a dependency between "Simulation" and "Action" was created and supplied the underlined keywords in Fig.6.

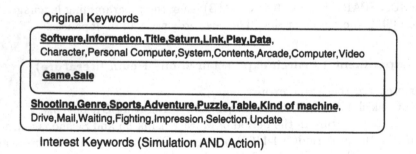

Interest Keywords (Simulation AND Action)

Fig. 6. Result of experiment about video games

By considering the user's interest in the words "Simulation" and "Action" related to sorts of games, the user could know other categories like "Shooting", "Sports", "Adventure" and "Puzzle". The user can get more pages about games using these keywords. Namely, this user has a specific interest in the genre of games, and our system caught it accurately.

---

[4] Video games are the games that game software links to the television by a exclusive hardware and one can play games on TV.

4.3   Experiment 3:Search expression about Japanese history

A user prepared the search expression
"(Aizu OR Satsuma OR Choshu) AND (Yoshinobu OR Youho OR Komei)"
to search for articles about "The Kinmon coup d'e tat [5] ", the incident in the
last part of the Edo period(1600-1867), but the user couldn't remember the word
"The Kinmon coup d'e tat". As a result of this input, two interest hypotheses
between "Aizu" and "Satsuma" and between "Yoshinobu" and "Komei" were
discovered and underlined keywords in Fig.7 were supplied as output of the
system.

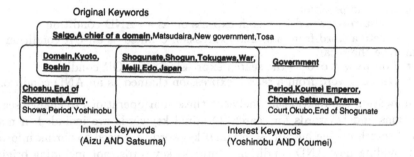

Original Keywords

Saigo,A chief of a domain,Matsudaira,New government,Tosa

**Domain,Kyoto,
Boshin**

**Shogunate,Shogun,Tokugawa,War,
Meiji,Edo,Japan**

Government

**Choshu,End of
Shogunate,Army**,
Showa,Period,Yoshinobu

**Period,Koumei Emperor,
Choshu,Satsuma,Drama**,
Court,Okubo,End of Shogunate

Interest Keywords
(Aizu AND Satsuma)

Interest Keywords
(Yoshinobu AND Koumei)

Fig. 7. Result of experiment about Japanese history

In Fig.7, the keyword "Kyoto" is supplied, since Kyoto is the place where
the Kinmon coup d'e tat took place, this keyword must be useful to get pages
about the incident. The keyword "Kyoto" is included in both types of keywords,
so that we can say that the user's interest helps focusing original keywords.

5   Refinement:Search Expression to get Web pages

5.1   Refinement of search expression to improve precision and recall

The purpose of our system is to support a user, who couldn't be content with
the results of searching Web pages using a search engine, to make next search
expression smoothly. This refinement of search expression is due to the discovery of new hypothesis because we can use keywords related to inferred user's
interests.

There are two criterion, precision and recall, to measure the accuracy of
search. Precision is the rate of pages matched with user's request of all output
pages and recall is the rate of output pages matched with user's request of all

---

[5] This was an incident that the domain of Choshu would overthrow the Edo government, but an army of the government, supported by the domain of Satsuma and
Aizu, won the fight. Yoshinobu Tokugawa was the last shogun, a general, of the Edo
government and Komei was the Emperor at the time.

pages the user wanted in the database. These criterions are described in the next equations using symbols A, the set of pages a user wants to get, and B, set of pages searched by the search engine.

$$Precision = \frac{|A \cap B|}{|B|} \qquad (2)$$

$$Recall = \frac{|A \cap B|}{|A|} \qquad (3)$$

$|\bullet|$ denotes the number of elements in the set. The following patterns improve precision and recall by refinement of the search expression.

- Improve the precision
  - Add a **new word** to search expression as an AND condition
  - Delete a word from a search expression chained as an OR condition
- Improve the recall
  - Add **new word** to the search expression as an OR condition
  - Delete a word from a search expression chained as an AND condition

We mention the relationship between these four operations to change search expressions and keywords for search. Original keywords are included in many pages of search engine output, so original keywords are used to shrink information by adding new AND conditions. Interest keywords, not including original keywords, are related to user's interests, so a user can get more wanted pages to by adding keyword as OR conditions. In addition to these, although there is a way of deleting keywords connected with AND and OR, a user decides which one to take by comparing search expression and output pages.

5.2 An example of refining search expression

We show an example of refinement for the last experiment about "The Kinmon coup d'e tat". In the first two experiments, the keywords we mentioned are generally used to improve Recall-value, and one can easily imagine the user can get more pages by adding those interest keywords. Therefore we describe improvement of Precision.

With the original search expression, 210 Web pages were searched but there was no page about the Kinmon coup d'e tat in the first 10 pages. Then the user examined the pages and noticed many unrelated pages mentioning Yoshinobu, and knew the new keyword "Kyoto" from the output of the system. Thus the user deleted Yoshinobu from search expression and added Kyoto to it as follows.

"(Aizu OR Satsuma OR Choshu) AND (Youho OR Komei) AND Kyoto"

Then the user input the refined search expression to the search engine. The result was narrowed down to 75 pages and the user could find the page titled "The Kinmon coup d'e tat" in the fifth [6]. We also calculated Precision, the pages including the words "The Kinmon coup d'e tat" as true. The value of precision changed from $21/210(= 0.1)$ to $11/75(= 0.147)$, so the precision improved and the pages were focused.

---

[6] on 1998/2/6 4:45

6   Conclusion

In this paper, we proposed a system which discovers user's hidden specific interests and which supplies keywords representing discovered interests to support search expression refinement. Discovery, we discribed in this paper, is not a learning to find general rules but a discovery of unknown events; That is to discover user's interests not expressed in search expression. Practically, the more complicated the world grows and the more various interests people have, it will be more difficult for people to express his or her interests in words. Therefore discovery system such as ours will be more effective in future.

References

1. Bruza, P. D. and Huibers, T. W. C.:"A Study of Aboutness in Information Retrieval", Artificial Intelligence Review, Vol.10, pp.381 – 407, (1996).
2. "Chasen home page":
   (URL) http://cactus.aist-nara.ac.jp/lab/nlt/chasen.html.
3. Chen, H., Schatz, B., Ng, T., Martinez, J., Kirchhoff, A. and Lin, C.: "A Parallel Computing Approach to Creating Engineering Concept Spaces for Semantic Retrieval: The Illinois Digital Library Initiative Project", IEEE Trans. Pattern Analysis and Machine Intelligence(PAMI), Vol.18, No.8, pp.771 – 782, (1996).
4. Fayyad, U., Piatetsky-S., G., and Smyth, P.: "From Data Mining to Knowledge Discovery in Databases", AI magazine, Vol.17, No.3, pp.37 – 54, (1996).
5. Furnas, G. W., Landauer, T. K., Gomez, L. M. and Dumais, S. T. : "The vocabulary problem in human-system communication", Communications of the ACM, Vol.30, No.11, pp.964 – 971, (1987).
6. A search engine:goo, (URL) http://www.goo.ne.jp/.
7. Krulwich, B.:"Learning User Interests Across Hetegenous Document Databases", AAAI Spring Symposium on Information Gathering from Heterogeneous, SS-95-08, pp.106 – 110, (1995).
8. Langley, P., Simon, H.A., et al:"Scientific Discovery - computational Explorations of the Creative Processes", MIT Press, (1997).
9. A search engine:Mondou,
   (URL) http://www.kuamp.kyoto-u.ac.jp/ labs/infocom/mondou/.
10. Ohsawa,Y. and Yachida, M.:"An Index Navigator for Understanding and Expressing User's Coherent Interest", Proc. of International Joint Conference of Artificial Intelligence, Vol.1, pp.722 – 729, (1997).
11. Salton ,G. and Buckey, C.:"Term-Weighting Approaches in Automatic Text Retrieval", Readings in Information Retrieval, pp.323 – 328, (1997).
12. Sunayama, W., Ohsawa, Y. and Yachida, M.: "Validating a Questionnaire Data Analysis System by Shrinking Iterative Questionnaires", IJCAI97, Validation, Verification & Refinement of AI Systems & Subsystems, Nagoya, pp.65 – 66, (1997).
13. A search engine:Yahoo, (URL) http://www.yahoo.co.jp/.

The Discovery of Rules from Brain Images

Hiroshi Tsukimoto and Chie Morita

Systems & Software Engineering Laboratory, Research & Development Center,
Toshiba Corporation, 70, Yanagi-cho, Saiwai-ku, Kawasaki 210, Japan

Abstract. As a result of the ongoing development of non-invasive analysis of brain function, detailed brain images can be obtained, from which the relations between brain areas and brain functions can be understood. Researchers are trying to heuristically discover the relations between brain areas and brain functions from brain images. As the relations between brain areas and brain functions are described by rules, the discovery of relations between brain areas and brain functions from brain images is the discovery of rules from brain images. The discovery of rules from brain images is a discovery of rules from pattern data, which is a new field different from the discovery of rules from symbolic data or numerical data. This paper presents an algorithm for the discovery of rules from brain images. The algorithm consists of two steps. The first step is nonparametric regression. The second step is rule extraction from the linear formula obtained by the nonparametric regression. We have to confirm that the algorithm works well for artificial data before the algorithm is applied to real data. This paper shows that the algorithm works well for artificial data.

1 Introduction

The discovery of rules from data is important, but many researchers deal with symbolic data or numerical data and few researchers deal with pattern data. We deal with the discovery of rules from pattern data, which is a new field different from the discovery of rules from symbolic data or numerical data.

There are many pattern data. Time series data are typical 1-dimensional pattern data. Images are typical 2-dimensional pattern data. There are many kinds of images such as remote-sensing images, industrial images and medical images. And there are many medical images such as brain images, lung images, and stomach images. Brain functions are the most complicated and there are a lot of unknown matters, and consequently the discovery of relations between brain areas and brain functions is a significant subject. Therefore, we deal with brain images.

Analysis of brain functions using functional magnetic resonance imaging(f-MRI), positron emission tomography(PET), magnetoencephalography(MEG) and so on are called non-invasive analysis of brain functions[3]. As a result of the ongoing development of non-invasive analysis of brain function, detailed brain images can be obtained, from which the relations between brain areas and brain functions can be understood, for example, the relation between a subarea and another subarea in the motor area and a finger movement.

Several brain areas are responsible for a brain function. Some of them are connected in series, and others are connected in parallel. Brain areas connected in series are described by "AND" and brain areas connected in parallel are described by "OR". Therefore, the relations between brain areas and brain functions are described by rules. The discovery of relations between brain areas and brain functions from brain images is the discovery of rules from brain images.

Researchers are trying to heuristically discover the rules from brain images. Several statistical methods, for example, principal component analysis, have been developed. However, the statistical methods can only present some principal areas for a brain function. They cannot discover rules. This paper presents an algorithm for the discovery of rules from brain images.

Many brain images are obtained from a subject. For example, brain images of different times or brain images of different tasks are obtained from a subject. Many brain images are obtained from many subjects. In the case of many subjects, some preprocessings are necessary, because brain shape varies among subjects. The algorithm discovers rules between brain areas and brain functions from many brain images.

Brain images can be divided into many meshes, for example, 100×100 matrix. Meshes are attributes. The activities of meshes are attribute values, and functions are classes. Then, brain images can be dealt with by supervised inductive learning. However, the usual inductive learning algorithms do not work well for brain images, because there are strong correlations between attributes(meshes). For example, a mesh has a strong correlation with the adjacent four meshes, that is, the upper mesh, the lower mesh, the left mesh and the right mesh. Usually, brain images are divided into small meshes, that is, there are many meshes(attributes). On the other hand, only a small number of samples can be obtained. Especially, when classes are some aspects of brain disorders, it is impossible to obtain many samples. For example, brain images are 100×100 matrices, that is, there are 10000 attributes, and only 100 samples have been obtained. That is, there are a small number of samples for inductive learning. Due to the two problems, that is, strong correlations between attributes and a small number of samples, the usual inductive learning algorithms[4] do not work well.

There are two solutions for the above two problems. The first one is the modification of the usual inductive learning algorithms. The other one is nonparametric regression. The modification of usual inductive learning algorithms will need a lot of effort. On the other hand, nonparametric regression has been developed for the above two problems. We will use nonparametric regression for the discovery of rules from brain images. The outputs of nonparametric regression are linear formulas, which are not rules. However, we have already developed a rule extraction algorithm from linear formulas.

The algorithm for the discovery of rules from brain images consists of two steps. The first step is nonparametric regression. The second step is rule extraction from the linear formula obtained by the nonparametric regression.

Before the algorithm is applied to real data, we have to confirm that the algorithm works well for artificial data and we also have to confirm how the algorithm works with too small samples or with too many noises. This paper shows that the algorithm works well for artificial data, while a typical inductive learning algorithm does not.

Section 2 briefly explains the outline of the discovery of rules from brain images. Section 3 briefly explains nonparametric regression. Section 4 explains the rule extraction algorithm. Section 5 describes experiments.

2   Rule extraction from brain images

Fig.1 shows a brain image. In Fig. 1, the brain image is divided into $6\times6(=36)$ meshes. Each mesh has a continuous value indicating activities. In this paper, for simplification, the value is Boolean, that is, on or off.

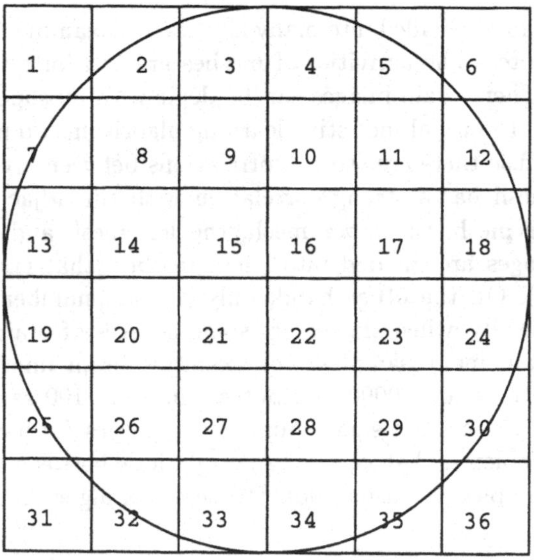

Fig. 1. Brain image

Assume that there are seven samples. Table 1 shows the data. In Table 1, 'on' means that the mesh is active and 'off' means that the mesh is inactive. Y in class shows that a certain function is on and N in class shows that a certain function is off.

Table 1. Data

sample	1	2	·36	class
S1	on	off	·off	Y
S2	on	on	·off	N
S3	off	off	·on	N
S4	off	on	·on	Y
S5	on	off	·off	N
S6	off	on	·on	N
S7	off	off	·on	Y

As stated in the introduction, attributes(meshes) in pattern data have strong correlations between adjacent meshes. In Fig. 1, for example, mesh 8 has a strong correlation with adjacent meshes, that is, mesh 2, mesh 7, mesh 9, and mesh 14.

3 Nonparametric regression

First, for simplification, the 1-dimensional case is explained[1].

3.1 1-dimensional nonparametric regression

Nonparametric regression is as follows: Let y stand for a dependent variable and t stand for an independent variable and let $t_j (j = 1, .., m)$ stand for measured values of t. Then, the regression formula is as follows:

$$y = \sum a_j t_j + e(j = 1, .., m),$$

where a_j are real numbers and e is a zero-mean random variable. When there are n measured values of y

$$y_i = \sum a_j t_{ij} + e_i (i = 1, .., n)$$

In usual linear regression, error is minimized, while, in nonparametric regression, error plus continuity or smoothness is minimized. When continuity is added to error, the evaluation value is as follows:

$$1/n \sum_{i=1}^{n} (y_i - \hat{y_i})^2 + \lambda \sum_{i=1}^{n} (\hat{y}_{i+1} - \hat{y_i})^2,$$

where $\hat{y}$ is an estimated value. The second term in the above formula is the difference of first order between the adjacent dependent variables, that is, the continuity of the dependent variable. λ is the coefficient of continuity. When λ is 0, the evaluation value consists of only the first term, that is, error, which means the usual regression. When λ is very big, the evaluation value consists of only the second term, that is, continuity, which means that the error is ignored and the solution $\hat{y}$ is a constant.

The above evaluation value is effective when the dependent variable has continuity, for example, the measured values of the dependent variable are adjacent in space or in time. Otherwise, the above evaluation value is not effective. When the dependent variable does not have continuity, the continuity of coefficients a_js is effective, which means that adjacent measured values of the independent variable have continuity in the influence over the dependent variables[2]. The evaluation value is as follows:

$$1/n \sum_{i=1}^{n} (y_i - \hat{y}_i)^2 + \lambda \sum_{j=1}^{m} (a_{j+1} - a_j)^2$$

When λ is fixed, the above formula is the function of a_i ($\hat{y}_i$ is the function of a_i). Therefore, a_is are determined by minimizing the evaluation value, and the optimal value of λ is determined by cross validation.

3.2   Calculation

Let $\mathbf{X}$ stand for $n \times m$ matrix. Let t_{ij} be an element of $\mathbf{X}$. Let $\mathbf{y}$ stand for a vector consisting of y_i. $m \times m$ matrix $\mathbf{C}$ is as follows:

$$\mathbf{C} = \begin{pmatrix} 1 & -1 & & \\ -1 & 2 & -1 & \\ & -1 & 2 & -1 \\ & & & \ddots \end{pmatrix}$$

Cross validation CV is as follows:

$$CV = n\tilde{\mathbf{y}}^t \tilde{\mathbf{y}}$$

$$\tilde{\mathbf{y}} = \mathbf{Diag}(\mathbf{I} - \mathbf{A})^{-1}(\mathbf{I} - \mathbf{A})\mathbf{y}$$

$$\mathbf{A} = \mathbf{X}(\mathbf{X}^t\mathbf{X} + (\mathbf{n} - 1)\lambda\mathbf{C})^{-1}\mathbf{X}^t,$$

where $\mathbf{DiagA}$ is a diagonal matrix whose diagonal components are $\mathbf{A}$'s diagonal components. The coefficients $\hat{\mathbf{a}}$ are as follows:

$$\hat{\mathbf{a}} = (\mathbf{X}^t\mathbf{X} + \mathbf{n}\lambda_o\mathbf{C})^{-1}\mathbf{X}^t\mathbf{y},$$

where λ_o is the optimal λ determined by cross validation.

3.3   2-dimensional nonparametric regression

In 2-dimensional nonparametric regression, the evaluation value for the continuity of coefficients a_{ij} is modified. In 1 dimension, there are two adjacent measured values, while, in 2 dimensions, there are four adjacent measured values. The evaluation value for the continuity of coefficients is not

$$(a_{i+1} - a_i)^2,$$

but the differences of first order between a mesh and the four adjacent meshes in the image. For example, in the case of mesh 8 in Fig.1, the adjacent meshes are mesh 2, mesh 7, mesh 9, and mesh 14, and the evaluation value is as follows:

$$(a_8 - a_2)^2 + (a_8 - a_7)^2 + (a_9 - a_8)^2 + (a_{14} - a_8)^2.$$

Consequently, in the case of 2 dimensions, $\mathbf{C}$ is modified in the way described above. When continuity is evaluated, four adjacent meshes are considered, while smoothness is evaluated, eight adjacent meshes are considered, for example, in the case of mesh 8 in Fig.1, the adjacent meshes are mesh 1, mesh 2, mesh 3, mesh 7,mesh 9, mesh 13, mesh 14 and mesh 15.

4   Rule extraction

4.1   The basic method

First, the basic method in the discrete domain is explained. The basic method is that linear formulas are approximated by Boolean functions. Note that the dependent variables in nonparametric regression should be normalized to [0,1]

Let (f_i) be the values of a linear formula. Let $(g_i)(g_i = 0$ or $1)$ be the values of Boolean functions. The basic method is as follows:

$$g_i = \begin{cases} 1(f_i \geq 0.5), \\ 0(f_i < 0.5). \end{cases}$$

This method minimizes Euclidean distance.

Fig. 2 shows a case of two variables. Crosses stand for the values of a linear formula and circles stand for the values of a Boolean function. $00, 01, 10$ and 11 stand for the domains, for example, 00 stands for $x = 0, y = 0$.
In this case, the values of the Boolean function $g(x, y)$ are as follows:

$$g(0,0) = 1, g(0,1) = 1, g(1,0) = 0, g(1,1) = 0.$$

Generally, let $g(x_1, ..., x_n)$ stand for a Boolean function, and let $g_i (i = 1, ..., 2^n)$ stand for values of a Boolean function, then the Boolean function is represented by the following formula:

$$g(x_1, ..., x_n) = \sum_{i=1}^{2^n} g_i a_i,$$

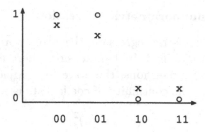

Fig. 2. Approximation

where $\sum$ is disjunction, and a_i is the atom corresponding to g_i, that is,

$$a_i = \prod_{j=1}^{n} e(x_j) \ (i = 1, ..., 2^n),$$

where

$$e(x_j) = \begin{cases} \overline{x_j}(e_j = 0), \\ x_j(e_j = 1), \end{cases}$$

where $\prod$ stands for conjunction, $\bar{x}$ stands for the negation of x, and e_j is the substitution for x_j, that is, $e_j = 0$ or 1. The above formula can be easily verified.

Therefore, in the case of Fig. 2, the Boolean function is as follows:
$g(x,y) = g(0,0)\bar{x}\bar{y} + g(0,1)\bar{x}y + g(1,0)x\bar{y} + g(1,1)xy,$
$g(x,y) = 1\bar{x}\bar{y} + 1\bar{x}y + 0x\bar{y} + 0xy,$
$g(x,y) = \bar{x}\bar{y} + \bar{x}y,$
$g(x,y) = \bar{x}.$

4.2   A polynomial algorithm

The method presented in Section 4.1 is exponential in computational complexity, and so the method is unrealistic when variables are many. A polynomial algorithm is presented in this subsection[5],[7] .

Expansion of linear functions by the atoms of Boolean algebra of Boolean functions
    Let

$$p_1 x_1 + ... + p_n x_n + p_{n+1}$$

stand for a linear function. Let

$$f_1 x_1 \cdots x_n + f_2 x_1 \cdots \overline{x_n} + ... + f_{2^n} \overline{x_1} \cdots \overline{x_n}$$

stand for the expansion by the atoms of Boolean algebra. Then,
$p_1 x_1 + ... + p_n x_n + p_{n+1} = f_1 x_1 \cdots x_n + f_2 x_1 \cdots \overline{x_n} + ... + f_{2^n} \overline{x_1} \cdots \overline{x_n}.$
f_i's are as follows[7]:

$$f_1 = p_1 + ... + p_n + p_{n+1},$$
$$f_2 = p_1 + ... + p_{n-1} + p_{n+1},$$

...

$$f_{2^{n-1}} = p_1 + p_{n+1},$$
$$f_{2^{n-1}+1} = p_2 + p_3 + ... + p_{n-1} + p_n + p_{n+1},$$

...

$$f_{2^n-1} = p_n + p_{n+1},$$
$$f_{2^n} = p_{n+1}.$$

The condition that $x_{i_1} \cdots x_{i_k} \overline{x}_{i_{k+1}} \cdots \overline{x}_{i_l}$ exists in the Boolean function after approximation.

The condition that $x_{i_1} \cdots x_{i_k} \overline{x}_{i_{k+1}} \cdots \overline{x}_{i_l}$ exists in the Boolean function after approximation is as follows:

$$\sum_{i_1}^{i_k} p_j + p_{n+1} + \sum_{1 \le j \le n, j \ne i_1,..,i_l, p_j < 0} p_j \ge 0.5.$$

Proof : Consider the existence condition of x_1 in the Boolean function after approximation. For simplification, this condition is called the existence condition. Because

$$x_1 = x_1 x_2 \cdots x_n \vee x_1 x_2 \cdots \overline{x}_n \vee ... \vee x_1 \overline{x}_2 \cdots \overline{x}_n,$$

the existence of x_1 equals the existence of the following terms:

$$x_1 x_2 \cdots x_n ,$$
$$x_1 x_2 \cdots \overline{x}_n,$$

...

$$x_1 \overline{x}_2 \cdots \overline{x}_n.$$

The existence of the above terms means that all coefficients of these terms $f_1, f_2, ..., f_{2^{n-1}}$ are greater than or equal to 0.5 (See 4.1). That is,

$MIN\{f_i\} \ge 0.5 (1 \le i \le 2^{n-1}).$

Because f_i's $(1 \le i \le 2^{n-1})$ are

$$f_1 = p_1 + ... + p_n + p_{n+1},$$
$$f_2 = p_1 + ... + p_{n-1} + p_{n+1},$$

...

$f_{2^{n-1}} = p_1 + p_{n+1},$

each $f_i (1 \le i \le 2^{n-1})$ contains p_1. If each p_j is non-negative, $f_{2^{n-1}} (= p_1 + p_{n+1})$ is the minimum because the other f_i's contain other p_j's, and therefore the other f_i's are greater than or equal to $f_{2^{n-1}} (= p_1 + p_{n+1})$. Generally, since each p_j is not necessarily non-negative, the $MIN\{f_i\}$ is f_i which contains all negative p_j. That is,

$MIN\{f_i\} = p_1 + p_{n+1} + \sum_{1 \le j \le n, j \ne 1, p_j < 0} p_j,$

which necessarily exists in $f_i (1 \le i \le 2^{n-1})$, because $f_i (1 \le i \le 2^{n-1})$ is

$p_1 + p_{n+1} +$ (arbitrary sum of $p_j (2 \le j \le n)$).

From the above arguments, the existence condition of x_1, $MIN\{f_i\} \ge 0.5$, is as follows:

$p_1 + p_{n+1} + \sum_{1 \le j \le n, j \ne 1, p_j < 0} p_j \ge 0.5.$

Since $p_1x_1 + ... + p_nx_n + p_{n+1}$ is symmetric for x_i, the above formula holds for other variables; that is, the existence condition of x_i is

$p_i + p_{n+1} + \sum_{1 \leq j \leq n, j \neq i, p_j < 0} p_j \geq 0.5$.

Similar discussions hold for $\overline{x}_i$, and so we have the following formula:

$p_{n+1} + \sum_{1 \leq j \leq n, j \neq i, p_j < 0} p_j \geq 0.5$.

Similar discussions hold for higher order terms $x_{i_1} \cdots x_{i_k} \overline{x}_{i_{k+1}} \cdots \overline{x}_{i_l}$, and so we have the following formula:

$\sum_{i_1}^{i_k} p_j + p_{n+1} + \sum_{1 \leq j \leq n, j \neq i_1, ..., i_l, p_j < 0} p_j \geq 0.5$.

Generation of DNF formulas

The algorithm generates terms using the above formula from the lowest order up to a certain order. A DNF formula can be generated by taking the disjunction of the terms generated by the above formula. A term whose existence has been confirmed does not need to be rechecked in higher order terms. For example, if the existence of x is confirmed, then it also implies the existence of xy, xz ,..., because $x = x \vee xy \vee xz$; hence, it is unnecessary to check the existence of $xy, xz, ...$ As can be seen from the above discussion, the generation method of DNF formulas includes reductions such as $xy \vee xz = x$. Let

$f = 0.65x_1 + 0.23x_2 + 0.15x_3 + 0.20x_4 + 0.02x_5$

be the linear function. The existence condition of $x_{i_1} \cdots x_{i_k} \overline{x}_{i_{k+1}} \cdots \overline{x}_{i_l}$ is

$\sum_{i_1}^{i_k} p_j + p_{n+1} + \sum_{1 \leq j \leq n, j \neq i_1, ..., i_l, p_j < 0} p_j \geq 0.5$.

In this case, each p_i is positive and $p_{n+1} = 0$; therefore the above formula can be simplified to

$\sum_{i_1}^{i_k} p_j \geq 0.5$.

For x_i, the existence condition is $p_i \geq 0.5$.

For $i = 1, 2, 3, 4, 5, p_1 \geq 0.5$,

therefore x_1 exists.

For x_ix_j, the existence condition is $p_i + p_j \geq 0.5$.

For $i, j = 2, 3, 4, 5, p_i + p_j < 0.5$,

therefore no x_ix_j exists.

For $x_ix_jx_k$, the existence condition is $p_i + p_j + p_k \geq 0.5$.

For $i, j, k = 2, 3, 4, 5, p_2 + p_3 + p_4 \geq 0.5$,

therefore $x_2x_3x_4$ exists.

Because higher order terms cannot be generated from x_5, the algorithm stops. Therefore, x_1 and $x_2x_3x_4$ exist and the DNF formula is the disjunction of these terms, that is,

$x_1 \vee x_2x_3x_4$.

Computational complexity of the polynomial algorithm

The computational complexity of generating the mth order terms is polynomial of nCm, that is, polynomial of n. Therefore, the computational complexity of generating DNF formulas from linear formulas is polynomial of n. Usual generations will be terminated up to a low order, because understandable propositions are desired. The error is small. The detailed explanation is found in [5], [7]

4.3   Extension to the continuous domain

Continuous domains can be normalized to [0,1] domains by some normalization method. So only [0,1] domains have to be discussed. First, we have to present a system of qualitative expressions corresponding to Boolean functions, in the [0,1] domain. The expression system is generated by direct proportion, reverse proportion, conjunction and disjunction. The direct proportion is $y = x$. The inverse proportion is $y = 1 - x$, which is a little different from the conventional one ($y = -x$), because $y = 1 - x$ is the natural extension of the negation in Boolean functions. The conjunction and disjunction will be also obtained by a natural extension. The functions generated by direct proportion, reverse proportion, conjunction and disjunction are called continuous Boolean functions, because they satisfy the axioms of Boolean algebra. For details, refer to [6]. In the domain [0,1], linear formulas are approximated by continuous Boolean functions. The basic method is the same as in the domain $\{0, 1\}$[5].

5   Experiments with artificial data

5.1   Experimental Conditions

The experimental conditions for images are as follows.

matrix :30×30 (900 (meshes)attributes)
attribute value :active/inactive
class(dependent variable) :positive/negative
sample number :42, which is small compared with 900 meshes(attributes).
rule : $(32 \lor 33) \land (100 \lor 101 \lor 102) \rightarrow$ positive
>   The numbers in the above rule mean that the meshes of the numbers are active. For example, the above rule can be interpreted as " If (mesh 32 is active or mesh 33 is active) and (mesh 100 is active or mesh 101 is active or mesh 102 is active), then the class value is positive, for example, some function is on.

The experimental conditions for sample generation are as follows.

positive samples : 21 positive samples are generated based on the above rule with noises. Noises are generated as follows: 100(or 200) numbers are selected at random from 1-900. 100 (or 200)meshes of the numbers are active. That is, the number of total active meshes is nearly 100 or 200.
negative samples : 21 negative samples are generated at random with the constraint that the number of total active meshes is 100 or 200. When a negative sample satisfies the above rule , the negative sample is deleted.
correlation between adjacent meshes : The meshes adjacent to active meshes are active with a probability. The probabilities are 0, 0.3,0.7, or 1.0
training sample and test sample : 42 samples are used for the training and other 42 samples are used for the test.
repetition : The above experiments are repeated 10 times and the accuracy is the average of 10 accuracies.

5.2   The results of C4.5

We applied C4.5[4] to the data in the preceding subsection. The accuracies of trees before pruning are shown in Table 2. t.a.m.n in Table 2 stands for total active mesh number. The probabilities in Table 2 means the probabilities of correlations between attributes, which was mentioned in the preceding subsection.

Table 2. C4.5 results

probability	0	0.3	0.7	1.0
t.a.m.n. 100	0.83	0.86	0.81	0.80
t.a.m.n. 200	0.80	0.82	0.75	0.77

The accuracies do not mainly depend on the number of the correlation probabilities, but depend on the number of noise meshes. This result shows that C4.5 does not work well, because C4.5 selects attributes based on the amount of information.

5.3   The results of the algorithm

This subsection presents the results of the algorithm. Table 3 shows the accuracies of the results with 100 noise meshes. Table 4 shows the accuracies of the results with 200 noise meshes

Table 3. Results(100)

probability	0	0.3	0.7	1.0
accuracy(linear formula)	0.87	0.92	0.93	0.97
accuracy(rule)	0.85	0.90	0.92	0.96

Table 4. Results(200)

probability	0	0.3	0.7	1.0
accuracy(linear formula)	0.84	0.90	0.92	0.94
accuracy(rule)	0.82	0.89	0.91	0.94

For example, when the probability is 0.7 and 100 noise meshes, the following rule has been obtained:

$$(32 \vee 33) \wedge (91 \vee 100 \vee 102) \to positive,$$

and $\lambda \simeq 10^6$.

The results show the following:

1. When the probabilities are 0, the accuracies are almost the same as the accuracies of C4.5. This result means that the algorithm does not work

well when the probabilities are 0, that is, there is no correlation between attributes.

2. When the probabilities are large, the accuracies are good. This result means that the algorithm works well when the probabilities are large, that is, there are strong correlations between attributes.

3. Accuracies of 100 noise meshes are slightly better than accuracies of 200 noise meshes. This result means that the accuracies of the algorithm is deteriorated when noises are large.

Consequently, the algorithm works well for the discovery of rules from artificial data. In the experiment, sample number is 42 and mesh number is 900. Since it is difficult to obtain samples, and small meshes are desired, experiments with a smaller number of samples are needed. In the experiment, the number of positive samples equals the number of negative samples. When classes are some aspects of brain disorders, it is difficult to obtain as many negative samples as positive samples, and so experiments with a small number of negative samples are needed.

6 Conclusions

This paper has presented an algorithm for the discovery of rules from brain images. The algorithm consists of two steps. The first step is nonparametric regression. The second step is rule extraction from the linear formula obtained by the nonparametric regression. We have to confirm that the algorithm works well for artificial data before the algorithm is applied to real data. This paper has shown that the algorithm works well for artificial data. Future work will include the application of the algorithm to real data.

References

1. Eubank, R.L.: Spline Smoothing and Nonparametric Regression. Marcel Dekker, New York (1988)
2. Miwa,T.:private communication (1998)
3. Posner, M.I., Raichle,M.E.: Images of Mind. W H Freeman & Co (1997)
4. Quinlan, J.R.: Induction of decision tree. Machine Learning 1 (1986) 81-106
5. Tsukimoto, H.: The discovery of logical propositions in numerical data. AAAI'94 Workshop on Knowledge Discovery in Databases (1994) 205-216
6. Tsukimoto, H.: On continuously valued logical functions satisfying all axioms of classical logic. Systems and Computers in Japan, Vol.25, No.12. (1994) 33-41
7. Tsukimoto,H., Morita,C., Shimogori,N.: An Inductive Learning Algorithm Based on Regression Analysis. Systems and Computers in Japan, Vol.28, No.3. (1997) 62-70

Instance Guided Rule Induction

Nobuhiro Yugami, Yuiko Ohta, and Seishi Okamoto

Fujitsu Laboratories, 2-2-1 Momochihama, Fukuoka, 814-8588, Japan

Abstract. This paper proposes a new supervised induction algorithm, IGR, that uses each training instances as a guide of rule induction. IGR learns a set of if-then rules by inducing a pseudo-optimun classification rule for each training instance. IGR weighs the induced rules by using the number of trianing instances covered by them and classifies new instances by majority voting with the weights. Experimental results with twenty datasets in UCI repository show IGR can induce more accurate classification rules than existing learning algorithms such as C4.5, AQ and LazyDT. The experiments also show that IGR does not generate too many rules even if it is applied to large problems.

1   Introduction

Supervised learning is one of the most important area in knowledge discovery in databases. Traditional supervised learning algorithms such as C4.5[15], CART[3] and AQ[14] induce a single hypothesis that works well on average for the whole of instance space. However, goodness (classification accuracy, or other criteria) for the whole instance space does not mean goodness for a particular part of the instance space. There may be a specific hypothesis that works quite well in a small part of the instance space but is poor outside the part. This suggests that it may be possible to achieve more accurate classification by combining such specific hypotheses.

LazyDT[9] is an interesting approach to learn such a specific hypothesis. It learns one if-then rule for classifying one new instance after the instance is given. Figure 1 shows a pseudo-code of LazyDT. When a target instance (the instance to be classified) is given, LazyDT tries to induce the best rule for it. The best rule means the rule covers the instance and covers maximum number of training instances. It starts from a rule with no precondition and continues to specialize a body of a rule until the body permits training instances belonging to a certain class. To specialize a rule, LazyDT estimates a test with a variant of information gain(see [9] in detail). LazyDT tries backtracking to search a better rule below the tests with sufficiently high estimation score (step 3 in figure 1). The experimental results showed that LazyDT outperformed C4.5 in many problems.

The most serious difficulty of LazyDT is its time complexity for classification. LazyDT starts learning after the test instance is given. Hence, its complexity for one classification equals to the complexity for inducing one rule. This complexity is not practical for large problems with many training instances represented by

procedure LazyDT(*target*, *Training*)
    inputs: a test instance, *target*, and a training set, *Training*.
    outputs: a class of *target* and its support size.

1.If all training instances in *Training* belong to a certain class c,
    then return the class c with support size $|Training|$.
2.Estimate all tests that accepts *target* by using weighted information gain.
3.**For each** test t whose estimation is higher than 90% of the highest gain
    call LazyDT(*target*, $\{i|i \in Training \land i$ satisfies $t\}$).
4.**Return** the result of recursive call with the largest support size.

Fig. 1. Pseudo-code of LazyDT

many attributes. Another difficulty of LazyDT is a lack of ability of pruning. As shown in figure 1, LazyDT specializes a rule until it covers only training instances belonging to a particular class and tends to generate over-specialized rules. This problem becomes serious in noisy problems.

This paper proposes a new learning algorithm, IGR, that resolves above difficulties of LazyDT and improves classification performance. In the following sections, we describe IGR in detail and show its ability by comparing with LazyDT, C4.5, AQ and TDDT-op[13].

2 IGR

2.1 Overview of IGR

This section describes IGR (Instance Guided Rule induction) and discusses how to avoid the difficulties of LazyDT pointed out in introduction. The first difficulty, slow classification, is essentially caused by lazy learning and the only solution is non-lazy learning that finishes learning before test instances are given. The problem is, then how to generate the best rule for a specific test instance, as LazyDT does, before the test instance is given. The basic idea of IGR for this problem is very simple. IGR learns the best rule for each training instance, instead of the test instance. If the training set is sufficiently large, we can expect that there exist training instances that are quite similar to any test instance and the best rules for them will also work well for the test instance.

IGR applies majority voting to avoid the second difficulty of LazyDT, a lack of pruning. When classifying a new instance, IGR applies all rules that cover the new instance and returns the majority class in the rules. Even if some rules are over-specialized and imply a wrong class(es), majority voting can decrease the effects of them. We will discuss the majority voting in IGR at the end of this section.

The over specialization not only leads wrong classification rules but also tends to generate "holes" of the instance space that are not covered by any rule and

procedure IGR(*Body*, *Targets*, *Training*)
    inputs: a conjunction of tests, *Body*, a set of target instance, *Targets*,
        and a training set, *Training*.
    output: a set of classification rules, *Rules*.

0.*Rules* := $\emptyset$.
1.If all training instances in *Training* belong to a certain class c,
    then return a rule set with one rule {If *Body* then class=c}.
2.Generate tests for specializing *Body* by grouping attribute values.
3.Estimate all tests by information gain.
4.**For each** target instance i in *Targets*,
    Selects the best test $t(i)$ with highest estimation for the class of i from
    the tests that accept i.
5.$T(t) := \{i|i \in Targets \wedge t(i) = t\}$.
6.**For each** test t s.t. $T(t) \neq \emptyset$
    $Rules = Rules \cup IGR(Body \wedge t, T(t), \{j|j \in Training \wedge j \text{ satisfies } t\})$.
7.**Return** *Rules*.

Fig. 2. Pseudo-code of IGR

the test instances in them can not be classified. To avoid this, IGR generates more rules than traditional learning algorithms such as C4.5 and AQ. Then, the rule set learned by IGR can cover most part of the instance space even if each rule is too specific and covers only a small part of the instance space.

The simplest approach to induce the best rule for each training instance is to apply LazyDT to all training instances. We don't adopt this approach because LazyDT is too slow to be applied to all training instances. There are three differences between IGR and LazyDT. First, IGR tries to induce rules for many target instances at once. Second, IGR does not try backtracking to search a better rule for a particular instance. It applies pure greedy search to induce a rule. These two differences are for reducing learning time. Finally, IGR can fix the conclusions(heads) of rules before learning because the conclusion of the rule for a particular training instance is the class of the instance. This makes it easy to select good tests for specialization.

Figure 2 shows a pseudo-code of IGR. IGR is a recursive algorithm and is called first as IGR($true, TSet, TSet$), where $TSet$ is the whole training set. IGR starts from a rule with no precondition and specializes it until the rule covers only training instances belonging to a certain class. To accelerate learning, IGR tries to specialize rules for many training instances at once. In figure 2, *Body* is a body of current rule and *Training* is a set of training instances that satisfy all tests in *Body*. *Targets* is a subset of *Training* and is a set of training instances for which the best rules should be induced by specializing *Body*. If all instances in *Training* belong to a particular class c, then IGR generates a rule "if *Body* then class c" as the best rule for the target instances in *Targets*. Otherwise,

IGR selects the best test for each target instance in *Targets*. For each test selected for at least one target instance, IGR specializes *Body* by the test and calls IGR recursively. For example, when IGR is called first and both of *Targets* and *Training* are the whole training set, IGR selects the first tests of bodies of best rules for all training instances. IGR then divides training instances by the selected test because the test estimation of the next tests can be done at once for the rules with the same body.

The number of rules induced by IGR is at most the number of training instances and IGR usually induces smaller number of rules because plural instances share one rule as their best rule. This happens when the condition of step 1 in figure 2 is satisfied and *Targets* involves plual target instances.

IGR is similar to AQ family[14] that also learns one rule for one seed instance. The most important difference between them is that IGR generates redundant rules to classify training instances. After learning one rule, AQ removes all positive instances covered by the rule and tries to induce rules for the remaining positive instances. Instead, IGR does not remove the instances covered by the rule and learns the best rules for all training instances. This means that IGR learns more rules than AQ. To classify training instances correctly, the rule set from AQ is sufficient and that of IGR involves redundant rules. However, this redundancy improves classification performance for new, unkonwn instances.

2.2 Test Selection

For general-to-special rule induction, how to select tests for specializing bodies of rules is one of key issues that decide classification performance of the induced rules. In the following discussion, we assume all attributes are nominal and numeric attributes are discretized. To specialize a rule, the simplest and the most traditional way is to restrict a value of a certain attribute to one of its possible values. However, this restriction tends to generate too specific rules when each attribute has many possible values[7]. Instead, IGR divides values of one attribute into two groups and uses a test that accepts values in one group. Note that the purpose of IGR is to generate a good rule for each training instance and the conclusion of the rule is the class of the target instance. To specialize a rule for a target instance belonging to a class c, IGR divides values of an attribute att whether the restriction of the attribute value increases the probability of c or not, and generates following two groups.

$$G_{pos}(c, att) = \{v | v \in Domain(att) \land P(c|(att = v) \land Body) > p(c|Body)\}$$

$$G_{neg}(c, att) = \{v | v \in Domain(att) \land P(c|(att = v) \land Body) \leq p(c|Body)\}$$

Then, IGR generates at most $2KL$ tests for specialization, where K is the number of classes and L is the number of attributes.

IGR uses a modified information gain as a test selection criterion. Let c and v be the class label and the value for attribute att of the target instance. The estimation of the test restricting values of attribute att to $G_{pos}(c, att)$ is

$$\begin{cases} IG(c, att) & \text{if } D(att = v) \geq AD(att, G_{pos}(c, att)) \\ IG(c, att) \times \frac{D(att=v)}{AD(att, G_{pos}(c, att))} & \text{if } D(att = v) < AD(att, G_{pos}(c, att)) \end{cases}$$

and the estimation for restricting values to $G_{neg}(c, att)$ is

$$\begin{cases} -IG(c, att) & \text{if } D(att = v) \geq AD(att, G_{neg}(c, att)) \\ -IG(c, att) \times \frac{D(att=v)}{AD(att, G_{neg}(c, att))} & \text{if } D(att = v) < AD(att, G_{neg}(c, att)) \end{cases}$$

where $IG(c, att)$ is the information gain for binary class, c or $\neg c$, gained by dividing $Training$ into two sets whether the value of att belongs to $G_{pos}(att, c)$ or $G_{neg}(att, c)$. $D(att = v)$ is the number of instances in $Training$ whose values for attribute att are v and $AD(att, G)$ be the average of $D(att = w)$ over attribute value $w \in G$. The estimation becomes positive (negative) if the test increases (decreases) the probability of class c. As a result, a test $att \in G_{neg}(c, att)$ is selected only when there is no attribute a for which the target instance's value belongs to $G_{pos}(c, a)$. The above definition also means that the estimation of the test depends on the target instance. The estimation becomes small when only a small number of training instances share the value of the test attribute att with the target instance. This is because the information gain is mainly caused by other values of att.

2.3   Rule weighting for majority voting

This subsection describes how IGR classifies a new instance. IGR uses a set of if-then rules for classification and causes two common problems, how to classify a new instance if the instance does not satisfy any rule and how to classify if the instance satisfies plural rules. In the former case, IGR returns a majority class in the whole training set. In the later case, IGR selects a class by majority voting with the following weight of rules.

$$Weight(r) = \sum_{i \in I(r)} \frac{1}{|n(i)|}$$

where $I(r)$ is a set of instances that satisfy a rule r and $n(i)$ be the number of rules that cover the training instance i. This definition means that the total contribution of one training instance to weights of rules is constant (1).

3   Experiment

In this section, we compare IGR with C4.5, AQ, LazyDT and TDDT-op[13] by using twenty datasets in UCI machine learning repository. AQ has many variants to induce a rule for one seed instance. To show how IGR improves classification performance by learning more rules than AQ, we applied the same tests for

Table 1. Charactaristics of datasets.

	classes	attributes	instances		classes	attribures	instances
breast	2	9	638	pima	2	8	768
cleveland	5	13	297	promoters	2	57	106
crx	2	15	653	satellite	7	36	6,435
dna	3	60	3,186	segment	7	19	2,310
glass	7	10	214	soy-large	19	35	307
hayes-roth	4	4	160	tic-tac-toe	2	9	958
iris	3	4	351	vehicle	4	18	846
krkp	2	36	3,196	voting	2	16	435
monks-1	2	6	432	waveform	3	21	5,000
monks-2	2	6	432	wine	3	13	178

specialization and the same test selection criterion used in IGR to AQ. TDDT-op is a multiple hypotheses learner based on decision trees and expands each internal node in decision trees with plural attributes. It has two parameters to control how many attributes are used to expand one internal node. We use the values for them in [13] that select seven test attributes at the top two levels, root node and its child nodes, and select one test attribute bellow them. These parameters generate about 50 times larger trees than traditional decision tree learners such as C4.5 and its classification performance is comparable to bagging[2] with 50 trials of C4.5.

Table 1 shows the domain characteristics of the datasets used in our experiments. IGR and LazyDT can not deal with numeric attributes directly and we discretized numeric attributes to five or ten intervals with constant population (five intervals for datasets with less than 500 instances and ten intervals for others). AQ also learned from discretized instances because it used the same tests for specialization with IGR as we described above. C4.5 and TDDT-op can deal with numeric attributes and they learned from original, non-discretized instances. We also removed instances with unknown attribute values because the current implimentation of IGR can not deal with them. To compare classification performance, we used 5-fold cross validation and averaged the results over 10 runs for each dataset, i.e. each algorithm was applied 50 times to each dataset.

Table 2 and 3 report the results of the experiments. Table 2 shows average classification accuracy and one standard deviation. Blanks for LazyDT means the dataset was too large to apply LazyDT. For example, LazyDT required more than ten minutes to classify one instance in satellite dataset (we used a sun workstation with 300MHz ultra-sparc) even if we implemented caching[9] for speeding up LazyDT. Table 3 shows the average number of rules induced by each algorithm except LazyDT that learns one rule for one test instance. For C4.5 and TDDT-op, we reported the number of leaf nodes as the number of rules.

IGR outperformed C4.5 with respect to classification accuracy in sixteen datasets and the difference was significant in eleven datasets (average of accu-

Table 2. Average accuracy and one standard deviation.

	C4.5	AQ	LazyDT	TDDT-op	IGR
breast	95.40±0.50	82.94±7.13	90.41±0.47	96.87±0.03	96.30±0.17
cleveland	47.41±1.55	55.03±0.88	56.16±1.01	55.17±1.47	55.91±1.46
crx	84.01±1.10	84.82±0.89	84.43±0.44	86.45±0.85	86.63±0.49
dna	93.71±0.23	83.52±3.38	–	94.63±0.29	96.07±0.29
glass	65.05±2.18	57.71±4.19	57.47±2.02	68.04±1.87	65.98±2.54
hayes-roth	74.31±2.35	70.06±3.80	77.00±2.60	72.00±2.41	83.69±3.44
iris	94.40±1.23	68.00±5.88	90.60±1.17	92.00±1.23	91.60±1.41
krkp	99.43±0.09	99.57±0.10	98.25±0.20	99.16±0.08	99.16±0.16
monks-1	91.90±1.30	85.67±9.20	92.94±2.26	95.95±1.83	98.03±0.84
monks-2	67.13±0.00	73.43±2.07	85.88±1.93	67.13±0.00	76.32±1.67
pima	71.58±1.64	71.05±1.20	69.02±0.92	73.88±0.55	74.15±0.91
promoters	77.17±5.14	51.70±8.33	55.66±1.83	91.60±2.83	88.49±2.88
satellite	85.23±0.31	73.51±0.58	–	88.61±0.26	86.23±0.14
segment	96.16±0.39	70.87±4.82	–	96.87±0.30	95.92±0.29
soy-large	88.73±1.39	69.64±2.72	79.45±1.73	88.05±1.25	88.93±1.32
tic-tac-toe	85.50±1.03	90.97±2.35	94.24±0.61	96.41±0.52	97.12±0.41
vehicle	71.26±1.15	56.28±2.79	55.58±0.91	71.02±0.76	70.14±0.99
voting	94.80±0.44	92.11±3.50	94.28±1.18	95.47±0.50	95.33±0.57
waveform	72.82±0.40	67.67±1.33	–	81.30±0.45	82.83±0.30
wine	93.60±1.16	60.34±9.45	71.35±1.76	95.06±1.42	95.90±0.99

racy difference was greater than its one standard deviation). Especially, IGR's accuracy was more than 8% higher than that of C4.5 in five datasets. IGR lost six datasets and the difference of accuracy was at most 3%. Consequently, IGR could induce much more accurate rules than C4.5.

By comparing with LazyDT, IGR lost only two datasets, monks-2 and cleveland. LazyDT performed poorly in small datasets such as promoters and wine. In such a small dataset, top-down (general to special) induction such as LazyDT and IGR, tends to induce too general rules. LazyDT classifies an instance with only one rule and is strongly affected by this difficulty. Instead, IGR can use plural rules for one classification and this ability decreases the effects of too general rules.

IGR also outperformed AQ in most datasets. IGR lost only one dataset, krkp, but IGR also achieved more than 99% accuracy. The poor results of AQ in datasets with numeric attributes, such as iris and wine, were mainly caused by the discretization of the attributes. Even if each interval involves many instances before learning, some of the intervals become empty, or involve few instances after specialization steps going on. This causes too specific rules and degrades the classification accuracy of induced rule set.

There was no significant difference between IGR and TDDT-op from the view point of classification accuracy. However, TDDT-op generated very large decision trees to achieve high classification accuracy. On average, it generated 53

Table 3. Average number of induced rules.

	C4.5	AQ	TDDT-op	IGR
breast	26	7	1,163	83
cleveland	77	25	3,246	139
crx	37	20	1,844	176
dna	103	31	8,601	323
glass	36	17	1,955	88
hayes-roth	15	8	359	16
iris	4	7	132	20
krkp	28	25	1,904	108
monks-1	33	8	1,372	79
monks-2	1	28	1	90
pima	95	32	3.661	282
promoters	12	4	783	23
satellite	441	97	18,738	1,722
segment	59	23	3,002	330
soy-large	40	20	2,140	91
tic-tac-toe	85	19	5,034	193
vehicle	121	27	5,671	365
voting	6	11	767	41
waveform	663	76	22,144	2,035
wine	6	4	490	33

times more rules than C4.5 and 16 times more rules than IGR. In other words, IGR scored comparable classification accuracy with only 6% of rules compared with TDDT-op.

Finally, we discuss IGR's time complexity. IGR was on average about three times slower than C4.5 and required about 4 minutes per cross validation fold (one run with 4/5 instances in the dataset) for satellite, the largest dataset in our experiments. IGR was much faster than TDDT-op. IGR was about ten times faster than TDDT-op in large datasets such as waveform and satellite. IGR is not a quite efficient algorithm but its learning speed is sufficiently fast to be applied to many real world problems.

4 Related Works

IGR learns if-then rules from training instances and there exists many related works to learn if-then rules. AQ family[14] is one of the most famous algorithms in this area. AQ starts from an empty rule set and continues to learn a if-then rule until all positive instances are covered by one of the induced rules. CN2[4]is based on similar strategy but it deals with noisy data based on the significance of the change of class distribution. RIPPER[5] also learns if-then rules and is quite faster than other algorithms.

Lazy learning[1] is an interesting approach to classification problems. Lazy learner is "lazy" because it does not start to learn until a test instance is given. Lazy learning has been mainly researched for nearest neighbor algorithms[6]. For example, Hasite and Tibshirani[11] proposed an algorithm to learn a local similarity measure that works quite well near the test instance. For learning logical rules, Friedman et al.[9] proposed LazyDT that learns an if-then rule when a new instance to be classified is given.

Bagging[2] and boosting[10][16] are recent interesting topics in supervised learning. IGR learns the best rules for the particular subspaces of the instance space by focusing on each training instance. Instead, bagging and boosting induce such rules by applying single learing algorithm to different trainig sets. Option decision trees[13] applies a different approach to learn the best rules for particular subspaces. It expands an internal node of decision trees with plural attributes and generates many trees at once. These algorithms significantly improves classification accuracy but tend to induce very large hypotheses.

Discretization of numeric attributes is an important technique for supervised learning. The current version of IGR applies the simplest discretization method, discretizing a domain of a numeric attribute to intervals with same population. There have been proposed more sophisticated algorithms for discretization. ChiMerge[12] is a bottom up discretization algorithm and uses statistical significance as an estimation function for intervals. Fayyad and Irani[8] proposed a discretization algorithm with entropy based heuristic.

5   Conclusions

This paper proposed a new supervised learning algorithm IGR. For each training instance, IGR learns the best rule with maximum support size, the number of training instances covered by the rule, and classifies a new instance with a set of induced rules.

The experimental results with UCI repository showed IGR learned much accurate rules than C4.5, AQ and LazyDT. IGR showed good performance even when LazyDT scored poor results. In our experiments, TDDT-op was the only algorithm with classification performance comparable to IGR. However, TDDT-op required about 16 times more rules than IGR to achieve the accuracy. These results show that IGR can learn accurate rule sets with reasonable size.

The current version of IGR has high ability for classification, but there are some problems to be tackled. The most important one is to apply a pruning procedure. IGR shows high classification ability without pruning as the experiments showed and the pruning will not improve the classification performance. The purpose of pruning for IGR is to reduce the number of rules without degrading classification accuracy.

The ability of IGR to find the optimum rule for each training instance, is quite useful for not only classification tasks but also knowledge discovery from databases. For example, IGR can help human experts to discover exceptional instances and rules. If the optimum rule for a certain training instance has large

support size, then the instance is typical in the database and such instances
and corresponding rules are trivial for domain experts. Instead, the instances
whose best rules have small support, are exceptional instances in the training
set and the instances and the corresponding rules may help the domain experts
to discover new knowledge.

References

1. Aha, D. W.: *Lazy Learning.* Kluwer academic publishers(1997)
2. Breiman, L.: Bagging Predictors. *Machine Learning, 24* (1996) 123–140
3. Breiman, L., Friedman, J. H., Olshen, R. A. and Stone, C. J.: Classification and
   Regression Trees. Wadsworth International Group(1984)
4. Clark, P. and Niblett, T.: The CN2 Induction Algorithm. *Machine Learning, 3*
   (1989) 261-283
5. Cohen, W.: Fast effective rule induction. *Proc. of the Twelfth International Conference on Machine Learning* (1995) 115–123
6. Dasarathy, B. V.: Nearest Neighbor(NN) Norms: NN Pattern Classification Techniques. *IEEE Computer Society Press*(1990)
7. Fayyad, U. M.: Branching on Attribute Values in Decision Tree Generation. In
   Proc. of the Twelfth National Conference on Artificial Intelligence (1994) 601–606
8. Fayyad, U. M. and Irani, K. B.: Multi-Interval Discretization of Continuous-Valued
   Attributes for Classification Learning. In *Proc. of the 13th International Joint
   Conference on Artificial Intelligence* (1993), 1022–1027
9. Friedman, J. H., Kohavi, R. and Yun, Y.: Lazy Decision Trees. In *Proc. of the
   13th National Conference on Artificial Intelligence* (1996) 717–724.
10. Freund, Y. and Schapire, R. E.: Experiments with a new Boosting Algorithm.
    In *Proc. of the Thirteenth International Conference on Machine Learning* (1996)
    148–156.
11. Hastie, T. and Tibshirani, R.: Discriminant Nearest Neighbor Classification. *IEEE
    Transactions on Pattern Analysis and Machine Learning, 18* (1995) 607–616
12. Kerber, R.: Discretization of numeric attributes. In *Proc. of the Tenth National
    Conference on Artificial Intelligence* (1992) 123-138
13. Kohavi, R. and Kunz, C.: Option Decision Trees with Majority Voting. In *Proc.
    of the Fourteenth International Conference on Machine Learning* (1997) 161–169.
14. Micalski, R. S. and Larson, J.: Incremental generation of vl1 hypotheses: the
    underlying methodology and the description of program AQ11. ISG 83-5, Dept. of
    Computer Science, Univ. of Illinois at Urbana-Champaign, Urbana(1983)
15. Quinlan, J. R.: *C4.5: Programs for Machine Learning.* Morgan Kaufmann Publishers (1993)
16. Quinlan, J. R.: Bagging, Boosting and C4.5. In *Proc. of the Thirteenth National
    Conference on Artificial Intelligence* (1996) 725–730

Learning with Globally Predictive Tests

Michael J. Pazzani

Department of Information and Computer Science
The University of California
Irvine, CA 92697
949-824-5888
email: pazzani@ics.uci.edu

Abstract. We introduce a new bias for rule learning systems. The bias only allows a rule learner to create a rule that predicts class membership if each test of the rule in isolation is predictive of that class. Although the primary motivation for the bias is to improve the understandability of rules, we show that it also improves the accuracy of learned models on a number of problems. We also introduce a related preference bias that allows creating rules that violate this restriction if they are statistically significantly better than alternative rules without such violations.

1 Introduction

A variety of rule learning systems have been developed that create rules to predict class membership of examples such as AQ15 [1], CN2 [2], ITRULE [3], C4.5-rules[4], FOIL [5], FOCL [6], Greedy3 [7], Ripper [8], and decision lists [9]. One commonly reported advantage of modeling predictive relationships with rules is the comprehensibility of the learned knowledge. Rule learners produce a set of learned rules of the form:

$$Test_1 \ \&...\& \ Test_n \rightarrow Class_i$$

where each test compares an attribute A_i to a value V_{ij} for that attribute. For nominal attributes, the possible tests include determining whether an attribute value of an example is equal to a particular value, is not equal to a particular value, or is a member of a set of values. For numerical values, the tests will determine whether an attribute value of an example is greater than or less than a particular value. Typically, the rules are ordered so that to classify an example, one predicts the class of the first rule whose antecedent is true. One common approach for ordering rules is an estimate of the accuracy of the rule (e.g., Quinlan [4]; Clark & Niblet [2]; Ali & Pazzani [10]).

Table 1 shows an example of some rules learned to screen infants for mild mental retardation [11] from a sample of over 4000 examples collected by the National Collaborative Perinatal Project of the National Institute of Neurological and Communicative Disorders and Stroke. The rules are relatively easy for an expert or novice to understand and could easily be applied by a person or a computer. However, the rules contain certain tests that are counter-intuitive and

Table 1. Some rules learned to screen infants for mild mental retardation.

```
IF the child has no emotional problems
AND the mother has normal IQ
THEN the risk is LOW

IF fetal distress is ascertained prior to or during labor
AND the mother's education level is less than 12 years
AND the mother smokes
THEN the risk is HIGH

OTHERWISE IF the child has no emotional problems
AND the mother's education level is at least 12 years
AND there were previous stillbirths
THEN the risk is LOW
```

puzzling to experts. In particular, the third rule predicts that there is low risk of mental retardation and contains a condition "there were previous stillbirths" that is normally thought of a risk factor for mental retardation. It is possible that this rule is a new medical finding for a sub-population of patients. However, before establishing such a claim, it is worthwhile to see if there are alternative models of the data that are equally predictive but do not require including such tests.

We present the following definition to facilitate the discussion of learning rules.

Definition 1 (Globally Predictive Test)

A test is globally predictive of $Class_i$ iff $P(Class_i|Test) > P(Class_i)$

Definition 2 (Locally Predictive Test)

A test is locally predictive of $Class_i$ in a Context iff $(Class_i|Test \ \& \ Context) > P(Class_i|Context)$ where Context is some Boolean combination of tests.

In this paper, we explore the implications of biasing rule learners to avoid using tests that are locally predictive of class memberships but are not globally predictive. A single rule that predicts class membership as a conjunction of globally predictive tests is an example of a simple causal schema: multiple necessary causes [12]. A set of such rules that enumerate alternative means of predicting class membership represents another simple causal schema: multiple sufficient causes. However, a rule that uses a test that is locally but not globally predictive is evoking a more complex causal schema in which there is an interaction among the variables. A predictive relationship involving such an interaction among variables is more difficult for people to learn from data [13]. We argue that to match the cognitive bias of human learners, knowledge discovery systems should avoid creating rules with locally predictive tests that are not globally predictive unless such tests are truly necessary to increase the accuracy of this model.

2   Background: Rule Learners

In this work, we will extend a rule learning system to implement the globally predictive bias. We will use FOCL [6] as a representative of this family of algorithms. FOCL is derived from Quinlan's [14] FOIL system. FOIL is designed to learn a set of rules that distinguish positive examples of a concept from negative examples.

FOIL operates by trying to find a rule that is true of as many positive examples as possible and no negative examples. It then removes the positive examples explained by that rule from consideration and finds another rule to account for other positive examples. It repeats this rule learning process until all of the positive examples are explained by some rule. Each rule can be viewed as a description of some subgroup of examples.

To learn an individual rule, FOIL first considers all possible rules consisting of a single test. It selects the best of these according to an information-gain heuristic that favors a test that is true of many positive examples and few negative examples. Next, FOIL specializes the rule using the same search procedure and information-based heuristic, considering how conjoining a test to the current rule would improve it by excluding many negative examples and few positives. This specialization process continues until the rule is not true of any negative examples, resulting in a single rule that is a conjunction of tests.

FOCL follows the same procedure as FOIL to learn a set of rules. However, it learns a set of rules for each class (such as low risk and high) enabling it to also deal with problems that have more than two classes. The rule learning algorithm is run once for each class, treating the examples of that class as positive examples and the examples of all other classes as negative examples. This results in a set of rules for each class. In this paper, we restrict our attention to a simple but effective procedure for converting a set of rules for each class into a single decision list such as that shown in Table 1. The learned rules are ordered by the Laplace estimate of each rule's accuracy [2] and the most frequent class is used as a default class.

When determining which test to add to the rule, FOCL (as well as other rule learners) considers tests in the context of the previous rules that were learned. The examples used to determine which test is best are those that are not true of any rule body that was learned previously and those that are true of the previous tests in the current rule. As a consequence, for all but the first test of the first rule, this family of algorithms can select a test that is locally predictive but not globally predictive. In the next section, we consider biasing rule learners to consider both the global and local predictability.

3   The Globally Predictive Test Bias

We experiment with two forms of the globally predictive test bias: a restriction bias and a preference bias. For the restriction bias, the procedure for selecting the best test is modified to exclude a test from consideration when learning a rule for $Class_i$ unless $P(Class_i|Test) > P(Class_i)$. The restriction bias therefore selects the globally predictive test that is best in the local context to add to a clause under consideration.

The preference bias prefers tests that are globally predictive. It will select a test that is not globally predictive if it is significantly better than the best locally predictive test that is globally predictive. First, the best test in the local context is found. If it is globally predictive, it is used in the rule. If it is not, the best test in the local context that is globally predictive is found. The two tests are then compared. If the globally predictive test is a significantly worse predictor in the local context than the test that is not globally predictive, the test that is not globally predictive is used in the rule. Otherwise, the test that is globally predictive is used. A χ^2 test is used to determine if there is a significant difference between the two tests. By default, if the probability that the two tests differ is greater than 0.75, then the locally but not globally predictive test is used. In our experiments, we determine the value of this probability parameter using cross-validation.

Whether the globally predictive test bias is useful in some domains is an empirical question. Clearly, the ability to have a locally but not globally predictive test will be useful in some problems such as those in which there are interactions among variables. However, this additional degree of freedom may be harmful in other domains resulting in inaccurate or confusing rules. In the next session, we investigate experimentally whether the globally predictive test restriction bias is useful.

3.1   Experiment 1: Restriction Bias

Here we report the results of running experiments on 16 problems selected from industrial and medical research projects at UCI and the UCI Repository of Machine Learning Databases [15]. Most of the data sets are available on the Internet at the UCI archive. The following additional data sets are used:

- Admissions: This database contains data on 312 high school students that were admitted to UC Irvine and the class label indicates whether the students enrolled at UCI. The attributes are scores on standardized tests and descriptive information such as age, gender, and residency.
- CERAD: Data collected by the Consortium to Establish a Registry for Alzheimer's Disease (CERAD). The particular problem of interest is to identify patients with early signs of dementia. The database contains 315 examples with the dementia status of each patient and the results of two commonly used cognitive tests for dementia screening, the Blessed Orientation, Memory and Concentration test and the Mini-Mental Status.
- FAQ: This is also data on screening for early signs of dementia collected by the UCI Center for Brain and Aging. The database contains 347 examples with the dementia status of each patient and the results of the Functional Activities Questionnaire.

- Retardation: This contains 4302 examples of newborn infants collected by the National Collaborative Perinatal Project of the National Institute of Neurological and Communicative Disorders and Stroke. The task is to determine whether an infant has mild mental retardation.
- Staging. This problem is to determine the severity of dementia of a patient from results of cognitive and neuropsychological tests. It contains 765 patient records.

We conducted an experiment in which we compared FOCL with the global predictive test restrictive bias to FOCL without this bias. The goal was to determine whether this bias is useful in practice. For each domain, we used paired ten-fold cross-validation of FOCL with and without this bias and computed the average accuracy for each of the databases. Table 2 also lists the average accuracy with and without the bias. We performed a paired t-test to determine whether there is a significant difference in using the bias on each data set. Figure 1 shows the difference in accuracy between using FOCL with the bias and using FOCL without the bias. Those problems in which a significant difference was found at the .05 level or greater are shown in black.

The results demonstrate that this bias results in a significant increase in accuracy on three data sets and a significant reduction on one. This shows that there are situations in which the extra freedom allowed by selecting a test that is locally predictive but not globally predictive is harmful. There are also situations in which the globally predictive test bias is harmful. The King-Rook-King-Pawn is an example of where the bias would not be expected to work well. This is a chess problem where the goal is to determine whether the white player with a king and rook can defeat a black player with a king and a pawn. The attributes in this problem correspond to features describing the locations of the pieces (e.g., the white king is in the last row). In this problem, it is the interaction among several features that determines whether white can win.

The CERAD data set is a particularly interesting illustration of the power of this bias. The attributes represent replies to questions designed to assess cognitive capabilities and those tests that are globally predictive of dementia represent incorrect answers to the questions. Those tests that are locally but not globally predictive of dementia are correct answers to questions. Although on a subsample of data they appear to be predictive of dementia, this is not a very reliable pattern when tested on unseen data.

Furthermore, rules that indicate that getting a question correct is a sign of dementia are puzzling to the experts in the domain. Others (e.g., Holte, Acker, & Porter [16], Pagallo & Haussler [7], Murphy & Pazzani [17], Vilalta, Blix, & Rendell [18]) have also reported on the problems associated with unreliably estimating descriptive statistics from small groups of examples and have proposed solutions based upon preventing examples from being partitioned into small groups. Here, we explore a different approach in which we reduce the hypothesis space to mitigate this problem. While the prior work has focused on improving the accuracy, we are motivated by improving the understandability of learned rules without reducing the accuracy.

Table 2. Databases used in the experiments and results of Experiment 1.

Problem	Classes	Without Bias	Restriction Bias
admissions	2	.696	.711
bupa	2	.664	.716 *
CERAD	2	.917	.949 *
colic	2	.827	.830
FAQ	2	.865	.870
glass	7	.674	.683
hepatitis	2	.800	.807
ion	2	.829	.838
krkp	2	.989	.971 *
mushrooms	2	.998	.999
pima	2	.724	.758 *
retardation	2	.701	.691
staging	4	.666	.671
voting	2	.936	.945
wine	3	.944	.950
wisc	2	.681	.705

The above discussion suggests that the globally predictive bias may aid in preventing overfitting. By requiring that tests be both globally and locally predictive, some unreliable tests may be eliminated from consideration. The previous experiment did not use any pruning method to reduce the effects of overfitting. In the next section, we report on an experiment designed to determine whether the globally predictive bias provides additional benefits when pruning.

3.2 Experiment 2: Restriction Bias with Pruning

The methodology used in this experiment is identical to the methodology used in Experiment 1. The only change is that reduced error pruning is used with FOCL both with and without the global predictive test restriction bias. Brunk & Pazzani [19] showed that reduced error pruning was more effective than the minimum description length heuristic used by FOIL on a variety of problems. Reduced error pruning operates by dividing the training data into two partitions. One partition (70% of the training data) is used to learn rules. The remaining training data is used in pruning. Two operators are used in pruning: deleting a rule and deleting the last test of a rule. These operators are applied to each rule learned and if a change improves the accuracy of the rules (as estimated on the pruning set), the change that results in the largest increase in accuracy is made permanent. Pruning is repeated until no change increases the accuracy of the learned rules. Figure 2 shows the difference in accuracy when reduced error pruning is used between FOCL with the global predictive test bias and FOCL without this bias. The black bars show significant differences in accuracy and positive values indicate that the bias was beneficial on that domain.

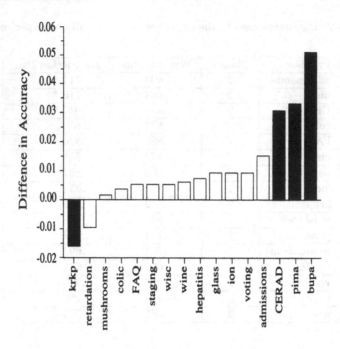

Fig. 1. Difference in accuracy between FOCL with the restriction bias and FOCL without this bias. Significant differences are shown in black. Positive values indicate that more accurate results are obtained when using the bias.

The results of these experiments indicate that the global predictive test bias provides an additional benefit over pruning. In 4 of the 16 problems, there is a significant increase in accuracy, while on two problems, there is a significant decrease. Furthermore, many of the increases in accuracy are greater than the largest decrease in accuracy. One possible reason that the bias provides benefits even when pruning is that pruning deletes tests that are unreliable but doesn't allow for the replacement of these tests with more reliable tests. In contrast, if the bias eliminates a test that is locally but not globally predictive, it finds a new test that is both globally and locally predictive to explain the data. Of course, this test is then also subject to the same pruning algorithm and is only retained if it is needed to increase accuracy on the pruning set.

Although the bias is useful on many problems, there are still some problems in which the bias is harmful. We would expect such a result with any bias for theoretical reasons (cf. Schaffer [20]) and with this particular bias we'd expect it to have problems when there are interactions among variables that make some variables locally but not globally predictive of class membership. In the next experiment, we relax the bias by preferring tests that are globally predictive.

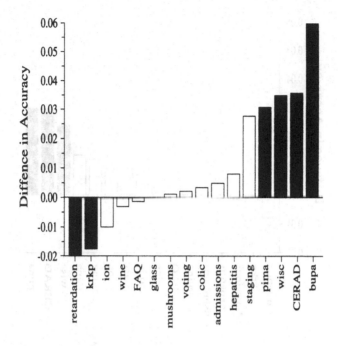

Fig. 2. Difference in accuracy between FOCL with the restriction bias and FOCL without this bias when using reduced error pruning.

3.3   Experiment 3: The globally predictive test preference bias

The globally predictive test bias is too restrictive for some domains. In this section, we explore a related preference bias. The preference will select a test that is not globally predictive if it is significantly better than the best locally predictive test that is globally predictive. In the experiments, a χ^2 test will be used to determine if there is a difference between the two tests. We use 5-fold cross validation to determine the best setting for the probability that there is a difference selecting from 0.05, 0.25, 0.5, 0.75 and 0.95. The experiment below is run using the same methodology as the previous two experiments. On each trial, the threshold for the χ^2 test is found by cross-validation on the training data before the global predictive test bias is compared to the accuracy of FOCL with this preference bias. The average difference in accuracy is plotted in Figure 3 for the 16 domains.

The results graphed in Figure 3 show that there are 3 domains in which the preference bias provides a significant increase in accuracy. Although there are decreases in accuracy, these are all less than one percent and none of these are significant. This suggests that the cross-validation test is generally effective at determining how large a difference is needed between the best locally but not globally predictive test and the best locally predictive test that is globally predictive to ignore the influence of the global predictiveness of a test.

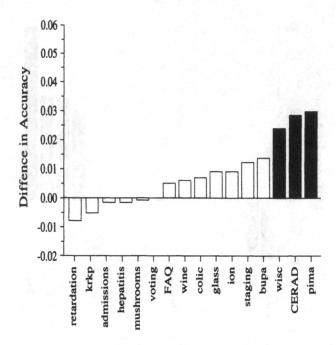

Fig. 3. Difference in accuracy between FOCL with the preference bias and FOCL without this bias.

An advantage of the preference bias over the restriction bias is that the preference bias does learn rules with tests that are locally predictive but not globally predictive. Such tests may represent important insights to convey to domain experts. However, unlike a system without any bias for globally predictive tests, the preference bias first ensures that there is not another alternative that is globally predictive. As a consequence, it includes fewer such tests in the rule, making it easier for an expert to verify that a useful interaction among variables has been found.

4   Discussion

In previous work (Pazzani, Mani & Shankle [21]), we addressed the problem of learning algorithms including counterintuitive tests in rules by having an expert provide "monotonicity constraints". For nominal variables, a monotonicity constraint is expert knowledge that indicates that a particular value makes class membership more likely. For numeric variables, a monotonicity constraint indicates whether increasing or decreasing the value of the variable makes class membership more likely. Lee, Buchanan, & Aronis [22] introduce similar expert constraints to the RL rule learner to make carcinogenicity more understandable and more accurate. Here, we show that much of the same effect could be achieved without consulting an expert by considering the global predictiveness of the training data. One advantage of the current approach is that it doesn't require an expert and may be applied more easily to many databases.

The expert monotonicity constraint bias was applied to the CERAD database. Pazzani, Mani & Shankle [21] report an accuracy of 90.7% using this constraint and 90.6% without. In contrast, C4.5 was 86.7% accurate, C4.5 rules was 82.6% accurate and a naive Bayesian classifier was 91.2%. The globally predictive test restriction bias obtained an accuracy of 94.4% on this database, substantially higher than the monotonicity constraint bias. There are two reasons for this difference in accuracy. First, one monotonicity constraint for a nominal value did not turn out to be globally predictive. This test was ignored when using monotonicity constraints but is frequently used with the global predictive test bias. Second, monotonicity constraints are not as specific as the global predictive test bias for numeric variables. In particular, a test includes both a comparison (such as greater than) and a specific numeric threshold. The global predictive test bias determines whether a test is globally predictive while the monotonicity constraint represents more general information about whether increasing values tend to make the class more likely. As a consequence, when using monotonicity constraints it is possible to have tests on numeric values that are locally but not globally predictive.

The globally predictive test bias represents a form of simplicity bias. However, in this case simplicity is not a syntactic property of the representation. Rather, it is a preference for a simple causal mechanism in which the influence of a variable on an outcome is not inverted in the context of other variables. That this bias is effective in increasing the accuracy of learned models is evidence that the databases commonly collected have such simple causal models. Similarly, the success of the bias may help to explain why replacing greedy searches for rules with more exhaustive searches (e.g., Rymon [23]; Webb [24]) has not been beneficial on most databases. Additional search would be useful to detect complex interactions among variables to find sets of locally but not globally predictive tests. However, if such situations are uncommon, the additional search is likely to overfit the data [25].

The globally predictive test bias, especially the restriction, could be viewed as a form of feature selection. For example, some approaches order variables by informativeness, a global criterion and select only the most informative prediction [26]. There are several differences however. First, we are selecting tests, not just variables. Second, we make this decision separately for each class. The most significant difference comes in the preference bias when we are favoring tests that meet these criteria but do not eliminate any test from consideration.

The original motivation of this work has been to improve expert acceptance of the results of knowledge discovery in databases. Experiments are in progress in which experts and novices judge the plausibility of rules learned with and without these global predictive constraints. Earlier experiments showed that experts preferred rules that obeyed monotonicity constraints and given the close relationship between monotonicity constraints and the global predictive test bias we are hopeful that the bias will prove useful in making the results of KDD more acceptable to experts.

5   Conclusions

We have explored the implications of biasing rule learners to create tests that are both globally and locally predictive of class membership. The results show that this bias improves the accuracy of learned models on a variety of domains. The knowledge discovery process is often viewed as an iterative process of modeling data with learning algorithms and changing the representation of the data or the parameters of the algorithm in an attempt to gain insight from the data. The global predictive test bias represents another tool in the toolkit that is intended to avoid overly complex models when simpler explanations of the data are possible.

Acknowledgements

We thank William Cohen for running RIPPER on synthetic data we developed to illustrate inverted influences. Comments by Pedro Domingos, Dennis Kibler, Cathy Kick, Richard Lathrop, Subramani Mani and the graduate students at UCI were helpful in clarifying the ideas presented in this paper.

References

1. Michalski, R., Mozetic, I., Hong, J., and Lavrac, N. (1986). The multi-purpose incremental learning system AQ15 and its testing application to three medical domains. *Proceedings of the 5th National Conference on Artificial Intelligence.* Philadelphia, PA: Morgan Kaufmann. 1041-1047.
2. Clark, P. and Niblett, T. (1989). The CN2 Induction Algorithm *Machine Learning,*3, 261- 284.
3. Goodman, R., and Smyth, P. (1989). The induction of probabilistic rule sets: the ITRULE algorithm, *Proceedings of the Sixth International Machine Learning Workshop,* (pp. 129- 132). Los Altos, CA: Morgan Kaufmann.
4. Quinlan, J.R. (1992). *C4.5: Programs for Machine Learning.* Los Altos, CA:Morgan Kaufmann.
5. Quinlan, J.R. (1990). Learning logical definitions from relations. *Machine Learning,* 5, 239-266.
6. Pazzani, M., and Kibler, D. (1992). The utility of knowledge in inductive learning. *Machine Learning,* 9, 57-94.
7. Pagallo, G., and Haussler, D. (1990). Boolean feature discovery in empirical learning.
8. Cohen, W. (1995). Fast effective rule induction. In *Proceedings of the Twelfth International Conference on Machine Learning,* Lake Tahoe, California.
9. Rivest, R. (1987). Learning decision lists. *Machine Learning,* 2:229 - 246.
10. Ali, K., and Pazzani, M. (1993). HYDRA: A noise-tolerant relational concept learning algorithm. *The International Joint Conference on Artificial Intelligence,* Chambery, France

11. Mani, M., McDermott, S., and Pazzani, M. (1997). Generating Models of Mental Retardation from Data with Machine Learning. *Proceedings IEEE Knowledge and Data Engineering Exchange Workshop* (KDEX-97), p. 114-119, IEEE Computer Society.
12. Kelley, H. (1983). The process of causal attribution. *American Psychologist*, 107-128.
13. Pazzani, M., and Silverstein, G. (1990). Feature selection and hypothesis selection: Models of induction. *Proceedings of the Twelfth Annual Conference of the Cognitive Science Society*, (pp. 221-228). Cambridge, MA: Lawrence Erlbaum.
14. Quinlan, J.R. (1990). Learning logical definitions from relations. *Machine Learning*, 5, 239-266.
15. Merz, C.J., and Murphy, P.M. (1998). UCI Repository of machine learning databases [http://www.lcs.uci.edu/ mlearn/MLRepository.html]. Irvine, CA: University of California, Department of Information and Computer Science. *Machine Learning* 5(1):71- 100.
16. Holte, R. Acker, L. and Porter, B. (1989). Concept learning and the problem of small disjuncts. *Proceedings International Joint Conference on Artificial Intelligence*. pp. 813- 818.
17. Murphy, P., and Pazzani, M. (1991). ID2-of-3: Constructive induction of m-of-n discriminators for decision trees. *Proceedings of the Eighth International Workshop on Machine Learning* (pp. 183-187). Evanston, IL:Morgan Kaufmann.
18. Vilalta, R., Blix, G. and Rendell, L. (1997). Global Data Analysis and the Fragmentation problem in Decision Tree Induction 9th European Conference on Machine Learning. *Lecture Notes in Artificial Intelligence*, Vol. XXX. Springer-Verlag, Heinderberg, pp 312- 326. *Workshop on Machine Learning* (pp. 183-187). Evanston, IL: Morgan Kaufmann.
19. Brunk, C., and Pazzani, M. (1991). An investigation of noise tolerant relational learning algorithms. *Proceedings of the Eighth International Workshop on Machine Learning* (pp. 389-391). Evanston, IL: Morgan Kaufmann.
20. Schaffer, C. (1994). A conservation law for generalization Proceedings of the 11th International Conference of Machine Learning, New Brunswick. Morgan Kaufmann.
21. Pazzani, M., Mani, S., and Shankle, W. R. (1997). Comprehensible knowledge-discovery in databases. In M. G. Shafto and P. Langley (Ed.) *Proceedings of the Nineteenth Annual Conference of the Cognitive Science Society*, pp. 596-601. Lawrence Erlbaum.
22. Lee, Y. Buchanan, B., and Aronis, J. (in press). Knowledge-Based Learning in Exploratory Science: Learning Rules to Predict Rodent Carcinogenicity. *Machine Learning*.
23. Rymon, R. (1993). An SE-based characterization of the induction problem. *Proceedings of the 10th International Conference of Machine Learning*, (pp. 268-275). Amherst, MA: Morgan Kaufmann.
24. Webb. G. (1993) Systematic search for categorical attribute-value data-driven machine learning. *Proceedings of the Sixth Australian Joint Conference on Artificial Intelligence*, (pp. 342-347). Melbourne: World Scientific.
25. Quinlan, J. R., and Cameron-Jones, R. (1995). Oversearching and layered search in empirical learning. *Proceedings Fourteenth International Joint Conference on Artificial Intelligence*, (pp. 1019-24). Morgan Kaufmann, Montreal.
26. Chen, M., Han, J., and Yu, P. (1996) Data Mining: An Overview from a Database Perspective. *IEEE Transactions on Knowledge and Data Engineering*, Vol. 8, No. 6.

Query-Initiated Discovery of Interesting Association Rules

Jongpil Yoon[1] and Larry Kerschberg[2]

[1] Sookmyung W. University, Computer Science Department, Seoul, 140-742, Korea
jyoon@sookmyung.ac.kr
[2] George Mason University, Information and Software Systems Engineering Dept.,
Fairfax, VA 22030-4444, USA
kersch@gmu.edu

Abstract. The approach presented in this paper is to discover association rules based on a user's query. Of the many issues in rule discovery, relevancy, interestingness, and supportiveness of association rules are considered in this paper. For a given user query, a database can be partitioned into three views: a positively-related-query view, a negatively-related-query view, and an unrelated-query view.
We present a methodology for data mining and rule discovery that incorporates pattern extraction from the three types of query views with pattern spanning to enlarge the scope of a pattern and its derived association rule. The rule discovery process involves several interrelated steps: 1) pattern extraction from both positively- and/or negatively-related query views, 2) pattern association across attributes to enhance the semantics of patterns, while performing 3) pattern spanning within an attribute domain to enhance the supportiveness of the resulting pattern.
The contributions of the paper includes the specification and development of the data mining method and associated tool that combines the operations of association and spanning on query views to derive semantically interesting patterns. These patterns can then be used in decision making because the patterns were mined from the user's original hypothesis as expressed by the query.

1 Introduction

A number of studies [1, 4, 8–10, 13] have concentrated on developing methods of rule discovery from databases. Although association rule discovery is becoming increasingly important, the discovered association patterns may not be semantically rich enough to assist users in understanding the application domain or in making decisions based on the discovered knowledge. It is partly because association patterns are not interesting or relevant to user's intentions with respect to the application domain, and partly because they only take into account the item sets or attribute values, but not their frequency (or occurrence). In this paper, we propose a viable and efficient strategy for discovering interesting patterns in large databases, and thus extend the work reported in previous research on this subject.

The interesting, relevancy, and supportiveness of patterns discovered from the databases are discussed in this paper.

- *Interestingness*. Interestingness as unexpectedness can be defined in general [5, 11]. Discovered patterns are interesting if they are unexpected.
- *Relevancy*. So many patterns are discovered, but not all of them are related to a user's concerns or needs. Some patterns, if discovered from a point of view corresponding to a users' query, are related to the user's concerns. A user query is a way of specifying a user's intention to a database.
- *Supportiveness* and *Confidence*. Discovered patterns can be used for applications only when they exceed the given confidence thresholds. The higher the supportiveness of a discovered pattern, the more applicable it may be. The higher the confidence of a discovered pattern is, the more it can be asserted for that database.

With these three pattern features in mind, we describe correlations among them. Knowledge discovery and data mining in databases (KDD) aims to extract patterns that are not previously known. Patterns are of interest if they are unexpected and are also useful to users. They should be relevant to a user's query as well. See the following example.

EXAMPLE 1.1: For example in a soil database, suppose that a farmer expects (or queries for) information about soil composition ratio amongst calcium, iron, and sodium as they relate to higher yield. A pattern as to what soil composition ratios produce higher yield is relevant to the user's expectation (or an answer in typical database processing), but it may not be interesting because it is pre-expected (asked) or known. On the other hand, a pattern as to which customers want bigger fruits may be interesting because it is not expected. However, it is not at all relevant to the user's query. Increasing the interestingness of a pattern may result in decreasing its relevancy, and vice versa. Our goal in this paper is then to extract patterns which are both related to and interesting to the user, based on the initial query. For the example, a pattern which characterizes those composition ratios that do not lead to higher yield is both related and interesting.

To achieve our goal, we classify a database into the three views according to a user's query. Consider a query posed to a table in a database. The first view is the answer (query) from the table. The second view is the remaining of the tuples which are not in the answer. Finally the third view consists of the other tables in the database, not specifies in the query. These views are explained later in detail in Section 3. The key idea is that patterns extracted from the (first and) second view(s) are both interesting and relevant to the query issued by a user.

The remainder of this paper is organized as follows: Section 2 reviews related work. Section 3 describes background knowledge, and defines the relationships amongst the properties of relevancy, interestingness, and supportiveness of patterns. Section 4 describes pattern spanning for a higher support. Section 5 describes our method for mining patterns by extracting, associating, and spanning patterns, using examples from the soil database. Finally, the contributions of this paper and future work are described in Section 6.

2   Related Work

Rule discovery in databases originated from finding *functional dependencies* in a database. Functional dependencies depict functional relationships among attribute values and can be viewed as constraints on the set of legal database relationships among data items. By counting the number of records that have the particular attribute values, data sets can be grouped and populated in a table.

A way of counting attribute values has been developed [3]. Based on counting attribute values, those attributes are associated to constitute a pattern. Associating patterns from databases have been considered by [1, 2]. The terms "support" and "confidence" are used to characterize a pattern. As association rules are specified over a collection of attributes, their support will decrease. Associating patterns may continue until the minimum support is reached. In contrast to this approach, we develop a method of *associating patterns being specified over maximally conjuncted attributes and also spanning values of those attributes to increase their support*.

Statistical methods for data mining have been introduced [7].

Classification functions have been developed using a *training data set* based on decision trees [1, 6]. The classification function, an interval, is generated by traversing in the decision (or classification) tree from the root to the leaves which are labeled with a particular group label. Numerical data (or categorical data) are also considered to extract patterns from them. In our paper, we also consider categorical data in order to span numeric patterns.

It is conceivable that the space of all possible combinations of attribute values and classes is too large to perform an exhaustive analysis. To resolve this problem, a practical learning process has employed heuristics to guide the discovery of useful rules [12]. The notion of interestingness of discovered rules has been introduced [5, 11]. In our approach, we use a user query to express an initial user hypothesis or interest, and then begin mining the query result for more interesting and relevant knowledge.

3   Preliminaries

Pattern Model

A *predicate* is a boolean function in the form of $(x\Theta X)$, where where x is a variable in the form of "attribute_name," X is an attribute value (i.e., X is in the domain of x), and Θ denotes a comparison operator ($>, =, \neq, \geq$,etc). We call an equality predicate $(x = X)$ a *point pattern*. As an illustration of a point pattern, (weight $= 7$) represents "weight is 7." A conjunction of predicates about numerical attribute values is defined as a range. For example, (weight $\geq$ 7) $\wedge$ (weight $\leq$ 9) specifies that the "weight is between 7 and 9." We call this type of a predicate a *range pattern*.

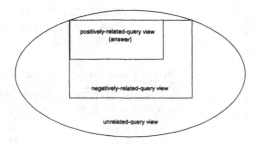

Fig. 1. Three Views for a Query

Query Related Views

A database query specifies a query condition, conveys a user's expectation to a database, and then its answer is a database subset that satisfies the query condition. As previously defined for a pattern, a query condition is specified in a conjunction of predicates. In SQL, a query specification is of the form "SELECT attributes FROM table names WHERE a condition." While processing a query, the tables specified in the FROM part of a query are joined.

Definition 1. (Query Related Views).

- Positively-Related Query View (PRQV). This view is precisely those tuples that satisfy the SQL query Q, $PRQV(Q) = Q(\mathcal{D}) = \pi_{A_1,..,A_k}(\sigma_{condition}(T_1 \bowtie .. \bowtie T_l))$, where A and T denote respectively attribute and table.
- Negatively-Related Query View (NRQV). This view represents the complement of the PR-query view with respect to the natural join of the tables specified in Q. The relational algebra specification is $NRQV(Q) = \pi_{A_1,..,A_k}(T_1 \bowtie .. \bowtie T_l) - PRQV(Q) = \pi_{A_1,..,A_k}[(T_1 \bowtie .. \bowtie T_l) - (\sigma_{condition}(T_1 \bowtie .. \bowtie T_l))] = \pi_{A_1,..,A_k}[(\sigma_{\neg condition}(T_1 \bowtie .. \bowtie T_l))]$. The arity of the attributes of both the PR- and NR-query views is the same.
- Unrelated Query View (URQV). This view consists of the all tables, and attributes not involved in the query specification. This is represents the complement of the combination of the PR- and NR-query views within the database $\mathcal{D}$. $URQV(Q) = \mathcal{D} - [PROQV(Q) \cup NRQV(Q)] = \mathcal{D} - [\pi_{A_1,..,A_k}(T_1 \bowtie .. \bowtie T_l)]$. Notice that we do not use this view in this paper because it is not relevant to the user based on the query.

The three views are depicted in Figure 1. The entire database of instances is depicted by an oval, the PR- and NR-query views are depicted by the inner- and outer-rectangles, respectively, while the UR-query views is represented by those instances not in the PR- or NR-query views.

If a query answer is identical to an entire database (all tables in a database), then there is no NR-query view nor UR-query view. Rather, there is only a PR-query view. If an SQL query specifies in its FROM part some of table names, and its answer contains all instances populated in those tables, then there is no

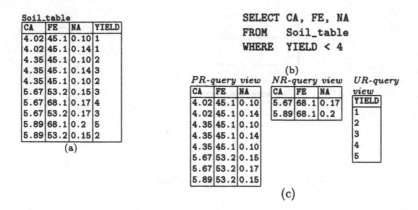

Fig. 2. Soil Table Example and Query Views

NR-query view, but there may be the other two views. Similarly, if a query has an empty answer and if the target tables are not empty, then there must be an NR-query view, but no PR-query view. An UR-query view is always generated, unless a query is specified over the entire database.

For example, consider the table as in Figure 2 (a). The query in Figure 2 (b) is posed to list calcium CA, iron FE, and sodium NA information from the table Soil_table if the yield is strictly less than 4. As the query is processed, the PR-query, NR-query, and UR-query views are generated as in Figure 2 (c).

Most previous KDD approaches consider PR-query views only, while our approach considers both PR-query and NR-query views.

Confidence of Patterns

Consider a rule $A \Longrightarrow B$. The confidence is defined as $P(\{A, B\})/P(\{A\})$, where $\{A, B\}$ is an itemset concept defined in supermarket basket examples [1]. A pattern k for query condition q has $\mathsf{count}(\sigma_{k \wedge q}(T))/\mathsf{count}(\sigma_k(T))$ of confidence, where count is an aggregation function on the database table T. We simply write it $P(q|k) = P(k, q)/P(k) = \mathsf{count}(\sigma_{k \wedge q}(T))/\mathsf{count}(\sigma_k(T))$. The confidence of the pattern k is 1.0 only if the pattern k is extracted from the PR-query view only, meaning that the predicate k is not satisfied within NR- or UR-query views. Similarly, the confidence of the pattern k' is $P(\neg q|k') = P(k', \neg q)/P(k')$ $= \mathsf{count}(\sigma_{k' \wedge \neg q}(T))/\mathsf{count}(\sigma'_k(T)) = 1.0$ if k' is extracted from the NR-query view.

Support of Patterns

For an itemset $\{A, B\}$, the support is $P(\{A, B\})$ [1]. Given a pattern k for a query condition q extracted from a PR-query view, its support is $\mathsf{count}(\sigma_{k \wedge q}(T))/\mathsf{count}(T)$ where count is an aggregation function on the database table T. The support is simply written $P(k, q) = \mathsf{count}(\sigma_{k \wedge q}(T))/\mathsf{count}(T)$, If the pattern k' is extracted from an NR-query view, its support is $P(k', \neg q) = \mathsf{count}(\sigma_{k' \wedge \neg q}(T))/\mathsf{count}(T)$.

Interest of Patterns

We base our definition of "interestingness" on those of Silberschatz [11], Fagin [5], and Brin [3].

As defined in [3]: the interest of A and B in a supermarket basket example is defined as $P(A,B)/P(A)P(B)$ for the rule in itemsets $A \implies B$. We also employ this definition: The interest of a pattern k for a query condition q extracted from a PR-query view is $P(k,q)/P(k)P(q) = \text{count}(\sigma_{k \wedge q}(T))/\text{count}(\sigma_k(T))\text{count}(\sigma_q(T))$. The interest of a pattern k' extracted from an NR-query view is $P(k', \neg q)/P(k')P(\neg q) = \text{count}(\sigma_{k' \wedge \neg q}(T))/\text{count}(\sigma_{k'}(T))\text{count}(\sigma_{\neg q}(T))$. In the following theorem, we investigate formally under what conditions the pattern extracted from NR-query view is more interesting than the pattern extracted from PR-query view.

Theorem 1. (Interesting pattern Generation). Patterns with higher support can be generated from the NR-query view, thereby leading to the discovery of more interesting patterns. This is particularly the case when the confidence of the $NRQV(Q) <<$ confidence of the $PRQV(Q)$.

Proof: Suppose that there is a smaller NR-query view as compared to a PR-query view, and the pattern support from an NR-query view is higher than the one from a PR-query view. The difference of the pattern interest between the NR-query view and the PR-query view is:

$P(k', \neg q)/P(k')P(\neg q) - P(k,q)/P(k)P(q) = [P(q)P(k', \neg q) - P(k,q)(1 - P(q))]/P(k)P(q)(1 - P(q)) \geq 0$

because the given conditions, a fewer NR-query view $P(q) > P(\neg q)$ and a higher support $P(k', \neg q) > P(k,q)$. Hence, under the given conditions, the theorem is proved.

The main goal of the KDD process proposed in this paper is to generate more interesting patterns from databases.

4 Pattern Spanning for Higher Support

A point pattern can be spanned with another point pattern or a range pattern so long as a given confidence threshold is satisfied. Consider a given confidence threshold c, a PR-query view T_p and an NR-query view T_n. Notice that $T = T_p \cup T_n$. Two point patterns $(v = V_i)$ and $(v = V_j)$ extracted from a PR-query view may be spanned to generate the range pattern $(v \geq V_i) \wedge (v \leq V_j)$ so long as the confidence $[\text{count}(\sigma_{(v \geq V_i) \wedge (v \leq V_j)}(T_p))/ \text{count}(\sigma_{(v \geq V_i) \wedge (v \leq V_j)}(T))] \geq c$, where $V_i \leq V_j$. If a point pattern $(v = V_i)$ and a range pattern $(v \geq V_j) \wedge (v \leq V_k)$, both of which are extracted from a PR-query view, are to be spanned, then the range pattern $(v \geq V_i) \wedge (v \leq V_k)$ can be obtained so long as the confidence $[\text{count}(\sigma_{(v \geq V_i) \wedge (v \leq V_k)}(T_p))/ \text{count}(\sigma_{(v \geq V_i) \wedge (v \leq V_k)}(T))] \geq c$, where $V_i \leq V_j$. Likewise, patterns extracted from an NR-query view may also be spanned.

Theorem 2. (Spanning for Higher Support). As patterns are spanned to generate a range pattern, the support of the spanned range pattern increases.

Proof: Trivial

In order to span patterns, the attribute values must be in a sort order. If attribute values are not numerical values, categorical attributes[1] can be used to create an order. A pattern can be spanned by another pattern if both patterns are specified over the same attributes. In order to span efficiently, we introduce a frequency graph, in which the y-axis denotes ordered attribute values, and the x-axis is the frequency of occurrence for the corresponding attribute values. The PR-query view attribute values frequencies are depicted on the right side of the x-axis and the NR-query view attribute values are depicted on the left side. A frequency graph, like a histogram, depicts the frequencies of (categorical or numerical) attribute values.

EXAMPLE 5.1: As an illustration, the query in Figure 2(b) is posed to the table in Figure 2(a). Therefore, the PR-query view and NR-query view are obtained as in Figure 2(c). Figure 3 depicts for an attribute NA the attribute values ordered along the y-axis or zero point of the horizontal. The PR-query view attribute values frequencies are depicted in the right side of the y-axis and the NR-query view attribute values are depicted in the left side. Depicted are the point patterns (NA=0.1), (NA=0.14) and (NA=0.15) extracted from a PR-query view, and the point patterns (NA=0.2) extracted from an NR-query view. Notice that (NA=0.17) is extracted from both the views.

Consider the spanning bar s1 in the figure. From the first two patterns, (NA=0.1) and (NA=0.14), the two patterns are spanned to be a range pattern $(NA \geq 0.1) \wedge (NA \leq 0.14)$. Consider the frequency, i.e., the number of instances in the range, is 5, and suppose the size of table, i.e., the total number of tuples is 10. Its support is then 0.5 by computing $P(k,q) = $ count $(\sigma_{k \wedge q}(T))/$count$(T)$ = count $(\sigma_{(NA \geq 0.1) \wedge (NA \leq 0.14) \wedge (YIELD < 4)}$ (Soil_table)) / count (Soil_table) = 5/10 = 0.5. Furthermore, spanning with (NA=0.15) by the spanning bar s2, the range pattern becomes a wider range $(NA \geq 0.1) \wedge (NA \leq 0.15)$ with the support of 0.7. We note that spanning increases the support factor. By spanning further as s3 in the figure, the pattern will be $(NA \geq 0.1) \wedge (NA \leq 0.17)$. Now, its support becomes 0.9 but its confidence lowers to 0.88 from 1.0. The reason for the confidence is that out of 9 instances within the range, only 8 instances are in a PR-query view, and therefore $P(q|k) = P(k,q)/P(k) = $ count$(\sigma_{k \wedge q}(T))$ / count$(\sigma_k(T)) = $ count$(\sigma_{(NA \geq 0.1) \wedge (NA \leq 0.17) \wedge (YIELD < 4)}$ (Soil_table)) / count $(\sigma_{(NA \geq 0.1) \wedge (NA \leq 0.17)}$ (Soil_table)) = 8/9 = 0.88. As spanning incorporates more patterns, then its support will be higher, but its confidence will be lower, especially if some instances pertain to an NR-query view.

5   Mining Patterns by Spanning Ahead

In our approach, rule discovery is performed based on a user query. For the query, the three views are generated: a PR-query view, an NR-query view, and a UR-query view. Patterns may be extracted from these views. There are three

---

[1] Alphabetical values are categorized and grouped in a certain order, for example, lexicographical order.

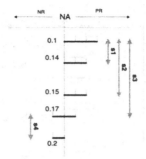

Fig. 3. A Frequency Graph of Patterns (NA= V_i) Extracted from Query Related Views

steps: 1) Pattern extraction, 2) Pattern association, and 3) Pattern spanning. First patterns are extracted and spanned with the adjacent patterns. Then steps 2 and 3 are repeated until all possible attributes or values have been exhausted while meeting given support and confidence thresholds.

5.1   Extraction of Database Patterns

An initial pattern is a point pattern of the form of (attribute = value). Like association rules extracted from large item sets [1], our pattern extraction method aims to extract as a pattern a conjunction of as many predicates as possible such that its support is within a given support threshold. The pattern extraction method proposed in this paper is as follows:

1. Construct two views, PR- and NR-query views, for a given query.
2. Generate patterns from each query view.
    A pattern (attribute = value) is represented as a predicate.
3. Constitute a maximal pattern by "spanning ahead" a pattern with adjacent patterns of the same attribute while meeting the given support threshold.
    The result will be a pattern, a so called *maximal pattern*. By maximal pattern we mean that a range pattern covering the maximal range of attribute values while meeting the minimum support threshold.

Now, let's apply the pattern extraction method to the database example as shown in Figure 2. Suppose the query Figure 2(b), SELECT CA, FE, NA FROM Soil_table WHERE YIELD < 4;, is posed to a database containing the table "Soil_table" as in Figure 2(a). The three views for the given query are obtained as shown in Figure 2(c). For the attributes, CA, FE, NA and YIELD from those views the frequency graphs are constructed as in Figure 4. For each attribute, there may be three views available, PR-, NR- and UR-query views are constructed (refer the left and right arrows on the top of the graph). We consider, however, PR- and NR-query views[2], with each attribute value and its frequency for each

---

[2] Notice that since patterns extracted from a UR-query view are not interesting and relevant to a given query, they are ignored in this paper.

#	point pattern	view	suppt	conf
1	(CA=4.02)	PR	0.2	1.0
2	(CA=4.35)	PR	0.3	1.0
3	(CA=5.67)	PR	0.3	0.667
4	(CA=5.67)	NR	0.3	0.667
5	(CA=5.89)	PR	0.2	0.5
6	(CA=5.89)	NR	0.2	0.5
7	(FE=45.1)	PR	0.5	1.0
8	(FE=53.2)	PR	0.3	1.0
9	(FE=68.1)	NR	0.2	1.0
10	(NA=0.1)	PR	0.3	1.0
11	(NA=0.14)	PR	0.2	1.0
12	(NA=0.15)	PR	0.2	1.0
13	(NA=0.17)	PR	0.2	0.5
14	(NA=0.17)	NR	0.2	0.5
15	(NA=0.2)	NR	0.1	1.0

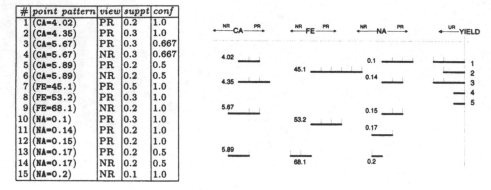

Fig. 4. Point Pattern extraction with Frequency Graph

#	maximal pattern from Figure 4	views	support	confidence	interest
1	(CA >= 4.02) ∧ (CA <= 5.67)	PR	0.8	0.875	0.11
2	(FE >= 45.1) ∧ (FE <= 53.2)	PR	0.8	1.0	0.125
3	(FE = 68.1)	NR	0.2	1.0	0.5
4	(NA >= 0.1) ∧ (NA <= 0.17)	PR	0.9	0.889	0.11

Fig. 5. Patterns Extracted from PR-/NR-Query Views

PR- or NR-query view. From each frequency graph depicted as in Figure 4, of course, patterns can be extracted. For example, for the attribute iron FE, the patterns (FE = 45.1) and (FE = 53.2) are extracted from the PR-query view, and the pattern (FE = 68.1) is extracted from the NR-query view. Currently, their supports are 0.5, 0.3, and 0.2, respectively.

In Figure 4, given the 0.2 support and 0.85 confidence thresholds, a pattern (CA = 4.02) is spanned with the adjacent pattern (CA = 4.35) to constitute a new pattern (CA >= 4.02) ∧ (CA <= 4.35). Its support increases from 0.2 to 0.5, and confidence is still 1.0. As discussed in the previous section, its interestingness is 0.11 because 7/8×8=0.11. By further spanning with the adjacent pattern (CA = 5.67), the new pattern will be (CA >= 4.02) ∧ (CA <= 5.67). Then its support again increases to 0.8, while its confidence decreases to 0.875. Spanning stops because further spanning violates the confidence threshold. Some examples of the patterns from the first step are listed in Figure 5. Notice that the interestingness of the pattern #3 extracted from the NR-query view is higher than all other patterns extracted from the PR-query view.

5.2 Association of Database Patterns

A pattern can be associated with another pattern by joining between attributes. As in market basket data, patterns at association level 1 are extracted over a single attribute. At level 2, every pair of patterns may be associated as long as the given minimum support is met. In the same manner, a pattern can be associated with one more pattern to obtain a higher level pattern.

#	associated pattern from Figure 5	views	support	confidence	interest
1	(CA >= 4.02) ∧ (CA <= 5.67) ∧ (FE >= 45.1) ∧ (FE <= 53.2)	PR	0.7	1.0	0.125
2	(CA >= 4.02) ∧ (CA <= 5.67) ∧ (NA >= 0.1) ∧ (NA <= 0.17)	PR	0.8	0.875	0.097
3	(FE >= 45.1) ∧ (FE <= 53.2) ∧ (NA >= 0.1) ∧ (NA <= 0.17)	PR	0.8	1.0	0.125
4	(CA >= 5.67) ∧ (CA <= 5.89) ∧ (FE = 68.1)	NR	0.2	1.0	0.5
5	(FE = 68.1) ∧ (NA >= 0.17) ∧ (NA <= 0.2)	NR	0.2	1.0	0.5

Fig. 6. Pattern Association from Maximal Patterns at Level 1

The pattern association is possible by joining a pattern over one attribute with another pattern over another attribute. This is similar to extracting association rules as in [1]. However our approach distinguishes between two different underlying views (PR- and NR-query views) and deal not only with attribute values but also their frequencies. In previous approaches, association patterns were extracted from PR-query views only, whereas our approach considers both PR- and NR-query views. Attribute values were regarded to be of equal frequency, whereas our approach takes *actual frequencies* into account.

The pattern (CA >= 4.02) ∧ (CA <= 5.67) is associated with another pattern (FE >= 45.1) ∧ (FE <= 53.2), shown in Figure 5, to constitute a new pattern (CA >= 4.02) ∧ (CA <= 5.67) ∧ (FE >= 45.1) ∧ (FE <= 53.2) This new association pattern is shown in Figure 6. Note that its support decreases to 0.7, while its confidence increases to 1.0. Similarly, the pattern (CA >= 4.02) ∧ (CA <= 5.67) is associated with the other pattern (NA >= 0.1) ∧ (NA <= 0.17) to constitute a new pattern (CA >= 4.02) ∧ (CA <= 5.67) ∧ (NA >= 0.1) ∧ (NA <= 0.17). Then its support decreases to 0.8, and its confidence increases to 0.875 in the PR-query view as well. However, when the pattern (FE = 68.1) is associated with the other pattern (NA >= 0.17) ∧ (NA <= 0.2) to constitute a new pattern (FE = 68.1) ∧ (NA >= 0.17) ∧ (NA <= 0.2), its support decreases to 0.2, while its confidence increases to 1.0 in the NR-query view. Some examples of the patterns are listed in Figure 6.

5.3 Spanning of Database Patterns

Those patterns associated as in the previous section may now also be spanned. Given two (associated) patterns, if they are specified over the same attribute but different values, as long as support and confidence thresholds are met they can be spanned to represent a range (or conjunction of predicates). In this way, spanned patterns may have a higher support and therefore they become ready to be associated at higher levels as long as the support is met.

For example, the pattern (CA >= 4.02) ∧ (CA <= 5.67) ∧ (FE >= 45.1) ∧ (FE <= 53.2) whose support is 0.7 as in Figure 6 can be spanned with another pattern (CA = 5.89) ∧ (FE >= 45.1) ∧ (FE <= 53.2) which does not appear in Figure 6 due to lower support than the given threshold to generate to a new pattern (CA >= 4.02) ∧ (CA <= 5.89) ∧ (FE >= 45.1) ∧ (FE <= 53.2) in the PR-query view. Then its support increases to 0.8, while its confidence is still 1.0 in the view. Some examples of the patterns are listed in Figure 7. Spanning and associating methods will be discussed in more detail.

242   Jongpil Yoon and Larry Kerschberg

#	spanned pattern from Figure 6	views	support	confidence	interest
1	(CA >= 4.02) ∧ (CA <= 5.89) ∧ (FE >= 45.1) ∧ (FE <= 53.2)	PR	0.8	1.0	0.125
2	(CA >= 4.02) ∧ (CA <= 5.89) ∧ (NA >= 0.1) ∧ (NA <= 0.17)	PR	0.9	0.88	0.11
3	(FE >= 45.1) ∧ (FE <= 53.2) ∧ (NA >= 0.1) ∧ (NA <= 0.2)	PR	0.8	1.0	0.125
4	(CA >= 4.02) ∧ (CA <= 5.89) ∧ (FE = 68.1)	NR	0.2	1.0	0.5
5	(FE = 68.1) ∧ (NA >= 0.1) ∧ (NA <= 0.2)	NR	0.2	1.0	0.5

Fig. 7. Spanning from Associated Patterns of Level 1

#	associated pattern from Figure 7	vs	supt	conf	int
1	(CA >= 4.02) ∧ (CA <= 5.89) ∧ (FE >= 45.1) ∧ (FE <= 53.2) ∧ (NA >= 0.1) ∧ (NA <= 0.17)	PR	0.8	1.0	0.125
2	(CA >= 4.02) ∧ (CA <= 5.89) ∧ (FE = 68.1) ∧ (NA >= 0.1) ∧ (NA <= 0.2)	NR	0.2	1.0	0.5

Fig. 8. Pattern Association at Level 2

In a level of association, we may obtain patterns with higher support unless spanning the patterns. By spanning patterns in each association level, we may generate such representationally powerful patterns that appeared in Figure 7.

5.4   Iteration between Association and Spanning

The association and spanning steps are iterated until no more attributes are available to meet the given support and confidence thresholds. While iterating through these two steps, redundant patterns should be eliminated. From the previous spanned patterns shown in Figure 7, more patterns can be associated as in Figure 8. In this example, Figure 8 contains the range patterns, similarly called largest itemsets in [1].

6   Conclusion

This paper has presented a novel approach to knowledge discovery from databases that uses a user query to guide the discovery process. The query result is used to construct positively-related and negatively-related query views which are then used to extract patterns that will lead to both interesting and relevant patterns.

This work extends other research in association rule discovery by combining association and spanning operations so as to maximize patterns for association patterns while maintaining high support and confidence levels.

Acknowledgement

This research was supported in part by non-direct research fund, Korean Science and Engineering Research Foundation grant number 96-0101-08-01-3, and by a grant from the Idaho National Engineering and Environmental Laboratory grant number 01-E-0729.

References

1. Rakesh Agrawal, Tomasz Imielinski, and Arun Swami. Mining association rules between sets of items in large databases. In Sushil Jajodia, editor, *Proc. ACM SIGMOD Intl. Conf. on Management of Data*, pages 207–216, Washington D.C., 1993.

2. S. Brin, R Motwani, and C Silverstein. Beyond market baskets: Generalizing association rules to correlations. In Peckman, editor, *Proc. ACM SIGMOD Intl. Conf. on Management of Data*, pages 265–276, Tucson, Arizona, 1997.

3. S. Brin, R Motwani, J. Ullman, and S. Tsur. Dynamic itemset counting and implication rules. In Peckman, editor, *Proc. ACM SIGMOD Intl. Conf. on Management of Data*, pages 255–264, Tucson, Arizona, 1997.

4. Y. Cai, N. Cercone, and J. Han. An attribute-oriented approach for learning classification rules from relational databases. In *the Sixth Int'l Conf. of Data Engineering*, pages 281–288, Los Angeles, Calif, 1990.

5. R. Fagin. Finite model theory - a personal perspective. In S. Abiteboul and P. Kanellakis, editors, *Proc. Int'l Conf on Database Theory*, 1990.

6. T. Fukuda, Y. Morimoto, and S. Morishita. Constructing efficient decision trees by using optimized numeric association rules. In *Proc. Intl. Conf. on Very Large Data Bases*, pages 146–155, Bombay, India, 1996.

7. C. Glymour, D. Madigan, D. Pregibon, and P. Smyth. Statistical themes and lessons for data mining. *Data Mining and Knowledge Discovery*, 1(1):11–28, 1997.

8. J. Han. Towards on-line analytical mining in large databases. *ACM Sigmod Record*, 27(1):97–107, 1998.

9. J. Han, Y. Cai, and N. Cercone. Data-driven discovery of quantitative rules in relational databases. *IEEE Transactions on Knowledge and Data Engineering*, 5(1):29–40, 1993.

10. K. Kaufman, R. Michalski, and L. Kerschberg. Mining for knowledge in databases: Goals and general description of the INLEN system. *Proc. IJCAI-89 Workshop on Knowledge Discovery in Databases*, pages 158–172, 1989.

11. A. Silberschatz and A. Tuzhilin. What makes patterns interesting in knowledge discovery systems. *IEEE Transactions on Knowledge and Data Engineering*, 8(6):970–974, 1996.

12. B. G. Silverman, M. R. Hieb, and T. M. Mezher. Unsupervised discovery in an operational control setting. In Gregory Piatetsky-Shapiro and William J. Frawley, editors, *Knowledge Discovery in Databases*, pages 432–448. MIT Press, 1991.

13. Jong P. Yoon and Larry Kerschberg. A framework for knowledge discovery and evolution in databases. *IEEE Transactions on Knowledge and Data Engineering*, 5(6):973–978, December 1993.

Boosting Cost-Sensitive Trees

Kai Ming Ting and Zijian Zheng

School of Computing and Mathematics
Deakin University,
Vic 3168, Australia
{kmting,zijian}@deakin.edu.au

Abstract. This paper explores two techniques for boosting cost-sensitive trees. The two techniques differ in whether the misclassification cost information is utilized during training. We demonstrate that each of these techniques is good at different aspects of cost-sensitive classifications. We also show that both techniques provide a means to overcome the weaknesses of their base cost-sensitive tree induction algorithm.

1 Introduction

Cost-sensitive classification deals with situations where different types of incorrect prediction cost differently. The cost of incorrect predictions is more important than the number of incorrect predictions in many real world applications. For example, in medical diagnosis, the cost committed in diagnosing someone as healthy when one has a life-threatening disease is usually considered to be much higher than another type of error—of diagnosing someone as ill when one is in fact healthy. The former type of error is more serious than the latter type of error.

Recently, Ting (1998) introduces a method to induce cost-sensitive trees directly from training data. The intuition is to have *unequal* initial weights which reflect the (given) costs of misclassifications. This effectively influences the learner to focus on instances which have high misclassification costs. This method can be easily adapted to the standard tree induction procedure, which effectively converts the standard procedure that seeks to minimize *the number of errors*, regardless of cost, to a procedure that seeks to minimize *the number of errors with high weight/cost*. Ting (1998) shows that C4.5CS, an adaptation of the standard algorithm C4.5 (Quinlan, 1993) by initial weight modification, performs very well with substantially fewer high cost errors and lower total misclassification costs in two-class problems. This result indicates that the standard tree induction algorithm is a good algorithm, in its own right, for cost-sensitive tree induction. However, C4.5CS has two weaknesses: (1) it requires the conversion of cost matrix to cost vector which hampers its performance in multi-class datasets, and (2) it might not perform well in terms of total misclassification costs in datasets with highly skewed class distribution.

Boosting has been shown to be an effective method of combining multiple models in order to enhance the predictive accuracy of a single model (Quinlan,

1996; Freund & Schapire, 1996). Because of the use of the instance weight modification, boosting appears to be a natural extension to the single cost-sensitive tree induction of C4.5CS. Boosting also offers a potential method to overcome the current weaknesses of C4.5CS. Another interesting issue is how the cost can be incorporated into the boosting procedure.

In this paper, we investigate these issues by exploring two techniques of boosting C4.5CS. The first technique is to employ C4.5CS as the base learning algorithm in the ordinary boosting procedure. The second technique is a variant of ordinary boosting which utilizes the misclassification cost information during the induction of multiple cost-sensitive trees. We call the first method **UBoost** (for **Boost**ing with **U**nequal initial instance weights), and the second method **Cost-UBoost** (for **UBoost** with **Cost**-sensitive adaptation). We conduct empirical evaluation to assess the performance of UBoost and Cost-UBoost with comparison to C4.5CS.

The next section describes the procedures used in UBoost and Cost-UBoost. Section 3 describes the cost matrix and the method of conversion to cost vector used in this paper. Section 4 reports experiments with C4.5CS, UBoost, and Cost-UBoost. Section 5 discusses some related issues. Section 6 describes related work, and we summarize our findings in the final section.

2   UBoost and Cost-UBoost

Here, UBoost is implemented by maintaining a weight for each training example (Quinlan, 1996) rather than drawing a succession of independent samples from the original examples (Freund & Schapire, 1996). The key differences between UBoost and the ordinary boosting procedure (Freund & Schapire, 1996) are that (1) the latter uses *equal* initial weights, whereas the former employs *unequal* initial weights which enable cost-sensitive inductions (Ting, 1998); and (2) the former uses the minimum expected cost criterion (see Equation (5) below) to select a prediction class during classification, instead of the maximum weight criterion.

Similar to the ordinary boosting procedure, UBoost induces multiple individual classifiers in sequential trials. At the end of each trial, the vector of weights is adjusted to reflect the importance of each training example for the next induction trial. This adjustment effectively increases the weights of misclassified examples. These weights cause the learner to concentrate on different instances in each trial and so lead to different classifiers. Finally, the individual trees are combined through voting to form a composite classifier. The UBoost procedure is shown as follows. Note that the weight adjustment formulas in step (iii) below are from a new version of boosting (Schapire, Freund, Bartlett, & Lee, 1997).

UBoost procedure: Given a training set $\mathcal{T}$ containing N examples, $w_k(n)$ denotes the weight of the nth example at the kth trial. The weight of every class j instance at $k = 1$ is initialized as follows (Ting, 1998).

$$w_1(n) = w^j = C^j \frac{N}{\sum_i C^i N^i}, \tag{1}$$

where w^j is the initial weight of a class j instance, and C^j is the cost of misclassifying a class j instance, and N^j is the number of class j instances. $N/\sum_i C^i N^i$ is a normalizing term such that $\sum_j w^j N^j = N$.

In each trial $k = 1, \ldots, K$, the following three steps are carried out.

(i) A decision tree T_k is constructed by using C4.5 from the training set under the weight distribution w_k.

(ii) $\mathcal{T}$ is classified using the decision tree T_k. Let $d(n) = 1$ if the nth example in $\mathcal{T}$ is classified incorrectly; $d(n) = 0$ otherwise. The error rate of this tree, ϵ_k, is defined as:

$$\epsilon_k = \frac{1}{N} \sum_n w_k(n) d(n). \tag{2}$$

If $\epsilon_k \geq 0.5$ or $\epsilon_k = 0$, then all $w_k(n)$ is reset using bootstrap sampling, i.e., $w_k(n)$ is set zero and incremented one unit every time instance n is selected in the sampling with replacement process to select N samples. Then, the process continues from step (i).

(iii) The weight vector $w_{(k+1)}$ for the next trial is created from w_k as follows:

$$w_{(k+1)}(n) = w_k(n) \frac{exp(-\alpha_k(-1)^{d(n)})}{z_k}, \tag{3}$$

where the normalizing term z_k and α_k are defined as follows.

$$z_k = 2\sqrt{(1 - \epsilon_k)\epsilon_k}, \qquad \alpha_k = \frac{1}{2} ln((1 - \epsilon_k)/\epsilon_k). \tag{4}$$

After K trials, the decision trees $T_1, \ldots, T_K$ are combined to form a single composite classifier. Given an example, the final classification of the composite classifier relies on the votes of all the individual trees. The vote of the tree T_k is worth α_k units. Since we use the expected misclassification cost to select the predicted class, the voting is not simply summing up the vote of every individual tree. Instead, the following computation is performed.

Let $t_k(x)$ be the leaf of the tree T_k where the example x falls into, and $W^i(t_k(x))$ be the total weight of class i examples in $t_k(x)$. The expected misclassification cost for class j with respect to the example x and the composite classifier consisting of trees $T_1, \ldots, T_K$ is given by:

$$EC^j(x) \propto \sum_i^I \sum_k^K \alpha_k W^i(t_k(x)) cost(i, j), \tag{5}$$

where $cost(i, j)$ is the misclassification cost of classifying a class i example as class j; and I is the total number of classes.

To classify a new example x, $EC^j(x)$ is computed for every class. The example x is assigned to class j with the smallest value for $EC^j(x)$. That is, $EC^j(x) < EC^{j'}(x)$ for all $j' \neq j$.

From the description above, it can be seen that UBoost only utilizes the mis-classification cost information during induction of the first tree through initial weight setting (Equation (1)) and during classification through the computation of the expected misclassification cost (Equation (5)). Its classifier induction process from $k = 2$ onwards does not employ the cost information.

One can modify the UBoost procedure so that the weights of misclassified examples are updated according to the costs associated with these misclassifications. Thus, each subsequent tree is cost-sensitive. Based on this idea, UBoost is modified to create a variant: **Cost-UBoost**. Cost-UBoost uses the same procedure as UBoost except the weight adjustment process in step (iii). We assume a unity condition $cost(i, j) \geq 1, \forall i \neq j$ (see details in the next section); and the weight adjustment is re-defined as follows.

$$w_{(k+1)}(n) = \frac{w'_{(k+1)}(n)}{\sum_n w'_{(k+1)}(n)} N, \tag{6}$$

$$w'_{(k+1)}(n) = \begin{cases} cost(actual(n), predicted(n)), & \text{if } actual(n) \neq predicted(n); \\ w_k(n), & \text{otherwise.} \end{cases} \tag{7}$$

Equation (7) replaces an instance's weight with the misclassification cost if it is misclassified by the current tree T_k; otherwise the current weight is retained for the next trial. Equation (6) performs normalization to ensure $\sum_n w_{k+1}(n) = N$. The normalization is important because there is no reason to alter the size of training set, which is equivalent to the sum of all training instance weights, while the individual instance weights are adjusted to reflect the relative importance of instances for making future prediction with respect to cost-sensitive classification.

During the classification stage, Cost-UBoost also uses Equation (5) for selecting the class with the minimum expected cost except that each individual tree in Cost-UBoost is worth 1 unit for voting, that is, $\alpha_k = 1$.

We denote the base line algorithm as C4.5CS, which induces a tree using the initial weight distribution w_1, computed using Equation (1). As UBoost and Cost-UBoost, C4.5CS also employs the same formulae for selecting the prediction class, in which $K = 1$ and $\alpha_k = 1$. Note that the tree induced by C4.5CS is exactly the same as the first tree in both UBoost and Cost-UBoost.

3   Cost Matrix and Cost Vector

In a classification task of J classes, the misclassification costs can be specified in a cost matrix of size $J \times J$. The off-diagonal entries contain the costs of misclassifications; and on the diagonal lie the costs for correct classifications which are zero in this case since our main concern here is total misclassification costs of a (composite) classifier.

Each entry of a cost matrix, denoted as $cost(i, j)$, is the cost of misclassifying a class i instance as belonging to class j. In all cases, $cost(i, j) = 0.0$, for $i = j$.

A cost matrix must be converted to a cost vector C^i in order to use Equation (1) for instance-weighting. In this paper, we employ the form of conversion suggested by Breiman, Friedman, Olshen, & Stone (1984):

$$C^i = \sum_j^J cost(i,j). \qquad (8)$$

In our experiments, without loss of generality, we impose a unity condition—$cost(i,j) \geq 1, \forall i \neq j$, and at least one $cost(i,j) = 1.0$. In two-class datasets, one of the two off-diagonal entries must be 1 and the other more than 1. The only reason to have this unity condition or normalization is to allow us to measure the number of *high cost errors*, which is defined as the number of misclassification errors that have costs more than 1.0.

Ting (1998) has noted that the cost matrix to cost vector conversion works well with the single cost-sensitive tree induction when there are only two classes. But it might be inappropriate when there are more than two classes because it collapses $J \times J$ numbers to J. In order to investigate the potential problem due to this conversion, we explicitly divide the experimental datasets into two groups: two-class and multi-class. Any performance discrepancy between these two groups is likely to be due to this conversion.

4   Experiments

In this section, we empirically evaluate UBoost and Cost-UBoost by comparing with C4.5CS. Twenty-two natural datasets from the UCI machine learning repository (Merz & Murphy, 1997) are used in the experiments. This test suite covers a wide variety of different datasets with respect to dataset size, the number of classes, the number of attributes, and types of attribute.

Ten 10-fold cross-validations (Breiman *et al.*, 1984) are carried out in each domain, except in the Waveform domain where 100 pairs of training set of size 300 and test set of size 5000 are randomly generated. In each run, the same cost matrix is employed in training and testing. All reported results are averaged over 100 runs.

A cost matrix for each dataset is randomly generated for each experimental run, except the Heart(Statlog) and GermanCredit datasets. In the latter cases, the costs (i.e., $cost(1,2) = 1.0$ and $cost(2,1) = 5.0$) specified in Michie, Spiegelhalter, & Taylor (1994) are used. In other datasets, the costs in the off-diagonal entries are any randomly generated integer between 1 and 10 in each matrix, satisfying the unity condition.

We use two measures to evaluate the performance of the algorithms employed for cost-sensitive classification. The first measure is the *total cost of misclassifications* made by a classifier on a test set (i.e., $\sum_m cost(actual(m), predicted(m))$). The second measure is the *number of high cost errors*. It is the number of misclassifications associated with costs higher than 1 made by a classifier on a test set. Note that the lowest misclassification cost is 1 in a normalized cost matrix (or a cost matrix with the unity condition). A good cost-sensitive classifier should have low total misclassification cost or small number of high cost errors.

4.1 Comparison of C4.5CS, UBoost, and Cost-UBoost

The parameter K controlling the number of classifiers generated in both UBoost and Cost-UBoost is set at 10 for all experiments. It is interesting to see the performance improvement that can be gained by a single order of magnitude increase in computation. All C4.5CS parameters have their default values as in C4.5, and only pruned trees are used.

Table 1. Comparison of C4.5CS, UBoost and Cost-UBoost

Datasets	C4.5CS		UBoost vs C4.5CS		Cost-UBoost vs C4.5CS		Cost-UBoost vs UBoost	
	cost	#hce	cost ratio	#hce ratio	cost ratio	#hce ratio	cost ratio	#hce ratio
Echocardiogram	7.1	0.58	.92	.21	.91	.47	.98	2.25
Hepatitis	5.4	0.40	1.09	.37	.94	.87	.86	2.33
Heart(Statlog)	9.7	0.73	1.06	.29	.88	.48	.83	1.67
Heart	13.7	1.31	.90	.24	.83	.51	.92	2.09
Horse	18.2	0.43	1.05	.23	.86	1.98	.83	8.50
Credit	26.0	0.93	1.09	.63	.76	1.46	.70	2.31
Breast-W	9.9	0.92	.94	.48	.71	.54	.75	1.14
Diabetes	34.8	1.88	1.04	.40	.93	.76	.89	1.91
German	30.5	0.16	.98	.44	.99	.31	1.00	.71
Euthyroid	22.9	1.52	1.39	.66	.88	1.39	.64	2.11
Hypothyroid	10.2	0.68	1.44	.69	.74	1.15	.51	1.66
Coding	935.6	27.36	.98	.26	.96	.50	.98	1.94
Mean			1.07	.41	.86	.87	.82	2.38
Lymphography	15.5	3.26	.89	1.29	.88	1.15	.99	.89
Glass	35.5	5.92	.72	.86	.75	.87	1.05	1.02
Waveform	6920.6	1154.27	.64	.74	.64	.74	1.01	1.00
Soybean	28.6	5.23	.90	.90	.75	.89	.83	.99
Annealing	35.1	6.98	.59	.56	.66	.69	1.13	1.23
Vowel	114.0	19.14	.52	.67	.55	.70	1.05	1.04
Splice	95.3	17.14	.97	.90	.81	.81	.83	.91
Abalone	676.1	121.74	.93	.93	.86	.85	.93	.92
Nettalk(s)	459.8	92.61	.79	.96	.74	.84	.93	.88
Satellite	464.3	76.02	.68	.95	.66	.86	.97	.90
Mean			.76	.88	.73	.84	.97	.98

Table 1 shows the misclassification costs and the number of high cost errors of C4.5CS. The ratios for the two measures for the pair-wise comparison among C4.5CS, UBoost, and Cost-UBoost are presented in the last three columns. A ratio of less than 1 for UBoost vs C4.5CS, for example, represents an improvement due to UBoost. The mean ratios over the twelve two-class problems and the ten multi-class problems are also shown.

In terms of total misclassification costs, UBoost achieves a mean reduction of 24% over C4.5CS in the multi-class problems, but it has a mean increase of 7% in the two-class problems. Cost-UBoost reduces the misclassification costs of C4.5CS in all datasets, with a mean reduction of 27% in the multi-class problems and 14% in the two-class problems. Using a two-tailed pairwise sign test, UBoost in the multi-class problems and Cost-UBoost in all datasets are superior to C4.5CS at a significance level better than 1%. Compared with UBoost, Cost-UBoost has lower costs in all but one two-class problems, with a mean reduction of 18%. Both UBoost and Cost-UBoost make comparable misclassification costs in the multi-class problems.

In terms of the number of high cost errors, UBoost improves C4.5CS dramatically in the two-class problems, achieving a mean 59% reduction over C4.5CS; but only a mean reduction of 12% in the multi-class problems. In comparison to C4.5CS, Cost-UBoost achieves a mean reduction of 13% in the two-class problems and 16% in the multi-class problems. Using a two-tailed pairwise sign test, UBoost in all datasets and Cost-UBoost in the multi-class problems are superior to C4.5CS at a significance level better than or equal to 1%. Comparing Cost-UBoost directly to UBoost, the former has the number of high cost errors 2.38 times larger than the latter in the two-class problems. Note that one single domain, Horse, makes a significant contribution to this increase. Both UBoost and Cost-UBoost have comparable number of high cost errors in the multi-class problems.

In summary, in two-class problems, UBoost is a better boosting procedure to minimize the number of high cost errors, and Cost-UBoost is better to minimize the total misclassification costs. In multi-class problems, both UBoost and Cost-UBoost perform comparably in either of the two measures.

4.2   Why different performance between two-class and multi-class problems?

In this section, we attempt to investigate the reason why there exists a difference of relative performance of UBoost and Cost-UBoost between two-class and multi-class problems. As noted in Section 3, the performance difference between two-class and multi-class problems could be due to the conversion of cost matrix to cost vector. One way to test this hypothesis is to have a form of cost matrix for multi-class problems such that it closely resembles the form of cost matrix for two-class problems. If the relative performance of UBoost versus Cost-UBoost is similar when presented with the new form of cost matrix for multi-class problem and the cost matrix for two-class problems, then we accept this hypothesis; otherwise we reject it.

One such possible form of cost matrix can be defined as: $cost(i, j) > 1$ only for a single value of $j = J$; and $cost(i, j \neq J) = 1$ for all $i \neq j$. We call this Type A cost matrix. Examples of this cost matrix and the more general Type G cost matrix, together with a two-class cost matrix are shown in Table 2. Type G is used in the last section.

Table 2. Examples of two types of cost matrix for multi-class problems with comparison to that for a two-class problem.

	Two-class		Type A			Type G	
	i		i			i	
	1 2	1	2	3	1	2	3
	1 0.0 6.0	1 0.0	3.0	7.0	1 0.0	3.0	2.0
j	2 1.0 0.0	2 1.0	0.0	1.0	2 1.0	0.0	7.0
		3 1.0	1.0	0.0	3 10.0	5.0	0.0

Table 3. Mean ratios for Cost-UBoost/UBoost on the effect of cost matrix types

Cost Matrix Type (Two/Multi-class)	cost ratio	#hce ratio
Type G (Multi-class)	.97	.98
Type A (Multi-class)	.95	3.49
(Two-class)	.82	2.38

Table 3 shows the summary results of using Type A cost matrix in multi-class problems. The previous results from Table 1 are also listed for ease of comparison.

In terms of the number of high cost errors, the relative performance of UBoost and Cost-UBoost, while using Type A cost matrix in multi-class problems, closely resembles that in two-class problems. UBoost performs significantly better than Cost-UBoost in reducing the number of high cost errors. This shows that the conversion of cost matrix to cost vector does play a major part in causing the different performance between two-class and multi-class problems.

In terms of total misclassification costs, however, the relative performance improves only slightly in favor of Cost-UBoost while using Type A cost matrix instead of Type G cost matrix. This shows that UBoost does not perform worse than Cost-UBoost in multi-class problems as in two-class problems (see Table 1).

4.3 Do UBoost and Cost-UBoost help to overcome C4.5CS' weaknesses?

We demonstrate the weaknesses of C4.5CS reported by Ting (1998), by comparing C4.5CS to C4.5c in this section. C4.5c is a version of C4.5 which employs the minimum expected cost criterion as in Equation (5) with $K = 1$ and $\alpha_k = 1$. The only difference between C4.5c and C4.5CS is that the former uses *equal* initial instance weights and the later uses *unequal* initial instance weights.

Table 4. Mean ratios of C4.5c, UBoost, and Cost-UBoost to C4.5CS

	Two-class		Multi-class		Euthyroid		Hypothyroid	
	cost ratio	#hce ratio	cost ratio	#hce ratio	cost ratio	#hce ratio	cost ratio	#hce ratio
C4.5c/C4.5CS	1.27	6.63	.99	.96	.92	1.96	.80	1.59
UBoost/C4.5CS	1.07	.41	.76	.88	1.39	.66	1.44	.69
Cost-UBoost/C4.5CS	.86	.87	.73	.84	.88	1.39	.74	1.15

Table 4 shows the mean ratios of C4.5c, UBoost, and Cost-UBoost to C4.5CS for both the total misclassification costs and the number of high cost errors. The third column shows the first weakness of C4.5CS in multi-class problems where it performs slightly worse than C4.5c. The last two columns show the results in the Euthyroid and Hypothyroid datasets which have highly skewed class distribution—they have default accuracies 90.7% and 95.2% respectively. In both datasets, C4.5CS performs worse than C4.5c in terms of total misclassification costs.

The results clearly show that both UBoost and Cost-UBoost provide a means to improve the performance of C4.5CS in multi-class problems. In datasets with highly skewed class distribution, one should employ Cost-UBoost, instead of UBoost, to improve the performance of C4.5CS in terms of total misclassification costs. On the other hand, in terms of the number of high cost errors, one should use UBoost instead.

5   Discussion

Although the conversion of cost matrix to cost vector contributes to the different performance between two-class and multi-class problems, it does not hamper the performance of either UBoost or Cost-UBoost in multi-class problems, as it does to C4.5CS. This shows the robustness of using multiple models.

The success of Cost-UBoost suggests that boosting's "success lay not in its specific form but its adaptive sampling property where increased weight was placed on those instances more frequently misclassified" (Breiman, 1996). This finding is consistent with Breiman's who uses a different form of adaptive sampling from Equation (3) for classification problems.

Ting (1998) shows that transforming C4.5 to C4.5CS is converting a procedure that seeks to minimize *the number of errors* to a procedure that seeks to minimize *the number of high cost errors*. Here we show that employing UBoost with C4.5CS as its base induction algorithm further minimize the number of high cost errors. The reason why this is more obvious in two-class problems than in multi-class problems is due to the use of conversion of cost matrix to cost vector, described in Section 4.2.

For UBoost, α_k may be dropped from Equation (5). Our experiments show that it makes little difference whether α_k is set to 1 or use it as suggested by Schapire *et al.* (1997). The mean ratios for UBoost($\alpha_k = 1$)/UBoost over twenty-two datasets are 1.04 and .96 for total misclassification costs and the number of high cost errors, respectively.

Ting & Zheng (1998) describe a related work applied to a situation where cost changes very often. In this situation, one might want to retain only one single (composite) classifier, i.e., perform only one induction, and use the same (composite) classifier for every cost change. The situation demands the boosting procedures to begin with equal weights. An experiment comparing UBoost (Cost-UBoost) with equal and unequal weights reveals that the result is in favor of the later in terms of the number of high cost errors. The mean ratios for

Boost(equal weight)/UBoost over twenty-two datasets are .96 and 1.11 for total misclassification costs and the number of high cost errors, respectively. Similarly, the mean ratios for Cost-Boost(equal weight)/Cost-UBoost are 1.00 and 1.30.

As with any other methods which employ multiple models, one loses comprehensibility of the induced trees with UBoost and Cost-UBoost.

6   Related Work

Cost-UBoost is inspired by the instance weight adjustment method in decision tree learning for boosting (Quinlan, 1996) and for effective cost-sensitive tree induction (Ting, 1998). The principle of using the minimum expected cost criterion for cost-sensitive classification is described by Michie et al. (1994).

There are some research on the induction of a single cost-sensitive tree. Breiman et al. (1984), Knoll, Nakhaeizadeh, & Tausend (1994), Pazzani, Merz, Murphy, Ali, Hume, & Brunk (1994), Webb (1996), and Ting (1998) investigate methods of incorporating variable misclassification costs into the processes of tree generation, tree pruning, and tree specialization for cost-sensitive classifications. Furthermore, there are a few decision tree learning algorithms that consider the costs of tests, such as EG2 (Núñez, 1991), CS-ID3 (Tan, 1993), and IDX (Norton, 1989). Turney (1995) studies the effect of test cost using a genetic algorithmic search in decision tree induction, under the condition of equal misclassification costs (i.e., $cost(1,2) = cost(2,1)$). All the above systems induce a single cost-sensitive tree. None of them has explored multiple models for cost-sensitive classifications.

Recent research in boosting has provided promising results from both theoretical and empirical aspects (Freund & Schapire, 1996; Quinlan, 1996; Schapire et al., 1997). Nevertheless, boosting has not been applied to cost-sensitive classifications yet, with the exception of Ting & Zheng (1998).

7   Summary

This paper has investigated practical issues in relation to the use of boosting for cost-sensitive classifications. We have explored two techniques for boosting cost-sensitive trees. One is UBoost—the ordinary boosting which begins with unequal initial instance weights, together with the minimum expected cost criterion in the classification stage. This first technique does not consider misclassification cost during induction of classifiers, except the first tree, and only use the cost information during classification stage by employing the minimum expected cost criterion to select the predicted class.

The second technique is Cost-UBoost, a variant of UBoost, which takes the advantage of the available misclassification cost information during training. This makes the boosting procedure more sensitive to the cost of misclassification.

Experimental results show that both UBoost and Cost-UBoost can significantly reduce the misclassification cost and the number of high cost errors of a single cost-sensitive tree in multi-class problems. In two-class problems, UBoost is a better choice in reducing the number of high cost errors; but in terms of misclassification cost, Cost-UBoost is a better choice than UBoost. We demonstrate that the different performance between two-class and multi-class problems is largely due to the use of cost matrix to cost vector conversion.

We also shows that both UBoost and Cost-UBoost can be employed to overcome the weaknesses of C4.5CS, the base cost-sensitive tree induction algorithm in both techniques, in multi-class problems and in datasets with highly skewed class distribution.

8   References

Breiman, L. (1996), Bias, variance, and arcing classifiers, *Technical Report 460*, Department of Statistics, University of California, Berkeley, CA.

Breiman, L., J.H. Friedman, R.A. Olshen, & C.J. Stone (1984), *Classification And Regression Trees*, Belmont, CA: Wadsworth.

Freund, Y. & R.E. Schapire (1996), Experiments with a new boosting algorithm, in *Proceedings of the Thirteenth International Conference on Machine Learning*, pp. 148-156, Morgan Kaufmann.

Knoll, U., G. Nakhaeizadeh, & B. Tausend (1994), Cost-sensitive pruning of decision trees, in *Proceedings of the Eighth European Conference on Machine Learning*, pp. 383-386. Berlin, Germany: Springer-Verlag.

Merz, C.J. & P.M. Murphy (1997), *UCI Repository of machine learning databases* [http://www.ics.uci.edu/~mlearn/MLRepository.html]. Irvine, CA: University of California, Department of Information and Computer Science.

Michie, D., D.J. Spiegelhalter, & C.C. Taylor (1994), *Machine Learning, Neural and Statistical Classification*, Ellis Horwood Limited.

Norton, S.W. (1989), Generating better decision trees, in *Proceedings of the Eleventh International Joint Conference on Artificial Intelligence*, pp. 800-805, Morgan Kaufmann.

Núñez, M. (1991), The use of background knowledge in decision tree induction, *Machine Learning, 6*, pp. 231-250.

Pazzani, M., C. Merz, P. Murphy, K. Ali, T. Hume, & C. Brunk (1994), Reducing misclassification costs, in *Proceedings of the Eleventh International Conference on Machine Learning*, pp. 217-225, Morgan Kaufmann.

Quinlan, J.R. (1993), *C4.5: Program for Machine Learning*, Morgan Kaufmann.

Quinlan, J.R. (1996), Bagging, boosting, and C4.5, in *Proceedings of the 13th National Conference on Artificial Intelligence*, pp. 725-730, AAAI Press.

Schapire, R.E., Y. Freund, P. Bartlett, & W.S. Lee (1997), Boosting the margin: A new explanation for the effectiveness of voting methods, in *Proceedings of the Fourteenth International Conference on Machine Learning*, pp. 322-330. Morgan Kaufmann.

Tan, M. (1993), Cost-sensitive learning of classification knowledge and its applications in robotics, *Machine Learning, 13*, pp. 7-33.

Ting, K.M. (1998), Inducing cost-sensitive trees via instance-weighting, to appear in *Proceedings of The Second European Symposium on Principles of Data Mining and Knowledge Discovery*, Springer-Verlag.

Ting, K.M. & Z. Zheng (1998), Boosting trees for cost-sensitive classifications, *Proceedings of the Tenth European Conference on Machine Learning*, LNAI-1398, pp. 190-195, Berlin: Springer-Verlag.

Turney, P.D. (1995), Cost-sensitive classification: Empirical evaluation of a hybrid genetic decision tree induction algorithm, *Journal of Artificial Intelligence Research, 2*, pp. 369-409.

Webb, G.I. (1996), Cost-sensitive specialization, in *Proceedings of the 1996 Pacific Rim International Conference on Artificial Intelligence*, pp. 23-34, Springer-Verlag.

On the Boosting Algorithm for Multiclass Functions Based on Information-Theoretic Criterion for Approximation

Eiji Takimoto and Akira Maruoka

Graduate School of Information Sciences, Tohoku University
Aoba 05, Aramaki, Aoba, Sendai, 980-8579 Japan
{t2, maruoka}@ecei.tohoku.ac.jp

Abstract. We consider the boosting technique that can be directly applied to the classification problem for multiclass functions. Although many boosting algorithms have been proposed so far, all of them are essentially developed for binary classification problems, and in order to handle multiclass classification problems, they need the problems reduced somehow to binary ones. In order to avoid such reductions, we introduce a notion of the pseudo-entropy function G that gives an information-theoretic criterion, called the conditional G-entropy, for measuring the loss of hypotheses. The conditional G-entropy turns out to be useful for defining the weakness of hypotheses that approximate, in some way, to a multiclass function in general, so that we can consider the boosting problem without reduction. We show that the top-down decision tree learning algorithm using G as its splitting criterion is an efficient boosting algorithm based on the conditional G-entropy. Namely, the algorithm intends to minimize the conditional G-entropy, rather than the classification error. In the binary case, our algorithm is identical to the error-based boosting algorithm proposed by Kearns and Mansour, and our analysis gives a simpler proof of their results.

1 Introduction

Boosting is a technique of finding a hypothesis with high accuracy by combining many *weak* hypotheses, which is only moderately accurate. In the binary classification problem, a weak hypothesis is usually defined to have error slightly less than $1/2$. Here, the goodness of a hypothesis is measured by error, i.e., the probability of misclassification. As to error, the probability just $1/2$ is considered to be worst because it can be achieved by random guessing. So is was surprising when Schapire [6] first gave a boosting algorithm that combines many weak hypotheses only slightly better than random guesses to construct a master hypothesis that approximates a target function with high accuracy. Since then many boosting algorithms have been proposed and extensively studied both in practice and theory [1–4, 7], but all of them are essentially based on error and require weak learning algorithms to produce hypotheses with error less than $1/2$.

More surprisingly, in some case, we can boost accuracy about a target function from hypotheses with error just 1/2. In fact, Natarajan gave a boosting algorithm from weak learning algorithms that are guaranteed to produce hypotheses with error only slightly below 1 (so, including just 1/2) but being *one-sided* [5]. It turned out that the hypotheses with one-sided error always have positive *mutual information* about a target function whenever the error is below 1, where the mutual information is based on Shannon entropy. So, it is natural to measure the goodness of a hypothesis in terms of the amount of information, concerning a target, in the hypothesis. Based on this observation, Takimoto, Tajika and Maruoka provided a framework in which the mutual information is used to measure how well the hypothesis approximates the target [9].

This information-based criterion for approximation turns out to be very useful for defining the weakness of hypotheses especially in multiclass classification setting for the following two reasons, where a target function f takes values in the set of classes $Y = \{1, \dots, N\}$ for $N \geq 2$. First of all, the mutual information is defined between any two multiclass functions, and secondly the weak hypothesis is naturally defined to have non-zero mutual information. In the error-based criterion, on one hand, the bound 1/2 cannot be relaxed to a larger bound, say, $1 - 1/N$, which is the error by a random guess of N labels. This is because the distribution over the instance space might give positive weights only on instances on which f takes either value in $\{1, 2\} \subseteq Y$, yielding essentially a binary classification problem. However, it seems too strong to require a weak learning algorithm to produce error less than 1/2 even if the distribution makes f take all values in Y with equal probability. In this case, a hypothesis with error less than $1 - 1/N$ would contain positive information on f.

In this paper, we investigate information-based boosting algorithms for multiclass classification problems. Although the error-based boosting algorithms proposed so far can also be applied to multiclass functions, they need the problems reduced somehow to the binary classification problems [3, 8]. As long as we stand on the error-based criterion, the bound cannot be greater than 1/2 anyway, and so the reduction to the binary setting would be unavoidable.

The first non-trivial information-based boosting algorithm is the work due to Kearns and Mansour [4], although it was not explicitly stated. They analyzed the performance of top-down algorithms for growing decision trees, such as C4.5 and CART, in order to give justification to empirically successful heuristics, and they showed that the algorithms are actually error-based boosting algorithms in the binary classification setting. The top-down algorithm uses a function G called the *splitting criterion* to decide which classification rule the internal node should be labeled with. In particular, C4.5 uses the Shannon entropy function as the splitting criterion. Then, under our interpretation, it can be viewed as an *information-based* boosting algorithm. We generalize the result for multiclass classification problems. First we extend the notion of entropy and introduce a function G called a *pseudo-entropy*. Using G, we define the criterion called the conditional G-entropy for measuring the loss of a hypothesis. Many functions including Shannon entropy function can be used as pseudo-entropy. We show that for any pseudo-entropy G, the top-down decision tree learning algorithm using G as its splitting criterion is an information-based boosting algorithm. Then, we choose a particular function G for later analysis. As a by-product, our analysis gives a simpler proof of the result by Kearns and Mansour.

2   Information-theoretic criterion for approximation

Let X denote an instance space and Y a finite set of labels. We assume $Y = \{1, \ldots, N\}$ with $N \geq 2$. Throughout the paper, we fix a target function $f : X \to Y$. Let D denote a probability distribution over X. We consider a criterion for approximation to f under D by means of a function $h : X \to Z$, where Z is a finite set but possibly different from Y. In the special case that $Y = Z$, we typically measure the loss (the badness of approximation) of h in terms of the probability of misclassification, i.e., $\Pr_D(f(x) \neq h(x))$. We call this measure the error of h. In this paper, we measure the loss of h from an information-theoretic view point. To do so, we extend the notion of entropy and introduce a function $G : [0,1]^N \to [0,1]$ having three properties:

1. There exist constants $a > 0$ and $0 < b \leq 1$ such that for any $(q_1, \ldots, q_N) \in [0,1]^N$ with $\sum_i q_i = 1$,

$$\min_{1 \leq i \leq N} (1 - q_i) \leq a\, G(q_1, \ldots, q_N)^b.$$

2. For any $(q_1, \ldots, q_N) \in [0,1]^N$ with $\sum_i q_i = 1$,

$$G(q_1, \ldots, q_N) = 0 \quad \Leftrightarrow \quad q_i = 1 \text{ for some } 1 \leq i \leq N.$$

3. G is concave.

We call such a function G a *pseudo-entropy function*. Note that Shannon entropy function

$$H_n(q_1, \ldots, q_N) = -\frac{1}{\log N} \sum_{i=1}^{N} q_i \log q_i$$

is also a pseudo-entropy function. In the following, G is assumed to be an arbitrary pseudo-entropy function.

Interpreting $G(q_1, \ldots, q_N)$ as uncertainty of labels under the probability distribution $(q_1, \ldots, q_N)$ over Y, we can define notions of the entropy, the conditional entropy and the mutual information based on G.

Definition 1. For $1 \leq i \leq N$, let $q_i = \Pr_D(f(x) = i)$. Then the *G-entropy of f with respect to D*, denoted $H_D^G(f)$, is defined as

$$H_D^G(f) = G(q_1, \ldots, q_N).$$

Then, we can naturally define the loss of h based on the conditional G-entropy of f given h, which is interpreted as uncertainty of f remaining after receiving the value of h.

Definition 2. For $1 \leq i \leq N$ and $z \in Z$, let

$$q_{i|z} = \Pr_D(f(x) = i|h(x) = z).$$

Then, the *conditional G-entropy of f given h with respect to D*, denoted $H_D^G(f|h)$, is defined as

$$H_D^G(f|h) = \sum_{z \in Z} \Pr_D(h(x) = z)G(q_{1|z}, \ldots, q_{N|z}),$$

and the *mutual G-information between f and h with respect to D*, denoted $I_D^G(f;h)$, is defined as

$$I_D^G(f;h) = H_D^G(f) - H_D^G(f|h).$$

Note that since G is concave, $I_D^G(f,h) \geq 0$ always holds, that is, any function h brings non-negative amount of information on f.

Now we show how the conditional G-entropy is related to the error. Since the range of h may be different from that of f, we have to specify a rule that transforms the value of h to an element in Y, so that we can define the error of h. For such a rule, it is natural to use the maximum likelihood estimator. More precisely, let the function $M : Z \to Y$ be defined as

$$M(z) = \arg \max_{1 \leq i \leq N} q_{i|z},$$

where $q_{i|z} = \Pr_D(f(x) = i|h(x) = z)$. Then, we can define the error of h to be $\Pr_D(M(h(x)) \neq f(x))$. By definition, we have

$$\Pr_D(M(h(x)) \neq f(x)) = \sum_{z \in Z} \Pr_D(h(x) = z)\Pr_D(M(z) \neq f(x)|h(x) = z)$$

$$= \sum_{z \in Z} \Pr_D(h(x) = z) \min_{1 \leq i \leq N}(1 - q_{i|z})$$

$$\leq \sum_{z \in Z} \Pr_D(h(x) = z)aG(q_{1|z}, \ldots, q_{N|z})^b.$$

Here we used Property 1 of pseudo-entropy function. By the concavity of the function $g(x) = x^b$,

$$\Pr_D(M(h(x)) \neq f(x)) \leq \sum_{z \in Z} \Pr_D(h(x) = z)aG(q_{1|z}, \ldots, q_{N|z})^b$$

$$\leq a \left(\sum_{z \in Z} \Pr_D(h(x) = z)G(q_{1|z}, \ldots, q_{N|z}) \right)^b$$

$$= a \, H_D^G(f|h)^b.$$

Moreover, the inequalities above together with Property 2 says that $\Pr_D(M(h(x)) \neq f(x)) = 0$ if and only if $H_D^G(f|h) = 0$. These imply that, in order to find a function h having small error, it suffices to minimize the conditional G-entropy of f given h. In what follows we fix an arbitrary pseudo-entropy function G. So we omit the superscript G and simply write $H_D(f)$, $H_D(f|h)$ and $I_D(f;h)$ to denote the G-entropy, the conditional G-entropy and the mutual G-information, respectively.

3 Weak learning based on the conditional G-entropy

In this section, we give notions of weak learning algorithms based on the conditional G-entropy. Throughout the paper, we only focus on weak learning and boosting with respect to an empirical distribution. Let S be a sequence of training examples $(\langle x_1, f(x_1) \rangle, \ldots, \langle x_m, f(x_m) \rangle)$, where each instance x_i belongs to X. A weak learning algorithm is given as input a sequence of training examples S along with a distribution D over S, i.e., a distribution over the multiset $\{x_1, \ldots, x_m\}$. Given such input, the weak learning algorithm chooses a function $h : X \to Z$ as its weak hypothesis. By a weak hypothesis we mean a function h with the conditional G-entropy $H_D(f|h)$ bounded above by a non-trivial value, so that h contains non-zero information on f. In particular, we consider a *relative bound* of the conditional G-entropy, which depends on D. Namely, a weak hypothesis is defined to have small conditional G-entropy *relative to* the G-entropy of the target function.

Definition 3 (Weak learning). A *weak learning algorithm* is an algorithm that, when given any sequence of examples S and any probability distribution D over S, produces a function h satisfying

$$H_D(f|h) \le (1 - \gamma) H_D(f)$$

for some positive constant $\gamma > 0$.

4 A boosting algorithm

Now we give an *information-based* boosting algorithm that minimizes the conditional G-entropy rather than the error. A boosting algorithm is given as input a sequence of examples S and accuracy parameter $\varepsilon > 0$. At each trial, it computes a probability distribution D over S and feeds D along with S to a weak learning algorithm available, and then it receives from the weak learning algorithm a hypothesis with not too large conditional G-entropy with respect to D. After an appropriate number of trials, it combines these weak hypotheses somehow to make a master hypothesis h with $H_U(f|h) \le \varepsilon$, where U denotes the uniform distribution over S. As mentioned before, small $H_U(f|h)$ implies small empirical error. Although this does not necessarily guarantee small generalization error, a standard argument shows that the generalization error of a hypothesis is not much larger than its empirical error if the complexity of the hypothesis is not too large (see e.g. [3]).

Kearns and Mansour showed that for binary classification problems, the top-down decision tree learning algorithm using a pseudo-entropy G as the splitting criterion is actually an error-based boosting algorithm [4]. We show that their algorithm is actually an information-based boosting algorithm for multiclass classification problems.

Let H denote a hypothesis class of the weak learning algorithm. Let T be any decision tree whose internal nodes are labeled with functions in H. Let *leaves*(T)

denote the set of leaves of T. Then T naturally defines the function that maps any instance $x \in X$ to a leaf in $leaves(T)$. For each $\ell \in leaves(T)$, w_ℓ denotes the probability that an instance x reaches leaf ℓ, and $q_{i|\ell}$ denotes the probability that $f(x) = i$ given that x reaches ℓ. Here, these probabilities are taken under the uniform distribution over S and so easy to compute. Note that, unlike the usual decision tree, we don't need the classification values labeled with the leaves of T. Then, the conditional G-entropy of f given T can be represented as

$$H_U(f|T) = \sum_{\ell \in leaves(T)} w_\ell G(q_{1|\ell}, \ldots, q_{N|\ell}). \tag{1}$$

The top-down algorithm we examine makes local modifications to the current tree T in an effort to reduce $H_U(f|T)$. Namely, at each local change, the algorithm chooses a leaf $\ell \in leaves(T)$ and $h \in H$ somehow and replaces ℓ by h. The tree obtained in this way is denoted by $T(\ell, h)$. Here the leaf ℓ becomes an internal node labeled with h and creates $|Z|$ child leaves. (So, the total number of leaves increases by $|Z| - 1$.) Now we show how the pair of ℓ and h is chosen at each local change. The algorithm simply chooses ℓ that maximizes $w_\ell G(q_{1|\ell}, \ldots, q_{N|\ell})$. Then it calculates the examples S_ℓ that reach the leaf ℓ and runs the weak learning algorithm **WeakLearn** available by feeding S_ℓ together with the uniform distribution D_ℓ over S_ℓ. Finally the top-down algorithm receives h that **WeakLearn** produces. We give the details of the algorithm in Figure 1. With respect to this distribution D_ℓ, $q_{i|\ell} = \Pr_{D_\ell}(f(x) = i)$ holds for any $1 \le i \le N$, and so we have $H_{D_\ell}(f) = G(q_{1|\ell}, \ldots, q_{N|\ell})$. Moreover, it is not hard to see that

$$H_U(f|T) - H_U(f|T(\ell, h)) = w_\ell(H_{D_\ell}(f) - H_{D_\ell}(f|h)) = w_\ell I_{D_\ell}(f; h). \tag{2}$$

This implies that $H_U(f|T)$ decreases if h has positive information on f with respect to the *filtered* distribution D_ℓ. Now we state our main result.

TopDown(S, ε)
begin
    Let T be the single-leaf tree.
    $\tau := (1/\varepsilon)^{|Z|/\gamma}$;
    for τ **times do begin**
        $\ell := \arg \max_{\ell \in leaves(T)} w_\ell G(q_{1|\ell}, \ldots, q_{N|\ell})$;
        $S_\ell := \{\langle x, f(x)\rangle \in S | x \text{ reaches leaf } \ell\}$;
        Let D_ℓ be the uniform distribution over S_ℓ.
        $h := $ **WeakLearn**(S_ℓ, D_ℓ);
        $T := T(\ell, h)$;
    end;
end.

Fig. 1. Algorithm **TopDown**(S, ε)

Theorem 1. *Let* WeakLearn *be a weak learning based on G-entropy and let γ be the associated constant. Then, for any sequence S of examples of f and any $\varepsilon > 0$, algorithm* **TopDown**(S, ε) *produces T with $H_U(f|T) \leq \varepsilon$.*

Proof. Let T be the tree at the beginning of the t-th iteration of algorithm **TopDown**(S, ε), and let H_t denote the conditional G-entropy $H_U(f|T)$. Note that the number of leaves of T is $(t-1)(|Z|-1)+1 \leq t|Z|$. So by equation (1), there must exist a leaf ℓ such that

$$w_\ell G(q_{1|\ell}, \ldots, q_{N|\ell}) = w_\ell H_{D_\ell}(f) \geq H_t/(t|Z|),$$

where D_ℓ is the uniform distribution over S_ℓ. Since h that **WeakLearn**(S_ℓ, D_ℓ) produces is a weak hypothesis, equation (2) implies

$$H_t - H_{t+1} \geq w_\ell I_{D_\ell}(f; h) \geq \gamma w_\ell H_{D_\ell}(f) \geq \gamma H_t/(t|Z|),$$

or equivalently

$$H_{t+1} \leq \left(1 - \frac{\gamma}{t|Z|}\right) H_t \leq e^{-\frac{\gamma}{t|Z|}} H_t.$$

Since $H_1 \leq 1$, we have

$$H_{\tau+1} \leq e^{-\frac{\gamma}{|Z|} \sum_{t=1}^{\tau} 1/t} \leq e^{-\frac{\gamma}{|Z|} \ln \tau} = \varepsilon$$

as desired.   □

5   A pseudo-entropy function

In the previous section, we showed that boosting is possible whenever we have a weak learning algorithm based on the conditional G-entropy for *any* pseudo-entropy function G. However, it is not clear what G is useful for weak learning. In particular, what G makes the condition of weak learning really weak so that we can easily find a weak learning algorithm based on the G. In this section, we consider the weak learnability under some restricted class of distributions, and we give a particular G as a useful pseudo-entropy function that guarantees that weak learning in this setting implies distribution-free weak learning.

5.1   The binary classification case

In the binary case where $|Y| = 2$, Kearns and Mansour observed that the top-down decision tree learning algorithm shows a good performance of error-based boosting for pseudo-entropy G having a strong concavity, and by this analysis they suggested $G(q_1, q_2) = \sqrt{q_1 q_2}$ that gives an efficient boosting algorithm. Moreover, Schapire and Singer showed that their AdaBoost algorithm works best when weak hypotheses are guaranteed to have not too large conditional G-entropy with $G(q_1, q_2) = \sqrt{q_1 q_2}$ rather than not too large error, provided that the hypotheses are based on a partitioning of the domain like a decision tree [8].

In the information-based setting as well, this choice of G turns out to have a nice property that significantly weaken the condition of weak learning. Namely, we have an equivalent definition of weak learning algorithms even if we restrict D within a class of *balanced* probability distributions. A distribution D is balanced if $q_1 = q_2 = 1/2$, where $q_i = \mathrm{Pr}_D(f(x) = i)$ for $i \in Y = \{1, 2\}$.

Proposition 1. *Assume that $G(q_1, q_2) = \sqrt{q_1 q_2}$ is used for the pseudo-entropy function. If there exists a weak learning algorithm with respect to balanced distributions, then there exists a distribution-free weak learning algorithm.*

Proof. Assume we have a weak learning algorithm A with respect to balanced distributions. Using A, we construct a distribution-free weak learning algorithm B. Suppose that B is given as input a probability distribution D over a sequence of examples $S = (\langle x_1, f(x_1)\rangle, \ldots, \langle x_m, f(x_m)\rangle)$. First B calculates $q_i = \mathrm{Pr}_D(f(x) = i)$ for $i = 1, 2$. Note that q_i is easy to compute since $q_i = \sum_{j: f(x_j) = i} D(x_j)$. Without loss of generality, we can assume that $G(q_1, q_2) \neq 0$. Now we define the balanced distribution D' induced by D as follows. For any $1 \le j \le m$, let $D'(x_j) = D(x_j)/(2q_i)$, where $i = f(x_j)$. Clearly the distribution D' is balanced. Then, B runs A with the examples S along with the distribution D' and receives h that A produces. Since A is a weak learning algorithm with respect to balanced distributions, h must satisfy

$$H_{D'}(f|h) \le (1 - \gamma) H_{D'}(f) \tag{3}$$

for some $\gamma > 0$.

Now we estimate $H_{D'}(f|h)$ in terms of probabilities with respect to D. For $i \in Y$ and $z \in Z$, let $w_z = \mathrm{Pr}_D(h(x) = z)$ and $q_{i|z} = \mathrm{Pr}_D(f(x) = i|h(x) = z)$. An easy calculus gives that

$$w'_z = \mathrm{Pr}_{D'}(h(x) = z) = \frac{w_z}{2}\left(\frac{q_{1|z}}{q_1} + \frac{q_{2|z}}{q_2}\right)$$

and

$$q'_{i|z} = \mathrm{Pr}_{D'}(f(x) = i|h(x) = z) = \frac{q_{i|z}/q_i}{q_{1|z}/q_1 + q_{2|z}/q_2}.$$

Using these equations, we have

$$H_{D'}(f|h) = \sum_{z \in Z} w'_z G(q'_{1|z}, q'_{2|z})$$

$$= \sum_z \frac{w_z}{2}\left(\frac{q_{1|z}}{q_1} + \frac{q_{2|z}}{q_2}\right)\left[\left(\frac{q_{1|z}/q_1}{q_{1|z}/q_1 + q_{2|z}/q_2}\right)\left(\frac{q_{2|z}/q_2}{q_{1|z}/q_1 + q_{2|z}/q_2}\right)\right]^{1/2}$$

$$= \sum_z \frac{w_z}{2} \frac{\sqrt{q_{1|z} q_{2|z}}}{\sqrt{q_1 q_2}} = \frac{H_D(f|h)}{2H_D(f)} = \frac{H_{D'}(f)H_D(f|h)}{H_D(f)}.$$

The last equality comes from the fact that $H_{D'}(f) = \sqrt{(1/2)(1/2)} = 1/2$. By inequality (3), we can conclude $H_D(f|h) \le (1 - \gamma) H_D(f)$, which implies that h is a weak hypothesis with respect to D. $\qquad\square$

It is not hard to see that any error-based weak hypothesis h with $\Pr_D(f(x) \neq h(x)) \leq 1/2 - \gamma$ immediately implies an information-based weak hypothesis with $H_D(f|h) \leq (1 - \gamma^2/2)H_D(f)$ with respect to any balanced distribution. Therefore, Proposition 1 says that any error-based weak learning algorithm can be transformed to an information-based weak learning algorithm in the distribution-free setting. Because we have an information-based boosting algorithm and an information-based strong learning algorithm is also an error-based strong learning algorithm, the overall transformations involved gives an error-based boosting algorithm. Here we have a much simpler proof of the main result of Kearns and Mansour.

5.2   The multiclass classification case

For the multiclass classification problem, we also give a pseudo-entropy function so that the class of balanced distributions suffices to be considered for weak learning. Our choice is a natural extension of the one given for the binary case. Namely,

$$G(q_1, \ldots, q_N) = \sum_{i=1}^{N} g(q_i),$$

where $g(x) = \sqrt{x(1-x)}$. Clearly G is a pseudo-entropy function.

First we give a definition of balanced distributions for the multiclass case.

Definition 4 (Balanced distributions). Let D be a probability distribution over X and q_i denote $\Pr_D(f(x) = i)$ for $i \in Y = \{1, \ldots, N\}$. The distribution D is said to be *balanced* if there exists a subset $Y' \subseteq Y$ such that $q_i = 1/|Y'|$ if $i \in Y'$ and $q_i = 0$ otherwise. We call $|Y'|$ the *degree* of the distribution.

Note that the G-entropy of f with respect to a balanced distribution with degree k is $\sqrt{k-1}$.

Now we show that a weak learning algorithm that is guaranteed to work well for balanced distributions can be transformed to a weak learning algorithm that works well for any distributions. The next lemma is used for the later analysis.

Lemma 1. *Let $q_1, \ldots, q_N$ be non-negative real numbers with $\sum_{i=1}^{N} q_i = 1$. Then, there exist at least two i's satisfying $g(q_i) \geq G(q_1, \ldots, q_N)/(2N)$.*

Proof. We assume without loss of generality that $q_1 \geq \cdots \geq q_N$. Note that $g(q_1) \geq \cdots \geq g(q_N)$. So, it suffices to show that $g(q_2) \geq G(q_1, \ldots, q_N)/(2N)$. Since the function g is concave, we have

$$\sum_{i \geq 2} g(q_i) \geq g\left(\sum_{i \geq 2} q_i\right) = g(1 - q_1) = g(q_1),$$

which implies that $G(q_1, \ldots, q_N) = g(q_1) + \sum_{i \geq 2} g(q_i) \geq 2g(q_1)$. Therefore,

$$g(q_2) \geq \frac{G(q_1, \ldots, q_N) - g(q_1)}{N - 1} \geq \frac{G(q_1, \ldots, q_N)}{2N}.$$

$\square$

Proposition 2. *If there exists a weak learning algorithm with respect to balanced distributions, then there exists a distribution-free weak learning algorithm.*

Proof. Assume we have a weak learning algorithm A with respect to balanced distributions. Using A, we construct a distribution-free weak learning algorithm B. Suppose that B is given as input a probability distribution D over $S = (\langle x_1, f(x_1) \rangle, \ldots, \langle x_m, f(x_m) \rangle)$. First B calculates $q_i = \Pr_D(f(x) = i)$ for $1 \leq i \leq N$. Let $Y' = \{i \in Y | g(q_i) \geq G(q_1, \ldots, q_N)/(2N)\}$. By Lemma 1, we have $k = |Y'| \geq 2$. Now we define the balanced distribution D' induced by D as follows. For any $1 \leq j \leq m$, let $D'(x_j) = D(x_j)/(kq_i)$ if $i = f(x_j) \in Y'$ and $D'(x_j) = 0$ otherwise. Then, B runs A with the examples S along with the distribution D' and receives h that A produces. Since A is a weak learning algorithm with respect to balanced distributions, h must satisfy $H_{D'}(f|h) \leq (1 - \gamma)H_{D'}(f)$ for some $\gamma > 0$.

Now we estimate $H_{D'}(f|h)$ in terms of probabilities with respect to D. For $i \in Y$ and $z \in Z$, let $w_z = \Pr_D(h(x) = z)$ and $q_{i|z} = \Pr_D(f(x) = i|h(x) = z)$. As shown in the proof of Proposition 1, we have

$$w'_z = \Pr_{D'}(h(x) = z) = \frac{w_z}{k} \sum_{l \in Y'} \frac{q_{l|z}}{q_l}$$

and

$$q'_{i|z} = \Pr_{D'}(f(x) = i|h(x) = z) = \frac{q_{i|z}/q_i}{\sum_{l \in Y'}(q_{l|z}/q_l)}$$

for $i \in Y'$ and $q'_{i|z} = 0$ for $i \notin Y'$. Using these equations, we have

$$H_{D'}(f|h) = \sum_{z \in Z} w'_z G(q'_{1|z}, \ldots, q'_{N|z})$$

$$= \sum_z \frac{w_z}{k} \sum_{i \in Y'} \left[\frac{q_{i|z}}{q_i} \sum_{l \in Y' \setminus \{i\}} \frac{q_{l|z}}{q_l} \right]^{1/2}.$$

Since $H_{D'}(f) = \sqrt{k-1}$, this is equivalent to

$$\sum_z w_z \sum_{i \in Y'} \left[\frac{q_{i|z}}{q_i} \sum_{l \in Y' \setminus \{i\}} \frac{q_{l|z}}{q_l} \right]^{1/2} \leq (1 - \gamma)k\sqrt{k-1}. \tag{4}$$

Let ϵ_i be the real number satisfying $\sum_z w_z g(q_{i|z}) = (1 - \epsilon_i)g(q_i)$ for $i \in Y'$. Note that since $\sum_z w_z q_{i|z} = q_i$ and g is concave, we must have $0 \leq \epsilon_i \leq 1$. Now we introduce the probability distribution over Z such that each $z \in Z$ is chosen with probability w_z. Then, we claim that the probability that $q_{i|z}$ is too small relative to q_i is upper-bounded in terms of ϵ_i.

(Claim) For any $i \in Y'$ and $\delta > 0$,

$$\Pr_z \left(q_{i|z} \leq q_i(1 - \delta) \right) = \sum_{z : q_{i|z} \leq q_i(1-\delta)} w_z \leq \frac{8\epsilon_i}{\delta^2}.$$

Proof of Claim. First we consider the case where $q_i \leq 1/2$. Using a Taylor's expansion analysis for g, we can show that $g(x) \leq g(q_i) + g'(q_i)(x - q_i)$ for any $0 \leq x \leq 1$, and especially for $x \leq q_i$, $g(x) \leq g(q_i) + g'(q_i)(x - q_i) + \frac{g''(q_i)}{2}(x - q_i)^2 = g(q_i) + g'(q_i)(x - q_i) - \frac{1}{8g(q_i)^3}(x - q_i)^2$. Therefore, we have

$$\sum_z w_z g(q_{i|z}) = \sum_{z:q_{i|z} \leq q_i} w_z g(q_{i|z}) + \sum_{z:q_{i|z} > q_i} w_z g(q_{i|z})$$

$$\leq \sum_z w_z \left(g(q_i) + g'(q_i)(q_{i|z} - q_i) \right) - \frac{1}{8g(q_i)^3} \sum_{z:q_{i|z} \leq q_i} w_z(q_{i|z} - q_i)^2$$

$$\leq g(q_i) - \frac{1}{8g(q_i)^3} \sum_{z:q_{i|z} \leq q_i(1-\delta)} w_z(q_{i|z} - q_i)^2$$

$$\leq g(q_i) - \frac{(q_i\delta)^2}{8g(q_i)^3} \Pr_z \left(q_{i|z} \leq q_i(1 - \delta) \right)$$

$$\leq g(q_i) \left(1 - \frac{\delta^2}{8} \Pr_z \left(q_{i|z} \leq q_i(1 - \delta) \right) \right).$$

Since $\sum_z w_z g(q_{i|z}) = (1 - \epsilon_i)g(q_i)$, the claim holds.

Next we consider the case where $q_i > 1/2$. Since the function $\sqrt{x}$ is concave, we have

$$\sum_z w_z g(q_{i|z}) \leq \sqrt{\sum_z w_z q_{i|z}(1 - q_{i|z})} = \sqrt{q_i - \sum_z w_z q_{i|z}^2}.$$

So,

$$q_i - \sum_z w_z q_{i|z}^2 \geq (1 - \epsilon_i)^2 q_i(1 - q_i) \geq (1 - 2\epsilon_i)q_i(1 - q_i),$$

which implies

$$2\epsilon_i q_i(1 - q_i) \geq \sum_z w_z(q_{i|z} - q_i)^2 \geq \sum_{z:q_{i|z} \leq q_i(1-\delta)} w_z(q_{i|z} - q_i)^2$$

$$\geq (q_i\delta)^2 \Pr_z \left(q_{i|z} \leq q_i(1 - \delta) \right).$$

Since $q_i > 1/2$, we have

$$\Pr_z \left(q_{i|z} \leq q_i(1 - \delta) \right) \leq \frac{2\epsilon_i(1 - q_i)}{q_i \delta^2} \leq \frac{2\epsilon_i}{\delta^2}.$$

End of Proof of Claim

Due to the claim above, inequality (4) can be rewritten as

$$(1 - \gamma)k\sqrt{k - 1} \geq \sum_z w_z \sum_{i \in Y'} \left[\frac{q_{i|z}}{q_i} \sum_{l \in Y' \setminus \{i\}} \frac{q_{l|z}}{q_l} \right]^{1/2}$$

$$\geq \Pr_z \left(\forall i \in Y', q_{i|z} \geq q_i(1 - \gamma/2)\right) k \left[(k-1)(1-\gamma/2)^2\right]^{1/2}$$

$$\geq \left(1 - (32/\gamma^2) \sum_{i \in Y'} \epsilon_i\right)(1 - \gamma/2)k\sqrt{k-1}$$

$$\geq \left(1 - \left((32/\gamma^2) \sum_{i \in Y'} \epsilon_i + \gamma/2\right)\right) k\sqrt{k-1}.$$

So, $\sum_{i \in Y'} \epsilon_i \geq \gamma^3/64$. This implies that there exists an $l \in Y'$ such that $\epsilon_l \geq \gamma^3/(64N)$. Since $l \in Y'$ and so $g(q_l) \geq G(q_1, \ldots, q_N)/(2N) = H_D(f)/(2N)$, we have

$$H_D(f|h) = \sum_i \sum_z w_z g(q_{i|z}) \leq \sum_{i \neq l} g(q_i) + (1 - \epsilon_l)g(q_l)$$

$$= \sum_i g(q_i) - \epsilon_l g(q_l) \leq H_D(f) - \left(\gamma^3/(64N)\right)\left(H_D(f)/(2N)\right)$$

$$= H_D(f)(1 - \gamma^3/(128N^2)).$$

Therefore, h is also a weak hypothesis with respect to D. □

References

1. Y. Freund. Boosting a weak learning algorithm by majority. *Information and Computation*, 121(2):256–285, 1995.
2. Y. Freund and R. E. Schapire. Game theory, on-line prediction and boosting. In *Proceedings of the 9th Workshop on Computational Learning Theory*, pages 325–332, 1996.
3. Y. Freund and R. E. Schapire. A decision-theoretic generalization of on-line learning and an application to boosting. *Journal of Computer and System Sciences*, 55(1):119–139, 1997.
4. M. Kearns and Y. Mansour. On the boosting ability of top-down decision tree learning algorithms. In *Proceedings of the 28th Annual ACM Symposium on Theory of Computing*, pages 459–468, 1996.
5. B. K. Natarajan. *Machine Learning: A Theoretical Approach*. Morgan Kaufmann, San Mateo, 1991.
6. R. Schapire. The strength of weak learnability. *Machine Learning*, 5(2):197–227, 1990.
7. R. Schapire, Y. Freund, P. Bartlett, and W. S. Lee. Boosting the margin: A new explanation for the effectiveness of voting methods. In *Proceedings of the 14th International Workshop on Machine Learning*, pages 322–330, 1997.
8. R. Schapire and Y. Singer. Improved boosting algorithms using confidence-rated predictions. In *Proceedings of the 11th Workshop on Computational Learning Theory*, 1998.
9. E. Takimoto, I. Tajika, and A. Maruoka. Mutual information gaining algorithm and its relation to PAC-learning. In *Proceedings of the 5th Conference on Algorithmic Learning Theory*, pages 547–559, 1994.

The Continuous–Function Attribute Class in Decision Tree Induction

Michael Boronowsky*

Center for Computing Technology, University of Bremen, FB3,
P.O.Box 33 04 40, D-28334 Bremen, e-mail: `michaelb@informatik.uni-bremen.de`

Abstract. The automatic extraction of knowledge from data gathered from a dynamic system is an important task, because continuous measurement acquisition provides an increasing amount of numerical data. On an abstract layer these data can generally be modeled as continuous functions over time. In this article we present an approach to handle continuous–function attributes efficiently in decision tree induction, if the *entropy minimalization heuristics* is applied.

It is shown how time series based upon continuous functions could be preprocessed if used in decision tree induction. A proof is given, that a *piecewise linear approximation* of the individual time series or the underlying continuous functions could improve the efficiency of the induction task.

1 Introduction

Dynamic systems generally have properties which can be observed over time. This can be the dynamic behavior of particular *system variables* of the system but it can also be the assignment of a particular *state* to the system which changes over time. Let $\mathcal{F}$ be the set of continuous functions over time or their mathematical transformations (e.g. derivations with respect to the time), modeling the system variables behavior over time of a certain system S. Further let C be a function of time t such that $C(t) \in \mathcal{C}$ and $\mathcal{C}$ is the set of *discrete* states where S can reside in. Below in the text *states* will be denoted as *classes*, so that $\mathcal{C}$ is the set of classes which could be assigned to S, and the set of continuous functions $\mathcal{F}$ will be referred to as *system functions*.

An interesting task is to extract rules to describe the observations of the system behavior – e.g. to find hypotheses to describe under which conditions a particular class from $\mathcal{C}$ is assigned to S.

One possible method to achieve this is the use of decision tree or rule induction, realized for example in the systems *CART* [3], *ID3* [14], *CN2* [5] and *C4.5* [15]. In this case the system functions $\mathcal{F}$ are forming an own attribute class – the *continuous–function attributes*. Generally continuous–function attributes can be handled as normal continuous–valued attributes – especially if the system functions are already sampled by data acquisition. The observations at discrete time

---

* Research supported by FNK–Forschungsförderung University of Bremen.

points t in a time interval $[t_{start}, t_{end}]$ of the system functions $f_j(t)$ with $f_j \in \mathcal{F}$ and the assigned classes $C(t) \in \mathcal{C}$ could be used to build an example set.

To preserve the characteristics of the system functions it is necessary to sample the continuous functions with high density, thus the time between adjoined examples must be short. But the shorter the time the greater the example set and the more data points must be handled in induction.

The handling of continuous–valued attributes in induction relies often on discrete attribute values of an appropriate attribute, described e.g. in [3] as *binary–split* on the ordered attribute values of one attribute of the example set. Improvements of the discretization of continuous attributes in machine learning has been the subject of a great deal of research e.g. [4, 13, 8, 7] or see [6, 16] for a good overview.

To the best of our knowledge no research is done on discretization of continuous attributes when the attribute values are coming from a continuous function. In this article we introduce a method to handle continuous–function attributes in decision tree induction based on piecewise–linear approximation of the system functions. Furthermore, a proof is presented that the piecewise–linear approximation of the system functions can increase the efficiency of decision tree induction, when the *entropy minimalization heuristics* is applied.

This proof can be understood as an extension of the analytical studies presented in [8]. In that article a method is introduced to handle continuous–valued attributes efficiently, when induction algorithms – based upon the information entropy minimalization heuristics – are applied. One important result of that work is the insight that the search for the best binary split that minimizes the entropy measure by choosing a single cut point can be restricted to the so called *boundary points*. But a real speed–up with this method is only achieved if a certain precondition holds: one particular value of an attribute must be assigned to a single class in the example set, otherwise additional boundary points have to be generated. In [7] the term *mixed block* is introduced to describe this case where different class labels are attached to identical attribute values of one attribute. Each attribute value of this kind forms an own mixed block where the borders of each block are boundary points.

In general it happens that more classes are assigned to identical attribute values – especially if huge data sets are used for induction – and it is desirable to achieve more universality, thus to be also efficient if identical attribute values are assigned to more than one class. Therefore, we have developed an alternative method to handle continuous–function attributes more efficiently – reducing the points to be tested while determining the minimal entropy – even if more classes are assigned to identical attribute (function) values.

In [2] a concrete application of the the continuous–function attribute class is introduced. In that article decision tree induction based on continuous function attributes is used to extract rules out of the measurements of a technical system concerning the dynamic of the observed system.

Using the classification introduced in [16] the work presented here could be understood as a *local, supervised (non–class–blind), non–parameterized and hard* discretization method of continuous attributes in the special case of continuous–function attributes.

2 Decision Tree Induction from Continuous–Function Attributes

If continuous–function attributes are handled as normal continuous–valued attributes the system functions have to be transformed into an appropriate discrete representation. Even if the system functions are already sampled the question arises, whether the data points could be reduced by abstraction.

When continuous–function attributes are used it seems useful to base the induction on time intervals rather than on discrete samples of the functions. If for example a system function f in a particular time interval belongs to a class $c \in C$, could this on the one hand be expressed by a number of samples of $f(t)$ in this time interval. On the other hand f could be abstracted by the duration of the class membership in this time interval. However, this requires an induction algorithm that can cope with the interval representation of the examples.

As mentioned above we want to apply the entropy minimalization heuristics. Compared with *traditional* approaches *(e.g. ID3, C4.5 etc.)* the handling of continuous–function attributes differs in two important aspects:

1. The entropy minimalization heuristics must be based upon time intervals and not on discrete objects.
2. When a split point has been found the entire example set must be partitioned according to this split point. The split is also based upon time intervals rather than on discrete objects.

In the following sections both aspects will be examined, the main focus being on the *time–based entropy minimalization heuristics*.

2.1 The Time–Based Entropy Minimalization Heuristics

The system functions $\mathcal{F}$ are going to be used in decision tree induction, e.g. to find rules describing the behavior of a particular system in a time interval $[t_{start}, t_{end}]$. The main idea of our approach is to approximate each system function $f \in \mathcal{F}$ in the considered time interval – within a given error ϵ – by a piecewise–linear function $\tilde{\mathbf{f}}$. Compared with the original function f the approximation $\tilde{\mathbf{f}}$ generally can be represented with a smaller number of points.

Let $\mathcal{T}_{\tilde{\mathbf{f}}} = \{t_1, t_2, \ldots, t_m\}$, $t_i < t_{i+1}$, $t_i \in [t_{start}, t_{end}]$ – be the set of *interesting time points*[1] – so that $\tilde{\mathbf{f}}(t)$, $t \in [t_i, t_{i+1})$ is represented by the individual linear functions $\tilde{f}_i(t) = m_i t + b_i$. Each function in $\mathcal{F}$ has its own set of interesting

---

[1] For readability in the following $\mathcal{T}_{\tilde{\mathbf{f}}}$ is denoted as $\mathcal{T}$.

time points. The individual functions $\tilde{f}_i$ are partially defined over the domain of $\tilde{\mathbf{f}}$, $domain(\tilde{f}_i) = [t_i, t_{i+1})$, $t_i, t_{i+1} \in \mathcal{T}$, $domain(\tilde{f}_i) \cap domain(\tilde{f}_j) = \emptyset$, $i \neq j$ and $\bigcup_{i=1}^{n} domain(\tilde{f}_i) \cup \{t_{end}\} = domain(\tilde{\mathbf{f}})$, whereby $\tilde{\mathbf{f}}$ is interpreted as a set of linear functions, such that $\tilde{f} \in \tilde{\mathbf{f}}$ and $|\tilde{\mathbf{f}}| = n$.

The piecewise–linear approximation of a function or a set of data points has been investigated by many researchers, e.g. [12, 1, 9]. In [11] an $O(n)$–algorithm is introduced, which finds – within a given error ϵ – a piecewise–linear function fit with the least number of corners for a set of data points considering the Chebyshev error.

A classifier which is learned from the individual $\tilde{\mathbf{f}}(t)$ and $C(t)$ in the time interval $[t_{start}, t_{end}]$ should also classify $f(t)$ correct in this time interval. To achieve this, it is assumed (but also supported by a empirical study) that the following additional time points $t_k \in [t_{start}, t_{end}]$ have to be elements of $\mathcal{T}$:

1. the start and end point t_{start}, t_{end} of the time interval,
2. the t_k where the class assignment – the value of $C(t_k)$ – changes, thus the borders of a particular, time–dependent class membership,
3. the t_k where the derivation of $f(t)$ with respect to time t is 0, thus the extreme values of f in the considered time interval. This way f is divided into monotonous segments.

Due to these additional time points a single class $c \in \mathcal{C}$ can be assigned to function $\tilde{f}$ over its domain and a function $class(\tilde{f}) \in \mathcal{C}$ can be defined, which specifies these class.

The approximation must be chosen in such a way that $\tilde{\mathbf{f}}(t_k) = f(t_k)$. To realize the abstraction of the $f \in \mathcal{F}$ the time points listed above should be determined first and then the functions can be approximated between these time points by piecewise–linear functions. Figure 1 illustrates this.

Calculating the Entropy The approximation $\tilde{\mathbf{f}}$ is used to calculate the entropy of f with regard to the error ϵ, if f is split at a particular threshold y along the ordinate axis. The calculation of the entropy is based upon time rather than on discrete objects, therefore a number of functions $time_c$ are defined – one for each class $f(t)$, $t \in [t_{start}, t_{end}]$ is assigned to. The functions are used to calculate for how much time the function $\tilde{\mathbf{f}}$ belongs to the appropriate class $c \in \mathcal{C}$ when the function value $\tilde{\mathbf{f}}(t)$ is below threshold y.

The function $time_c(y)$ of an approximation $\tilde{\mathbf{f}}$ is defined as follows

$$time_c(y) = \sum_{\tilde{f} \in Low_c(y)} duration_{Low}(\tilde{f}) + \sum_{\tilde{f} \in Dec_c(y)} duration_{Dec}(\tilde{f}, y) + \sum_{\tilde{f} \in Inc_c(y)} duration_{Inc}(\tilde{f}, y)$$

$Low_c(y)$ is the set of linear functions $\tilde{f}$, such that $Low_c(y) = \{\tilde{f} \in \tilde{\mathbf{f}} \mid \forall_{t \in domain(\tilde{f})} y \geq \tilde{f}(t) \land c = class(\tilde{f})\}$. Thus the complete ranges of all functions in $Low_c(y)$ are lower or equal to threshold y and class c is assigned to all these functions. In

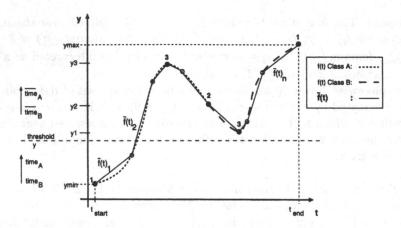

Fig. 1. System function f and schematic example of an approximation $\tilde{f}$. The nodes are numbered according to the enumeration list; the light nodes are the corners of the piecewise–linear functions

this case the duration of the class membership $duration_{Low}(\tilde{f})$ is calculated as the difference of the maximal and minimal value of the domain of the individual functions $\tilde{f}$, hence $duration_{Low}(\tilde{f}) = max(domain(\tilde{f})) - min(domain(\tilde{f}))$.

The set of linear functions $Dec_c(y) = \{\tilde{f} \in \tilde{\mathbf{f}} \mid \exists_{t \in domain(\tilde{f})} y = \tilde{f}(t) = mt + b \wedge m < 0 \wedge c = class(\tilde{f})\}$ contains the *decreasing* functions which are intersected by threshold y and $Inc_c(y) = \{\tilde{f} \in \tilde{\mathbf{f}} \mid \exists_{t \in domain(\tilde{f})} y = \tilde{f}(t) = mt + b \wedge m > 0 \wedge c = class(\tilde{f})\}$ contains the *increasing* functions which are intersected by y all belonging to class c.

The duration of the intersected functions with respect to threshold y is $duration_{Inc}(\tilde{f}, y) = 1/m \, [y - \tilde{f}(min(domain(\tilde{f})))]$, $\tilde{f}(t) = mt + b$ for the increasing functions and $duration_{Dec}(\tilde{f}, y) = -1/m \, [y - \tilde{f}(max(domain(\tilde{f})))]$ for the decreasing ones. Thus both functions $duration_{Inc}(\tilde{f}, y)$ and $duration_{Dec}(\tilde{f}, y)$ are linear functions with respect to y.

All individual functions $time_c(y)$ share some fundamental properties:

- $time_c(y)$ is a piecewise–linear function, monotonously increasing but not always continuously,
- $time_c(y) = 0$ for $y \leq y_{min}$, where y_{min} is the absolute minimum value of $f(t)$, $t \in [t_{start}, t_{end}]$,
- $time_c(y) = tmax_c$ for $y \geq y_{max}$, where y_{max} is the absolute maximum value of $f(t)$, $t \in [t_{start}, t_{end}]$. $tmax_c$ is the total duration – in the considered time interval – where class c is assigned to the $\tilde{f}$.

The function $time_c(y)$ calculates the time of a class membership when $\tilde{\mathbf{f}}(t) \leq y$, furthermore a number of functions $\overline{time}_c(y) = tmax_c - time_c(y)$ are defined to calculate the duration of a particular class membership when $\tilde{\mathbf{f}}(t) > y$. This is also illustrated in figure 1 and in figure 2, whereby the relation between the two figures is slightly simplified[2].

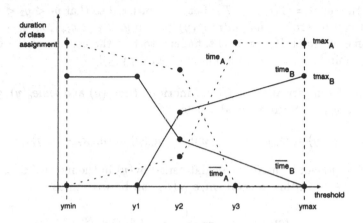

Fig. 2. Example of the function plots of the individual $time_c$ and $\overline{time}_c$ functions

The functions $time_c$ and $\overline{time}_c$ and the constants $tmax_c$ are used to calculate the class entropy if $f \in \mathcal{F}$ is split at a certain threshold value, with respect to the approximation error ϵ. In the following the formulas for the calculation of the entropy are specified, they can be understood as the *continuous version* of the formulas used in $C4.5$ [15]. A threshold y splits f in a lower and a higher part. Let $\mathcal{C}$ be the set of the assigned classes, then the function

$$info(y) = -\sum_{i \in \mathcal{C}} \frac{time_i(y)}{\sum_{k \in \mathcal{C}} time_k(y)} \ln\left(\frac{time_i(y)}{\sum_{k \in \mathcal{C}} time_k(y)}\right)$$

calculates the information content of the lower part and the function

$$\overline{info}(y) = -\sum_{i \in \mathcal{C}} \frac{\overline{time}_i(y)}{\sum_{k \in \mathcal{C}} \overline{time}_k(y)} \ln\left(\frac{\overline{time}_i(y)}{\sum_{k \in \mathcal{C}} \overline{time}_k(y)}\right)$$

for the higher part. The total entropy – when f is split at threshold y – is calculated with

$$entropy(y) = \frac{\sum_{i \in \mathcal{C}} time_i(y)\, info(y) + \sum_{i \in \mathcal{C}} \overline{time}_i(y)\, \overline{info}(y)}{\sum_{i \in \mathcal{C}} tmax_i}.$$

The threshold value where $entropy(y)$, $y_{min} \le y \le y_{max}$ has got its absolute minimum is considered to be the best split point of f.

---

[2] The correct functions consists of some more corners, whereby the functions are still linear between them.

Finding the Minimal Entropy The determination of the minimal entropy can be still an expensive operation. But we have found out that the chosen abstraction of f has got a very interesting property. The best split point of f with respect to the error ϵ – thus the y where $entropy(y)$ becomes minimal – can be found in the set $S = \{\tilde{\mathbf{f}}(t) \mid t \in \mathcal{T}\}$. Let S be ordered so that $y_1 < y_2 < \cdots < y_n$, $y_i \in S$. It can be proven that $entropy(y)$ within $y_i < y < y_{i+1}$, $y_i, y_{i+1} \in S$ cannot become minimal. Hence, it is sufficient to test the corners of the functions $time_c(y)$ to find the absolute minimum of $entropy(y)$!

Due to the linearization of f the functions $time_c(y)$ and $\overline{time}_c(y)$, $y_i < y < y_{i+1}$ and $y_i, y_{i+1} \in S$ can be modeled as

$$g_c(y) = m_c\,y + b_c \qquad \text{and} \qquad h_c(y) = tmax_c - g_c(y)\,.$$

$g_c(y)$ and $h_c(y)$ can be used to calculate $entropy(y)$ in the interval $y_i < y < y_{i+1}$ and $y_i, y_{i+1} \in S$. In this interval $entropy(y)$ can be written as

$$entropy_*(y) = \frac{1}{\sum_{i \in C} tmax_i}\,(A(y) + B(y))\,, \tag{1}$$

whereby $A(y)$ and $B(y)$ are defined as

$$A(y) = \left[\sum_{i \in C} g_i(y)\right]\left[-\sum_{i \in C} \frac{g_i(y)}{\sum_{k \in C} g_k(y)} \ln\left(\frac{g_i(y)}{\sum_{k \in C} g_k(y)}\right)\right]$$

$$= -\left[\sum_{i \in C} g_i(y) \ln(g_i(y))\right] + \left[\sum_{i \in C} g_i(y) \ln\left(\sum_{k \in C} g_k(y)\right)\right] \tag{2}$$

$$B(y) = \left[\sum_{i \in C} h_i(y)\right]\left[-\sum_{i \in C} \frac{h_i(y)}{\sum_{k \in C} h_k(y)} \ln\left(\frac{h_i(y)}{\sum_{k \in C} h_k(y)}\right)\right]$$

$$= -\left[\sum_{i \in C} h_i(y) \ln(h_i(y))\right] + \left[\sum_{i \in C} h_i(y) \ln\left(\sum_{k \in C} h_k(y)\right)\right]\,. \tag{3}$$

We want to show that $entropy_*(y)$ cannot have a real minimum as extreme value. Therefore the second derivation of $entropy_*(y)$ with respect to y may not get positive, thus we must prove that

$$entropy''_*(y) \le 0\,.$$

The second derivation of $entropy_*(y)$ is

$$entropy''_*(y) = \frac{1}{\sum_{i \in C} tmax_i}\,(A''(y) + B''(y))\,. \tag{4}$$

In the following we are going to prove that $A''(y)$ and $B''(y)$ never can become positive. Because of this and the fact that $\sum_{i \in C} tmax_i$ generally is positive we

can conclude that $entropy''_*(y)$ also can not become positive.
First we will prove

$$A''(y) \leq 0 . \tag{5}$$

Taking the first derivation of equation (2) with respect to y leads to

$$A'(y) = - \left[\sum_{i \in C} g'_i(y) \ln (g_i(y)) \right] - \left[\sum_{i \in C} g'_i(y) \right]$$
$$+ \left[\sum_{i \in C} g'_i(y) \ln \left(\sum_{k \in C} g_k(y) \right) \right] + \left[\sum_{i \in C} \frac{\sum_{k \in C} g'_k(y)}{\sum_{k \in C} g_k(y)} g_i(y) \right] . \tag{6}$$

Using the derivation $g'_c(y) = m_c$ then equation (6) can be simplified to

$$A'(y) = - \left[\sum_{i \in C} m_i \ln (g_i(y)) \right] + \left[\sum_{i \in C} m_i \ln \left(\sum_{k \in C} g_k(y) \right) \right] . \tag{7}$$

The second derivation of $A(y)$ is

$$A''(y) = - \left[\sum_{i \in C} m_i \frac{g'_i(y)}{g_i(y)} \right] + \left[\sum_{i \in C} m_i \frac{\sum_{k \in C} g'_k(y)}{\sum_{k \in C} g_k(y)} \right]$$
$$= - \left[\sum_{i \in C} \frac{m_i^2}{g_i(y)} \right] + \frac{\left(\sum_{i \in C} m_i \right)^2}{\sum_{i \in C} g_i(y)} . \tag{8}$$

Recall that we want to prove equation (5), hence – using equation (8) – we must prove

$$\frac{\left(\sum_{i \in C} m_i \right)^2}{\sum_{i \in C} g_i(y)} \leq \left[\sum_{i \in C} \frac{m_i^2}{g_i(y)} \right] . \tag{9}$$

Due to the properties of $time_c$ we know that $g_c(y)$ never becomes negative, thus equation (9) can be rewritten as

$$\left(\sum_{i \in C} m_i \right)^2 \leq \left[\sum_{i \in C} \frac{m_i^2}{g_i(y)} \right] \left[\sum_{i \in C} g_i(y) \right] \tag{10}$$

which is equivalent to

$$\left[\sum_{i \in C} \frac{m_i}{\sqrt{g_i(y)}} \sqrt{g_i(y)} \right]^2 \leq \left[\sum_{i \in C} \left(\frac{m_i}{\sqrt{g_i(y)}} \right)^2 \right] \left[\sum_{i \in C} \left(\sqrt{g_i(y)} \right)^2 \right] . \tag{11}$$

Choosing $a_i(y)$ and $b_i(y)$ appropriate, substitution leads to

$$\left[\sum_{i \in C} a_i(y) b_i(y) \right]^2 \leq \left[\sum_{i \in C} a_i(y)^2 \right] \left[\sum_{i \in C} b_i(y)^2 \right] , \tag{12}$$

which is exactly the *Cauchy–Schwarz Sum Inequality* [10] and proven to be correct. Hence, equation (5) is also proven to be true.

Now we must show that

$$B''(y) \leq 0 . \tag{13}$$

Due to the structural similarity of (2) and (3) the first derivation of (3) has got the same structure as equation (6). Using $h'_c(y) = -m_c$ the first derivation of $B(y)$ can be written as

$$B'(y) = \left[\sum_{i \in C} m_i \ln \left(h_i(y) \right) \right] - \left[\sum_{i \in C} m_i \ln \left(\sum_{k \in C} h_k(y) \right) \right] . \tag{14}$$

The second derivation of $B(y)$ is

$$B''(y) = - \left[\sum_{i \in C} \frac{m_i^2}{h_i(y)} \right] + \frac{\left(\sum_{i \in C} m_i \right)^2}{\sum_{i \in C} h_i(y)} \tag{15}$$

which has got the same structure as equation (8). Substitute $h_i(y)$ by $g_i(y)$, equations (9) ... (12) also proves equation (13) to be correct[3].
Hence, it is true that $entropy''_*(y)$ never becomes positive in the considered interval. It is well possible that $entropy''_*(y)$ becomes zero and then the $entropy_*(y)$ remains constant for the interval. In this case it could happen that only the border values considered as minimum, whereas strictly speaking $entropy_*(y)$ also has this value for the whole interval. Because only a single split point is needed and no qualitative difference exists between the potential split points, it is not problematic that one of the border values is chosen as split point.

If the individual functions $time_c(y)$ and $\overline{time}_c(y)$ are modeled by linear functions, we have shown that under certain preconditions no real minimum can occur in $entropy_c(y)$. Since $time_c(y)$ and $\overline{time}_c(y)$ are generally only linear in the interval $y_i < y < y_{i+1}$ and $y_i, y_{i+1} \in S$, a real minimal value for $entropy(y)$ can only occur at the borders of the interval.

To emphasize this once again, this is also valid if identical function values of f_j are assigned to different classes and certainly a great difference to the approach introduced in [8], which only increases efficiency in the cases where a class assignment to a particular attribute value is not inconsistent[4]. These attribute values can be handled by Fayyad's and Irani's approach, but only in the conventional way, by calculating the entropy for the splits between adjacent inconsistent attribute values.

But it is possible to combine our approach to reduce the points to be tested minimizing $entropy(y)$ with Fayyad's and Irani's method [8]. The $y_i \in S$ which are assigned to a single class and which are no *boundary points* in the sense of

---

[3] Using the property of $\overline{time}_c(y)$, that this function never becomes negative for $y_{min} \leq y \leq y_{max}$ and thus $h_c(y)$ also cannot become negative in this interval.

[4] Inconsistent with respect to the considered attribute and not for the total example set.

Fayyad and Irani could be removed from S, because this points cannot minimize entropy(y).

The points to be tested as minimum of entropy(y) are related to the corners of the piecewise–linear function $\tilde{f}$. Thus the quality of the approximation of f has got direct influence on the considered split points. It could happen that the induced decision tree has not minimal depth for the average of the examples. The influence of the approximation quality on the generated decision tree will have to be the subject of further research. However, our empirical study has shown that the induction algorithm always generates a correct decision tree, which classifies $f(t)$ correctly in the considered time interval.

2.2 Performing a Split

Once the function $f \in \mathcal{F}$ with the best split point y has been determined, all functions in $\mathcal{F}$ must be partitioned according to the split of f at y. The split is performed in two steps:

1. First two sets T_{low} and T_{high} are built. T_{low} contains the time intervals where f is below or equal y and T_{high} contains the time intervals the time intervals when f is above y.
2. The time intervals in these two sets are used to split all functions $\mathcal{F}$ in lower and upper function parts in such a way that the functions during the time intervals from T_{low} (T_{high}) are in their lower (upper) parts, respectively.

A split intersects the individual functions in $\mathcal{F}$ resulting in a number of border values at the intersection points. The determination of these border values should be based upon the original system functions rather than on their abstraction. This is necessary to guarantee the properness of the induced decision tree. Further, the abstraction of the functions should be updated, so that the border of an interval at an intersection point is represented by the exact function value. After the split the time–based entropy minimalization heuristics can be used to determine the following *best* split point. The total process is repeated until no further best split point can be found.

3 Conclusion

In this paper a new attribute class for decision tree induction is introduced – the continuous–function attribute. We have developed an entropy based heuristics which is based upon time intervals rather than on discrete objects. This seems to be an appropriate approach if continuous functions are used for induction. The piecewise–linear approximation of the continuous functions is used as an abstraction of the individual functions. Due to this approximation the application of the entropy minimalization heuristics could be realized efficiently. The proof which was given here shows that only particular points have to be tested as minimum of the entropy minimalization heuristics. Thus the costs for the determination of the best cut point, with respect to a certain approximation error could be reduced this way.

The influence of the chosen piecewise–linear abstraction on the generated decision tree is going to be the subject of further research, because the immediate relation between abstraction and considered split points is evident.

References

1. Baines, M. J. (1994). Algorithms for Optimal Piecewise Linear and Constant L_2 Fits to Continuous Functions with Adjustable Nodes in One and Two Dimensions. *Mathematics of Computation*, 62(206):645–669.
2. Boronowsky, M. (1998). Automatic Measurement Interpretation of a Physical System with Decision Tree Induction. In *Proceedings of the International Symposium on Intelligent Data Engineering and Learning in Hong Kong, China*. Springer.
3. Breiman, L., Friedman, J. H., Olshen, R. A., and Stone, C. J. (1984). *Classification and Regression Trees*. Wadsworth, Belmont.
4. Catlett, J. (1991). On changing continuous attributes into ordered discrete attributes. In Kodratoff, Y., editor, *Proceedings of the European Working Session on Learning*, pages 164–178.
5. Clark, P. and Niblett, T. (1989). The CN2 Induction Algorithm. *Machine Learning*, 1:261–282.
6. Dougherty, J., Kohavi, R., and Sahami, M. (1995). Supervised and Unsupervised Discretitzation of Continuous Features. In *Machine Learning: Proceedings of the 12th International Conference*, pages 194–202. Morgan Kaufmann.
7. Elomaa, T. and Rousu, J. (1998). General and Efficient Multiplitting of Numerical Attributes. *Machine Learning*. to appear.
8. Fayyad, U. M. and Irani, K. B. (1992). On the Handling of Continuous–Valued Attributes in Decision Tree Generation. *Machine Learning*, 8:87–102.
9. Goodrich, M. T. (1995). Efficient Piecewise–Linear Function Approximation Using the Uniform Metric. *Discrete & Computational Geometry*, 14:445–462.
10. Gradshteyn, I. S. and Ryzhik, I. M. (1979). *Table of integrals, series, and products*. Academic Press, 5th. edition. page 1092.
11. Hakimi, S. L. and Schmeichel, E. F. (1991). Fitting Polygonal Functions to a Set of Points in the Plane. *CVGIP: Graphical Models and Image Processing*, 53(2):132–136.
12. Imai, H. and Iri, M. (1988). Polygonal Approximation of a Curve – Formulations and Algorithms. In G.T.Toussaint, editor, *Computational Morphology*, pages 71–86. Elsevier Science Publishers B.V.
13. Kerber, R. (1992). Chimerge: Discretization of Numeric Attributes. In *Proceedings of the 10th National Conference on Artificial Intelligence*, pages 123–127.
14. Quinlan, J. R. (1986). Induction of Decision Trees. *Machine Learning*, 1.
15. Quinlan, J. R. (1993). *C4.5 Programs for Machine Learning*. Morgan Kaufmann.
16. Susmaga, R. (1997). Analyzing Discretizations of Continuous Attributes Given a Monotonic Discrimination Function. *Intelligent Data Analysis*, 1(3).

Feature Transformation and Multivariate Decision Tree Induction

Huan Liu and Rudy Setiono

School of Computing, National University of Singapore, Singapore 119260.
{liuh,rudys}@comp.nus.edu.sg

Abstract. Univariate decision trees (UDT's) have inherent problems of *replication, repetition,* and *fragmentation.* Multivariate decision trees (MDT's) have been proposed to overcome some of the problems. Close examination of the conventional ways of building MDT's, however, reveals that the fragmentation problem still persists. A novel approach is suggested to minimize the fragmentation problem by separating hyperplane search from decision tree building. This is achieved by *feature transformation.* Let the initial feature vector be $\mathbf{x}$, the new feature vector after feature transformation T is $\mathbf{y}$, i.e., $\mathbf{y} = T(\mathbf{x})$. We can obtain an MDT by (1) building a UDT on $\mathbf{y}$; and (2) replacing new features $\mathbf{y}$ at each node with the combinations of initial features $\mathbf{x}$. We elaborate on the advantages of this approach, the details of T, and why it is expected to perform well. Experiments are conducted in order to confirm the analysis, and results are compared to those of C4.5, OC1, and CART.

1 Introduction

Among many algorithms for univariate decision tree (UDT) induction, C4.5 [18] is one that is widely used and well received. The basic algorithm is divide-and-conquer. Let the classes be $\{C_1, C_2, ..., C_k\}$. To construct a decision tree from dataset D, we go through one of these steps: (1) if D contains one or more examples, all belonging to a single class C_j. Stop; (2) if D contains no example, the most frequent class at the parent of this node is chosen as the class. Stop; or (3) if D contains examples belonging to a mixture of classes, information gain is then used as a heuristic to split D into partitions (branches) based on the values of a *single feature*. Start recursion on each branch to build subtrees. C4.5 also devised sophisticated methods to deal with continuous features, overfitting, and so on [18]. Note that at each decision node, a univariate (single feature) test is performed to partition the data. Researchers noticed that this univariate test leads to (1) the replication problem - subtrees that are replicated in a UDT; (2) the repetition (or repeated testing) problem - features that are repeatedly tested (more than once) along a path in a UDT; and (3) the fragmentation problem - data is gradually partitioned into small fragments [16, 3]. Replication and repetition always imply fragmentation, but fragmentation may occur without any replication or repetition if many features need to be tested [6]. Thus reduced is the UDT's capability of expressing concepts succinctly, which renders many concepts either difficult or impossible to express by UDT's. In addition, the univariate tests in tree building are also found sensitive to noise and minor variations in the data [20]. Multivariate decision trees (MDT's) can overcome these problems. In building an MDT, each test can be based on one or more features, i.e., a multivariate test. For example, the replication problem can be solved by testing Boolean combinations of the features; the repeated testing problem can be handled by forming linear combinations of the features. In [16,3], it is shown that these combinations improve accuracy and reduce tree size.

2   Classical Way of Building MDT's and Its Limitations

Multivariate decision tress have been proposed to overcome the replication and repetition problems. In essence, an MDT is built in a way similar to building a UDT with one difference - performing univariate or multivariate tests. In building a UDT, at each test, we need to select the most promising feature. Many heuristic measures were proposed, such as impurity [2], information gain [17], and more in [14]; in building an MDT, however, we need to find the most promising hyperplane (a set of features is needed). It is known that selecting the best set of features for test at a node is an NP-complete problem [9]. In order to find an "optimal" hyperplane for a multivariate test, researchers offered various suggestions such as Regression [2], Linear Programming [1,3], Randomization [15], and Simulated Annealing [9]. In the following, we examine the conventional procedure of MDT building with reference to UDT building.

2.1   Common aspects of the MDT building

Many issues in building MDT's are the same as those for UDT's. Both types of trees are built from labeled examples. In finding a linear combination of initial features for an MDT test, however, we encounter some issues that are not found in a UDT. Testing on a single feature is equivalent to finding an axis-parallel hyperplane to partition the data, while testing on multiple features we try to find an oblique hyperplane. The latter is significantly more complicated than the former. All existing methods of building an MDT can be described in the following procedure (the conventional way):

1. Select features that should be included in a linear combination;
2. Find the coefficients of a multivariate test that result in the best partition of the examples;
3. Build the MDT recursively; and
4. Prune the MDT to avoid overfitting.

In Step 1, a UDT only examines at maximum N features where N is the number of features describing a dataset, but an MDT needs to look at $\sum_{i=1}^{M} \binom{N}{i}$, where M is the best number of features for the current oblique hyperplane. A UDT does not require Step 2 which is also a very time-consuming step [9].

Basically, we need to find an optimal orientation for the hyperplane in an M-dimensional space. In order to avoid exhaustive search for M features and corresponding coefficients, researchers offered some solutions. Breiman et al. [2] suggested a perturbation algorithm that modifies one coefficient of the hyperplane at a time. Murthy et al. [15] implemented OC1 that uses a randomized algorithm to avoid local minima in modifying coefficients (and several coefficients can be modified at a time). Brodley and Utgoff [3] reviewed more algorithms for finding the coefficients of a hyperplane (e.g., linear programming, a recursive least squares procedure, the Pocket algorithm, and the thermal training procedure). Heath et al [9] employed a simulated annealing algorithm. All in all, various solutions have been proposed for finding optimal coefficients, and no absolute conclusion on which solution is the best [3]. Step 3 is the same for both MDT's and UDT's with an MDT having fewer recursions since it induces a simpler tree in general (though every node is more complicated). Step 4 is also common to both types of trees. The difference is that an MDT node contains more information (an oblique hyperplane) compared to a UDT node (an axis-parallel hyperplane). Extra care should be taken in pruning an MDT node.

2.2 Limitations of the conventional procedure

From the head-to-head comparison above, we realize (1) it is indeed *a complicated process* to induce an MDT following *exactly* the way in which a UDT is built; (2) in this process, simulated annealing, randomization algorithms, or heuristics have to be used for selecting a split point at node in an MDT to avoid exhaustive search; (3) searching for optimal hyperplanes has to be repeated as many times as the number of multivariate test nodes. For example, if a final MDT has 10 internal nodes, that means ten oblique hyperplanes have been formed; (4) the fragmentation problem persists. The last point is obvious since hyperplane search is intertwined with tree expansion as in univariate decision tree induction. For example, if the data is partitioned equally at each test, the size of the data for a test at a new level k (which is 0 at the tree root) decreases by a factor of 2^k. Sample shortage can cause inappropriate estimation for methods such as regression, neural networks, and simulated annealing. Because a hyperplane (multivariate test) is generally more powerful in partitioning data into uniform classes than a univariate test is, powerful multivariate tests have to be based on small fragments of data due to the fragmentation problem.

3 A Novel Perspective

The conventional way of MDT building has limitations. We propose here a different perspective. As we know that hyperplane search at each multivariate test could be complicated and time-consuming, can we separate hyperplane search from tree induction? The separation will also avoid finding hyperplanes with data fragments. Our proposal is to build an MDT via *feature transformation*.

Let the initial feature vector be $\mathbf{x} = (x_1, x_2, ..., x_N)$, the new feature vector after applying feature transformation T is $\mathbf{y} = (y_1, y_2, ..., y_M)$, where $\mathbf{y} = T(\mathbf{x})$. We obtain a linear MDT by (1) building a UDT with $\mathbf{y}$; and (2) replacing new feature vector $\mathbf{y}$ at each node with the combinations of initial feature vector $\mathbf{x}$.

Algorithm BMDT

Input: dataset D with feature vector $\mathbf{x}$
Output: an MDT
BMDT (D)
   1. Trans $=$ buildTrans(D) with $\mathbf{y} = T(\mathbf{x})$;
   2. $D_Y =$ Trans(D) /*D_Y with feature vector $\mathbf{y}$*/;
   3. $Tree_Y =$ createUDT(D_Y);
   4. MDT $=$ Repl$(Tree_Y)$ with $\mathbf{y} = T(\mathbf{x})$

---

buildTrans() creates a data transformer (**Trans**) that can change the original data D into D_Y which is described by feature vector $\mathbf{y}$. In the process, we also obtain feature transformation T. createUDT() induces a univariate decision tree, $Tree_Y$. Repl() replaces all y_i in $Tree_Y$ with combinations of x_j's. By transforming features from $\mathbf{x}$ to $\mathbf{y}$, we can indeed create an MDT about $\mathbf{x}$ by inducing a UDT of $\mathbf{y}$. Steps 2 and 4 in BMDT are straightforward. The important components of BMDT are (1) buildTrans() and (2) createUDT(). Our choices are a feedforward neural network (NN) for buildTrans() and C4.5 for createUDT().

Transformation via a neural network

The tool for transformation is a standard 2-layer feedforward neural network [19] which has the input, hidden, and output units. The first layer of weights is formed by the input and hidden units, and the second layer by the hidden and output units. At each layer, the units are fully connected. The weighted inputs are fed into hidden units and summed. A squashing function (e.g., logistic) is normally applied at hidden units, thus an *activation value* is obtained. The activation values are then fed to the output units where the values are summed at each output unit, and the class labels are determined. The typical issues about such NN's are (1) how many hidden units should be used; and (2) input/output coding of the data. In BMDT, we simply use an excessive number of hidden units (e.g., if there are N inputs (features), use $N/2$ hidden units), and rely on a pruning algorithm [8, 21], to find the sufficient and necessary number of hidden units for good performance from the neural net. After pruning, the number of hidden units is usually much smaller than that of the input units. The input and output coding schemes are standard, for example, using normalized values for continuous data, binary coding for nominal data, and thermometer coding for ordinal discrete data.

In transforming features using a neural network, the remaining hidden units are our new features $\{y_1, y_2, ..., y_M\}$. The activation values of an original example form a new example in the feature space $\mathbf{y}$. As such, we obtain transformed data (D_Y in BMDT) as well as transformed features. Two more issues remain: (1) when should the network training and pruning stop (i.e., achieving sufficiently high accuracy without overfitting the data); and (2) how to find the combination of initial features $\{x_1, x_2, ..., x_N\}$ for each new feature in **y**. The solution for (1) exists, i.e., by employing cross validation during network training and pruning. The solution for (2) is found by understanding how an activation value is obtained in the net. We will discuss it now.

Rewriting univariate tests in x

Let's look at how an activation value y_j is obtained. If we use the hyperbolic tangent function (δ) at the hidden units, we have

$$y_j = \delta(\sum_{i=1}^{N} w_{ij}x_i + \tau_j) \qquad (1)$$

where w_{ij} are weights on the connections from input i to hidden unit j, and τ_j is the bias for hidden unit j. Note that in (1) δ is a squashing function which normalizes new features in **y**; and it is a one-to-one mapping function. The complexity thus introduced is that the transformation is not linear any more. However, this does not prevent us from building a linear MDT. After a UDT is induced from D_Y, a typical test is "$y_j < c$" and c is some threshold value determined by tree induction. Applying equation (1), we obtain an equivalent form of the test, "$\delta(\sum_{i=1}^{N} w_{ij}x_i) < c$". Taking the inverse δ, the test is "$(\sum_{i=1}^{N} w_{ij}x_i) < \delta^{-1}(c)$", a linear inequality involving the initial features **x**. A linear MDT is thus built. It is obvious that in this context, a nonlinear MDT and a linear MDT are equivalent. We choose the linear form since it is simpler in representation, faster to test, and easier for us to compare with the results of other linear MDT builders such as OC1 and CART.

Why use a neural network?

A feature transformer should satisfy the following conditions: (1) it generates new features; (2) new features preserve information; (3) it compresses information so there are fewer new features than original ones; (4) reverse transformation is possible; and (5) it is simple and standard. Among many choices (see related work), we select a standard two-layer feedforward network for the task. This is because it satisfies all the conditions above. After the network training and pruning, hidden units are new features (**y**). Neural network research [23, 22] demonstrates that hidden units are a powerful compressor of input information. Excessive hidden units can be pruned away without impairing the network's capability. More often than not, pruning can improve its performance by avoiding overfitting caused by excessive hidden units - hence, after condensing and pruning, hidden units (the new features) preserve the information represented by input units (the original features); the number of hidden units is usually much smaller than that of input units. As is shown earlier, it is straightforward to rewrite a multivariate test of y_i by a combination of original **x**.

Analysis

Before we discuss empirical study, some analysis can help us to know why this novel approach deserves such a study and is worthy of our effort. With BMDT, training just one neural network has made it possible to find all hyperplanes required for a multivariate decision tree, instead of searching for as many hyperplanes as the number of multivariate test nodes in a decision tree. The power of an MDT lies in optimal or near optimal hyperplanes, the search of such a hyperplane can, however, be intractable [9]. If possible, we should minimize the number of such search.

All training data is used in neural network training and pruning to transform original features x to new features y. New features y are then used in tree induction. This two-phase procedure effectively decouples hyperplane search from tree induction, unlike the classical way of MDT building in which a new hyperplane is searched as a new test node is required, and every new hyperplane has to be formed with less data (i.e., the fragmentation problem). In other words, the feature transformation approach (BMDT) extracts hyperplane search out of tree induction so that the fragmentation problem is minimized.

Pruning multivariate decision tree is not as straightforward as in univariate decision tree pruning. Overfitting does occur in MDT's as pointed out in [3], which is partially due to sample shortage that causes inappropriate estimation for methods such as regression, neural networks, and simulated annealing, but pruning the entire node with a multivariate test may result in even more classification errors. Brodley and Utgoff suggested that in such a case, one should try to reduce the error by eliminating some features from the multivariate test, in other words, no node is pruned. Overfitting is handled in the proposed approach through pruning in a univariate decision tree. No special treatment is needed. In addition, inappropriate estimation caused by sample shortage is limited because the new approach minimizes the fragmentation problem.

4   Empirical Study and Discussion

Our analysis above shows that the proposed approach should perform well. The analysis will be further confirmed via experiments. In particular, we wish to verify whether (1) BMDT can create MDT's; and (2) MDT's constructed by BMDT are comparable to, if not better than, those by other methods, such as CART, OC1, and C4.5, in terms of tree size and predictive accuracy. We select datasets commonly used and publicly available from the UC Irvine data repository [13] with different data types (nominal, continuous, and mixed), additional two data sets (Par12a and Mux12) are taken from [27]. They are summarized in Table 1. For nominal attributes, their values are binarized. Doing so is not biased against C4.5 since researchers [24] observed that this would improve univariate decision trees' accuracy. Five classification algorithms are used in the experiments, they are BMDT, CART, OC1, C4.5, and NN (a standard 2-layer feedforward neural network with pruning). CART is chosen because it is the very first one that proposed multivariate tests; OC1 is the latest one and publicly available, Murthy et al. [15] also compared OC1 with other multivariate decision tree induction algorithms with favorable results. C4.5 is chosen due to its reputable performance and used here for reference as base average tree size and average accuracy.

Results are presented in Tables 2 and 3. Also included are the results of a two-tailed t-test. The P-values are calculated for C4.5, OC1, and CART against BMDT based on 10-fold cross validation. Included also are the results of the

Name	#Data	#A	Type
1 Monk1	432	17	binary
2 Monk2	432	17	binary
3 Monk3	432	17	binary
4 Par12a	1024	12	binary
5 Mux12	1024	12	binary
6 TicTacToe	958	27	binary
7 Vote	300	48	binary
8 BreastCancer	699	9	continuous
9 Bupa	345	6	continuous
10 Ionosphere	351	34	continuous
11 Iris	150	4	continuous
12 Pima-diabetes	768	8	continuous
13 Sonar	208	60	continuous
14 Australian	690	14	mixed
15 HeartDisease	297	13	mixed
16 Housing	506	13	mixed

Table 1. Dataset Summary. #Data - data size, Type - attribute type, and #A - number of attributes.

neural networks (NN) based on which BMDT builds MDT's. We want to understand whether the difference between every pair is statistically significant. If it is significant, we further check which result is better by looking at the two average values. We observe the following from experiments:

Predictive accuracy (Table 2): First of all, BMDT's accuracy is not significantly different from NN's. However, BMDT performs significantly differently from C4.5 in 7 out of 16 cases. For all the 7 cases, BMDT's accuracy is better except for one (Mux12). When the results of BMDT are statistically different from those of OC1 and CART, BMDT always achieves higher average accuracy. The numbers of cases for which BMDT achieves significantly higher average accuracy over OC1 and CART are 6 and 4 out of 16 cases, respectively.

Tree size (Table 3): BMDT builds trees that are significantly smaller than those of C4.5's in all cases. In 7 cases, trees created by BMDT are significantly different from those of OC1's, in which two of the OC1's trees are smaller. In 9 cases, trees by BMDT are significantly different from those of CART's, in which only one of CART's trees is smaller.

An example: The dataset is Pima-diabetes. In Table 3, it is seen that C4.5 creates a UDT with average tree size of 122.4 nodes, BMDT builds an MDT with average tree size of 3 nodes. That means the MDT has one root and two leaves (class labels). Hence, there is only one oblique hyperplane at the root. Choosing one out of ten MDT's to show how it is obtained, we have the following: (1) after pruning, the network contains one hidden unit and 4 out of the 8 original inputs, i.e., after transformation, the new feature vector y is one dimensional in the new space; (2) building a UDT over the feature y, we have "if $y \leq 0$ then class 0, else class 1"; (3) inverse transformation and the network weights of the first layer

Name	BMDT	NN	C4.5	OC1	CART
Monk1	100	100	100	100	100
(P-value)		-	-	-	-
Monk2	100	100	93.7*	99.77	99.31
(P-value)		-	**(0.0025)**	(0.3434)	(0.3434)
Monk3	100	100	100	100	97.22*
(P-value)		-	-	-	**(0.0133)**
Par12a	100	100	100	99.49*	99.68
(P-value)		-	-	**(0.0109)**	0.1355
Mux12	99.6	99.39	100.0	99.44	99.32
(P-value)		(0.6324)	**(0.0982)**	(0.6822)	(0.3826)
TicTacToe	98.0	97.71	93.8*	91.75*	86.33*
(P-value)		(0.1801)	**(0.0001)**	**(0.0034)**	**(0.0001)**
Vote	96.3	96.33	97.3	93.10*	95.17
(P-value)		(0.7583)	(0.1944)	**(0.0422)**	(0.4073)
B-Cancer	96.1	96.0	95.3	94.99	94.71
(P-value)		(0.7583)	(0.2423)	(0.4033)	(0.2777)
Bupa	68.8	68.18	61.1*	66.09	65.22
(P-value)		(0.5565)	**(0.0475)**	(0.5289)	(0.3443)
Ionosphere	88.1	88.34	90.0	88.32	86.61
(P-value)		(0.5986)	(0.4249)	(0.9376)	(0.6827)
Iris	95.3	91.33	94.0	96.00	94.00
(P-value)		(0.3005)	(0.6975)	(0.7914)	(0.6510)
Pima	76.4	76.32	70.9*	72.40*	72.40*
(P-value)		(0.8996)	**(0.0094)**	**(0.0362)**	**(0.0362)**
Sonar	86.1	86.55	85.5	85.58	87.02
(P-value)		(0.3521)	(0.8519)	(0.9524)	(0.8482)
Australian	85.7	86.52	84.2	83.04*	86.09
(P-value)		0.3955	(0.2014)	**(0.0922)**	(0.7988)
Heart	82.8	83.15	72.0*	73.74*	77.78
(P-value)		0.7944	**(0.0046)**	**(0.0068)**	0.1335
Housing	85.8	85.96	82.0*	82.41	80.83*
(P-value)		0.5912	**(0.0256)**	(0.2625)	**(0.0470)**

Table 2. 10-fold cross validation results - accuracy (%). P-values in bold mean that results are significantly different from BMDT's. * indicates that BMDT is significantly better.

(input to hidden units) give the oblique hyperplane (in original attributes $\mathbf{x}$):
if $5.67x_1 + 1.28x_2 + 2.81x_6 + 44.27x_7 \leq 316.20$ then class 0, else 1. In contrast to the UDT (109 nodes and each node with one attribute) built on the initial features $\mathbf{x}$ created by C4.5, the MDT has only three nodes and an oblique hyperplane with four attributes. In this case, the MDT has the simplest possible tree but a more complex node. Another point to note is that for this particular fold cross validation, UDT's accuracy is 64.94%, while MDT's is 76.62%.

We summarized the results with the following discussion.

- We can expect that BMDT performs well in general. This is because it overcomes the replication and repetition problems by still constructing multivariate decision trees, as well as minimizes the fragmentation problem through feature transformation. The latter cannot be achieved by the traditional way of constructing MDT's. Via feature transformation, BMDT circumvents the difficult problem of choosing M out of N features every time when a multi-

Name	BMDT	C4.5	OC1	CART
Monk1	9.8	14.8*	7.0	9.4
(P-value)		**(0.0001)**	**(0.0013)**	(0.5594)
Monk2	8.0	87.6*	5.4	7.4
(P-value)		**(0.0001)**	(0.1174)	(0.7418)
Monk3	10.0	15.0*	3.0	3.0
(P-value)		**(0.0001)**	**(0.0001)**	**(0.0001)**
Par12a	11.0	31.0*	11.4	27.8*
(P-value)		**(0.0001)**	(0.7472)	**(0.0899)**
Mux12	33.6	212.0*	29.6	63.0*
(P-value)		**(0.0001)**	(0.4846)	**(0.0028)**
TicTacToe	5.6	71.8*	17.6*	20.8*
(P-value)		**(0.0001)**	**(0.0787)**	**(0.0185)**
Vote	3.0	6.6*	3.2	3.2
(P-value)		**(0.0001)**	(0.3434)	(0.3434)
B-Cancer	3.2	22.2*	5.0	5.8*
(P-value)		**(0.0001)**	(0.1105)	**(0.0533)**
Bupa	5.8	79.2*	9.2	18.6
(P-value)		**(0.0001)**	(0.3961)	(0.1495)
Ionosphere	3.0	24.8*	8.6*	21.4*
(P-value)		**(0.0001)**	**(0.0552)**	**(0.0005)**
Iris	5.0	8.2*	5.0	5.2
(P-value)		**(0.0011)**	-	(0.3434)
Pima	3.0	122.4*	11.4*	26.2*
(P-value)		**(0.0001)**	**(0.0196)**	**(0.0722)**
Sonar	3.0	31.0*	8.2*	16.4*
(P-value)		**(0.0001)**	**(0.0056)**	**(0.0002)**
Australian	5.4	68.7*	4.8	6.2
(P-value)		**(0.0001)**	(0.8226)	(0.7571)
Heart	3.0	51.9*	9.2*	7.4*
(P-value)		**(0.0001)**	**(0.0566)**	**(0.0008)**
Housing	9.6	53.8*	4.2	8.8
(P-value)		**(0.0001)**	(0.1510)	(0.8640)

Table 3. 10-fold cross validation results - tree size. P-values in bold and * have the same meanings as in Table 2.

variate test node is created; tree pruning also becomes simpler. A feedforward neural network preserves and condenses information. Network pruning plays a key role in removing irrelevant information and in avoiding overfitting data. Referring to Tables 2 and 3, we notice that better average accuracy obtained by BMDT is usually due to a tree with fewer nodes.

- It is clear in Table 2 that average accuracy rates of BMDT and NN are the same except for one case that BMDT's is better but the difference is not statistically significant. Is there any advantage to have BMDT over NN? This question brings us back to the old question whether we need a decision tree when a neural network is present, or vice versa. It boils down to an issue of explicitness of learned representations (rules versus weights). The learned representation of a decision tree framework is explicit and allows one to examine the rules and explain how a prediction is made, while trained neural networks are largely black boxes[1]. As we see, the explicitness requires

---

[1] There are efforts at opening these black boxes [25, 7, 22].

extra work in BMDT. If no explanation for prediction is needed, rather than BMDT, NN should be the first choice due to its high accuracy.

– Adding multivariate tests, in a conventional way of constructing MDT's [15, 3], decreases the number of nodes in the tree, but increases the number of features tested per node. It is not the case for BMDT: owing to feature transformation, testing at a node is still univariate, but the number of nodes in the tree decreases. Recall that the dimensionality of y is usually much smaller than that of x, this is because y corresponds to the hidden units after network pruning and x to the input units.

– Although oblique trees are usually small in terms of nodes in a tree, each node is more complicated than a node in an axis-parallel tree. Hence, there is a tradeoff to consider between the two types of trees: a large tree with simple nodes vs. a small tree with complicated nodes. This also indicates that MDT's are a complimentary alternative to UDT's. Adopting which type of trees depends on the problems (i.e., data) at hand.

– How can one include unordered features in a linear combination test? Different solutions have been suggested. In CART, Breiman et al [2] suggested to form a linear combination using only the ordered features. Another solution is to map each multi-valued unordered feature to m numeric features, one for each observed value of the feature [26]. This is equivalent to the binary coding mentioned earlier when the neural net input coding was discussed.

5   Related Work

In order to overcome the replication problem and alleviate the fragmentation problem, researchers have suggested various solutions. [16] proposed compound boolean features; [12, 28] employed constructive induction to construct new features defined in terms of existing features;   [27] tried global data analysis - assessing the value of each partial hypothesis by recurring to all available training data. However, all these solutions are mainly designed for boolean data. The approach proposed in this work can handle all types of data (namely nominal, continuous, or mixed).

Feature transformation can be achieved by other forms. One popular approach is principle component analysis (PCA). The basic idea is to calculate eigenvalues of the covariance matrix of the data with x and select eigenvectors whose values are greater than a threshold (e.g., 0.9) to form a transformation matrix with M (the number of new features) rows and N (the number of original features) columns. The big disadvantage of PCA is that it cannot take advantage of the class information, though it is available. By discarding those eigenvectors with eigenvalues less than the threshold, PCA cannot fully preserve the input information.

Feature transformation may also be linked to techniques such as discretization [5, 11] and subset selection [10, 4]. Discretization transforms continuous data into discrete one. Subset selection, as its name implies, chooses a subset of original features. Although both techniques can reduce the dimensionality of the data, it is clear that neither new nor compound features are generated.

6 Summary

Univariate decision trees inherit the problems of replication, repetition, and fragmentation. Multivariate decision trees can overcome some of these problems. Close examination of the conventional way of building an MDT reveals that the fragmentation problem persists. Further investigation shows that the intertwining of hyperplane search and tree induction is the major cause since the conventional way of the MDT building follows exactly the way in which a univariate decision tree is built. We propose to separate hyperplane search from tree induction via feature transformation. A neural network is employed to transform original features to new features that preserve and compress information in data. It is chosen because of its many advantages over other methods. Instead of searching hyperplanes as the tree grows, applying feature transformation to the MDT building is, in effect, finding all the hyperplanes before tree induction. In this way, all training data is used for hyperplane search so that the fragmentation problem is minimized. We experimented with the new approach (BMDT) to prove the benefits of using this approach. The reduction of tree nodes from UDT's to MDT's is significant with equal or better accuracy - this indicates the effectiveness of BMDT in attacking the problems of replication, repetition, and fragmentation. The important conclusion is that it is possible and worthwhile to separate hyperplane search from tree induction; and doing so also makes building a multivariate decision tree simpler.

Acknowledgments

Thanks to S. Murthy and S. Salzberg for making OC1 publicly available. Thanks also to J. Yao, M. Dash, X. Ma for helping obtain the results of OC1 and CART reported in this paper.

References

1. K.P. Bennett and O.L. Mangasarian. Neural network training via linear programming. In P.M. Pardalos, editor, *Advances in Optimization and Parallel Computing*, pages 56–67. Elsevier Science Publishers B.V., Amsterdam, 1992.
2. L. Breiman, J.H. Friedman, R.A. Olshen, and C.J. Stone. *Classification and Regression Trees*. Wadsworth & Brooks/Cole Advanced Books & Software, 1984.
3. C.E. Brodley and P.E. Utgoff. Multivariate decision trees. *Machine Learning*, 19:45–77, 1995.
4. M. Dash and H. Liu. Feature selection methods for classifications. *Intelligent Data Analysis: An International Journal*, 1(3), 1997. http://www-east.elsevier.com/ida/free.htm.
5. U.M. Fayyad and K.B. Irani. Multi-interval discretization of continuous-valued attributes for classification learning. In *Proceedings of the Thirteenth International Joint Conference on Artificial Intelligence*, pages 1022–1027. Morgan Kaufmann Publishers, Inc., 1993.

6. J.H. Friedman, R. Kohavi, and Y. Yun. Lazy decision trees. In *Proceedings of the Thirteenth National Conference on Artificial Intelligence*, pages 717–724, 1996.
7. L. Fu. *Neural Networks in Computer Intelligence*. McGraw-Hill, 1994.
8. B. Hassibi and D.G. Stork. Second order derivatives for network pruning: Optimal brain surgeon. *Neural Information Processing Systems*, 5:164–171, 1993.
9. D. Heath, S. Kasif, and S. Salzberg. Learning oblique decision trees. In *Proceedings of the Thirteenth International Joint Conference on AI*, pages 1002–1007, France, 1993.
10. K. Kira and L.A. Rendell. The feature selection problem: Traditional methods and a new algorithm. In *Proceedings of the Tenth National Conference on Artificial Intelligence*, pages 129–134. Menlo Park: AAAI Press/The MIT Press, 1992.
11. H. Liu and R. Setiono. Chi2: Feature selection and discretization of numeric attributes. In J.F. Vassilopoulos, editor, *Proceedings of the Seventh IEEE International Conference on Tools with Artificial Intelligence, November 5-8, 1995*, pages 388–391, Herndon, Virginia, 1995. IEEE Computer Society.
12. C Matheus and L. Rendell. Constructive induction on decision trees. In *Proceedings of International Joint Conference on AI*, pages 645–650, August 1989.
13. C.J. Merz and P.M. Murphy. UCI repository of machine learning databases. http://www.ics.uci.edu/~mlearn/MLRepository.html. Irvine, CA: University of California, Department of Information and Computer Science, 1996.
14. John Mingers. An empirical comparison of selection measures for decision-tree induction. *Machine Learning*, 3:319–342, 1989.
15. S Murthy, S. Kasif, S. Salzberg, and R. Beigel. Oc1: Randomized induction of oblique decision trees. In *Proceedings of AAAI Conference (AAAI'93)*, pages 322–327. AAAI Press / The MIT Press, 1993.
16. G. Pagallo and D. Haussler. Boolean feature discovery in empirical learning. *Machine Learning*, 5:71–99, 1990.
17. J.R. Quinlan. Induction of decision trees. *Machine Learning*, 1(1):81–106, 1986.
18. J.R. Quinlan. *C4.5: Programs for Machine Learning*. Morgan Kaufmann, 1993.
19. D.E. Rumelhart, J.L. McClelland, and the PDP Research Group. *Parallel Distributed Processing*, volume 1. Cambridge, Mass. The MIT Press, 1986.
20. I.K. Sethi. Neural implementation of tree classifiers. *IEEE Trans. on Systems, Man, and Cybernetics*, 25(8), August 1995.
21. R. Setiono. A penalty-function approach for pruning feedforward neural networks. *Neural Computation*, 9(1):185–204, 1997.
22. R. Setiono and H. Liu. Understanding neural networks via rule extraction. In *Proceedings of International Joint Conference on AI*, 1995.
23. R. Setiono and H. Liu. Analysis of hidden representations by greedy clustering. *Connection Science*, 10(1):21–42, 1998.
24. J.W. Shavlik, R.J. Mooney, and G.G. Towell. Symbolic and neural learning algorithms: An experimental comparison. *Machine Learning*, 6(2):111–143, 1991.
25. G.G. Towell and J.W. Shavlik. Extracting refined rules from knowledge-based neural networks. *Machine Learning*, 13(1):71–101, 1993.
26. P.E. Utgoff and C.E. Brodley. An incremental method for finding multivariate splits for decision trees. In *Machine Learning: Proceedings of the Seventh International Conference*, pages 58–65. University of Texas, Austin, Texas, 1990.
27. R. Vilalta, G. Blix, and L. Rendell. Global data analysis and the fragmentation problem in decision tree induction. In M. van Someren and G. Widmer, editors, *Machine Learning: ECML-97*, pages 312–326. Springer-Verlag, 1997.
28. J. Wnek and R.S. Michalski. Hypothesis-driven constructive induction in AQ17-HCI: A method and experiments. *Machine Learning*, 14, 1994.

Formal Logics of Discovery and Hypothesis Formation by Machine

Petr Hájek* and Martin Holeňa

Institute of Computer Science, Academy of Sciences
182 07 Prague, Czech Republic
hajek@uivt.cas.cz, martin@uivt.cas.cz

Abstract. The following are the aims of the paper: (1) To call the attention of the community of Discovery Science to certain existing formal systems for DS developed in Prague in 60's till 80's suitable for DS and unfortunately largely unknown. (2) To illustrate the use of the calculi in question on the example of the GUHA method of hypothesis generation by computer, subjecting this method to a critical evaluation in the context of contemporary data mining. (3) To stress the importance of Fuzzy Logic for DS and inform on the present state of mathematical foundations of Fuzzy Logic. (4) Finally, to present a running research program of developing calculi of symbolic fuzzy logic for DS and for a fuzzy GUHA method.

1 Introduction

The term "logic of discovery" is admittedly not new: let us mention at least Popper's philosophical work [42], Buchanan's dissertation [4] analyzing the notion of a logic of discovery in relation to Artificial Intelligence and Plotkin's paper [41] with his notion of a logic of discovery as a logic of induction plus a logic of suggestion. In relation to data mining one has to mention the concept of exploratory data analysis, as elaborated by Tukey [50]. Can there be a formal (symbolic) logic of discovery? And why should it be developed? The answer is yes, various formal calculi can be and have been developed. And the obvious *raison d'être* for them (besides their purely logical importance) is that the computer can understand, process and (sometimes) evaluate formulas of a formal language, which is important for discovery as a cognitive activity studied by AI and Discovery Science (DS). The present paper has the following aims: (1) To call the attention of the DS community to certain existing formal systems for DS developed in Prague in 60's till 80's – not just for some reasons of priority but since we find them natural, suitable for DS and unfortunately largely unknown. (2) To illustrate the use of the calculi in question on the example of the GUHA method of hypothesis generation by computer, subjecting this method to

---

* Partial support of the grant No. A1030601 of the Grant Agency of the Academy of Sciences of the Czech Republic is acknowledged. The authors thank to D. Harmancová for her help in preparing the text of this paper.

a critical evaluation in the context of contemporary data mining. (3) To stress the importance of Fuzzy Logic (and, more generally Soft Computing) for DS and inform on the present state of mathematical foundations of Fuzzy Logic. (4) Finally, to present a running research program of developing calculi of symbolic fuzzy logic for DS and for a fuzzy GUHA method.

2   Calculi of the logic of discovery

We refer here on calculi whose syntax and semantics is fully elaborated in the monograph [16]. Since there exists a survey paper [15] (which we would like to recommend to the reader) we shall be rather sketchy (see also [10]).

Distinction is made between observational and theoretical languages. Formulas of an observational language are used to speak about the data; formulas of a theoretical language on a universe not directly being at our disposal. Examples: in my sample of 200 persons, 120 have positive outcome of test T (observational). In my universe of discourse, the probability of positive outcome of the test is higher than 50% (theoretical).

Data are finite structures. For simplicity, think of a data structure as of a rectangular matrix whose rows correspond to objects and columns correspond to values of a variate. Each variate has a *name* (X, TEST etc.) and domain from which the values are taken (real, integer, numbers 1 to 20, etc.). We may also have some distinguished subsets of the domain named attributes, e.g. the interval $\langle 10, 20 \rangle$ for age (teenagers). Formulas may be built from atoms if the form $X{:}A$ (read X is A, e.g. age is teenager) using logical connectives – such formulas are called *open formulas*. Given data **M**, it is clear what we mean saying that an object m *satisfies* on open formula φ; $Fr_{\mathbf{M}}(\varphi)$ is the frequency of φ in **M**, i.e. the number of objects in **M** satisfying φ. For a pair φ, ψ of open formulas we have four frequencies $a = Fr_{\mathbf{M}}(\varphi \& \psi)$, $b = Fr_{\mathbf{M}}(\varphi \& \neg \psi)$, $c = Fr_{\mathbf{M}}(\neg \varphi \& \psi)$, $d = Fr_{\mathbf{M}}(\neg \varphi \& \neg \psi)$ ($\neg$ being negation, & being conjunction). The quadruple (a, b, c, d) is the *fourfold table* of (φ, ψ).

Generalized quantifiers are used to get sentences, i.e. formulas expressing properties of the data as whole, e.g. for φ, ψ open formulas, $(\forall x)\varphi$ means "all objects satisfy φ", (Majority $x)\varphi$ means "more than 50% objects satisfy φ", $(\exists x)\varphi$ means "at least one object satisfies φ" etc. (Many$_p x)$ $(\psi | \varphi)$, written also $\varphi \sqsupset_p \psi$ means p-many x satisfying φ satisfy ψ, i.e. $Fr(\varphi \& \psi)/Fr(\varphi) \geq p$ etc.

Semantics of a unary quantifier q is given by its *truth function* (also called *associated function*) Tr_q assigning to each column vector of zeros and ones (the course of values of a formula) 0 or 1. For example, $Tr_{Majority}(V) = 1$ iff the column V contains more 1's than 0's. Similarly for a binary quantifier (like "Many...are..."), but now V is a matrix consisting of two column vectors of 0's and 1's.

One of most distinguishing features of the described approach is a *tight connection between logic and statistic*.[1] The key idea of that connection is to view

---

[1] A connection of logic and statistics obeying this schema was, in the 1970s, one of the main contributions of the GUHA approach to exploratory data analysis [11, 16,

each data matrix, used for evaluating observational sentences, as a realization of a random sample. Consequently, the truth function of a generalized quantifier, composed with random samples with values in its domain, is a random variable. Since random variables expressible as a composition of a function of many variables with multidimensional random samples are often used as test statistics for *testing statistical hypotheses*, it is possible to cast statistical tests in the framework of generalized quantifiers. In the most simple case of dichotomous data matrices, this can be accomplished for example as follows: Let M_D be a two-column matrix of zeros and ones the rows of which contain evaluations, in given data, of some pair of open formulae (φ, ψ). Thus all those evaluations are viewed as realizations of independent two-dimensional random vectors, all having the same distribution D. Suppose that D is known to belong to the set $\mathcal{D}$ described by the nonsingularity condition $p_{\psi|\varphi} \in (0,1)$, where $p_{\psi|\varphi}$ is the conditional probability corresponding to D of ψ being satisfied conditioned on φ being satisfied. The parametrizability of $\mathcal{D}$ by $p_{\psi|\varphi}$ makes it possible to express also a null hypothesis $D \in \mathcal{D}_0$ by means of $p_{\psi|\varphi}$. In particular, given $\alpha, \theta \in (0,1)$, the following statistical test can be considered: *test the null hypothesis $p_{\psi|\varphi} \leq \theta$ using a test statistic $\sum_{i=a}^{a+b} \binom{a+b}{i} \theta^i (1-\theta)^{a+b-i}$, and the critical region $(0, \alpha)$.* This leads to a binary quantifier $\beth_\theta^!$ called *likely implication* with the threshold θ, whose truth function is defined, for each natural k and each matrix $M \in \{0,1\}^{k,2}$, as follows:

$$Tr_{\beth_\theta^!}(M) = 1 \text{ iff } \sum_{i=a}^{a+b} \binom{a+b}{i} \theta^i (1-\theta)^{a+b-i} \leq \alpha.$$

Thus the quantifier $\beth_\theta^!$ captures the fact that the test leads to rejecting $p_{\psi|\varphi} \leq \theta$ at the significance level α.

Dually, we may consider a quantifier $\beth_\theta^?$ (suspicious implication) capturing the fact that a particular test does not reject $p_{\psi|\varphi} \geq \theta$. It is important that both quantifiers belong to an infinite family of implicational (multitudinal) quantifiers defined by simple monotonicity conditions (and containing also very simple quantifiers like *many...are...*. All quantifiers $\beth$ from this family share some important logical properties (e.g. $(\varphi_1 \& \varphi_2) \beth \psi$ implies $\varphi_1 \beth (\neg \varphi_2 \vee \psi)$). See [16] for details; for recent developments see [45, 46].

In this way we get observational logical calculi with interesting formal properties – a particular branch of *finite model theory* as logical foundations of database theory. Sentences of an observational language express interesting *patterns* that may be recognized in given data. In contradistinction to this, *theoretical sentences* are interpreted in possibly infinite structures, not directly accessible. They may express properties of *probability, possibility, (in)dependence etc.* The corresponding calculi have been elaborated, also using the notion of a generalized quantifier. *Modal logic* is relevant here; theoretical structures are defined as parametrized by "possible worlds" and e.g. the probability of φ is defined as

21–23]. Two decades later, such a connection became an important flavour of data mining [31, 32, 55, 56].

the probability of the set of all possible worlds in which φ is true. See the references above for details and note that there is important literature on probability quantifiers, notably [30] and, in our context, [11].

Inductive inference is the step from an observation (expressed by a sentence α of an observational language) to a theoretical sentence Φ, given some theoretical frame assumption Frame. The rationality of such step is given by the fact that assuming Frame, if Φ were false then we could prove that the observation α is unlikely (in some specified sense), i.e.

$$\text{Frame}, \neg\Phi \vdash \text{unlikely}(\alpha).$$

This is a starting point for various formal developments, including (but not identical with) statistical hypothesis testing.

3 The GUHA method and data mining

The development of this method of exploratory data analysis started in mid-sixties by papers by Hájek, Havel and Chytil [14]. Even if the original formalism appears simple-minded today, the principle formulated there remains the same till today: to use means of formal logic to let the computer generate all hypotheses interesting with respect to a research task and supported by the data. In fact the computer generates interesting observational sentences rather than hypotheses (theoretical sentences); but the observational sentences correspond to theoretical sentences via a rule of inductive inference as above and, in addition, they are interesting as statements about the data themselves, in particular if the data are immensely large.

This general program may be realized in various forms; the main form that has been implemented (repeatedly) and practically used is the GUHA package ASSOC for generating hypotheses on associations and high conditional probabilities via observational sentences expressing symmetric associations $\varphi \sim \psi$ (φ, ψ are mutually associated) or multitudinal (implicational) associations of the form $\varphi \sqsupset \psi$ (many objects having φ have ψ).[2] Here φ, ψ are open formulas. Each formally true sentence of a desired form is output (saved) together with the corresponding fourfold table and values of several characteristics. Here we cannot go into details; the corresponding theory is found in the monograph [16] and the paper [12], a report on a PC-implementation is in [17]. See also [20].[3]

---

[2] A new implementation of a generalization of ASSOC, called FFTM (four-fold table miner) has been prepared by J. Rauch.

[3] The following is an example of a sentence found true in the data of the paper [29] dealing with interdependencies among cytological prognostic factors and estimation of patient risk:

(cath-d: >65.00 and ps2: >10.00 and grad: 2-3) $\sqsupset$ stadium: N-M

with the table (28,3,244,229), i.e. there were 31 patients (from 504) satisfying the antecedent formula; 28 of them satisfied "stadium N-M".

At the time when the theoretical principles of the GUHA approach were developed, data analysts typically dealt with tens to hundreds of objects, thousands being already an exception. Future databases of a size breaking the terabyte limit will increasingly often contain data about the whole population, thus making the inference from a sample to the population in principle superfluous. Indeed, sample-based methods are used in data mining mostly for efficiency reasons [1, 33, 35, 36, 49, 53, 54]. But notice that an inductive inference from frame assumptions and an observation to a theoretical sentence is not an inference from a sample to the population. Therefore, it remains fully justified even if the observation is based on data covering the whole population, i.e. even under the conditions of data mining.

GUHA has been developed as a method of exploratory data analysis but has been rather rarely used in practice until recently, when the availability of a PC version of ASSOC and the ubiquity of performant personal computers has made its sophisticated algorithms easily accessible even for occasional users needing to tackle realistically-sized problems. Indeed, since 1995 nearly a dozen GUHA applications have been reported, some of them being listed in [20].

Though these applications are increasingly various, including such areas as pharmacy, linguistics or musicology, we shall mention here only two applications in *medicine* and *economy*.

In the *biochemistry of cancer cells,* hypotheses generated by GUHA helped to elucidate some of the mechanisms driving the rise of metastases of a primary tumor [29, 37]. The rise of metastases depends on the metabolism of the cells and on the exchange of different substances between cells and their environment. Substances entering the metabolism of the cells are interconnected through various dependences. The GUHA approach contributed to finding such dependences. It also supported the researchers in deriving decision rules for more extensive use of tumor markers in the treatment of oncological patients.

In *financial market analysis,* GUHA has been recently applied to time series from foreign exchange terminal quotations of exchange rates between USD and three other leading currencies (GBP, DEM, JPY). Those three time series were scanned to find jumps, i.e. changes above a prescribed threshold. For each jump, a number of indicators were computed, characterizing either the jump itself (time, direction, interval since the previous jump) or the whole time series at the moment of the jump (e.g., activity, volatility, density and trend of recent jumps, interval of identical trend). Then GUHA was used to generate hypotheses about relationships between various indicators of all three time series. The generated hypotheses served as a basis for a new method of predicting near future jumps and their directions. This method was tested on data from a following time period, and the results were found promising [38, 39].

Finally let us mention an ongoing application of GUHA in chemistry done in Czech-Japanese cooperation [18], [19].

Compared with modern data mining methods, GUHA lacks a thorough coupling to the database technology. Actually, such a coupling was under development in the 1980s [43]. However, it was oriented exclusively towards network databases relying on the Codasyl proposal [6], that time still commercially the most successful kind of databases. As Codasyl databases became obsolete, that development has been abandoned. To couple GUHA to relational and object-oriented databases remains a task for the future.

In spite of that difference, we feel that GUHA fully deserves to be considered an early example of data mining. This opinion can be justified from multiple points of view.

Purpose. In this respect, GUHA has several features typical for data mining [7, 8, 26, 48]:

- search for relationships hidden in the data,
- limiting the search to relationships interesting according to some predefined criteria,
- focus on relationships that can not be found in a trivial way (e.g., that could not be found through SQL queries),
- automating the search as far as possible,
- optimization to avoid blind search whenever possible.

Methods. GUHA is similar to some modern data mining approaches in employing logic for the specification of and navigation through the hypotheses space, while employing data analysis, in particular statistical methods, for the evaluation of hypotheses in that space. Moreover, that similarity goes even further, covering also the main kinds of statistical methods employed for the evaluation, namely statistical hypotheses testing, most often in the context of contingency tables [5, 9, 31, 32, 55, 56].

Scope. GUHA relates, in particular, to *mining association rules.* Indeed, if $A = \{A_1, \ldots, A_m\}$ is the set of binary attributes in a database of size k, and if $X, Y \subset A, X \cap Y = \emptyset$, then the association rule $X \Rightarrow Y$ is significant in the database (according to [1, 2, 27, 28, 40, 47, 54]) if and only if the GUHA sentence

$$\bigwedge_{i \in X} A_i \sqsupset_{B,p} \bigwedge_{i \in Y} A_i$$

holds for the $k \times m$ dichotomous data matrix formed by the values of the attributes from A. Here, $\sqsupset_{B,p}$ is a *founded* version of the generalized quantifier $\sqsupset_p$ mentioned in section 2 (version requiring the frequence a to be at least as large as a predefined *base* $B \in \mathcal{N}$, see also [12, 17]). Moreover, there is a very simple relationship between the parameters p and B of that quantifier, and the *support* $s \in (0, 1)$ and *confidence* $c \in (0, 1)$ of the above association rule:

$$p = c \ \& \ B = k \cdot s.$$

In addition, several concepts pertaining to mining association rules have some counterpart in GUHA:

- Mining rules with *item constraints* [47] can be covered by GUHA using *relativized sentences* with filtering conditions [12, 17].

- The notion of a *frontier/border set*, crucial for efficient finding of all large/frequent itemsets [35, 54], is closely related to the GUHA concept of *prime sentences* [16].
- The gap between association rules and functional dependencies known from databases [2, 34] can be partially bridged in GUHA by means of *improving literals* [12, 16].

4   Impact of Soft Computing

Soft computing is a relatively new name for a branch of research including fuzzy logic, neural networks, genetic and probabilistic computing.[4] Here we contemplate on soft exploratory data analysis or soft data mining in GUHA-style and in general.

First on fuzzy logic. This is admittedly a fashionable term with several meanings. Following Zadeh we shall distinguish between FL_w (fuzzy logic in wide sense) and FL_n (fuzzy logic in narrow sense), the former being practically everything dealing with fuzziness, thus synonymous with fuzzy set theory (also in wide sense). In the narrow sense, fuzzy logic is just the study of some calculi of many-valued logic understood as logic of graded truth; Zadeh stresses that the agenda of fuzzy logic differs from the agenda of traditional many-valued logic and includes entries as generalized quantifiers (usually, many etc.), approximate reasoning and similar. In the last period of development fuzzy logic (FL_n) has been subjected to a serious mathematical and logical investigation resulting, among other works, in the monograph [13]. It has turned out that, on the one hand, calculi of fuzzy logic (based on the notion of a triangular norm) admit classical investigation concerning axiomatizability, completeness, question of complexity etc., both for propositional and predicate logic, and, on the other hand several entries of Zadeh's agenda can be analyzed in terms of *deduction* in appropriate theories. The main aim of that book is to show that fuzzy logic is (can be) a real fully-fledged logic. This does not contradict the fact that fuzzy logic in wide sense has many extra-logical aspects. But mathematical foundations of fuzzy logic may be understood as an integral part of mathematical foundations of Soft Computing – and the paradigm of Soft Computing is obviously relevant for the intended development of Discovery Science.

5   Fuzzy logic of discovery and fuzzy GUHA

Needless to say, fuzzy logic in the wide sense has been repeatedly used in Data Analysis; see e.g. [3] Sect. 5.5 for a survey. It is very natural to ask how can the

---

[4] Let us quote from Zadeh, the father of fuzzy set theory and fuzzy logic [52]: *"The guiding principle of soft computing is: exploit the tolerance for imprecision, uncertainty, partial truth, and approximation to achieve tractability, robustness, low solution cost and better rapport with reality. One of the principal aims of soft computing is to provide a foundation for the conception, design and application of intelligent systems employing its member methodologies symbiotically rather than in isolation."*

methods and results of fuzzy logic in the narrow sense be applied to the calculi of logic of discovery as sketched above; thus what are fuzzy observational and theoretical languages of DS. This should clearly not be a self-purpose fuzzification: First, the typical observational quantifiers used in GUHA are *associational* (φ, ψ are associated, positively dependent in the data) or *multitudinal* (many φ's are ψ's). Until now it has been always defined in some crisp way, using a parameter (p-many etc.). But it is much more natural to understand them in a frame of a fuzzy logic, at least in two kinds of systems:

(a) open observational formulas are crisp as before (like "age is $\langle 10 - 20 \rangle$" – yes or no), but quantified observational formulas are fuzzy, e.g. the truth value of (Many x)φ in M can be the relative frequence of φ in M.

(b) Also atomic observational formulas are fuzzy, i.e. the attributes are fuzzy, e.g. "age is young" where "young" is a fuzzy attribute with a given fuzzy truth function on numerical values of age. For both variants, [13] contains foundations; but elaboration remains a research task.

Second, fuzzy hypothesis testing may be developed in the framework of fuzzy generalized quantifiers in the FL_n sense, mentioned in the preceding section. Fuzziness can enter a statistical test mainly in the following ways:

(i) The data analyst has only a vague idea about the null hypothesis to test. In that case, the set $\mathcal{D}_0 \subset \mathcal{D}$, considered in Section 2, should be replaced by an appropriate fuzzy set $\tilde{D}$ on $\mathcal{D}$. For example, the set $(0, \theta \rangle$ determining the null hypothesis for the parameter $p_{\psi|\varphi}$ should by replaced by a fuzzy set on $(0, 1)$.

(ii) The data analyst has only a vague idea about the critical region to use for the test. Then a fuzzy set on $(0, 1)$ should be used instead of the interval $(0, \alpha \rangle$. This corresponds to the situation when the data analyst is not sure about the significance level α to choose.

In our opininon, especially the fuzzification of the tested null hypotheses is highly relevant for exploratory data analysis and data mining. In fact, exploratory analysis and data mining are typically performed in situations when only very little is known about the distribution of the random variates that generated the data. Consequently, it is very difficult to specify precisely the set $\mathcal{D}_0$ determining the tested null hypothesis, e.g., to choose a precise value of the threshold θ in our example.

Recently, statistical tests with fuzzy null hypotheses have been intensively studied in the context of GUHA generalized quantifiers $\sqsupset^!$ and $\sqsupset^?$ mentioned in section 2 [24, 25]. The investigations were mainly intended for the fuzzy hypotheses paraphrased as "$p_{\psi|\varphi}$ is low" (replacing $p_{\psi|\varphi} \leq \theta$) in the case of the quantifier $\sqsupset^!$, and "$p_{\psi|\varphi}$ is high" (replacing $p_{\psi|\varphi} \geq \theta$) in the case of $\sqsupset^?$. However, actually a much more general setting of nonincreasing / nondecreasing linguistic quantifiers (in the sense introduced by Yager in [51]) has been used.

A number of important results concerning the fuzzy-hypotheses generalizations $\sqsupset^!_\sim$, $\sqsupset^?_\sim$ of the quantifiers $\sqsupset^!$, $\sqsupset^?$, respectively, have been proven in [25]. Logical theory of the fuzzy quantifiers "the probability of ... is high" and "the conditional probability of ... given ... is high" is elaborated in [13].

All this seems to be a promising research domain. Let us add two details: (a) languages for large data concerning large event sequences should be developed, i.e. the ordering of objects in the data matrix is relevant and expressible in the language. Some rudimentary beginnings can be found in [16].

(b) For processing extremely large data sets by a GUHA-like procedure, there are good possibilities of parallelization.

6 Conclusion

This paper is a kind of position paper; we have offered formal logical foundations (partly old and forgotten, partly new and under development) for a certain direction (branch) of DS, namely logic of discovery as hypothesis formation relevant to data mining.

This is based on fully fledged formal calculi with exactly defined syntax and semantics in the spirit of modern mathematical logic. We have stressed the paradigm of soft computing, in particular of fuzzy logic together with its strictly logical foundations. In our opinion the surveyed kind of Logic of Discovery can be found a valuable contribution to Discovery Science as intended.

References

1. AGRAWAL, R., MANNILA, H., SRIKANT, R., TOIVONEN, H., AND VERKAMO, A. Fast discovery of association rules. In *Advances in Knowledge Discovery and Data Mining*, U. Fayyad, G. Piatetsky-Shapiro, P. Smyth, and R. Uthurusamy, Eds. AAAI Press, Menlo Park, 1996, pp. 307–328.
2. AGRAWAL, R., AND SRIKANT, R. Fast algorithms for mining association rules. In *Proceedings of the 20th International Conference on Very Large Data Bases* (1994).
3. ALTROCK C. VON *Fuzzy Logic and Neurofuzzy Applications Explained*. Prentice Hall PTR UpperSaddle River, NJ 1995.
4. BUCHANAN, B.G. *Logics of Scientific Discovery*. Stanford AI Memo no. 47, Stanford University 1966.
5. CHATFIELD, C. Model uncertainty, data mining and statistical inference. *Journal of the Royal Statistical Society. Series A 158* (1995), 419–466.
6. CODASYL DATA BASE TASK GROUP. DBTG report. Tech. rep., ACM, 1971.
7. FAYYAD, U., PIATETSKY-SHAPIRO, G., AND SMYTH, P. From data mining to knowledge discovery: An overview. In *Advances in Knowledge Discovery and Data Mining*, U. Fayyad, G. Piatetsky-Shapiro, P. Smyth, and R. Uthurusamy, Eds. AAAI Press, Menlo Park, 1996, pp. 1–36.
8. FRAWLEY, W., PIATETSKY-SHAPIRO, G., AND MATHEUS, C. Knowledge discovery in databases: An overview. In *Knowledge Discovery in Databases*, G. Piatetsky-Shapiro and W. Frawley, Eds. AAAI Press, Menlo Park, 1991, pp. 1–27.
9. GLYMOUR, C., MADIGAN, D., PREGIBON, D., AND SMYTH, P. Statistical inference and data mining. *Communications of the ACM 39* (1996), 35–41.
10. HÁJEK, P. On logics of discovery. In *Math. Foundations of Computer Science; Lect. Notes in Comp. Sci. vol. 32* (1975), Springer, pp. 30–45.
11. HÁJEK, P. Decision problems of some statistically motivated monadic modal calculi. *Int. J. for Man-Machine studies 15* (1981), 351–358.

12. HÁJEK, P. The new version of the GUHA procedure ASSOC (generating hypotheses on associations) – mathematical foundations. In *COMPSTAT 1984 – Proceedings in Computational Statistics* (1984), pp. 360–365.

13. HÁJEK, P. *Metamathematics of fuzzy logic*. Kluwer, 1998.

14. HÁJEK, P., HAVEL, I., AND CHYTIL, M. The GUHA-method of automatic hypotheses determination. *Computing 1* (1966), 293–308.

15. HÁJEK, P., AND HAVRÁNEK, T. On generation of inductive hypotheses. *Int. J. for Man-Machine studies 9* (1977), 415–438.

16. HÁJEK, P., AND HAVRÁNEK, T. *Mechanizing hypothesis formation (mathematical foundations for a general theory)*. Springer-Verlag, Berlin-Heidelberg-New York, 1978.

17. HÁJEK, P., SOCHOROVÁ, A., AND ZVÁROVÁ, J. GUHA for personal computers. *Computational Statistics and Data Analysis 19* (1995), 149–153.

18. HÁLOVÁ, J., ŽÁK, P., ŠTROUF, O. QSAR of Catechol Analogs Against Malignant Melanoma by PC-GUHA and CATALYST^{TM} software systems, poster, VIII. Congress IUPAC, Geneve (Switzerland) 1997. *Chimia 51* (1997), 532.

19. HÁLOVÁ, J., ŠTROUF, O., ŽÁK, P., SOCHOROVÁ, A., UCHIDA, N., YUZUVI, T., SAKAKIBAVA, K., HIROTA, M.: QSAR of Catechol Analogs Against Malignant Melanoma using fingerprint descriptors, *Quant. Struct.-Act. Relat. 17* (1998), 37–39.

20. HARMANCOVÁ, D., HOLEŇA, M., AND SOCHOROVÁ, A. Overview of the GUHA method for automating knowledge discovery in statistical data sets. In *Procedings of KESDA '98 – International Conference on Knowledge Extraction from Statistical Data* (1998), M. Noirhomme-Fraiture, Ed., pp. 39–52.

21. HAVRÁNEK, T. The approximation problem in computational statistics. In *Mathematical Foundations of Computer Science '75; Lect. Notes in Comp. Sci. vol. 32* (1975), J. Bečvář, Ed., pp. 258–265.

22. HAVRÁNEK, T. Statistical quantifiers in observational calculi: an application in GUHA method. *Theory and Decision 6* (1975), 213–230.

23. HAVRÁNEK, T. Towards a model theory of statistical theories. *Synthese 36* (1977), 441–458.

24. HOLEŇA, M. Exploratory data processing using a fuzzy generalization of the GUHA approach. In *Fuzzy Logic*, J. Baldwin, Ed. John Wiley and Sons, New York, 1996, pp. 213–229.

25. HOLEŇA, M. Fuzzy hypotheses for GUHA implications. *Fuzzy Sets and Systems 98* (1998), 101–125.

26. HOLSHEIMER, M., AND SIEBES, A. Data mining. The search for knowledge in databases. Tech. rep., CWI, Amsterdam, 1994.

27. HOUTSMA, M., AND SWAMI, A. Set-oriented mining of association rules. Tech. rep., IBM Almaden Research Center, 1993.

28. KAMBER, M., HAN, J., AND CHIANG, J. Using data cubes for metarule-guided mining of multi-dimensional association rules. Tech. rep., Simon Fraser University, Database Systems Research Laboratory, 1997.

29. KAUŠITZ, J., KULLIFFAY, P., PUTEROVÁ, B., AND PECEN, L. Prognostic meaning of cystolic concentrations of ER, PS2, Cath-D, TPS, TK and cAMP in primary breast carcinomas for patient risk estimation and therapy selection. To appear in *International Journal of Human Tumor Markers*.

30. KEISLER, U. J. Probability quantifiers. In *Model-Theoretic Logics*, J. Barwise and S. Feferman, Eds. Springer-Verlag, New York, 1985, pp. 539–556.

31. KLÖSGEN, W. Efficient discovery of interesting statements in databases. *Journal of Intelligent Information Systems 4* (1995), 53–69.
32. KLÖSGEN, W. Explora: A multipattern and multistrategy discovery assistant. In *Advances in Knowledge Discovery and Data Mining*, U. Fayyad, G. Piatetsky-Shapiro, P. Smyth, and R. Uthurusamy, Eds. AAAI Press, Menlo Park, 1996, pp. 249–272.
33. LIN, D.I. AND KEDEM, Z. Pincer search: A new algorithm for discovering the maximum frequent set. In *Proceedings of EDBT'98: 6th International Conference on Extending Database Technology* (1998).
34. MANNILA, H., AND RÄIHÄ, K. Dependency inference. In *Proceedings of the 13th International Conference on Very Large Data Bases* (1987), pp. 155–158.
35. MANNILA, H., TOIVONEN, H., AND VERKAMO, I. Efficient algorithms for discovering association rules. In *Knowledge Discovery in Databases*, U. Fayyad and R. Uthurusamy, Eds. AAAI Press, Menlo Park, 1994, pp. 181–192.
36. MUELLER, A. Fast sequential an parallel algorithms for association rule mining: A comparison. Tech. rep., University of Maryland – College Park, Department of Computer Science, 1995.
37. PECEN, L., AND EBEN, K. Non-linear mathematical interpretation of the oncological data. *Neural Network World*, 6:683–690, 1996.
38. PECEN, L., PELIKÁN, E., BERAN, H., AND PIVKA, D. Short-term fx market analysis and prediction. In *Neural Networks in Financial Engeneering* (1996), pp. 189–196.
39. PECEN, L., RAMEŠOVÁ, N., PELIKÁN, E., AND BERAN, H. Application of the GUHA method on financial data. *Neural Network World 5* (1995), 565–571.
40. PIATETSKY-SHAPIRO, G. Analysis and presentation of strong rules. In *Knowledge Discovery in Databases*, G. Piatetsky-Shapiro and W. Frawley, Eds. AAAI Press, Menlo Park, 1991, pp. 229–248.
41. PLOTKIN, G. D. A further note on inductive generalization. *Machine Intelligence 6* (1971), 101–124.
42. POPPER, K. R. *The Logic of Scientific Discovery*. Hutchinson Publ. Group Ltd., London, 1974.
43. RAUCH, J. Logical problems of statistical data analysis in data bases. In *Proceedings of the Eleventh International Seminar on Data Base Management Systems* (1988), pp. 53–63.
44. RAUCH, J. Logical Calculi for knowledge discovery in databases, In: (Komarowski, Żytkov, ed.) *Principles of Data Mining and Knowledge Discovery*, Lect. Notes in AL, vol. 1263, Springer-Verlag 1997.
45. RAUCH, J. *Classes of four-fold table quantifiers*, accepted for PKDD'98.
46. RAUCH, J. *Four-fold table predicate calculi for Discovery Science*, poster, this volume.
47. SRIKANT, R., VU, Q., AND AGRAWAL, R. Mining association rules with item constraints. In *Proceedings of the Third International Conference on Knowledge Discovery and Data Mining KDD-97* (1997).
48. TELLER, A., AND VELOSO, M. Program evolution for data mining. *International Journal of Expert Systems 8* (1995), 216–236.
49. TOIVONEN, H. *Discovery of Frequent Patterns in Large Data Collections*. PhD thesis, University of Helsinki, 1996.
50. TUKEY, J. W. *Exploratory Data Analysis*. Addison-Wesley, Reading, 1977.
51. YAGER, R. On a semantics for neural networks based on fuzzy quantifiers. *International Journal of Intelligent Systems 7* (1992), 765–786.

52. ZADEH, L. A. What is soft computing? (Editorial). *Soft Computing 1* (1997), 1.
53. ZAKI, M., PARATHASARATHY, S., LI, W., AND OGIHARA, M. Evaluation of sampling for data mining of association rules. In *Proceedings of the 7th International Workshop on Research Issues in Data Engineering* (1997), pp. 42–50.
54. ZAKI, M., PARATHASARATHY, S., OGIHARA, M., AND LI, W. New parallel algorithms for fast discovery of association rules. *Data Mining and Knowledge Discovery 1* (1997), 343–373.
55. ZEMBOWICZ, R., AND ŻYTKOV, J. From contingency tables to various forms of knowledge in databases. In *Advances in Knowledge Discovery and Data Mining*, U. Fayyad, G. Piatetsky-Shapiro, P. Smyth, and R. Uthurusamy, Eds. AAAI Press, Menlo Park, 1996, pp. 329–352.
56. ŻYTKOV, J., AND ZEMBOWICZ, R. Contingency tables as the foundation for concepts, concept hierarchies and rules: The 49er system approach. *Fundamenta Informaticae 30* (1997), 383–399.

Finding Hypotheses from Examples by Computing the Least Generalization of Bottom Clauses

Kimihito Ito[1] and Akihiro Yamamoto[2]

[1] Department of Electrical Engineering
[2] Division of Electronics and Information Engineering
and
Meme Media Laboratory
Hokkaido University
N 13 W 8, Sapporo 060-8628 JAPAN
{itok,yamamoto}@meme.hokudai.ac.jp

Abstract. In this paper we propose a new ILP method Bottom Reduction. It is an extension of Bottom Generalization for treating multiple examples. Using Bottom Reduction we can find all hypotheses which subsume all of given examples relative to a background theory in Plotkin's sense. We do not assume any restriction on the head predicates of the examples. Applying Bottom Reduction to examples which have a predicate different from those of other examples, we can reduce the search space of hypotheses. We have already implemented a simple learning system named BORDA based on Bottom Reduction on a Prolog system, and we present, in this paper, some techniques in its implementation.

1 Introduction

In this paper we propose a new Inductive Logic Programming method, *Bottom Reduction*, which finds hypotheses from multiple examples represented as definite clauses, and show how it should be implemented as a program.

We are now intending to contribute to Discovery Science with Inductive Logic Programming (ILP, for short) [6]. ILP is a research area where various methods have been investigated for finding new rules from given examples and a background theory in logical manners. Formally, given examples $E_1, \ldots, E_n$ and a background theory B, an ILP method finds hypotheses H such that

$$B \wedge H \vdash E_1 \wedge \ldots \wedge E_n. \tag{1}$$

The foundations of ILP methods are usually Logic Programming theories. Since a relational database can be regarded as a logic program, we expect that we can apply ILP methods to discovering knowledge from large scale databases.

We focus on an ILP method Bottom Generalization proposed by one of the authors [11–13, 15]. Bottom Generalization was developed for *learning from entailment*, which means the case $E_1, \ldots, E_n$ are definite clauses in the formula

(1). Various methods other than Bottom Generalization were also developed for learning from entailment, e.g. Inverse Entailment [5], Saturation with generalization [9], V^noperator [3]. Bottom Generalization has the following two good properties which none of these methods have:

- Its correctness and completeness are given.
- Its relation to abduction and deduction is formally shown.

Bottom Generalization can be regarded as a combination of SOLDR-derivation from a goal and bottom-up evaluation of a logic program. SOLDR-resolution is a modification of SOL-resolution developed by Inoue [2], and can be regarded as abduction. Since bottom-up evaluation is deduction, Bottom Generalization can be regarded as an amalgamation of abduction and deduction. Moreover, SOLDR-derivation is a natural extension of SLD-derivation, and we can implement SOLDR-derivation by a simple extension of Prolog meta-interpreter.

No ILP system which adopts Bottom Generalization had not been constructed, however, before we implemented BORDA system expressed in this paper. This is due to the following two problems. Firstly Bottom Generalization cannot work for multiple examples. Bottom Generalization was aimed to generate hypotheses which can explain *one* given example with respect to a background theory. When we treat more than one example, we must develop a new method for finding hypotheses. The second problem is that the number of bottom clauses or the length of each bottom clause may be infinite. In order to construct the ILP system using Bottom Generalization, we must develop some methods to solve both of them.

In order to solve the first problem we introduce *Bottom Reduction*. It is an extension of Bottom Generalization obtained by replacing inverse subsumption with the least common generalization [7]. It is important in the research of ILP methods that we justify them with the semantics for logic programs. We have already pointed out that the logical background of some ILP methods is not sufficient to justify them [14]. We show in this paper that Bottom Reduction is correct as a hypothesis generating method and complete in the sense that every hypothesis generated by the method subsumes all of the given examples relative to a background theory. We solve the second problem on Bottom Generalization by restricting the syntax of the background theory. The restrictions are well-known in Logic Programming theories and Inductive Logic Programming theories. By using these two solutions, we have implemented a simple learning system named BORDA.

This paper is organized as follows: In the following section we illustrate Bottom Reduction with a simple example. In Section 3 we prepare some basic definitions and notations and give a survey of our previous works. In Section 4 we give a formal definition of Bottom Reduction and show its logical properties. In Section 5 we illustrate an implementation method and discuss how the bottom evaluation should be terminated in a finite time. In the last section we describe concluding remarks.

2   An Overview of Bottom Reduction Method

Let us assume that the following definite program B is given as a background theory:

$$B = \begin{matrix} (s(a) \leftarrow) \\ \wedge\ (s(X) \leftarrow t(X)) \\ \wedge\ (p(f(X)) \leftarrow r(X)) \\ \wedge\ (q(g(X)) \leftarrow r(X)). \end{matrix}$$

Consider that the following clauses E_1 and E_2 are given as positive examples:

$$E_1 = p(f(a)) \leftarrow,$$
$$E_2 = q(g(b)) \leftarrow t(b).$$

Our learning method Bottom Reduction looks for definite clauses H such that H subsumes both E_1 and E_2 relative to B in Plotkin's sense.

At first the method constructs, by using Bottom Generalization, highly specific hypotheses from one given example and a background theory. We call the hypotheses *bottom clauses*. ¿From E_1 the following two bottom clauses are constructed:

$$C_1 = p(f(a)) \leftarrow s(a),$$
$$C_2 = r(a) \leftarrow s(a).$$

¿From the property of bottom clauses, any clause H which subsumes E_1 relative to B iff H subsumes either C_1 or C_2. Since H must also subsume E_2 relative to B, it subsumes either of the bottom clauses of E_2:

$$D_1 = q(g(b)) \leftarrow s(a), t(b), s(b),$$
$$D_2 = r(b) \leftarrow s(a), t(b), s(b).$$

At the next step, Bottom Reduction makes the least generalization of each pair of a bottom clause of E_1 and one of E_2, under the subsumption order $\succeq$. We get the following four clauses, due to the selection of bottom clauses:

$$\begin{matrix} H_{11} = lgg(C_1, D_1) = \leftarrow s(a), s(X), \\ H_{12} = lgg(C_1, D_2) = \leftarrow s(a), s(X), \\ H_{21} = lgg(C_2, D_1) = \leftarrow s(a), s(X), \\ H_{22} = lgg(C_2, D_2) = r(X) \leftarrow s(a), s(X). \end{matrix}$$

Because the target of learning is a definite program, we throw away H_{11}, H_{12}, and H_{21}. Then it holds that

$$H_{22} \succeq C_2 \text{ and } H_{22} \succeq D_2,$$

that is, H_{22} is a definite clause which subsumes both of E_1 and E_2 relative to B. We will later show that any definite clause which subsumes H_{22} subsumes both of E_1 and E_2 relative to B.

For any definite clause H obtained by applying inverse subsumption to H_{22}, e.g. $r(X) \leftarrow s(X)$, it holds that $B \wedge H \vdash E_1 \wedge E_2$.

3 Preliminaries

3.1 Basic Definitions and Notations

A *definite program* is a finite conjunction of definite clauses. A *definite clause* is a formula of the form

$$C = \forall X_1 \ldots X_k (A_0 \vee \neg A_1 \vee \ldots \vee \neg A_n)$$

and a *goal clause* is of the form

$$D = \forall X_1 \ldots X_k (\neg A_1 \vee \ldots \vee \neg A_n)$$

where $n \geq 0$, A_i's are all atoms, and $X_1, \ldots, X_k$ are all variables occurring in the atoms. The atom A_0 is the head of C. A *Horn clause* is either a definite clause or a goal clause. We represent the formulas in the form of implication:

$$C = A_0 \leftarrow A_1, A_2, \ldots, A_n,$$
$$D = \leftarrow A_1, A_2, \ldots, A_n.$$

For the definite clause C, C^+ and C^-, respectively, denote a definite clause $A_0 \leftarrow$ and a goal clause $\leftarrow A_1, A_2, \ldots, A_n$.

The *complement* of C is a formula

$$\neg(C\sigma_C) = (\neg A_0 \wedge A_1 \wedge \ldots \wedge A_n)\sigma_C,$$

where σ_C is a substitution which replaces each variable in C with a Skolem constant symbol. We sometimes write σ instead of σ_C when it causes no ambiguity. For a definite clause C, note that $\neg(C\sigma^+)$ is equivalent to a goal clause

$$\leftarrow A_0\sigma$$

and $\neg(C\sigma^-)$ is equivalent to a definite program

$$P = (A_1\sigma \leftarrow) \wedge \ldots \wedge (A_k\sigma \leftarrow)$$

because $C\sigma$ contains no variable.

In our discussion, we often use a covering function T_P for a logic program P, it is very famous in Logic Programming theories [4].

3.2 Relative Subsumption and Bottom Clauses

Let H and E be definite clauses. If there is a substitution θ such that every literal in $H\theta$ occurs in E, we say H *subsumes* E and write $H \succeq E$. Given a definite clauses E, *inverse subsumption* generates a definite clause H such that $H \succeq E$.

When a definite program B is given, we say H *subsumes* E *relative to* B and write $H \succeq E$ (B) if there is a (possibly non-definite) clause F such that

$$B \models \forall Y_1 \ldots Y_n (E' \leftrightarrow F')$$

and H subsumes F, where E' and F' are obtained by removing universal quantifiers from E and F, respectively, and $Y_1, \ldots, Y_n$ are all variables occurring in E' and F'.

Proposition 1 *Let C and D be definite clauses and let B be a definite program. If $C \succeq D$ (B) then $B \wedge C \vdash D$.*

A clause C is a *generalization of D under subsumption order* if $C \succeq D$, and *generalization of D under relative subsumption order* if $C \succeq D$ (B).

The *bottom set* for E relative to B is a set of literals

$$\mathrm{Bot}(E, B) = \{L \mid L \text{ is a ground literal and } B \wedge \neg(E\sigma) \models \neg L\}.$$

A *bottom clause* for E relative to B is a definite clause which is a disjunction of some literals in $\mathrm{Bot}(E, B)$. The set of all bottom clauses is denoted by $\mathrm{BOTC}(E, B)$.

In our previous works, we showed the relation between bottom clauses and relative subsumption.

Theorem 1 ([12, 13]) *Let B be a definite program and E a definite clause such that $B \not\models E$. A hypothesis H subsumes E relative to B iff H contains no Skolem constants and subsumes some bottom clause for E relative to B.*

3.3 How to Generate Bottom Clauses

We gave in [11, 15] a method with which every atom can be generated in the bottom set for E relative to B. At first note that the formula $B \wedge \neg(E\sigma)$ in the definition of bottom sets is a conjunction of a definite program $P = B \wedge \neg(E^-\sigma)$ and a ground goal clause $G = \neg(E^+\sigma)$.

The following theorem shows that the set of all negative literals in $\mathrm{Bot}(E, B)$ can be obtained by generating the set $T_P \uparrow \omega$.

Theorem 2 ([15]) *Let A be a ground atom, P a definite program, and G a goal. Suppose that $P \wedge G$ is consistent. Then $P \wedge G \models A$ iff $P \models A$.*

Every positive literal in $\mathrm{Bot}(E, B)$ can be obtained by SOLDR-resolution. An SOLDR-derivation can be formalized in the same way as an SLD-derivation, with assuming that an atom in a goal is selected according to a *computation rule* R and that variables in each input clause is *standardized apart* ([4]).

An *SOLDR-derivation* of (P, G) consists of a finite sequence of quadruples $\langle G_i, F_i, \theta_i, C_i \rangle$ $(i = 0, 1, 2, \ldots, n)$ which satisfies the following conditions:

1. G_i and F_i are goal clauses, θ_i is a substitution, and C_i is a variant of a clausal formula in P the variables of which are standardized apart by renaming.
2. $G_0 = G$ and $F_0 = \square$.
3. $G_n = \square$ and $F_n = F$.
4. For every $i = 0, \ldots, n-1$, if $G_i =\leftarrow A_1, \ldots, A_k$, $F_i =\leftarrow B_1, \ldots, B_h$, and A_m is the atom selected from G by R, then one of the following three holds:
   (a) (Resolution) θ_i is an mgu of C_i^+ and A_m,
   $G_{i+1} =\leftarrow (A_1, \ldots, A_{m-1}, M_1, \ldots, M_l, A_{m+1}, \ldots, A_k)\theta_i$, and $F_{i+1} = F_i\theta_i$, where $C_i^- = \leftarrow M_1, \ldots, M_l$.
   (b) (Simple Skip) $h = 0$(that is, $F_i = \square$), θ_i is an identity substitution, $G_{i+1} =\leftarrow A_1, \ldots, A_{m-1}, A_{m+1}, \ldots, A_k$, and $F_{i+1} =\leftarrow A_m$.

(c) (Skip with Reduction) $h = 1(F_i = \leftarrow B)$, θ_i is an mgu of B and A_m, $G_{i+1} = \leftarrow (A_1, \ldots, A_{m-1}, A_{m+1}, \ldots, A_k)\theta_i$, and $F_{i+1} = \leftarrow B\theta_i$.

The goal F_n is called the *consequence* of the SOLDR-derivation. The skip operation was firstly invented for SOL-resolution [2]. If such a consequence $\leftarrow A$ is found, we say it is *SOLDR-derivable* from (P, G). We put the set

$$SOLDR(P, G) = \{\leftarrow A \mid \leftarrow A \text{ is SOLDR-derivable from } (P, G)\}.$$

Theorem 3 ([11, 15]) *If $P \wedge G$ is consistent, $\leftarrow A \in SOLDR(P, G)$ is equivalent to $P \wedge G \models \neg A$ for any ground atom A.*

Corollary 1 *For a definite program B and a definite clause E such that $B \not\models E$,*

$$\text{Bot}(E, B) = \{L \mid \neg L \in SOLDR(P, G) \cup T_P \uparrow \omega\}$$

where $P = B \wedge \neg(E^-\sigma)$ and $G = \neg(E^+\sigma)$.

4   The Bottom Reduction Method

Let C and D be definite clauses. A clause H is a *common generalization* of C and D if $H \succeq C$ and $H \succeq D$. It is the *least common generalization* (the *least generalization*, for short) if $H' \succeq H$ for any common generalization H' of C and D. We denote the common least generalization of C and D by $lgg(C, D)$. It was shown in [7] that $lgg(C, D)$ exists for any C and D.

The following proposition holds.

Proposition 2 *For a definite clause C and D, the following two are equivalent:*

1. *The predicate occurs in head of C is the same as that of D.*
2. *$lgg(C, D)$ is a definite clause.*

4.1   Definition and Justification of Bottom Reduction

Now we give the formal definition of Bottom Reduction method.

Definition 1 Let B be a definite program, E_1 and E_2 be definite clauses such that $B \not\models E_1 \vee E_2$. A definite clause H is obtained by *Bottom Reduction* method from E_1 and E_2 relative to B if H subsumes $lgg(C, D)$ where $C \in \text{BOTC}(E_1, B)$ and $D \in \text{BOTC}(E_2, B)$.

The logical justification of Bottom Reduction is given by the following theorem.

Theorem 4 *Suppose that $B \not\models E_1 \vee E_2$. Then a definite clause H is obtained by Bottom Reduction from E_1 and E_2 relative to B iff $H \succeq E_1$ (B) and $H \succeq E_2$ (B).*

The only-if-part of Theorem 4 is proved as the corollary of the next lemma.

Lemma 1 *Suppose that $B \not\models E_1 \vee E_2$. For any definite clauses $C \in \mathrm{BOTC}(E_1, B)$ and $D \in \mathrm{BOTC}(E_2, B)$,*

$$lgg(C, D) \succeq E_1\ (B)\ and\ lgg(C, D) \succeq E_2\ (B).$$

Proof By Theorem 1 it holds that $C \succeq E_1\ (B)$ for any $C \in \mathrm{BOTC}(E_1, B)$. Since $lgg(C, D) \succeq C$, we get $lgg(C, D) \succeq E_1\ (B)$ from the definition of the relative subsumption. In the same way, we can show that $lgg(C, D) \succeq E_2\ (B)$. □

Corollary 2 *Let $C \in \mathrm{BOTC}(E_1, B)$, $D \in \mathrm{BOTC}(E_1, B)$, and $H \succeq lgg(C, D)$. Then $H \succeq E_1\ (B)$ and $H \succeq E_2\ (B)$.*

The next lemma proves the if-part of Theorem 4.

Lemma 2 *Assume that $H \succeq E_1\ (B)$, that $H \succeq E_2\ (B)$, and that $B \not\models E_1 \vee E_2$. Then $H \succeq lgg(C, D)$ for some $C \in \mathrm{BOTC}(E_1, B)$ and $D \in \mathrm{BOTC}(E_2, B)$.*

Proof From the assumptions of the lemma, there exist $C \in \mathrm{BOTC}(E_1, B)$ and $D \in \mathrm{BOTC}(E_2, B)$ such that $H \succeq C$ and $H \succeq D$ by Theorem 1. Then $H \succeq lgg(C, D)$. □

4.2   The Advantage of Bottom Reduction

The advantage of Bottom Reduction is that we give no restriction to the head predicates of given examples. The head predicate of a definite clause is the predicate which occurs in the positive literal in it. Most of other ILP methods assume (sometimes implicitly) that all examples should share one head predicate. We now show that removing this restriction reduces the search space of hypotheses.

Let us consider the following definite program B as a background theory:

$$B = \begin{array}{l} (mortal(X) \leftarrow birds(X)) \\ \wedge\ (mortal(X) \leftarrow fishes(X)) \\ \wedge\ (mortal(X) \leftarrow mammals(X)) \\ \wedge\ (has\_navel(X) \leftarrow mammals(X)). \end{array}$$

Now assume that the following definite clause E_1 and E_2 are given as examples:

$$E_1 = mortal(socrates) \leftarrow human(socrates),$$
$$E_2 = mortal(plato) \leftarrow human(plato).$$

The set $\mathrm{BOTC}(E_1, B)$ consists of four clauses:

$$C_1 = birds(socrates) \leftarrow human(socrates),$$
$$C_2 = fishes(socrates) \leftarrow human(socrates),$$
$$C_3 = mammals(socrates) \leftarrow human(socrates),$$
$$C_4 = mortal(socrates) \leftarrow human(socrates).$$

The set $BOTC(E_2, B)$ also consists of four clauses:

$$D_1 = birds(plato) \leftarrow human(plato),$$
$$D_2 = fishes(plato) \leftarrow human(plato),$$
$$D_3 = mammals(plato) \leftarrow human(plato),$$
$$D_4 = mortal(plato) \leftarrow human(plato).$$

By making $lgg(C_i, D_j)$ $(i, j = 1, \ldots, 4)$, we get 16 clauses. The definite clauses in them are

$$H_1 = birds(X) \leftarrow human(X),$$
$$H_2 = fishes(X) \leftarrow human(X),$$
$$H_3 = mammals(X) \leftarrow human(X),$$
$$H_4 = mortal(X) \leftarrow human(X).$$

Note that each H_i is obtained by replacing constants in C_i and D_i with a variable X. In this case we have as many hypotheses as bottom clauses of E_1.

Next consider that the following E_2' is given instead of E_2.

$$E_2' = has\_navel(aristotle) \leftarrow human(aristotle)$$

Then $BOTC(E_2', B)$ consists of two clauses:

$$D_1' = has\_navel(aristotle) \leftarrow human(aristotle),$$
$$D_2' = mortal(aristotle) \leftarrow human(aristotle).$$

The hypothesis we get by applying the least generalization method is

$$H_5 = mammals(X) \leftarrow human(X).$$

This case shows that we have less hypotheses when we have examples whose predicate symbol are different from those of others.

5   An ILP System: BORDA

In this section, we describe an ILP system called BORDA based on **Bottom Reduction Algorithm**. BORDA is implemented on a Prolog system.

5.1   Restriction on Background Theory

In general the number of bottom clauses may be infinite. The length of a bottom clause may not be bounded. In Order to keep the termination of BORDA, we restrict a background theory to generative and weakly reducing definite program.

Definition 2 A clause $A \leftarrow B_1, \ldots, B_n$ is *generative* if every variable in A appears in $B_1, B_2, \ldots$, or B_n. A program P is *generative* if every clause in P is generative.

Bottom Reduction Algorithm for finitely many bottom clauses

input: two definite clauses E_1 and E_2, a definite program B
output: H which subsumes both of E_1 and E_2
begin

        Generate bottom clauses $C_1, C_2, \ldots, C_n$ of E_1 relative to B;
        Generate bottom clauses $D_1, D_2, \ldots, D_m$ of E_2 relative to B;
        for $1 \le i \le n$ **do**
            for $1 \le j \le m$ **do**
                if the head predicate of C_i is the same as one of D_j **then**
                    $H:=lgg(C_i, D_j)$;
                    output H;
                endif
            endfor
        endfor
end.

Fig. 1. A Bottom Reduction Algorithm for finitely many bottom clauses.

Definition 3 A clause $A \leftarrow B_1, \ldots, B_n$ is *weakly reducing* if $\|A\| \ge \|B_i\|$ for every $i = 1, \ldots, n$, where $\|A\|$ is the total number of function symbols, constant symbols and variables occurring in A. A program P is *weakly reducing* if any clause in P is weakly reducing.

Let B be a definite program given as a background theory and E be a definite clause given as an example. We denote the cardinality of a set S as $|S|$.

The number of bottom clauses is $|SOLDR(P, G)|$, and the length of each bottom clause is $|T_P \uparrow \omega|$, where $P = B \wedge \neg(E^-\sigma)$ and $G = \neg(E^+\sigma)$. With restricting the background theory to weakly reducing, $SOLDR(P, G)$ is a finite set. Since the set $T_P \uparrow \omega$ may be infinite in general. We use $T_P \uparrow n$ for some integer n instead of $T_P \uparrow \omega$. In spite of this approximation the system is sound from Corollary 2. Since P is generative, $T_P \uparrow n$ is a finite set.

With restricting the background theory to generative and weakly reducing, both the number of bottom clauses and the length of each bottom clause are finite. Then we can adopt the algorithm shown Figure 1 to find a definite clause H which subsume both C and D relative to B. Note that if two definite clauses C, D have distinct head predicates, then $lgg(C, D)$ is not a definite clause, due to Proposition 2.

5.2 An Implementation of Bottom Reduction

In order to implement Bottom Reduction Method, we need an implementation of SOLDR-resolution, a program which generates the set $T_P \uparrow \omega$, and a program for Plotkin's least generalization algorithm for clauses.

As shown in Section 3.3, we make bottom clauses using SOLDR-derivation and generating $T_P \uparrow \omega$. Because SOLDR-derivation is natural extension of SLD-

```
bottomReduction(E1,E2,B,H):-
    botc(E1,B,C), botc(E2,B,D),
    same_head_pred(C,D), lgg(C,D,H).

botc(E,B,C):-
    skolemize(E,[EHead,:-|EBody]),
    add(EBody,B,P),!, soldr([EHead],P,[CHead]), tp(P,10,CBody),
    unskolemize([CHead,:-|CBody],C).

soldr([],_Prog,[]).
soldr([G|Gs],Prog,F):-
    copy_term(Prog,Prog2), member([G,:-|Bs],Prog2),
    append(Bs,Gs,NewG), soldr(NewG,Prog,F).
soldr([G|Gs],Prog,[G]):-
    soldr(Gs,Prog,[]).
soldr([F|Gs],Prog,[F]):-
    soldr(Gs,Prog,[F]).
```

Fig. 2. A prolog program of Bottom Reduction Method

derivation, we can implement SOLDR-derivation by simple extension of the Prolog meta-interpreter.

Figure 2 shows the main part of the implementation code for BORDA. Please refer some textbooks [1, 10] or manuals about details of the programming language Prolog.

The predicate $soldr(G, P, F)$ is a meta-interpreter of SOLDR-derivation. Consequences F is obtained with SOLDR-derivation from a goal G on a definite program P. The definition of soldr/3 uses three intermediate predicates. The predicate copy_term/2 is one of the built-in predicates which Prolog systems support, and calculates variants of given term. The predicate append/3 is for concatenating two lists, member/2 is for selecting an element from a given list.

The call $botc(C, B, H)$ generates an element $H \in \mathrm{BOTC}(C, B)$. In its definition, predicates skolemize/2, add/3, soldr/3, tp/3, and unskolemize/2 are used. The predicate skolemize/2 replaces variables in a given clause with Skolem constants and unskolemize/2 does its converse. The call $add(Goal, B, P)$ generates a definite program P inserting all literals appearing in $Goal$ into the background theory as facts. The call $tp(P, N, M)$ computes the Herbrand model $M = T_P \uparrow N$.

Finally the call $bottomReduction(E_1, E_2, B, H)$ generates a clause H which subsumes both of C and D relative to B. The call $same\_head\_pred(C, D)$ checks whether the predicate symbol of head of C is the same as one of D. Using backtracking mechanism of Prolog, the predicate bottomReduction/4 searches

C_i and D_j whose heads have the same predicate symbol, and outputs $H = lgg(C_i, D_j)$. If such H is found, then H subsumes both E_1 and E_2 relative to B.

We illustrate an example run of BORDA. Theoretically a clause $H = r(A) \leftarrow s(a), s(A)$ is derived from examples $E_1 = p(f(a))$ and $E_2 = q(g(Y)) \leftarrow t(Y))$ under a background theory

$$B = \begin{matrix} & (s(a) \leftarrow) \\ \wedge & (s(X) \leftarrow t(X)) \\ \wedge & (p(f(X)) \leftarrow r(X)) \\ \wedge & (q(g(X)) \leftarrow r(X)) \end{matrix}$$

using Bottom Reduction. The program BORDA works H as follows:

```
| ?-E1= [p(f(a)),:-],
    E2= [q(g(Y)),:-,t(Y)],
    B = [ [s(a),:-],
          [s(X),:-,t(X)],
          [p(f(X)),:-,r(X)],
          [q(g(X)),:-,r(X)] ],
    bottomReduction(E1,E2,B,H).

H = [r(_A),:-,s(a),s(_A)] ?
yes
```

6   Conclusion

In this paper, we gave a new ILP method, Bottom Reduction, which is an extension of Bottom Generalization, and showed its theoretical background. We also explained BORDA, which is a Prolog implementation of Bottom Reduction.

The theoretical background is based on the fact that we can obtain every clause which subsumes given examples relative to a background theory, by computing the least generalization of the bottom clauses for the examples. We treated only the least generalization of two bottom clauses, but it would be quite easy to treat the least generalization of arbitrarily many bottom clauses. It is also be very easy to allow more than two examples.

We illustrated that applying Bottom Reduction to examples which have a predicate different from those of other examples contributes to the reduction of search space. Though learning from examples with various head predicates should be included in the framework of ILP, it seems quite strange that the advantage we shown in Section 4.2 had never been discussed explicitly before. We have to examine other ILP systems whether they are applicable to such learning. By restricting background theory to generative and weakly reducing definite programs, BORDA can generate hypotheses in finite time. It is in our future research plan to investigate other restrictions.

Acknowledgments

This work has been partly supported by Grant-in-Aid for Scientific Research No.10143201 from the Ministry of Education, Science and Culture, Japan.

References

1. Flach, P.: Simply Logical –Intelligent Reasoning by Example–, Wiley (1994).
2. Inoue, K.: Linear Resolution for Consequence Finding. Artificial Intelligence, 56 (1992) 301–353
3. Jung, B.: On Inverting Generality Relations. In Proceedings of the 3rd International Workshop on Inductive Logic Programming (1993) 87–101
4. Lloyd, J. W.: Foundations of Logic Programming : Second, Extended Edition. Springer - Verlag (1987)
5. Muggleton, S.: Inverse entailment and Progol. New Generation Computing, Vol.13, No.3-4 (1995) 245–86
6. Muggleton, S.(Ed.): Inductive Logic Programming . Academic Press, the APIC Series, No 38 (1992)
7. Plotkin, G. D.: A Note on Inductive Generalization. Machine Intelligence 5 (1970) 153-163
8. Plotkin, G. D.: A Further Note on Inductive Generalization. Machine Intelligence, 6 (1971) 101–124
9. Rouveirol, C.: Extensions of Inversion of Resolution Applied to Theory Completion. In:Muggleton,S.(ed.):Inductive Logic Programming,Academic Press (1992) 63–92.
10. Sterling, L., Shapiro, E.: The Art of Prolog. The MIT Press,Cambridge MA (1986)
11. Yamamoto, A.: Representing Inductive Inference with SOLD-Resolution. In Proceedings of the IJCAI'97 Workshop on Abduction and Induction in AI (1997) 59–63
12. Yamamoto, A.: Which Hypotheses Can Be Found with Inverse Entailment? In Proceedings of the Seventh International Workshop on Inductive Logic Programming (LNAI 1297) (1997) 296 – 308, The extended abstract is in Proceedings of the IJCAI'97 Workshop on Frontiers of Inductive Logic Programming (1997) 19–23
13. Yamamoto, A.: An Inference Method for the Complete Inverse of Relative Subsumption. To appear in New Generation Computing (1998)
14. Yamamoto, A.: Revising the Logical Foundations of Inductive Logic Programming Systems with Ground Reduced Programs. To appear in New Generation Computing (1998)
15. Yamamoto, A.: Using Abduction for Induction based on Bottom Generalization. To appear in: A. Kakas and P. Flach (eds.): Abductive and Inductive Reasoning : Essays on their Relation and Integration (1998).

On the Completion of the Most Specific Hypothesis Computation in Inverse Entailment for Mutual Recursion

Koichi Furukawa

Graduate School of Media and Governance, Keio University
5322 Endo, Fujisawa, Kanagawa 252, JAPAN
furukawa@sfc.keio.ac.jp

Abstract. In this paper, we introduce a complete algorithm for computing the most specific hypothesis (MSH) in Inverse Entailment when the background knowledge is a set of definite clauses and the positive example is a ground atom having the same predicate symbol as that of the target predicate to be learned.

Muggleton showed that for any first order theory (background knowledge) B and a single clause (a positive example) E, the MSH can be computed by first computing all ground (positive and negative) literals which logically follow from $B \wedge \neg E$ and negating their conjunction. However, Yamamoto gave a counter example and indicated that Muggleton's proof contains error. Furukawa gave a sufficient condition to guarantee the above algorithm to compute the MSH. Yamamoto defined a class of problems where the algorithm computes the MSH. In this paper, we extend the MSH computation algorithm to ensure that it computes the MSH correctly under the condition described above.

1 Introduction

Recently, inductive logic programming (ILP) has been paid a big attention as a framework of supervised learning in first order logic. Decision tree making systems such as ID3[8], CART[1], and C4.5[9] are well known as supervised learning systems in propositional logic. An advantage of ILP over these systems is its capability of utilizing background knowledge. On the other hand, there is a trade-off between the expressiveness and its performance.

Muggleton proposed a method to solve this problem based on **inverse entailment**[6]. Let B be background knowledge, E a positive example, and H a hypothesis. Then, H should satisfy the condition $B \wedge H \models E$ when B alone cannot explain E. Muggleton developed a very efficient algorithm to compute the best hypothesis given background knowledge and a set of positive examples, and implemented a system called Progol[6]. Progol's algorithm for computing the **most specific hypothesis (MSH)**, which any hypothesis H satisfying $B \wedge H \models E$ entails, is based on the "fact" that if we restrict both a hypothesis and a positive example to be a single clause, then the MSH can be computed by first computing all ground (positive and negative) literals which logically follows from $B \wedge \neg E$ and then negating their conjunction. We call this algorithm *Algorithm IE*.

However, Yamamoto[10] disproved the above "fact" by showing a counter example.

Furukawa [2] gave a sufficient condition for *Algorithm IE* to compute the correct MSH. The condition is "background knowledge must be in clausal form consisting of definite clauses which do not contain any negative literals having the target predicate symbol." This defines the case when *Algorithm IE* cannot compute MSH: that is, the case when the entire program includes mutual recursive clauses part of which are defined in the background knowledge and the rest by the newly introduced hypothesis. For this reason, we call this mutual recursive case.

Yamamoto[11] identified the class of problems for which *Algorithm IE* can compute the right MSHs in terms of SB-Resolution proposed by Plotkin[7].

This paper proposes a complete algorithm for computing the MSH when the background knowledge is a set of definite clauses and the positive example is a ground atom having a predicate to be learned.

The outline of this paper is as follows. In Section 2, we give terminology to be used later. In Section 3, we introduce a **model constraining clause** which plays an essential role for building a complete algorithm for computing the MSH. In section 4, we define a complete algorithm for computing MSHs under the above condition. Finally, we conclude the paper and give the future work.

2   Preliminaries

This section introduces terminology, concepts and lemmas needed later to introduce our algorithm.

2.1   Terminology and Lemmas in Logic

Definition 1. *A clausal form which consists of only definite clauses is called a* **definite clausal form.**

Definition 2. *Let S be a set of all conjuncts P_i in a given conjunction $P_1 \wedge P_2 \wedge \ldots$. Then a conjunction which consists of a subset of S is called a* **subconjunction** *of the original conjunction.*

Definition 3. *When an atom P has a vector of variables $\mathbf{X}$, we represent the atom as $P\{\mathbf{X}\}$.*

2.2   Terminology in Inductive Logic Programming
Logical setting of ILP

Components in the logical setting of ILP are positive examples, background knowledge and hypotheses generated from them. We assume that background knowledge B, a hypothesis H, and a positive example E are all in clausal form. Then the following relations hold:

1. For a positive example E and background knowledge B, $B \not\models E$ holds.
2. A hypothesis H should satisfy $B \wedge H \models E$.

1. states that background knowledge B alone cannot explain a positive example E, whereas 2. states that B together with H can explain E.

Definition 4. *The second formula in the logical setting of ILP can be transformed to*

$$B \wedge \neg E \models \neg H.$$

We call this formula the **inverse entailment** *of the original formula.*

The inverse entailment expression suggests a possibility that a hypothesis can be deduced from background knowledge and a positive example.

Definition 5. *We denote the conjunction of all (positive and negative) ground literals which are true in all Herbrand models of $B \wedge \neg E$ by $\neg bot(B, E)$.*

The aim of this paper is to define the bottom of the subsumption lattice for searching the best hypothesis. We will derive it by modifying $bot(B, E)$. Later we treat the disjunction $bot(B, E)$ as a set as usual in clausal theory.

Definition 6. *A predicate which represents the target concept of learning is called* **a target predicate.** *Positive examples are represented by a target predicate.*

3 Model Constraining Clauses

In this section, we define **model constraining clauses** and its related theorem which play an essential role for building a complete algorithm for computing the MSH.

Definition 7. *We define a* **model constraining clause** *inductively.*

1. *A clause in the background knowledge B containing a negative disjunct of the target predicate is a model constraining clause.*
2. *A clause obtained by applying a folding operation with respect to a model constraining clause to a clause in B is a model constraining clause.*
3. *Only clauses obtained by 1 and 2 are model constraining clauses.*

In this section, we assume that model constraining clauses have only one body literal having the target predicate. The treatment for more general cases will be discussed later in the next section.

A model constraining clause constrains possible models of $B \wedge \neg E$. Let $Head(\mathbf{X}) \leftarrow Body(\mathbf{X})$ be a model constraining clause. Then, any Herbrand model of $B \wedge \neg E$ satisfies one of the following conditions: (1) it does not contain

$Body(\mathbf{b})$ or (2) it contains both $Head(\mathbf{b})$ and $Body(\mathbf{b})$, where $\mathbf{b}$ is a vector of symbols representing constants in Herbrand universe which is to be substituted to $\mathbf{X}$. This is why we call this kind of clauses as model constraining clauses. We divide the models of $B \wedge \neg E$ into two kinds by this model constraining clause as described above.

Note that the nature of symbols in $\mathbf{b}$ is very similar to that of Skolem constants in a sense that they represent some constants in the Herbrand universe. Only the purposes they are introduced are different from each other. We will give a new name for this symbol.

Definition 8. *A symbol is called as a* **Pseudo-Skolem (P-Skolem) constant** *when it represents any constant in the Herbrand universe. It can be substituted to any constant in the Herbrand universe.*

Definition 9. *Let* $C = Head\{\mathbf{X}\} \leftarrow Body\{\mathbf{X}\}$ *be a model constraining clause in* B *and a positive example be* E. *Then, two quasi-MSH* $bot^+(B, E)(\mathbf{b})$ *and* $bot^-(B, E)(\mathbf{b})$ *parameterized by a P-Skolem constant vector* $\mathbf{b}$ *corresponding to* $\mathbf{X}$ *are defined by:*

$$bot^+(B, E)(\mathbf{b}) = bot(B \backslash C, E) \cup \{Body\{\mathbf{b}\}\}$$
$$bot^-(B, E)(\mathbf{b}) = bot(B \backslash C, E) \cup \{\neg Head\{\mathbf{b}\}, \neg Body\{\mathbf{b}\}\}$$

Note that the signs of literals to be added are inverted because we compute $bot^*(B, E)$ instead of $\neg bot^*(B, E)$.

We first show the counter example given by Yamamoto[10] for which Algorithm IE does not work properly in order to explain the essential point of the problem.

Example 1

$$\left\{ \begin{array}{l} B_1 = even(0) \leftarrow \\ \quad\quad even(s(X)) \leftarrow odd(X) \\ E_1 = odd(s(s(s(0)))) \leftarrow \\ H_1 = odd(s(X)) \leftarrow even(X) \end{array} \right\}$$

Then,

$$\neg bot(B_1, E_1) = even(0) \wedge \neg odd(s(s(s(0))))$$

and thus

$$bot(B_1, E_1) = odd(s(s(s(0)))) \leftarrow even(0).$$

In this example, H_1 is a correct hypothesis because it explains E_1 together with B_1. However, it cannot be computed from $bot(B_1, E_1)$ since $H_1 \models bot(B_1, E_1)$ does not hold.

The flaw in Muggleton's proof of the completeness of the *Algorithm IE* comes from the fact that Skolemized ground literals in $\neg H$ can represent different ground literals in different models [2].

Let us consider this proof for Yamamoto [10]'s counter example. The negation of H_1 is expressed as:

$$\neg H_1 = \neg odd(s(a)) \wedge even(a),$$

where a is a Skolem constant.

The problem is that although H_1 satisfies the condition $B_1 \wedge H_1 \models E_1$, $\neg H_1$ is not a sub-conjunction of $\neg bot(B_1, E_1)$.

By considering the second clause (a model constraining clause) of B_1:

$$even(s(X)) \leftarrow odd(X),$$

it turns out that every model of $B_1 \wedge \neg E_1$ either contains $\{\neg odd(s(b))\}$ or $\{odd(s(b)), even(s(s(b)))\}$ where b is a P-Skolem constant for the variable X. In both cases, $\neg H_1 = \neg odd(s(a)) \wedge even(a)$ becomes true when $b = s(0)$ by simply assigning $a = s(s(0))$ and $a = 0$ respectively.

Thus, when $\neg H$ contains Skolem constants, $\neg H$ need not be a sub-conjunction of $\neg bot(B, E)$ even if it is true for all models of $B \wedge \neg E$.

We can avoid this problem by simply adding these literals containing P-Skolem constants which are true in both groups of models.

First, we compute two quasi-MSHs parameterized by a P-Skolem constant b corresponding to the variable X in the model constraining clause as follows:

$$bot^+(B_1, E_1)(b) = odd(s^3(0)); odd(b) \leftarrow even(0)$$
$$bot^-(B_1, E_1)(b) = odd(s^3(0)) \leftarrow even(0), odd(b), even(s(b))$$

Now, we consider those hypotheses which do not entail $bot(B_1, E_1)$. Since they must entail both $bot^+(B_1, E_1)(b)$ and $bot^-(B_1, E_1)(b)$, they must have $E_1 = odd(s^3(0))$ as their head and either $odd(b)$ or $even(s(b))$ as their body (by particularly considering the latter quasi-MSH).

Therefore, both

$$odd(s^3(0)) \leftarrow even(s(b))$$

and

$$odd(s^3(0)) \leftarrow odd(b)$$

are their candidates.

On the other hand, the hypotheses which do not entail $bot(B, E)$ in general must satisfy the following theorem.

Theorem 1. *Those hypotheses which do not entail $bot(B, E)$ must contain a goal literal calling to the model constraining clause, that is, a negative literal which is unifiable to the head literal of the model constraining clause.*

Proof. For a hypothesis which does not entail $bot(B, E)$, the proof of the example E is not of SB-Resolution and the hypothesis must appear more than once in the proof. Therefore, the hypothesis must contain a goal whose proof tree contains the goal predicate. Since we defined model constraining clauses inductively by applying all possible foldings with respect to the goal predicate, the goal's proof can be replaced by a call to one of the model constraining clauses. □

By this theorem, one of the candidate hypotheses is discarded and there remains only one clause

$$odd(s^3(0)) \leftarrow even(s(b))$$

as a candidate hypothesis.

Since we do not need any other literals than $\neg Head\{\mathbf{b}\}$, we redefine one of the quasi-MSHs as follows:

Definition 10. *Let* $C = Head\{\mathbf{X}\} \leftarrow Body\{\mathbf{X}\}$ *be a model constraining clause in* B *and a positive example be* E. *Then, a quasi-MSH* $bot^-(B, E)(\mathbf{b})$ *parameterized by a P-Skolem constant vector* $\mathbf{b}$ *corresponding to* $\mathbf{X}$ *is defined by:*

$$bot^-(B, E)(\mathbf{b}) = bot(B \backslash C, E) \cup \{\neg Head\{\mathbf{b}\}\}$$

4   A Complete Algorithm for Computing MSH

In this section, we introduce a complete algorithm, called Algorithm *Augmented Inverse Entailment in Mutual Recursion (AIEMR)*, for computing the most specific hypothesis when background knowledge is given by a definite clausal form and the positive example is a ground atom having a predicate to be learned.

First, we consider the case when there exists only one model constraining clause in background knowledge. More general case will be treated later in this section.

4.1   Single model-constraining-clause case

Algorithm *AIEMR*

1. Let B be an arbitrary definite clausal form having a model constraining clause $C = Head\{\mathbf{X}\} \leftarrow Body\{\mathbf{X}\}$, E a ground atom positive example and $MSH\ bot(B \backslash C, E)$.

2. We compute two quasi-MSHs parameterized by a P-Skolem constant vector $\mathbf{b}$ corresponding to the variable $\mathbf{X}$:

$$bot^+(B, E)(\mathbf{b}) = bot(B \backslash C, E) \cup \{Body\{\mathbf{b}\}\}$$
$$bot^-(B, E)(\mathbf{b}) = bot(B \backslash C, E) \cup \{\neg Head\{\mathbf{b}\}\}$$

3. Build an initial candidate hypothesis from $bot^-(B, E)(\mathbf{b})$:

$$E \leftarrow Head\{\mathbf{b}\}.$$

4. Repeat the following procedure as much as we need by using type information as a generator:
   (a) Instantiate the P-Skolem constant vector in the above initial candidate hypothesis by the generated constants of Herbrand universe and variablize the obtained expression.
   (b) Test the variablized candidate whether it entails the first quasi-MSH $bot^+(B, E)(\mathbf{b})$ or not.
   (c) Let the substitution to $\mathbf{b}$ for which the variablized candidate succeeded in the entailment test be θ. Update MSH by $MSH \cup \{\neg Head\{\mathbf{b}\}\theta\}$.

Note that the above "algorithm" is not guaranteed to terminate for two reasons. Firstly, the entailment test between two clauses is undecidable in general [4]. It is well known that the entailment test between clauses can be replaced by the subsumption test between them when they are not self-recursive [5]. Self-recursive case is then solved by applying the flattening-and-subsaturants technique [6]. Secondly, the entire loop may not terminate. In the next subsection, we will give a finite bound for the loop counts.

As a result of performing Algorithm $AIEMR$, a new literal $\neg Head\{b\}\theta$ is added to the original set of literals $bot(B\backslash C, E)$ for each successful loop in the step 4.

Let us apply this algorithm to Example 1. Here, we concentrate the problem of computing an appropriate substitution to **b**. We will discuss how much we should repeat the loop in the step 4 at the next subsection.

We successively instantiate the P-Skolem constant b in the initial candidate hypothesis

$$odd(s^3(0)) \leftarrow even(s(b))$$

by the elements of Herbrand universe, $0, s(0), s^2(0), \ldots$ generated by the type information program for the argument of $even$:

$$\left\{ \begin{array}{l} int(0). \\ int(s(X)) : -int(X). \end{array} \right\}$$

Then we variablize them. The results are as follows:

$$\left\{ \begin{array}{ll} b = 0 : & odd(s^2(Y)) \leftarrow even(Y) \\ b = s(0) : & odd(s(Y)) \leftarrow even(Y) \\ b = s^2(0) : & odd(Y) \leftarrow even(Y) \\ b = s^3(0) : & odd(Y) \leftarrow even(s(Y)) \\ \ldots & \ldots \end{array} \right\}$$

Note that the variable X in the above expressions corresponds to the Skolem constant a in $\neg H_1$.

Then, for each variablized candidate, we test whether it entails the first quasi-MSH $bot^+(B_1, E_1)(b)$ or not.

The first clause $odd(s^2(Y)) \leftarrow even(Y)$ does not entail $bot^+(B_1, E_1)(0)$. On the other hand, the second clause for $b = s(0)$, $odd(s(Y)) \leftarrow even(Y)$, entails it. Then, we equate MSH to $bot(B_1, E_1) \cup \{\neg even(s(s(0)))\}$ and continue further computation for finding best hypothesis from this bottom clause in the same way as Progol.

Since Algorithm $AIEMR$ computes only those hypotheses which satisfy both of the quasi-MSHs, the correctness of the algorithm is trivial. The difficult part is its termination property. Before discussing this issue, we will show one more example of the algorithm.

Example 2

$$\left\{\begin{array}{l} B_2 : \ even(0) \leftarrow \\ \qquad even(s^2(0)) \leftarrow \\ \qquad even(s^3(X)) \leftarrow odd(X) \\ E_2 : \ odd(s^7(0)) \leftarrow \\ H_2 : \ odd(s(X)) \leftarrow even(X) \\ H_{2'} : \ odd(s^2(X)) \leftarrow even(X) \end{array}\right\}$$

From B_2 and E_2, we obtain

$$bot(B_2, E_2) = odd(s^7(0)) \leftarrow even(0), even(s^2(0)).$$

The model constraining clause of B_2 is:

$$even(s^3(X)) \leftarrow odd(X)$$

Therefore, we have the following quasi-MSHs:

$$bot^+(B_2, E_2)(b) = \\ odd(s^7(0)), odd(b) \leftarrow even(0), even(s^2(0)) \\ bot^-(B_2, E_2)(b) = \\ odd(s^7(0)) \leftarrow even(0), even(s^2(0)), even(s^3(b))$$

Now, we build the initial candidate hypothesis for this example from $bot^-(B_2, E_2)$ in the same way as Example 1:

$$odd(s^7(0)) \leftarrow even(s^3(b))$$

Next, by successively substituting $0, s(0), s^2(0), \dots$ to b and variablizing, we obtain:

$$\left\{\begin{array}{l} b = 0 : \qquad odd(s^4(Y)) \leftarrow even(Y) \\ b = s(0) : \quad odd(s^3(Y)) \leftarrow even(Y) \\ b = s^2(0) : odd(s^2(Y)) \leftarrow even(Y) \\ b = s^3(0) : odd(s(Y)) \leftarrow even(Y) \\ b = s^4(0) : odd(Y) \leftarrow even(Y) \end{array}\right\}$$

By applying the substitution $\{Y/0\}$ to the candidate hypothesis $odd(s^2(Y)) \leftarrow even(Y)$ for $b = s^2(0)$, we obtain $odd(s^2(0)) \leftarrow even(0)$ which turns out to entail $bot^+(B_2, E_2)(s^2(0))$ (however, this is not our intended solution!).

Similarly, we obtain a clause $odd(s^3(0)) \leftarrow even(s^2(0))$ by applying the substitution $\{Y/s^2(0)\}$ to the candidate hypothesis $odd(s(Y)) \leftarrow even(Y)$ corresponding to $b = s^3(0)$. This clause entails $bot^+(B_2, E_2)(s^3(0))$, which is our intended solution. The final MSH is given by:

$$MSH = odd(s^7(0)) \leftarrow even(0), even(s^2(0)), even(s^5(0)), even(s^6(0)).$$

In the above discussion, we assumed that the body of the constraining clause contains only one literal. Since bot^- contains only $\neg Head\{b\}$, it does not cause any change to the algorithm even if there is more than one body literal. Although we need to add all body literals in the model constraining clause to obtain bot^+, this does not cause any change to the algorithm.

4.2 Termination Proof of the Algorithm $AIEMR$

Now, we will consider the termination problem of Algorithm $AIEMR$. Since we finally replaced the entailment test by subsumption test in the algorithm, the only thing we need to do is to find an upper bound of the iteration number of the step 4.

First, we need to define the sizes of a term and an atom.

Definition 11. *The size of a term t, denoted by $|t|$, is defined recursively as follows:*

1. *The sizes of a constant symbol and an uninstantiated variable are both 0.*
2. *The size of an n-ary term $f(t_1, ..., t_n)$, $|f(t_1, ..., t_n)|$, is defined by $|t_1| + ... + |t_n| + 1$.*

Since a constant is usually regarded as a function with arity 0, the above definition looks unnatural. The reason why we define the size of a constant as 0 is that we are only concerned with the increase of the term size caused by the repeated application of funcitons.

Definition 12. *The size of an n-ary atom $p(t_1, ..., t_n)$ is $|t_1| + ... + |t_n|$.*

Note that the size of a term (and that of an atom) may change after variables are instantiated. Particularly, the following equation holds:

$$|P\{\mathbf{X}\}\theta| = |P\{\mathbf{Y}\}| + |\mathbf{X}\theta|$$

where $\mathbf{Y}$ is an unbound new name of $\mathbf{X}$.

In computing, for example, the candidate hypothesis $odd(s(Y)) \leftarrow even(Y)$ for $b = s^3(0)$ in Example 2, we first substitute $s^3(0)$ to b in the initial hypothesis $odd(s^7(0)) \leftarrow even(s^3(b))$, and then we variablize. After that, we unify the left hand expression $odd(s(Y))$ to $odd(b) = odd(s^3(0))$ and unify the right hand to a unit ground clause $even(s^2(0))$ in the background knowledge to test whether it entails $bot^+(B_2, E_2)(s^3(0))$ or not.

Assume, in general, that the hypothesis is given by $odd(s^n(Y)) \leftarrow even(Y)$. Then since the head of this clause is unified to $odd(b)$, the size of b is given by the sum of $n = |s^n(Y)|$ and the size of a term to be substituted to Y. Hence, the equation

$$|b| = n + |Y| \tag{1}$$

holds.

Since the hypothesis $odd(s^n(Y)) \leftarrow even(Y)$ is obtained from the initial hypothesis $odd(s^7(0)) \leftarrow even(s^3(b))$ by applying substitution to b and after that applying variablization, the differences of sizes of each side of the clauses are kept to be equal. Therefore, in the above case, the equation $n = 7 - (3 + |b|)$ holds.

Let the positive example be E, the head of the model constraining clause be $Head\{\mathbf{X}\}$. Then, in general, 7 is the size of the positive example E and 3 is the size $Head\{X\}$ of the head of the model constraining clause P. Thus, the above equation becomes

$$n = |E| - (|Head\{X\}| + |b|). \tag{2}$$

On the other hand, since Y is substituted to some ground literal in the background knowledge, the size does not exceed the maximum size N of terms appearing in background knowledge. That is,

$$|Y| \leq N \tag{3}$$

holds.

By summarizing the above observation, we obtain the following theorem.

Theorem 2. *Let the positive example be E, the head of the model constraining clause be $Head\{X\}$ and the maximum size of terms appearing in the background knowledge be N. Then, the size of the P-Skolem constant vector b in the initial candidate hypothesis $E \leftarrow Head\{b\}$ is bounded by*

$$|b| \leq (|E| - |Head\{X\}| + N)/2$$

Proof. The above inequality follows directly from (1), (2) and (3) by eliminating both n and $|Y|$. □

For the above *Example 2*, since $|E|=7$, $|Head\{X\}| = |even(s^3(X))| = 3$ and $N = 3$, we obtain $|b| \leq 3$.

4.3   Plural model-constraining-clauses case

We showed that there are two kinds of models when there is one model constraining clause. If we have more than one model constraining clauses, in principle, we need to consider all possible combinations of model classes. The number of model types would be exponential when the model constraining clauses increase. However, we can avoid this combinatorial explosion. Let

$$C = Head\{X\} \leftarrow Body\{X\}$$

be a model constraining clause. When we apply Algorithm *AIEMR* to this clause, then a new literal $\neg Head\{X\}$ is added to the original set of literals $bot(B \backslash C, E)$. However, literals to be added due to the given model constraining clause are not increased by the existence of other model constraining clauses. Therefore, we can incrementally add new literals by successively applying Algorithm *AIEMR* to each model constraining clause.

5   Conclusion

In this paper, we introduced a complete algorithm for computing the most specific hypothesis when the background knowledge is a set of definite clauses and the positive example is a ground atom having a predicate to be learned. We showed the correctness and termination of the algorithm.

We also gave brief sketch of a complete algorithm for computing the MSH in abductive ILP setting elsewhere ([3]).

The development of a complete algorithm for computing the MSH in more general cases is the future research problem to be solved.

Acknowledgment

The author would like to express his deep thanks to Dr. Muggleton who gave an insightful suggestion for pursuing this problem. The author also would like to thank Dr. Yamamoto for his helpful discussion on this problem.

References

1. Breiman, L., Friedman, J.H., Olshen, R.A. and Stone, C. Classification And Regression Trees, Belmont, CA: Wadsworth International Group, 1984.
2. Furukawa, K., Murakami, T., Ueno, K., Ozaki, T. and Shimaze, K. On a Sufficient Condition for the Existence of Most Specific Hypothesis in Progol, Proc. of ILP-97, Lecture Notes in Artificial Intelligence 1297, Springer, 157-164, 1997.
3. Furukawa, K. On the Completion of the Most Specific Hypothesis in Inverse Entailment for Mutual Recursion and Abductive ILP Setting, Proceedings of 32nd SIG-FAI, JSAI, March, 1998 (in Japanese).
4. Marcinowski, J. and Pacholski, L. Undecidability of the Horn-clause implication problem, Proc. of 33rd Annual Symposium on Foundations of Computer Science, 354-362, 1992.
5. Muggleton, S. Inverting Implication, Proc. of ILP92, 19-39, ICOT Technical Memorandom: TM-1182, 1992.
6. Muggleton, S. Inverse Entailment and Progol, New Generation Computing, Vol.13, 245-286, 1995
7. Plotkin, G.D. Automatic Method of Inductive Inference, PhD thesis, Edinburgh University, 1971.
8. Quinlan, J.R. Induction of decision trees,Machine Learning, 1(1), 81-106, 1986
9. Quinlan, J.R. C4.5: Programs for Machine Learning, Morgan Kaufmann, San Mateo, CA, 1993
10. Yamamoto, A. Improving Theories for Inductive Logic Programming Systems with Ground Reduced Programs. Technical Report, Forschungsbericht AIDA-96-19 FG Intellektik FB Informatik TH Darmstadt, 1996.
11. Yamamoto, A. Which Hypotheses Can Be Found with Inverse Entailment? Proc. of ILP-97, Lecture Notes in Artificial Intelligence 1297, Springer, 296-308, 1997.

Biochemical Knowledge Discovery Using Inductive Logic Programming*

Stephen Muggleton[1], Ashwin Srinivasan[2], and R.D. King[3] and M.J.E. Sternberg[4]

[1] Department of Computer Science, University of York
stephen@cs.york.ac.uk,
WWW home page: http://www.cs.york.ac.uk/ stephen
[2] Computing Laboratory, University of Oxford
ashwin@comlab.ox.ac.uk,
[3] Department of Computer Science, The University of Wales Aberystwyth
rdk@aber.ac.uk,
WWW home page: http://www.aber.ac.uk/ rdk
[4] Biomolecular Modelling Laboratory, Imperial Cancer Research Fund
m.sternberg@icrf.icnet.uk,
WWW home page: http://www.icnet.uk/bmm

Abstract. Machine Learning algorithms are being increasingly used for knowledge discovery tasks. Approaches can be broadly divided by distinguishing discovery of procedural from that of declarative knowledge. Client requirements determine which of these is appropriate. This paper discusses an experimental application of machine learning in an area related to drug design. The bottleneck here is in finding appropriate constraints to reduce the large number of candidate molecules to be synthesised and tested. Such constraints can be viewed as declarative specifications of the structural elements necessary for high medicinal activity and low toxicity. The first-order representation used within Inductive Logic Programming (ILP) provides an appropriate description language for such constraints. Within this application area knowledge accreditation requires not only a demonstration of predictive accuracy but also, and crucially, a certification of novel insight into the structural chemistry. This paper describes an experiment in which the ILP system Progol was used to obtain structural constraints associated with mutagenicity of molecules. In doing so Progol found a new indicator of mutagenicity within a subset of previously published data. This subset was already known not to be amenable to statistical regression, though its complement was adequately explained by a linear model. According to the combined accuracy/explanation criterion provided in this paper, on both subsets comparative trials show that Progol's structurally-oriented hypotheses are preferable to those of other machine learning algorithms.

---

* The results in this paper are published separately in [7, 16]

1   INTRODUCTION

Within the AI literature, the distinction between procedural and declarative knowledge was first introduced by McCarthy [8], though it is strongly related to Ryle's difference between "knowing how" and "knowing that" [15]. While procedural knowledge can often be conveniently described in algorithmic form, logical sentences are usually used to capture declarative knowledge.

Most of Machine Learning has been concerned with the acquisition of procedural knowledge. For instance, the nested if-then-else rules used in decision tree technology largely describe the flow of procedural control. As a consequence of this procedural bias, the emphasis in the testing methodology has been on predictive accuracy, while communicability, the primary hallmark of declarative knowledge, has largely been disregarded in practice. This is despite early and repeated recognition of the importance of comprehensibility in machine learning [9–11]. In certain domains client requirements dictate the need for inductive discovery of declarative knowledge. This is the case in the area of rational drug design (see James Black's Nobel Lecture reprinted in [1]).

A range of specialists are involved within the the pharmaceutical industry. These include computational chemists, molecular biologists, pharamcologists, synthetic and analytical chemists. The bottleneck in the process of drug design is the discovery of appropriate constraints to reduce the large number of candidate molecules for synthesis and testing. Since such constraints need to be used by synthetic chemists in the molecular design process, they must be stated in appropriately structural, and ideally 3-D terms. The constraints will describe both structural attributes which enhance medicinal activity as well as those which should be absent, owing to toxic side-effects. Such design-oriented constraints are declarative in nature.

For the development of such constraints, Inductive Logic Programming (ILP) [13], with its emphasis on the declarative representation of logic programs, is the obvious Machine Learning technique. This is not to say that other Machine Learning approaches cannot be used when, for instance, appropriate known structural properties have been encoded as molecular attributes. However, in industrial drug discovery tasks such 'indicator' attributes are typically either unknown, or only understood to a limited degree. Without such attributes, computational chemists usually apply linear regression to pre-computed bulk properties of the molecules involved. Common bulk properties include the octanol/water partition coefficient $log(P)$, and molecular reactivity as measured by the electronic property ϵ_{LUMO}. Although the derived regression equations can be used for testing candidate molecules within a molecular modelling package, they provide synthetic chemists with no structural clues as to how candidate molecules should be designed.

In this paper we use the problem of mutagenicity prediction to compare the ILP algorithm Progol [12] with various other state-of-the-art machine learning, neural and statistical algorithms. Mutagenicity is a toxic property of molecules related to carcinogenicity (see 2). The algorithms are compared both with and without provision of specially defined structural attributes. The derived hypotheses are compared both in terms of predictive accuracy, and in terms of declarative, structural description provided. Progol is capable of using low-level atom and bond relations to define structural properties not found in the original attribute language.

For the purposes of such a comparison it is not immediately obvious how one should define a performance criterion which combines both predictive accuracy and declarative explanation. Below we provide a working definition of such a combining function, appropriate for this problem domain.

Definition 1. Combination of Accuracy and Explanation. *If the predictive accuracies of two hypotheses are statistically equivalent then the hypothesis with better explanatory power will be preferred. Otherwise the one with higher accuracy will be preferred.*

Clearly no preference is possible in the case in which the hypotheses are equivalent in terms of both accuracy and explanation. In the experiments reported, hypotheses are judged to have "explanatory power" (a boolean property) if they provide insight for molecular design based on their use of structural descriptors, and not otherwise.

2   MUTAGENESIS

Mutagenic compounds cause mutations in the DNA sequence. These compounds are often, though not always, carcinogenic (cancer-producing). It is of considerable interest to the pharmaceutical industry to determine molecular indicators of mutagenicity, since this would allow the development of less hazardous new compounds. This paper explores the power of Progol to derive Structure-Activity Relations (SARs). A data set in the public domain is used, which was originally investigated using linear regression techniques in [4]. This set contains 230 aromatic and hetero-aromatic nitro compounds that were screened for mutagenecity by the widely used Ames test (with *Salmonella typhimurium* TA 98). The data present a significant challenge to SAR studies as they are chemically diverse and without a single common structural template.

In the course of the original regression analysis, the data were found to have two sub-groups, with 188 compounds being amenable to linear modelling by regression and the remainder being regression "unfriendly". The 188 compounds have also been examined with good results by neural-network techniques [17]. In keeping with attribute-based approaches to SAR, these studies rely on representing compounds in terms of global chemical properties, notably $log(P)$ and ϵ_{LUMO} (see previous section). Usually, some degree of structural detail is introduced manually in the form of logical indicator attributes to be used in the regression equation. For example, in [4] the authors introduce the variable I_1 to denote the presence of 3 or more fused rings, and I_a to denote the class "acenthrylenes". Regression and neural-network models have then been constructed using these 4 basic attributes, namely $log(P), \epsilon_{LUMO}, I_1$, and I_a.

3 EXPERIMENT

3.1 Hypothesis

The experimental hypothesis is as follows.

Hypothesis. The ILP program Progol has better performance – using the criterion described in Definition 1 – than the attribute-based algorithms embodied in linear regression, decision-trees and neural networks.

3.2 Materials

We first describe the algorithms used in the comparative trial. The description of Progol is somewhat more detailed, given that it is less likely to be as well known to the reader as the attribute-based learners.

Progol A formal specification for Progol is as follows.

- B is background knowledge consisting of a set of definite clauses $= C_1 \wedge C_2 \wedge$ ...
- E is a set of examples $= E^+ \wedge E^-$ where
  - E^+ or 'positive examples' $= e_1 \wedge e_2 \wedge \ldots$ are definite clause s and
  - E^- or 'negative examples' $= f_1 \wedge f_2 \wedge \ldots$ are non-definite Horn clauses.
- $H = D_1 \wedge D_2 \wedge \ldots$ is an hypothesised explanation of the examples in terms of the background knowledge.

Progol can be treated as an algorithm A such that $H = A(B, E)$ is an hypothesis from predefined language $\mathcal{L}$. The language $\mathcal{L}$ consists of legal forms of the predicates in B that can appear in H. Each D_i in H has the property that it can explain at least one positive example and the set H is consistent with the negative examples. That is, $B \wedge D_i \models e_1 \vee e_2 \vee \ldots$, $\{e_1, e_2, \ldots\} \subseteq E^+$ and $B \wedge H \wedge E^- \not\models \Box$. If more than one such H exists then Progol returns the first one in an arbitrary lexicographic ordering $\mathcal{O}(\mathcal{L})$. Figure 1 is a diagrammatic view of the space searched by a program like Progol when constructing the individual clauses D_i.

Progol is thus provided with a set of positive and negative examples together with problem-specific background knowledge B. The aim is to generate an hypothesis, expressed as a set of rules, which explains all the positive examples in terms of the background knowledge whilst remaining consistent with the negative examples. To achieve this, *Progol* 1) randomly selects a positive example e_i; 2) uses inverse entailment ([12]) to construct the most specific hypothesis $\perp(B, e_i)$ which explains e_i in terms of B; 3) finds a rule D_i which generalises $\perp(B, e_i)$ and which maximally compresses a set of entailed examples E_i; and 4) adds D_i to the hypothesis H and repeats from 1) with examples not covered so far until no more compression is possible. Compression is here defined as the difference, in total size of formula, between E_i and D_i. Compression formalises

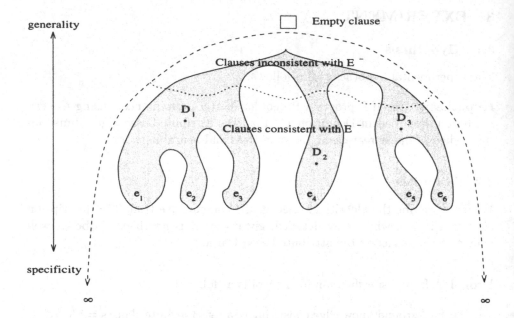

Fig. 1. The space of clauses searched by logic-based learning algorithms. The space enclosed by the parabolic broken line contains all clauses that can potentially be included in a theory. Clauses near the top are usually too general, that is, are inconsistent with the negative examples. Clauses become progressively more specific as one proceeds downwards from the empty clause. For example, given a set of positive examples, $E^+ = e_1 \wedge e_2 \wedge \ldots e_6$ and negative examples E^-, the shaded area shows the part of the space searched by Progol. In this space, the clauses D_i represent clauses that maximally compress the examples 'under' them in the search space. Progol ensures that it only searches those parts in the grey space.

the reduction of information provided by the rule in describing the data. Figure 1 illustrates the reduced, finite, search space of *Progol*. The reduction in the search space is such that, unlike most machine learning algorithms, it is generally feasible to find the optimal rule satisfying conditions (3) and (4) above in reasonable time (more detailed complexity arguments can be found in [12]).

The results in this paper were obtained using a Prolog implementation of Progol, called *P-Progol*. Details of obtaining and using a C implementation of Progol can be found in [12]. *P-Progol* is available by ftp access to *ftp.comlab.ox.ac.uk*. The relevant directory is *pub/Packages/ILP*. The implementation includes on-line documentation that clarifies the points of difference with the C version. The theory underlying both versions is the same and is described fully in [12]. However, differences in search strategies and pruning facilities imply that given the same language and resource restrictions, the two versions can compute different answers. For convenience, in the rest of this paper, we shall refer to *P-Progol* as Progol.

Attribute-based algorithms The regression was achieved using the Minitab package (Minitab Inc, Pennsylvania State University, Pa). The learning rule used by the neural technique relies on the back-propagation of errors. Changes in weight are calculated by solving a set of differential equations as described in [6]. This technique removes the need for learning-rate or momentum parameters [14]. The actual implementation of the back-propagation algorithm was supplied by J Hirst of the Imperial Cancer Research Fund. Finally, the procedure embodied by the CART algorithm [2] as implemented by the Ind package [3] is used to construct classification trees.

Data For this study, we consider the following characterisation for mutagenic molecules. All molecules whose mutagenic activity, as measured by the Ames test, is greater than 0 are termed "active". All others are taken to be "inactive". Figure 2 shows the distribution of molecules within these classes for the two different subsets identified in [4].

Subset	"Active"	"Inactive"	Total
"Regression friendly"	125	63	188
"Regression unfriendly"	13	29	42
All	138	92	230

Fig. 2. Class distribution of molecules

For each of the two subsets, the task is to obtain a characterisation of the "active" molecules. Unlike the attribute-based algorithms, Progol can employ an explicit 2-dimensional atom and bond representation of the molecules. This representation supports the encoding of generic chemical background knowledge. This is described in the next section.

ILP representation The obvious generic description of molecules is to use atoms and their bond connectivities. For the chemicals described here, these were obtained automatically using the molecular modelling program QUANTA[1]. For the experiments here, each chemical was entered manually via a molecular sketchpad. All atoms in the chemical are then automatically typed by QUANTA to consider their local chemical environment, an estimate of the electronic charge on each atom estimated, and connecting bonds are classified (as single, double, aromatic, etc.). It is worth noting here that the manual use of a sketchpad is clearly not mandatory – with appropriate software, the chemicals could have been extracted directly from a database and their structure computed. Further, the choice of QUANTA was arbitrary, any similar molecular modelling package would have been suitable.

---

[1] Distributed by Molecular Simulations Inc, USA

There is a straightforward translation of the QUANTA facts into Prolog facts suitable for Progol. The result is that each compound is represented by a set of facts such as the following.

    atm(127, 127_1, c, 22, 0.191)
    bond(127, 127_1, 127_6, 7)

This states that in compound 127, atom number 1 is a carbon atom of QUANTA type 22 with a partial charge of 0.191, and atoms 1 and 6 are connected by a bond of type 7 (aromatic). Again, this representation is not peculiar to the compounds in this study, and can be used to encode arbitrary chemical structures. With the *atm* and *bond* predicates as basis, it is possible to use background knowledge definitions of higher-level 2-dimensional chemical substructures that formalise concepts like methyl groups, nitro groups, 5- and 6-membered rings, aromatic rings, heteroaromatic rings, connected rings, etc. For example, the following Prolog program fragment can be used to detect benzene rings.

```
% benzene - 6 membered carbon aromatic ring
% recall that QUANTA type 7 indicates an aromatic bond
benzene(Drug,Ring_list) :-
        atoms(Drug,6,Atom_list,[c,c,c,c,c,c]),
        ring6(Drug,Atom_list,Ring_list,[7,7,7,7,7,7]).

% get N lexico-graphically ordered atoms and their elements
atoms(Drug,N,[Atom|Atoms],[Elem|Elems]) :-
...

% find 6 connected atoms and their bond types
ring6(Drug,[Atom1|List],[Atom1,Atom2,Atom4,Atom6,Atom5,Atom3],
        [Type1,Type2,Type3,Type4,Type5,Type6]) :-
        ...
```

A list of definitions used for the experiments is given in Appendix A.

3.3   Method

The comparative trials reported here were conducted under the following conditions.

1. Analysis of the two subsets of compounds in Figure 2 (188 and 42) were conducted separately.
2. For each subset, Progol and attribute-based algorithms were compared a) using the molecular properties $log(P)$ and ϵ_{LUMO} and b) using additional expert-derived structural indicator attributes I_1 and I_a. In both cases, Progol also had access to the definitions of the generic chemical properties described earlier. Owing to the lack of a relational description language, these properties could not be provided effectively to the attribute-valued algorithms, though they would be readily available in a real-world situation.

3. Predictive accuracy estimates were obtained using cross-validation. For the subset of 188 compounds, 10-fold cross-validation was employed, and for the 42 compound a leave-one-out procedure was used. Comparison of estimates use McNemar's test. Details are available in [16].

4. Explanatory power of Progol's hypotheses was compared to those of the attribute-based algorithm with highest predictive accuracy. Hypotheses were judged to have explanatory power if and only if they made use of structural attributes of molecules.

5. Relative performance was assessed in each case by combining accuracy and explanation using Definition 1.

3.4 Results

There are two cases to consider. These are analysis without (case $\bar{I}$) and with (case I) the indicator attributes $I_{1,a}$.

Accuracy Figure 3 tabulates the predictive accuracies of theories obtained on the subsets of 188 and 42 compounds for cases $\bar{I}$ and I. In the tabulation REG denotes linear regression, DT denotes decision-tree and NN a neural-network with 3 hidden units.

Algorithm	188		42	
	$\bar{I}$	I	$\bar{I}$	I
REG	0.85 (0.03)	0.89 (0.02)	0.67 (0.07)	0.67 (0.07)
DT	0.82 (0.02)	0.88 (0.02)	0.83 (0.06)	0.83 (0.06)
NN	0.86 (0.02)	0.89 (0.02)	0.64 (0.07)	0.69 (0.07)
Progol	0.88 (0.02)[1]	0.88 (0.02)[1]	0.83 (0.06)[2]	0.83 (0.06)[2]
Default class	0.66 (0.03)	0.66 (0.03)	0.69 (0.07)	0.69 (0.07)

Fig. 3. Predictive accuracy estimates of theories of mutagenicity. Accuracy is defined as the proportion of correct predictions obtained from cross-validation trials. Estimates for the 188 compounds are from a 10-fold cross-validation, and those for the 42 are from a leave-one-out procedure. The "Default class" algorithm is one that simply guesses majority class. Estimated standard error is shown in parentheses after each accuracy value. Superscripts on Progol results refer to the pairwise comparison in turn of Progol against the attribute-based programs. With the null hypothesis that Progol and its adversary classify the same proportions of examples correctly, superscript 1 indicates that the probability of observing the classification obtained is ≤ 0.05 for DT, and 2 that this probability is ≤ 0.05 for REG and NN only (but not DT).

It is evident that on the "regression-friendly" subset (188 compounds), Progol yields state-of-the-art results (at $P < 10\%$). It is also unsurprising that regression and neural-network perform well on this subset which had been identified as being amenable to a regression-like analysis. It is apparent that the predictive accuracy of Progol on the "regression-unfriendly" data is the same as that obtained using a decision-tree. Figure 4 summarises the comparison of Progol's predictive accuracy against the best attribute-based algorithm for the two subsets. For the 188 compounds, the best attribute-based algorithm is in turn NN (case $\overline{I}$) and REG (case I). For the 42 compounds, the best attribute-based algorithm is DT in both cases.

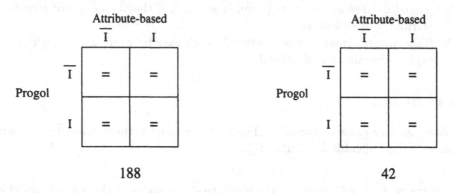

Fig. 4. Comparing the predictive accuracies of Progol and the best attribute-based algorithm. '=' means statistically equivalent at $P = 0.05$.

Explanation Appendix B lists the theories obtained by Progol and the attribute-based algorithm with the highest predictive accuracy – with the exception of case $\overline{I}$ on the 188 compounds. Here the best attribute-based algorithm is the neural-network and we do not list the weights assigned to each node. Figure 5 summarises the comparative assessment of the explanatory power of these theories, using the criterion in Step 4, Section 3.3.

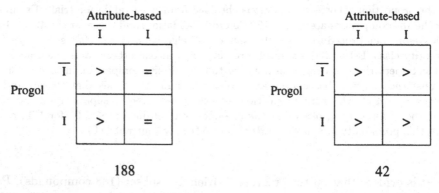

Fig. 5. Comparing the explanatory power of Progol and the best attribute-based algorithm. '=' means the hypotheses of both algorithms use structural attributes. '>' means Progol's hypothesis uses structural attributes and the other algorithm's hypothesis does not.

Performance The relative overall performance of Progol against the best attribute-based algorithm is obtained by combining the results in Figures 4, 5 using Definition 1. This is shown in Figure 6.

The experimental hypothesis holds in all but two of the eight cases. The expected case in real life drug design is the $\overline{I}, \overline{I}$ one, in which Progol outperforms all other algorithms tested.

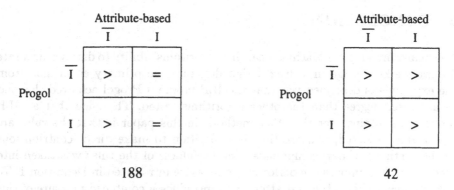

Fig. 6. Comparing the performance of Progol and the best attribute-based algorithm. The combination of accuracy and explanation from Figures 4 and 5 is according to Definition 1.

4 FUTURE WORK

A next step in representing chemical structure is to incorporate three-dimensional information for the compounds studied. We are currently considering a straightforward extension to the representation described in Section 3.2 to include the 3-dimensional co-ordinates for every atom in the molecule. These co-ordinates are then used to obtain a 3-dimensional co-ordinate specification for each structural group in Appendix A. This specifies the X,Y,Z co-ordinates of the nominal "centre" of the group. For example, the earlier Prolog program fragment to detect benzene rings is now modified as follows:

```
% benzene - 6 membered carbon aromatic ring centred at X,Y,Z
benzene(Drug,Atom_list,X,Y,Z) :-
    atoms(Drug,6,Atom_list,[c,c,c,c,c,c],Coordinates),
    ring6(Drug,Atom_list,Coordinates,Ring_list,[7,7,7,7,7,7],X,Y,Z).

}}
```

Euclidean distances between such group-centres within a molecule are easily calculable, and form the basis of a rudimentary reasoning about the 3-dimensional structure of the molecule. This approach is similar to recent work reported in [5], where Progol is used to identify a pharmacophore common to a series of active molecules. A pharmacophore is a 3-dimensional description of a molecule that consists of 3–5 functional groups and the distances between them (angles and other geometric properties may also be used). In [5] it is shown that in a blind-test Progol was able to identify the pharmacophore generally thought responsible for ACE inhibition.

5   CONCLUSION

This paper investigates Machine Learning algorithms' ability to discover accurate declarative knowledge in a drug design domain. The primary conclusion from the experiments described was that the ILP program Progol achieved this end to a greater degree than the other algorithms tested. The edge that an ILP program maintains over the other methods in this paper is that the rules are structurally oriented, and are thus in a position to make direct contributions to the synthesis of new compounds. The usefulness of the rules was taken into account in the comparative performance measure introduced in Definition 1. To our knowledge this is the first attempt to introduce a combined measure of this kind. It was only possible to define such a measure since the application domain dictates the way in which the rules would be used. One avenue of future research for knowledge discovery systems would be to use such a measure directly as a utility function in the hypotheses space search, thus maximising the chances of constructing accurate and comprehensible hypotheses. Such a function would be quite different in nature to the simple differential cost functions used in algorithms such as CART [11].

The ILP-constructed rules can be viewed as "knowledge-compilations" in the Michie sense – namely not just expressing the structural relationships that hold between response and measured attributes, but doing so in a form " ... that is meaningful to humans and evaluable in the head" (page 51, [11]).

Acknowledgements

This research was supported partly by the Esprit Basic Research Action ILP II (project 20237), the EPSRC grant GR/K57985 on 'Experiments with Distribution-based Machine Learning' and an EPSRC Advanced Research Fellowship held by Stephen Muggleton. Ashwin Srinivasan currently holds a Research Fellowship supported by the Nuffield Trust and Green College, Oxford. The authors are indebted to Donald Michie for his advice and interest in this work.

References

1. J. Black. Drugs from emasculated hormones: the principle of syntopic antagonism. *Bioscience Reports*, 9(3), 1989. Published in *Les Prix Nobel*, 1988. Printed in Sweden by Nostedts Tryckeri, Stockholm.

2. L. Breiman, J.H. Friedman, R.A. Olshen, and C.J. Stone. *Classification and Regression Trees.* Wadsworth, Belmont, 1984.

3. W. Buntine. Ind package of machine learning algorithms. Technical Report 244-17, Research Institute for Advanced Computer Science, NASA Ames Research Center, Moffett Field, CA 94035, 1992.

4. A.K. Debnath, R.L Lopez de Compadre, G. Debnath, A.J. Schusterman, and C. Hansch. Structure-Activity Relationship of Mutagenic Aromatic and Heteroaromatic Nitro compounds. Correlation with molecular orbital energies and hydrophobicity. *Journal of Medicinal Chemistry*, 34(2):786 – 797, 1991.

5. P. Finn, S. Muggleton, D. Page, and A. Srinivasan. Pharmacophore Discovery using the Inductive Logic Programming system Progol *Machine Learning*, 30:241 – 271, 1998.

6. C.W. Gear. *Numerical Initial Value Problems is Ordinary Differential Equations.* Prentice-Hall, Edgewood Cliffs, NJ, 1971.

7. R.D. King, S.H. Muggleton, A. Srinivasan, and M.J.E. Sternberg. Structure-activity relationships derived by machine learning: The use of atoms and their bond connectivities to predict mutagenicity by inductive logic programming. *Proc. of the National Academy of Sciences*, 93:438–442, 1996.

8. J. McCarthy. Programs with commonsense. In *Mechanisation of thought processes (v 1)*. Her Majesty's Stationery Office, London, 1959. Reprinted with an additional section in *Semantic Information Processing*.

9. R.S. Michalski. A theory and methodology of inductive learning. In R. Michalski, J. Carbonnel, and T. Mitchell, editors, *Machine Learning: An Artificial Intelligence Approach*, pages 83–134. Tioga, Palo Alto, CA, 1983.

10. D. Michie. The superarticulacy phenomenon in the context of software manufacture. *Proceedings of the Royal Society of London*, A 405:185–212, 1986. Reprinted in: The Foundations of Artificial Intelligence: a Sourcebook (eds. D. Partridge and Y. Wilks), Cambridge University Press, 1990.

11. D. Michie, D.J. Spiegelhalter, and C.C. Taylor, editors. *Machine Learning, Neural and Statistical classification.* Ellis-Horwood, New York, 1994.

12. S. Muggleton. Inverse Entailment and Progol. *New Gen. Comput.*, 13:245–286, 1995.

13. S. Muggleton and L. De Raedt. Inductive logic programming: Theory and methods. *Journal of Logic Programming*, 19,20:629–679, 1994.

14. A. J. Owens and D. L. Filkin. Efficient training of the back-propagation network by solving a system of stiff ordinary differential equations. In *Proceedings IEEE/INNS International Joint Conference of Neural Networks*, pages 381–386, Washington DC, 1989.

15. G. Ryle. *The Concept of Mind.* Hutchinson, 1949.

16. A. Srinivasan, S.H. Muggleton, R.D. King, and M.J.E. Sternberg. Theories for mutagenicity: a study of first-order and feature based induction. *Artificial Intelligence*, 85:277–299, 1996.

17. D. Villemin, D. Cherqaoui, and J.M. Cense. Neural network studies: quantitative structure-activity relationship of mutagenic aromatic nitro compounds. *J. Chim. Phys*, 90:1505–1519, 1993.

A. Some elementary chemical concepts defined in terms of the atom and bond structure of molecules

The following are Prolog definitions for some simple chemical concepts that can be defined directly using the atomic and bond structure of a molecule.

```
% In the following QUANTA bond type 7 is aromatic.

% Three benzene rings connected linearly
anthracene(Drug,[Ring1,Ring2,Ring3]) :-
        benzene(Drug,Ring1),
        benzene(Drug,Ring2),
        Ring1 @> Ring2,
        interjoin(Ring1,Ring2,Join1),
        benzene(Drug,Ring3),
        Ring1 @> Ring3,
        Ring2 @> Ring3,
        interjoin(Ring2,Ring3,Join2),
        \+ interjoin(Join1,Join2,_),
        \+ members_bonded(Drug,Join1,Join2).

% Three benzene rings connected in a curve
phenanthrene(Drug,[Ring1,Ring2,Ring3]) :-
        benzene(Drug,Ring1),
        benzene(Drug,Ring2),
        Ring1 @> Ring2,
        interjoin(Ring1,Ring2,Join1),
        benzene(Drug,Ring3),
        Ring1 @> Ring3,
        Ring2 @> Ring3,
        interjoin(Ring2,Ring3,Join2),
        \+ interjoin(Join1,Join2,_),
        members_bonded(Drug,Join1,Join2).


% Three benzene rings connected in a ball
ball3(Drug,[Ring1,Ring2,Ring3]) :-
        benzene(Drug,Ring1),
        benzene(Drug,Ring2),
        Ring1 @> Ring2,
        interjoin(Ring1,Ring2,Join1),
        benzene(Drug,Ring3),
        Ring1 @> Ring3,
        Ring2 @> Ring3,
        interjoin(Ring2,Ring3,Join2),
        interjoin(Join1,Join2,_).

members_bonded(Drug,Join1,Join2) :-
        member(J1,Join1),
        member(J2,Join2),
        bondd(Drug,J1,J2,7).


ring_size_6(Drug,Ring_list) :-
        atoms(Drug,6,Atom_list,_),
        ring6(Drug,Atom_list,Ring_list,_).

ring_size_5(Drug,Ring_list) :-
        atoms(Drug,5,Atom_list,_),
        ring5(Drug,Atom_list,Ring_list,_).

% benzene - 6 membered carbon aromatic ring
benzene(Drug,Ring_list) :-
        atoms(Drug,6,Atom_list,[c,c,c,c,c,c]),
        ring6(Drug,Atom_list,Ring_list,[7,7,7,7,7,7]).
```

```
carbon_5_aromatic_ring(Drug,Ring_list) :-
        atoms(Drug,5,Atom_list,[c,c,c,c,c]),
        ring5(Drug,Atom_list,Ring_list,[7,7,7,7,7]).

carbon_6_ring(Drug,Ring_list) :-
        atoms(Drug,6,Atom_list,[c,c,c,c,c,c]),
        ring6(Drug,Atom_list,Ring_list,Bond_list),
        Bond_list \== [7,7,7,7,7,7].

carbon_5_ring(Drug,Ring_list) :-
        atoms(Drug,5,Atom_list,[c,c,c,c,c]),
        ring5(Drug,Atom_list,Ring_list,Bond_list),
        Bond_list \== [7,7,7,7,7].

hetero_aromatic_6_ring(Drug,Ring_list) :-
        atoms(Drug,6,Atom_list,Type_list),
        Type_list \== [c,c,c,c,c,c],
        ring6(Drug,Atom_list,Ring_list,[7,7,7,7,7,7]).

hetero_aromatic_5_ring(Drug,Ring_list) :-
        atoms(Drug,5,Atom_list,Type_list),
        Type_list \== [c,c,c,c,c],
        ring5(Drug,Atom_list,Ring_list,[7,7,7,7,7]).

atoms(Drug,1,[Atom],[T]) :-
        atm(Drug,Atom,T,_,_),
        T \== h.
atoms(Drug,N1,[Atom1|[Atom2|List_a]],[T1|[T2|List_t]]) :-
        N1 > 1,
        N2 is N1 - 1,
        atoms(Drug,N2,[Atom2|List_a],[T2|List_t]),
        atm(Drug,Atom1,T1,_,_),
        Atom1 @> Atom2,
        T1 \== h.

ring6(Drug,[Atom1|List],[Atom1,Atom2,Atom4,Atom6,Atom5,Atom3],
        [Type1,Type2,Type3,Type4,Type5,Type6]) :-
        bondd(Drug,Atom1,Atom2,Type1),
        memberchk(Atom2,[Atom1|List]),
        bondd(Drug,Atom1,Atom3,Type2),
        memberchk(Atom3,[Atom1|List]),
        Atom3 @> Atom2,
        bondd(Drug,Atom2,Atom4,Type3),
        Atom4 \== Atom1,
        memberchk(Atom4,[Atom1|List]),
        bondd(Drug,Atom3,Atom5,Type4),
        Atom5 \== Atom1,
        memberchk(Atom5,[Atom1|List]),
        bondd(Drug,Atom4,Atom6,Type5),
        Atom6 \== Atom2,
        memberchk(Atom6,[Atom1|List]),
        bondd(Drug,Atom5,Atom6,Type6),
        Atom6 \== Atom3.


ring5(Drug,[Atom1|List],[Atom1,Atom2,Atom4,Atom5,Atom3],
        [Type1,Type2,Type3,Type4,Type5]) :-
        bondd(Drug,Atom1,Atom2,Type1),
        memberchk(Atom2,[Atom1|List]),
        bondd(Drug,Atom1,Atom3,Type2),
        memberchk(Atom3,[Atom1|List]),
        Atom3 @> Atom2,
        bondd(Drug,Atom2,Atom4,Type3),
        Atom4 \== Atom1,
        memberchk(Atom4,[Atom1|List]),
        bondd(Drug,Atom3,Atom5,Type4),
        Atom5 \== Atom1,
```

```
        memberchk(Atom5,[Atom1|List]),
        bondd(Drug,Atom4,Atom5,Type5),
        Atom5 \== Atom2.

nitro(Drug,[Atom0,Atom1,Atom2,Atom3]) :-
        atm(Drug,Atom1,n,38,_),
        bondd(Drug,Atom0,Atom1,1),
        bondd(Drug,Atom1,Atom2,2),
        atm(Drug,Atom2,o,40,_),
        bondd(Drug,Atom1,Atom3,2),
        Atom3 @> Atom2,
        atm(Drug,Atom3,o,40,_).

methyl(Drug,[Atom0,Atom1,Atom2,Atom3,Atom4]) :-
        atm(Drug,Atom1,c,10,_),
        bondd(Drug,Atom0,Atom1,1),
        atm(Drug,Atom0,Type,_,_),
        Type \== h,
        bondd(Drug,Atom1,Atom2,1),
        atm(Drug,Atom2,h,3,_),
        bondd(Drug,Atom1,Atom3,1),
        Atom3 @> Atom2,
        atm(Drug,Atom3,h,3,_),
        bondd(Drug,Atom1,Atom4,1),
        Atom4 @> Atom3,
        atm(Drug,Atom4,h,3,_).


% intersection(+Set1, +Set2, ?Intersection)

interjoin(A,B,C) :-
        intersection(A,B,C),
        C \== [].

bondd(Drug,Atom1,Atom2,Type) :-  bond(Drug,Atom2,Atom1,Type).

member(X,[X|_]).
member(X,[_|T]):-

connected(Ring1,Ring2):-
        Ring1 \= Ring2,
        member(Atom,Ring1),
        member(Atom,Ring2), !.
```

B. Theories obtained

Progol

On the subset of 188 compounds, the Progol theory for "active" molecules for case $\overline{I}$ is as follows.

```
active(A) :-
   logp(A,B), gteq(B,4.180).

active(A) :-
   lumo(A,B), lteq(B,-1.937).

active(A) :-
   logp(A,B), gteq(B,2.740), ring_size_5(A,C).
```

For case I, the rules obtained are as follows (again, ground facts are not shown).

```
active(A) :-
   ind1(A,1.000).
active(A) :-
   atm(A,B,o,40,-0.384).
active(A) :-
   atm(A,B,c,29,0.017).
active(A) :-
   atm(A,B,c,29,C), gteq(C,0.010).
active(A) :-
   atm(A,B,o,40,-0.389), bond(A,C,B,2), bond(A,D,C,1).
active(A) :-
   bond(A,B,C,1), bond(A,D,B,2), ring_size_5(A,E).
```

On the subset of 42 compounds, cases $\overline{I}$ and I yielded the description below.

```
active(A) :-
   bond(A,B,C,2), bond(A,D,B,1), atm(A,D,c,21,E).
```

Attribute-based algorithms

On the subset of 188 compounds, the best attribute-based algorithm (that is, with highest predictive accuracy) for case $\overline{I}$ is a neural-network. We refrain from reproducing the weights on the nodes here. For case I on the same subset, the best algorithm is regression. The equation obtained is below.

$$Activity = -2.94(\pm0.33) + 0.10(\pm0.08)logP - 1.42(\pm0.16)LUMO$$

$$- 2.36(\pm0.50)Ia + 2.38(0.23)I1 \tag{1}$$

On the subset of 42 compounds, the following decision-tree has the highest predictive accuracy.

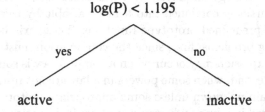

Computational Characteristics of Law Discovery Using Neural Networks

Ryohei Nakano[1] and Kazumi Saito[2]

[1] NTT Communication Science Laboratories,
2-4, Hikaridai, Seika, Soraku, Kyoto 619-0237 Japan
nakano@cslab.kecl.ntt.co.jp
[2] NTT R&D Management Department,
3-19-2, Nishi-shinjuku, Tokyo 163-8019 Japan
saito@rdh.ecl.ntt.co.jp

Abstract. Lately the authors have proposed a new law discovery method called RF5 using neural networks; i.e., law-candidates (neural networks) are trained by using a second-order learning algorithm, and an information criterion selects the most suitable from law-candidates. Our previous experiments showed that RF5 worked well for relatively small problems. This paper evaluates how the method can be scaled up, and analyses how it is invariant for the normalization of input and output variables. Since the sizes of many real data are middle or large, the scalability of any law discovery method is highly important. Moreover since in most real data different variables have typical values which may differ significantly, the invariant nature for the normalization of variables is also important.

1 Introduction

The discovery of a numeric law such as Kepler's third law $T = kr^{3/2}$ from a set of data is the central part of scientific discovery systems. After the pioneering work of the BACON systems [4, 5], several methods [7, 12, 13] have been proposed. The basic search strategy employed by these methods is much the same: two variables are recursively combined into a new variable by using multiplication, division, or some predefined prototype functions. These existing methods suffer from the following problems: first, since the combination must be done in order, a combinatorial explosion may occur when a complex law is sought for data with many variables. Second, when some powers in a law are not restricted to integers, the law may remain unknown unless some appropriate prototype functions such as $r^{3/2}$ are prepared in advance; however, a priori information is rarely available. Third, these methods are often criticized for their lack of noise tolerance since most real data contain noise [7, 12].

We believe a connectionist approach has great potential to solve the above problems. In order to directly learn a generalized polynomial form, a computational unit called a product unit has been proposed [2]. However, serious difficulties have been reported when using standard BP [9] to train networks containing these units [6]. Second-order methods employing nonlinear optimization techniques [3] such as quasi-Newton methods have been expected to overcome learning difficulties encountered by first-order methods such as BP. Thus, we have recently proposed a connectionist approach called RF5 [10] for discovering numeric laws, after inventing a second-order learning algorithm called BPQ [11]. Our previous experiments showed that RF5 worked well for relatively small problems each of which consists of at most hundreds of samples [10].

This paper evaluates how RF5 can be scaled up, and moreover analyses how it is invariant for the normalization of input and output variables. Since in many cases the size of real data is middle or large, it is highly important to know how a law discovery method can be scaled up. Moreover in most real data, different variables have typical values which may differ by orders of magnitude depending on the units for expression. Thus, it is also quite important to know how a method is invariant for the normalization of variables. After reviewing RF5 in Section 2, we evaluate the computational characteristics of RF5 in Section 3: the scalability and the invariance for variable normalization.

2   Law Discovery using Neural Networks

This section reviews our law discovery method called *RF5* (*Rule extraction from Facts version 5*), which employs the connectionist problem formalization, the BPQ algorithm, and the MDL criterion.

2.1   Problem Formalization

First a connectionist problem formalization for numeric law discovery [2] is explained. Let $\{(x_1, y_1), \cdots, (x_m, y_m)\}$ be a set of samples, where x_t is an n-dimensional input vector and y_t is a target value corresponding to x_t. In this paper, a class of numeric laws expressed as a generalized polynomial function

$$y_t = v_0 + \sum_{j=1}^{h} v_j \prod_{k=1}^{n} x_{tk}^{w_{jk}} \tag{1}$$

is considered, where each parameter v_j or w_{jk} is an unknown real number and h is an unknown integer. If a target law consists of a periodic or discontinuous function, we cannot exactly discover it when Eq. (1) is assumed. However, if the range of each input variable is bounded, such a law can be closely approximated to a generalized polynomial function with a finite number of terms. Hereafter, $(v_0, \cdots, v_h)^T$ and $(w_{j1}, \cdots, w_{jn})^T$ are expressed as v and w_j, respectively, where a^T means the transposed vector of a. In addition, a vector consisting of all parameters, $(v^T, w_1^T, \cdots, w_h^T)^T$ is simply expressed as Φ, and $N(= nh + h + 1)$ denotes the dimension of Φ.

Here we assume $x_{tk} > 0$; then, Eq. (1) is equivalent to

$$y_t = v_0 + \sum_{j=1}^{h} v_j \exp\left(\sum_{k=1}^{n} w_{jk} \ln x_{tk}\right). \tag{2}$$

Equation (2) can be regarded as the feedforward computation of a three-layer neural network where the activation function of each hidden unit is $\exp(s) = e^s$. Here h, w_j, and v denote the number of hidden units, the weights between the input units and hidden unit j, and the weights between the hidden units and the output unit respectively. Let the output value of hidden unit j for input x_t be

$$e_{tj} = \exp\left(\sum_{k=1}^{n} w_{jk} \ln x_{tk}\right),\tag{3}$$

and the output value of the output unit for x_t is described as

$$z(x_t; \Phi) = v_0 + \sum_{j=1}^{h} v_j e_{tj}.\tag{4}$$

A hidden unit defined by e_{tj} is called a *product unit* [2]. The discovery of numeric laws subject to Eq. (1) can thus be defined as a neural network learning problem to find the Φ that minimizes the sum-of-squares error function

$$f(\Phi) = \frac{1}{2} \sum_{t=1}^{m} (y_t - z(x_t; \Phi))^2.\tag{5}$$

2.2 BPQ Algorithm for Learning

In our early experiments and as reported in earlier studies [6], the problem of minimizing Eq. (5) turned out to be quite tough. Thus, in order to efficiently and constantly obtain good results, RF5 employs a new second-order learning algorithm called *BPQ* [11]; by adopting a quasi-Newton method [3] as a basic framework, the descent direction $\Delta\Phi$ is calculated on the basis of a partial BFGS update and a reasonably accurate step-length λ is efficiently calculated as the minimal point of a second-order approximation. In first-order learning algorithms which calculate the search direction as the gradient direction, a large number of iterations are often required until convergence. On the other hand, in existing second-order methods [3] which converge more quickly by using both gradient and curvature information, it is difficult to suitably scale up for large problems, and much computation is required for calculating the optimal step-length. BPQ can be reasonably scaled up by introducing a storage space parameter, and the computational complexity for calculating the optimal step-length is reasonably small, almost equivalent to that of gradient vector evaluation.

For the problem of minimizing Eq. (5), the partial BFGS update can be directly applied, while the basic procedure for calculating the optimal step-length λ must be slightly modified. In the step-length calculation, since λ is the single variable, we can express $f(\Phi + \lambda\Delta\Phi)$ simply as $\zeta(\lambda)$. Its second-order Taylor expansion is given as $\zeta(\lambda) \approx \zeta(0) + \zeta'(0)\lambda + \frac{1}{2}\zeta''(0)\lambda^2$. When $\zeta'(0) < 0$ and $\zeta''(0) > 0$, the minimal point of this approximation is given by

$$\lambda = -\frac{\zeta'(0)}{\zeta''(0)}.\tag{6}$$

Here, the method for coping with the other cases is exactly the same as described in [11]. For three-layer neural networks defined by Eq. (5), we can efficiently calculate both $\zeta'(0)$ and $\zeta''(0)$ as follows.

$$\zeta'(0) = -\sum_{t=1}^{m}(y_t - z(\boldsymbol{x}_t; \boldsymbol{\Phi}))z'(\boldsymbol{x}_t; \boldsymbol{\Phi}), \tag{7}$$

$$\zeta''(0) = \sum_{t=1}^{m}(z'(\boldsymbol{x}_t; \boldsymbol{\Phi})^2 - (y_t - z(\boldsymbol{x}_t; \boldsymbol{\Phi}))z''(\boldsymbol{x}_t; \boldsymbol{\Phi})). \tag{8}$$

2.3 Criterion for Model Selection

In general, for a given set of data, we cannot know the optimal number of hidden units in advance. Moreover, since the data is usually corrupted by noise, the law-candidate which minimizes Eq. (5) is not always the best one. We must thus consider a criterion to adequately evaluate the law-candidates discovered by changing the number of hidden units. In this paper, by assuming that the target output values are corrupted by Gaussian noise with a mean of 0 and an unknown standard deviation of σ, finding an adequate number of hidden units is formalized as a model selection problem of the maximum likelihood estimation problem. Thus, we adopt the MDL (Minimum Description Length) criterion [8] for this purpose. The MDL fitness value is defined by

$$\text{MDL} = \frac{m}{2}\log(\text{MSE}) + \frac{N}{2}\log m, \tag{9}$$

where MSE represents the value of the mean squared error defined by

$$\text{MSE} = \frac{1}{m}\sum_{t=1}^{m}(y_t - z(\boldsymbol{x}_t; \hat{\boldsymbol{\Phi}}))^2. \tag{10}$$

Here, $\hat{\boldsymbol{\Phi}}$ is a set of weights which minimizes Eq. (5), N is the number of parameters in $\boldsymbol{\Phi}$, and m is the number of samples.

3 Computational Characteristics of RF5

3.1 Scalability

The law discovery method RF5 was evaluated by using the following law, a modified version of Sutton and Matheus [13].

$$y = 2 + 3x_1^{-1}x_2^3 + 4x_3x_4^{1/2}x_5^{-1/3}. \tag{11}$$

In this problem, the total number of variables is 9 ($n = 9$), which means there exist irrelevant variables $x_6, \cdots, x_9$. Each sample is generated as follows: each value of variables $x_1, \cdots, x_9$ is randomly generated in the range of $(0, 1)$, and the corresponding value of y is calculated using Eq. (11). To evaluate RF5's scalability, the number of samples ranges from small to large ($m = 200$, $2K$, $20K$, $100K$). Moreover, to evaluate RF5's noise tolerance, we corrupted each value of y by adding Gaussian noise with a mean of 0 and a standard deviation of 0.1. The experiments were done on DEC alpha/600/333 workstation.

Remember RF5 employs a generalized polynomial model of Eq. (1) to fit data. As a law of Eq. (1), whether a constant term v_0 exists or not is far from trivial. For example, Kepler's third law $T = kr^{3/2}$ has no constant term, while Hagen-Rubens' law $R = 1 - 2(\mu/\sigma)^{0.5}$ has a constant term. Since usually we do not know such knowledge in advance, we have to try both with-v_0 and without-v_0 models; thus, in the experiments we tried both models. Note that for the without-v_0 model, the number of parameters decreases by one: $N = nh + h$.

In the experiments, we changed the number of hidden units from 1 to 3 ($h = 1, 2, 3$) and performed 10 trials for each of them. The initial values for w_j were independently generated according to a Gaussian distribution with a mean of 0 and a standard deviation of 1; the initial values for v_j were set to 0, but v_0 was initially set to the average output value of all training samples. The iteration was terminated when any of the following three conditions was met: the MSE value was sufficiently small, i.e.,

$$\frac{1}{m} \sum_{t=1}^{m} (y_t - z(\boldsymbol{x}_t; \boldsymbol{\Phi}))^2 < 10^{-8}, \tag{12}$$

the gradient vector was sufficiently small, i.e.,

$$\frac{1}{N} \|\nabla f(\boldsymbol{\Phi})\|^2 < 10^{-8}, \tag{13}$$

or the total CPU time exceeded T_{max} seconds, where T_{max} was changed according to the number of samples m: $T_{max} = 100$, 100, 300, 1000.

Table 1 shows the averages of the final MSE and MDL values, the numbers of iterations, and CPU times for both with-v_0 and without-v_0 models. In each model for any scale of data, the final MSE value was minimized when $h = 3$ although the difference between $h = 2$ and $h = 3$ was frequently very small. Note that for any scale of data, the final MDL value was minimized when $h = 2$ in the with-v_0 model, which means the correct number of hidden units $h = 2$ and the correct with-v_0 model were successfully found for any scale of data. When $h = 2$ in the with-v_0 model, all 10 trials converged to the global minimum for any scale of data. The law discovered by RF5 for $100K$ samples was

$$y = 2.000 + 3.001 x_1^{-1.000} x_2^{3.000} + 3.998 x_3^{1.000} x_4^{0.500} x_5^{-0.333}, \tag{14}$$

where the weight values were rounded off to the third decimal place. The law discovered by RF5 was almost identical to Eq. (11). This shows that RF5 is robust and noise tolerant to some degree for the above scale of data. Note that without preparing some appropriate prototype functions, existing numeric discovery methods cannot find such laws as described in Eq. (11). This point is an important advantage of RF5 over existing methods.

Table 1. Learning results for noisy data with a constant term

num. of samples	const. term	hidden units	final MSE value	final MDL value	num. of iterations	CPU time (sec)
	with	$h = 1$	2.783	131.5	203	0.1
	with	$h = 2$	0.010	-404.2	234	0.2
$m = 200$	with	$h = 3$	0.009	-388.0	1908	2.3
	without	$h = 1$	9.444	251.0	130	0.1
	without	$h = 2$	0.166	-126.4	143	0.1
	without	$h = 3$	0.010	-384.2	282	0.3
	with	$h = 1$	2.912	1110.6	1493	49.6
	with	$h = 2$	0.010	-4523.0	418	4.1
$m = 2K$	with	$h = 3$	0.010	-4490.2	469	5.8
	without	$h = 1$	14.077	2682.6	209	1.3
	without	$h = 2$	0.263	-1261.4	233	2.2
	without	$h = 3$	0.010	-4490.9	291	3.6
	with	$h = 1$	2.867	10588.9	770	300.5
	with	$h = 2$	0.010	-46000.1	230	45.8
$m = 20K$	with	$h = 3$	0.010	-45960.0	2158	290.5
	without	$h = 1$	12.422	25244.3	573	185.5
	without	$h = 2$	0.208	-15592.8	541	219.0
	without	$h = 3$	0.010	-45961.6	264	51.3
	with	$h = 1$	2.711	49932.6	659	1004.8
	with	$h = 2$	0.010	-230344.9	342	753.9
$m = 100K$	with	$h = 3$	0.010	-230295.6	1026	1004.7
	without	$h = 1$	12.147	124910.0	520	1005.3
	without	$h = 2$	0.204	-79269.5	389	1005.4
	without	$h = 3$	0.010	-230295.3	425	845.3

Figure 1 shows how the CPU time until convergence grows as the number of samples m increases for the case of $h = 2$ in the with-v_0 model. The figure indicates the CPU time grows in proportion to m, which means RF5 has very good scalability. The linear scalability is not so surprising because the complexity of BPQ is $nhm + O(Nm)$ [11], which is dominant in the complexity of RF5. Note that too simple models and too redundant models require more CPU time than the best model, as is shown in Table 1. However, the linear scalability still holds roughly. Thus, it was shown that RF5 has the desirable linear scalability for the growth of the number of samples.

Additionally, RF5 was evaluated by using the following law without a constant term.

$$y = 3x_1^{-1}x_2^3 + 4x_3x_4^{1/2}x_5^{-1/3}. \tag{15}$$

It is expected RF5 will select the without-v_0 model having $h = 2$ as the best model among all models generated by the combination of with/without-v_0 and $h = 1, 2, 3$. The conditions for the experiment were the same as the above. Table 2 shows the averages for $2K$ samples over 10 runs. The final MDL value was minimized when $h = 2$ in the without-v_0 model, which means the correct number of hidden units $h = 2$ and the correct without-v_0 model were successfully found. The law discovered by RF5 for $100K$ samples was

$$y = 2.999x_1^{-1.000}x_2^{3.000} + 4.000x_3^{1.000}x_4^{0.500}x_5^{-0.333}, \tag{16}$$

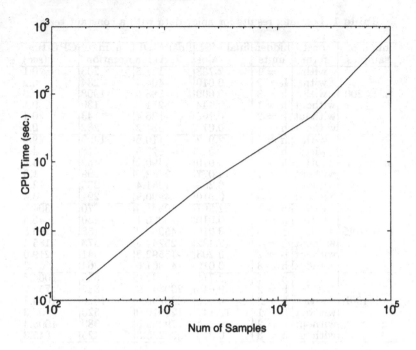

Fig. 1. Scalability of RF5 for noisy data with a constant model ($h = 2$)

where the weight values were rounded off to the third decimal place. The discovered law was almost identical to Eq. (15).

Table 2. Learning results for noisy data without a constant term

num. of samples	const. term	hidden units	final MSE value	final MDL value	num. of iterations	CPU time (sec)
	with	h = 1	2.558	981.0	542	19.2
	with	h = 2	0.010	-4529.5	243	2.2
m = 2K	with	h = 3	0.010	-4501.7	3851	46.7
	without	h = 1	4.988	1645.1	382	12.2
	without	h = 2	0.010	-4532.2	146	1.3
	without	h = 3	0.010	-4504.0	2442	28.5

3.2   Invariance for Variable Normalization

Many real data may consist of variables whose typical values differ by several orders of maginitude depending on the units for expression. Thus, it is quite important to know how RF5 is influenced by the normalization of variables. One of the most common forms of normalization is a linear transformation of input and output variables [1]. Since input variables are subject to the logarithmic transformation, we consider scaling transformations for them. Thus, we have the following normalization,

$$\tilde{x}_k = a_k x_k, \quad \tilde{y} = cy + d, \tag{17}$$

where we assume $a_k \neq 0$ and $c \neq 0$ to exclude meaningless normalization. Since the generalized polynomial modeling Eq. (1) of RF5 is nonlinear, it is not obvious whether RF5 is invariant for the above normalization. Here a modeling or method is called invariant when it shows the same generalization[1] capability even after the normalization. The optimal weights before normalization, v^* and w_j^*, satisfy the following:

$$\frac{\partial f}{\partial v_0} = -\sum_{t=1}^{m}(y_t - v_0^* - \sum_{j=1}^{h} v_j^* e_{tj}^*) = 0 \tag{18}$$

$$\frac{\partial f}{\partial v_j} = -\sum_{t=1}^{m}(y_t - v_0^* - \sum_{j'=1}^{h} v_{j'}^* e_{tj'}^*)e_{tj}^* = 0 \tag{19}$$

$$\frac{\partial f}{\partial w_{jk}} = -\sum_{t=1}^{m}(y_t - v_0^* - \sum_{j'=1}^{h} v_{j'}^* e_{tj'}^*)v_j^* e_{tj}^* \ln x_{tk} = 0, \tag{20}$$

where $e_{tj}^* = \exp(\sum_k w_{jk}^* \ln x_{tk})$. The optimal weights after normalization, $\tilde{v}^*$ and $\tilde{w}_j^*$, satisfy the following:

$$\frac{\partial f}{\partial \tilde{v}_0} = -\sum_{t=1}^{m}(\tilde{y}_t - \tilde{v}_0^* - \sum_{j=1}^{h} \tilde{v}_j^* \tilde{e}_{tj}^*) = 0 \tag{21}$$

$$\frac{\partial f}{\partial \tilde{v}_j} = -\sum_{t=1}^{m}(\tilde{y}_t - \tilde{v}_0^* - \sum_{j'=1}^{h} \tilde{v}_{j'}^* \tilde{e}_{tj'}^*)\tilde{e}_{tj}^* = 0 \tag{22}$$

$$\frac{\partial f}{\partial \tilde{w}_{jk}} = -\sum_{t=1}^{m}(\tilde{y}_t - \tilde{v}_0^* - \sum_{j'=1}^{h} \tilde{v}_{j'}^* \tilde{e}_{tj'}^*)\tilde{v}_j^* \tilde{e}_{tj}^* \ln \tilde{x}_{tk} = 0. \tag{23}$$

It is easily seen that if the weights before and after normalization satisfy the following linear relationship, the weights satisfy the above equations.

$$\tilde{v}_0^* = cv_0^* + d, \quad \tilde{v}_j^* = \frac{c}{q_j^*} v_j^*, \quad \tilde{w}_j^* = w_j^*, \tag{24}$$

where $q_j^* = \exp\left(\sum_k w_{jk}^* \ln a_k\right)$ and $\tilde{e}_{tj}^* = e_{tj}^* q_j^*$. Note that the input normalization $\tilde{x}_k = a_k x_k$ is absorbed by the weights $\{v_j^*\}$, the output normalization $\tilde{y} = cy + d$ is absorbed by the weights v_0^* and $\{v_j^*\}$, and the weights $\{w_j^*\}$ keep

---

[1] The goal of training is not to learn an exact representation of the training data itself, but rather to build a statistical model of the process which generates the data. How a trained model approximates to the true model is called *generalization* [1].

unchanged. Thus, the above normalization only multiplies the generalized polynomial function by c, and does not influence the generalization capability of RF5.

To confirm the above analysis, by using Eq. (11) and the normalization $\tilde{x}_k = 10x_k$, $\tilde{y} = 3y + 5$, the same experiments for $m = 100K$ were performed, except enlarged Gaussian noise with a mean of 0 and a standard deviation of 0.3. The law selected by RF5 was

$$y = 10.999 + 0.090x_1^{-1.000}x_2^{3.000} + 0.817x_3^{1.000}x_4^{0.500}x_5^{-0.333}, \qquad (25)$$

where the weight values were rounded off to the third decimal place. By following Eqs. (24) and (14), we have $\tilde{v}_0^* = 3 \times 2.000 + 5 = 11.000$, $q_j^* = 10^{(\sum_k w_{jk}^*)}$, $q_1^* = 10^{2.000} = 100$, $\tilde{v}_1^* = 3 \times 3.001/100 = 0.090$, $q_2^* = 10^{1.167} = 14.69$, $\tilde{v}_2^* = 3 \times 3.998/14.69 = 0.816$; these match Eq. (25) very well. Thus, it was shown that the normalization of Eq. (17) can be absorbed by the linear weight transformation of Eq. (24) and does not influence the generalization capability of RF5.

4   Conclusion

This paper evaluated how a law discovery method called RF5 can be scaled up, and analysed how it is invariant for the normalization of input and output variables. It was shown that RF5 has a linear scalability for the growth of data size, and is invariant for scaling transformations of input variables and/or a linear transformation of an output variable. In the future, we plan to do further experiments to evaluate RF5 using a wider variety of real problems.

References

1. C.M. Bishop. *Neural networks for pattern recognition*. Clarendon Press, Oxford, 1995.
2. R. Durbin and D. Rumelhart. Product units: a computationally powerful and biologically plausible extension. *Neural Computation*, 1(1):133–142, 1989.
3. P.E. Gill, W. Murray, and M.H. Wright. *Practical optimization*. Academic Press, 1981.
4. P. Langley. Bacon.1: a general discovery system. In *Proc. 2nd National Conference of the Canadian Society for Computational Studies of Intelligence*, pages 173–180, 1978.
5. P. Langley, H.A. Simon, G. Bradshaw, and J. Zytkow. *Scientific discovery: computational explorations of the creative process*. MIT Press, 1987.
6. L.R. Leerink, C.L. Giles, B.G. Horne, and M.A. Jabri. Learning with product units. In *Advances in Neural Information Processing Systems 7*, pages 537–544, 1995.
7. B. Nordhausen and P. Langley. A robust approach to numeric discovery. In *Proc. 7th Int. Conf. on Machine Learning*, pages 411–418, 1990.
8. J. Rissanen. *Stochatic complexity in statistical inquiry*. World Scientific, 1989.

9. D.E. Rumelhart, G.E. Hinton, and R.J. Williams. Learning internal representations by error propagation. In *Parallel Distributed Processing, Vol.1*, pages 318–362. MIT Press, 1986.

10. K. Saito and R. Nakano. Law discovery using neural networks. In *Proc. 15th International Joint Conference on Artificial Intelligence*, pages 1078–1083, 1997.

11. K. Saito and R. Nakano. Partial BFGS update and efficient step-length calculation for three-layer neural networks. *Neural Computation*, 9(1):239–257, 1997.

12. C. Schaffer. Bivariate scientific function finding in a sampled, real-data testbed. *Machine Learning*, 12(1/2/3):167–183, 1993.

13. R.S. Sutton and C.J. Matheus. Learning polynomial functions by feature construction. In *Proc. 8th Int. Conf. on Machine Learning*, pages 208–212, 1991.

Development of SDS2: Smart Discovery System for Simultaneous Equation Systems

Takashi Washio and Hiroshi Motoda

Institute of Scientific and Industrial Research, Osaka University
8-1 Mihogaoka, Ibaraki, Osaka, 567, Japan
{washio,motoda}@sanken.osaka-u.ac.jp

Abstract. SDS2 is a system to discover and identify the quantitative model consisting of simultaneous equations reflecting the first principles underlying the objective process through experiments. It consists of SSF and SDS, where the former is to discover the structure of the simultaneous equations and the latter to discover a quantitative formula of each complete equation. The power of SDS2 comes from the use of the complete subset structure in a set of simultaneous equations, the scale-types of the measurement data and the mathematical property of identity by which to constrain the admissible solutions. The basic principles, algorithms and implementation of SDS2 are described, and its efficiency and practicality are demonstrated and discussed with large scale working examples.

1 Introduction

Number of methods have been proposed to discover quantitative formulae of scientific laws from experimental measurements. Langley and others' BACON systems [1] are the most well known as a pioneering work. FAHRENHEIT [2] and ABACUS [3] are such successors that use basically similar algorithms to BACON in searching for a complete equation governing the measured data. Major drawbacks of the BACON family are their complexity in the search of equation formulae and the considerable amount of ambiguity in their results for noisy data [4][5]. To alleviate these difficulties, some later systems, e.g. ABACUS and COPER [6], utilize the information of the unit dimension of quantities to prune the meaningless terms. However, their applicability is limited only to the case where the quantity dimension is known.

Another difficulty of the conventional systems to discover a model of practical and large scale process is that such process is represented by multiple equations. Some of the aforementioned systems such as FAHRENHEIT and ABACUS can identify each operation mode of the objective process and derive an equation to represent each mode. However, many processes such as large scale electric circuits are represented by simultaneous equations. The model representation in form of simultaneous equations is essential to grasp the dependency structure among the multiple mechanisms in the processes [7][8]. An effort to develop a system called LAGRANGE has been made to automatically discover dynamical models represented by simultaneous equations[9]. However, it derives many redundant models in high computational complexity while the soundness of the solutions is not guaranteed.

The objective of this study is to develop a new scientific discovery system named "*SDS2 (Smart Discovery System 2)*" which is an extended version of our previous system "*SDS*" [10]. It overcomes the drawbacks of the conventional scientific discovery systems. The main extenstion of SDS2 is to discover a complex model of an objective process represented by a set of simultaneous equations from measured quantities in experiments. SDS2 has been developed based on some mathematical principles established in our past research[10][11][12]. It consists of "*SSF (Simultaneous System Finder)*" and the main body of SDS, where the former is to discover the structure of the simultaneous equations and the latter to discover a quantitative formula of each complete equation.

2 Outline of Principle and Algorithm of SSF

We set two assumptions on the objective process. One is that the objective process can be represented by a set of quantitative, continuous, complete and under-constrained simultaneous equations for the quantity ranges of our interest. Another is that all of the quantities in every equation can be measured, and all of the quantities except one dependent quantity can be controlled in every equation to their arbitrary values in the range under experiments while satisfying the constraints of the other equations. These assumptions are common in the past BACON family except the features associated with the simultaneous equations.

SSF requires a list of the quantities for the modeling of the objective process and their actual measurements. Starting from the set of control quantities having small cardinality, SSF experimentally tests if values of any quantities become to be fully under control. If such controlled quantities are found, the collection of the control quantities and the controlled quantities are considered to represent a mechanism determining the state of the objective process. The set of the control quantities and the controlled quantities is called a "*complete subset*". It reflects the dependency structure of the quantities in the simultaneous equations of the objective process as explained later in this section. SSF applies further experimental tests to derive new complete subsets while efficiently focusing the search space based on the complete subsets derived in the previous tests. After deriving all complete subsets, SSF reconstructs the simultaneous dependency structure of the quantities embeded in the objective process.

The principles of SSF are explained through a simple electric circuit consisting of two parallel resistances and a battery depicted in Fig. 1. One way of modeling this is by

$$V_1 = I_1 R_1 \ [1], V_2 = I_2 R_2 \ [2], V_e = V_1 \ [3] \text{ and } V_e = V_2 \ [4], \tag{1}$$

where R_1, R_2:two resistances, V_1, V_2:voltage differences across the resistances, I_1, I_2:electric current going through the resistances and V_e:voltage of the battery.

The same circuit can be modeled by another set of equations.

$$I_1 R_1 = I_2 R_2 \ [1], V_2 = I_2 R_2 \ [2], V_e = V_1 \ [3] \text{ and } V_e = V_2 \ [4]. \tag{2}$$

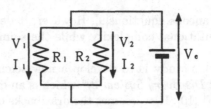

Fig. 1. An circuit of parallel resistances.

Both representations give correct behaviors of the circuit. The configuration of
the quantities in a set of simultaneous equations is represented by an *"incidence
matrix"* T where its rows $E = \{eq_i | i = 1, ...M\}$ correspond to the mutually
independent equations and its columns $Q = \{q_j | j = 1, ..., N\}$ to the quantities.
If the j-th quantity appears in the i-th equation, then the (i, j) element of T,
i.e., $T_{ij} = 1$, and otherwise $T_{ij} = 0$ [8]. When a subset consisting of n independent equations containing n undetermined quantities are obtained by exogenously
specifying the values of some extra quantities in the under-constrained simultaneous equations, the values of those n quantities are determined by solving the
equations in the subset. In terms of an incidence matrix, exogenous specification
of a quantity value corresponds to eliminating the column of the quantity. Under
this consideration, the following definition is introduced.

Definition 1 (complete subset). *Given an incidence matrix T, after applying elimination of a set of columns, $RQ(\subset Q)$, let a set of nonzero columns of
$T[CE, Q - RQ]$ be $NQ(\subseteq Q - RQ)$, where $CE \subseteq E$, and $T[CE, Q - RQ]$ is
a sub-incidence matrix for equations in CE and quantities in $Q - RQ$. CE is
called a "complete subset" of order n, if $|CE| = |NQ| = n$. Here, $| \bullet |$ stands for
the cardinality of a set.*

If we exogenously specify the values of V_e and R_1, the first, the third and the
forth rows of the matrix for Eq.(1) come to contain the three nonzero columns
of V_1, V_2 and I_1. Thus these equations form a complete subset of order 3, and
the three quantities are determined while the others, I_2 and R_2, are not. On
the other hand, if the identical specification on V_e and R_1 is made in the latter
model, no complete subset of order 3 is obtained, since every combination of
three rows in the matrix for Eq.(2) contains more than three nonzero columns.
In the real electric circuit, the validity of the consequence derived by the former
model is clear. The model having the incidence matrix which always derives
a valid interpretation in determining the quantities of an objective process is
named *"structural form"* in this paper.

　　The relation between the two models having an identical complete subset is
characterized by the following theorem.

Theorem 1 (invariance theorem). *Given a transform $f : U_E \to U_E$ where
U_E is the entire universe of equations. When CE is a complete subset of order n in T, $f(CE)$ is also a complete subset of order n, if $f(CE)$ for $CE \subset
U_E$ maintains the number of equations and the nonzero column structure, i.e.,*

$|CE| = |f(CE)|$ and $CQ = CQ_f$, where CQ_f is a set of nonzero columns in $T[f(CE), Q]$.

Various simultaneous equation formulae maintaining the equivalence of the quantitative relations and the dependency structure can be derived by limiting the transformation f to a quantitative one satisfying the *"invariance theorem"* such as substitution and arithmetic operation among equations. Our approach identifies only one specific form defined bellow.

Definition 2 (canonical form of a complete subset). *Given a complete subset* CE *of order* n, *the "canonical form" of* CE *is the form where all elements of the nonzero columns* CQ *in its incidence matrix* $T[CE, Q]$ *are* 1.

Because every admissible form is equivalent to every other, the identification of the canonical form is sufficient, and every other can be derived by applying each appropriate f to the form.

Though each complete subset represents a basic mechanism to determine the values of quantities in given simultaneous equations, some complete subsets are not mutually independent. For instance, the following four complete subsets can be found in the example of Eqs.(1).

$$\{[3], [4]\}(n = 2), \ \{[1], [3], [4]\}(n = 3), \ \{[2], [3], [4]\}(n = 3), \ \{[1], [2], [3], [4]\}(n = 4) \ (3)$$

The number in [ ] indicates each equation and n the order of the subset. They have many overlaps, and the complete subsets having higher orders represent the redundant mechanism with the lower subsets. Thus, the following definitions are introduced to decompose the internal structure of a complete subset.

Definition 3 (independent component of a complete subset). *The independent component* DE_i *of the complete subset* CE_i *is defined as*

$$DE_i = CE_i - \bigcup_{\substack{\forall CE_j \subset CE_i \\ \text{and } CE_j \in L}} CE_j.$$

The set of essential quantities DQ_i *of* CE_i *which do not belong to any other smaller complete subsets but are involved only in* CE_i *is also defined as*

$$DQ_i = CQ_i - \bigcup_{\substack{\forall CE_j \subset CE_i \\ \text{and } CE_j \in L}} CQ_j,$$

where CQ_i *is a set of nonzero columns of* $T(CE_i, Q)$.
The order δn_i *and the degree of freedom* δm_i *of* DE_i *are defined as*

$$\delta n_i = |DE_i| \text{ and } \delta m_i = |DQ_i| - |DE_i|.$$

In the example of Eq.(3), the three independent components are derived.

$$DE_1 = \{[3], [4]\} - \phi = \{[3], [4]\}, \quad \delta n_1 = 2 - 0 = 2,$$
$$DE_2 = \{[1], [3], [4]\} - \{[3], [4]\} = \{[1]\}, \quad \delta n_2 = 3 - 2 = 1,$$
$$DE_3 = \{[2], [3], [4]\} - \{[3], [4]\} = \{[2]\}, \quad \delta n_3 = 3 - 2 = 1. \quad (4)$$

Because each independent component DE_i is a subset of the complete subset CE_i, the nonzero column structure of DE_i also follows the invariance theorem. Consequently, the subset of the canonical from of CE_i is applicable to represent DE_i. Based on this consideration, the definition of the canonical form of the simultaneous equations is introduced.

Definition 4 (canonical form of simultaneous equations). *The "canonical form" of a set of simultaneous equations consists of the equations in $\cup_{i=1}^{b} DE_i$ where each equation in DE_i is represented by the canonical form in the complete subset CE_i, where b is the total number of DE_i.*

If the canonical form of simultaneous equations are experimentally derived to reflect the actual dependency structure among quantities, then the model must be a "*structural form*". Thus, the following terminology is introduced.

Definition 5 (structural canonical form). *If the canonical form of simultaneous equations is derived to be a "structural form", then the form is named "structural canonical form".*

Under our aforementioned assumption on the measurements and the controllability of quantities, a bottom up algorithm described in Fig. 2 has been developed and implemented into SSF. Starting from the set of control quantities C_{hi} having small cardinality, this algorithm applies the experiments to check if any quantities become to be controled. If the set of controlled quantities D_{hi} is found, the union of C_{hi} and D_{hi} is considered as a newly found complete subset CE_i. Then, based on the definition3, its DE_i, DQ_i, δn_i and δm_i are derived and stored. Once any new independent component is derived, only δm_i of the quantities in every DQ_i and the quantities which do not belong to any DQ_i so far found are used for control. The constraint of DQ_i does not miss any complete subset to search due to the monotonic lattice structure among complete subsets.

3   Outline of Principles and Algorithm of SDS

SDS uses the information of scale-types of quantities to discover the formula of each equation. The quantitative scale-types are interval, ratio and absolute scales[13]. Examples of the interval scale quantities are temperature in Celsius and sound tone where the origins of their scales are not absolute, and are changeable by human's definitions. Examples of the ratio scale quantities are physical mass and absolute temperature where each has an absolute origin. The absolute scale quantities are dimensionless quantities.

The properties of the quantities in terms of the scale-types yields "*scale-type constraint*" characterized by the following theorems [10].

Theorem 2 (Extended Buckingham Π-theorem). *If $\phi(x_1, x_2, x_3, ..., x_n) = 0$ is a complete equation, and if each argument is one of interval, ratio and absolute scale-types, then the solution can be written in the form*

$$F(\Pi_1, \Pi_2, ..., \Pi_{n-w}) = 0,$$

Input: A set of quantities $Q = \{q_k | k = 1, ..., N\}$ that should appear in the model of an objective process and $X = \{x_k | x_k = q_k,$ for all but directly controllable $q_k \in Q\}$.

Output: The sets of quantities in independent components DE, the sets of essential quantities in independent DQ and the orders of independent components N.

(S1) $DE = \phi$, $DQ = \phi$, $N = \phi$, $M = \phi$, $h = 1$ and $i = 1$.

(S2) *Choose $C_j \subset DQ_j \in DQ$ for some DQ_j and also $C_x \subseteq X$, and take their union $C_{hi} = ... \cup C_j \cup ... \cup C_x$ while maintaining $|C_j| \leq \delta m_j$ and $|C_{hi}| = h$. Control all $x_k \in C_{hi}, k = 1, ..., |C_{hi}|$ in the experiment.*

(S3) *Let a set of all quantities whose values are determined be $D_{hi} \subseteq (Q - C_{hi})$ where $D_{hi} \neq \phi$. Set $DE_{hi} = C_{hi} + D_{hi}$, $DQ_{hi} = DE_{hi} - \cup_{\substack{\forall DE_{h'i'} \subset DE_{hi} \\ DE_{h'i'} \in DE}} DE_{h'i'}$, $\delta n_{hi} = |D_{hi}| - \sum_{\substack{\forall DE_{h'i'} \subset DE_{hi} \\ DE_{h'i'} \in DE}} \delta n_{h'i'}$, and $\delta m_{hi} = |DQ_{hi}| - \delta n_{hi}$. If $\delta n_{hi} > 0$, then add DE_{hi} to the list DE, DQ_{hi} to the list DQ, δn_{hi} to the list N, δm_{hi} to the list M and $X = X - DQ_{hi}$.*

(S4) *If all quantities are determined, i.e., $D_{hi} = Q - C_{hi}$, then finish, else if no more C_{hi} where $|C_{hi}| = h$ exists, set $h = h + 1, i = 1$ and go to (S2), else set $i = i + 1$ and go to (S2).*

Fig. 2. Algorithm for finding structural canonical form

where n is the number of arguments of ϕ, w is the basic number of bases in $x_1, x_2, x_3, .., x_n$, respectively. For all i, Π_i is an absolute scale-type quantity.

Bases are such basic factors independent of the other bases in the given ϕ, for instance, as length $[L]$, mass $[M]$, time $[T]$ of physical unit and the origin of temperature in Celsius.

Theorem 3 (Extended Product Theorem). *Assuming primary quantities in a set R are ratio scale-type, and those in another set I are interval scale-type, the function ρ relating $x_i \in R \cup I$ to a secondary quantity Π is constrained to one of the following two:*

$$\Pi = (\prod_{x_i \in R} |x_i|^{a_i})(\prod_{I_k \subseteq I} (\sum_{x_j \in I_k} b_{kj}|x_j| + c_k)^{a_k})$$

$$\Pi = \sum_{x_i \in R} a_i \log |x_i| + \sum_{I_k \subseteq I} a_k \log(\sum_{x_j \in I_k} b_{kj}|x_j| + c_k) + \sum_{x_\ell \in I_g \subseteq I} b_{g\ell}|x_\ell| + c_g$$

where all coefficients except Π are constants and $I_k \cap I_g = \phi$.

These theorems state that any meaningful complete equation consisting only of the arguments of interval, ratio and absolute scale-types can be decomposed into an equation of absolute scale-type quantities having an arbitrary form and equations of interval and ratio scale-type quantities having specific forms. The former $F(\Pi_1, \Pi_2, ..., \Pi_{n-w}) = 0$ is called an "ensemble" and the latter $\Pi = \rho(x_1, x_2, x_3, ..., x_n)$ "regime"s.

Another constraint named "*identity constraint*" is also used to narrow down the candidate formulae [10]. The basic principle of the identity constraints comes by answering the question that "*what is the relation among Θ_h, Θ_i and Θ_j, if*

$\Theta_i = f_{\Theta_j}(\Theta_h)$ and $\Theta_j = f_{\Theta_i}(\Theta_h)$ are known?" For example, if $a(\Theta_j)\Theta_h + \Theta_i = b(\Theta_j)$ and $a(\Theta_i)\Theta_h + \Theta_j = b(\Theta_i)$ are given, the following identity equation is obtained by solving each for Θ_h.

$$\Theta_h \equiv -\frac{\Theta_i}{a(\Theta_j)} + \frac{b(\Theta_j)}{a(\Theta_j)} \equiv -\frac{\Theta_j}{a(\Theta_i)} + \frac{b(\Theta_i)}{a(\Theta_i)}$$

It is easy to prove that the admissible relation among the three is as follows.

$$\Theta_h + \alpha_1\Theta_i\Theta_j + \beta_1\Theta_i + \alpha_2\Theta_j + \beta_2 = 0 \tag{5}$$

The algorithm of SDS is outlined in Fig. 3. In (S1-1), SDS searches bi-variate relations having the linear form in IQ through data fitting. Similar bi-variate equation fitting to the data is applied in (S2-1) and (S2-3) where the admissible equations are power product form and logarithmic form respectively. If this test is passed, the pair of quantities is judged to have the admissible relation of "*Extended Product Theorem.*"

In (S1-2), triplet consistency tests are applied to every triplet of equations in IE. Given a triplet of the linear form equations in IE,

$$\bar{a}_{xy}x + y = b_{xy}, \bar{a}_{yz}y + z = b_{yz}, \bar{a}_{xz}x + z = b_{xz}, \tag{6}$$

the following condition must be met for the three equations to be consistent.

$$\bar{a}_{xz} = -\bar{a}_{yz}\bar{a}_{xy} \tag{7}$$

This condition is used to check if the triplet of quantities belong to an identical regime. SDS searches every maximal convex set MCS where each triplet of equations among the quantities in this set has passed the test. The similar test is applied to the quantities in RE in (S2-2).

Once all regimes are identified, new terms are generated in (S3-1) by merging these regimes in preparation to compose the ensemble equation. SDS searches bi-variate relations between two regimes Πs having one of the formulae specified in the equation set CE. The repertoire in CE governs the ability of the equation formulae search in SDS. Currently, only the two simple formulae of power product form and linear form are given in CE. Nevertheless, SDS performs very well in search for the ensemble equation. When one of the relations specified in CE is found, the pair of the regime Πs is merged into a new term. This procedure is repeated in couple for both product and linear forms until no new term generation becomes possible.

In (S3-2), the identity constraints are applied for further merging terms. The bi-variate least square fitting of the identity constraint such as Eq.(5) is applied to AQ. If all the coefficients except one are independent in a relation, the relation is solved for the unique dependent coefficient, and the coefficient is set to be the merged term of the relation. If all coefficients are independent in a relation, the relation is the ensemble equation. If such ensemble equation is not found, SDS goes back to the (S3-1) for further search.

4   Implementation of SDS2

The major function of SFF is to derive the structural canonical form of the simultaneous equations representing an objective process. However, SDS to discover a complete equation can not directly accept the knowledge of the structural canonical form for the discovery. Accordingly, some additional process to provide information acceptable for SDS is required to consistently implement the two parts into SDS2. First, the problem to derive quantitative knowledge of the

Input: A set of interval scale quantities IQ, a set of ratio scale quantities RQ and a set of absolute scale quantities AQ.

Output: The candidate models of the objective system AQ.

(S1-1) Apply bi-variate test for an admissible linear equation of interval scale to every pair of quantities in IQ. Store the resultant bi-variate equations accepted by the tests into an equation set IE and the others not accepted into an equation set NIE.

(S1-2) Apply triplet test to every triplet of associated bi-variate equations in IE. Derive all maximal convex sets MCSs for the accepted triplets, and compose all bi-variate equations into a multi-variate equation in each MCS. Define each multi-variate equation as a term. Replace the merged terms by the generated terms of the multi-variate equations in IQ. Let RQ = RQ + IQ.

(S2-1) Apply bi-variate test for an admissible equation of ratio scale to every pair of quantities in RQ. Store the resultant bi-variate equations accepted by the tests into an equation set RE and the others not accepted into an equation set NRE.

(S2-2) Apply triplet test to every triplet of associated bi-variate equations in RE. Derive all maximal convex sets for the accepted triplets, and compose all bi-variate equations into a multi-variate equation in each maximal convex set. Define each multi-variate equation as a term. Replace the merged quantities by the generated terms in RQ.

(S2-3) Apply bi-variate test for an admissible logarithmic equation between the linear forms of interval scale-type quantities and the other terms in RQ. Replace the terms in the resultant bi-variate equations accepted in the tests by the generated terms in RQ.

(S3) Let AQ = AQ + RQ. Given candidate formulae set CE, repeat steps (3-1) and (3-2) until no more new term become generated.

(S3-1) Apply bi-variate test of a formula in CE to every pair of the terms in AQ, and store them to AE. Merge every group of terms into a unique term respectively based on the result of the bi-variate test, if this is possible. Replace the merged terms with the generated terms of multi-variate equations in AQ.

(S3-2) Apply identity constraints test to every bi-variate equation in AE. Merge every group of terms into a unique term respectively based on the result of the identity constraints test, if they are possible. Replace the merged terms with the generated terms of multi-variate equations in AQ. Go back to step (2-1).

Fig. 3. Outline of SDS algorithm

simultaneous equations must be decomposed into subproblems to derive each equation individually. For the purpose, an algorithm to decompose the entire problem into such small problems is implemented. The values of the quantities within each independent component DE_i of a complete subset CE_i are mutually constrained, and have the order $\delta m_i = |DQ_i| - \delta n_i$ degree of freedom. Accordingly, the constraints within the independent component disable the bi-variate tests among the quantities of an equation in the structural canonical form, if the order δn_i is more than one. However, this difficulty is removed if the $(\delta n_i - 1)$ quantities are eliminated by the substitutiion of the other $(\delta n_i - 1)$ equations within the independent component. The reduction of the number of quantities by $(\delta n_i - 1)$ in each equation enables to control each quantities as if it is in a complete equation. This elimination of quantities is essential to enable the application of SDS which uses the bi-variate test. The reduction of quantities in equations provides further advantage, since the required amount of computaion in the equation search depends on the number of quantities. In addition, the smaller degree of freedom of the objective equation in the search introduces more robustness against the noise in the data and the numerical error in data fitting. The algorithm for the problem decomposition of SSF which minimizes the number of quantities involved in each equation is given in Fig. 4. This algorithm uses the list of the complete subsets and their order resulted in the algorithm of Fig.2. The quantities involved in each equation are eliminated by the equations in the other complete subset in (S2). In the next (S3), the quantities involved in each equation are eliminated by the other equation within the same complete subset, if the order of the subset is more than one. The quantities to be eliminated in (S2) and (S3) are selected by lexicographical order in the current SSF. This selection can be more tuned up based on the information of the sensitivity to noise and error of each quantities in the future.

> Input: *The lists DE, DQ and N obtained in the algorithm of Fig.2.*
> Output: *The list of quantities contained in a transformed equation*
> $\{DE_{ij}|i = 1, ..., |DE|, j = 1, ..., \delta n_i\}$.
> (S1) For $i = 1$ to $|DE|$ {
>   For $j = 1$ to $|DE|$ where $j \neq i$ {
>   If $DE_i \supset DE_j$ where $DE_i, DE_j \in DE$ {
> $DE_i = DE_i - DQ'_j$, where DQ'_j is arbitrally, and
> $DQ'_j \subset DQ_j \in DQ$ and $|DQ'_j| = \delta n_j$.}}}
> (S2) For $i = 1$ to $|DE|$ {
>   For $j = 1$ to δn_i {
> $DE_{ij} = DE_i - DQ_{ij}$, where DQ_{ij} is arbitrally, and
> $DQ_{ij} \subset DQ_i \in DQ$ and $|DQ_{ij}| = \delta n_i - 1$.}}

Fig. 4. Algorithm for minimization.

Another role of SSF for the equation discovery system is to teach how to control the quantities in the bi-variate experiments. The convntional SDS just tries to fix the values of all quantities except two during the bi-variate tests. However, such control of the objective process is impossible in case of the simultaneous equation system. SDS must be taught the quantities to control and the quantities determined in the process to appropriately arrange the experiments. SSF derives the information by applying a constraint propagation method to the knowledge of the structural canonical form. The algorithm is basically the same with the causal ordering [7].

5   Evaluation of SDS2

SDS2 has been implemented using a numerical processing shell named MATLAB [14]. The performance of SDS2 has been evaluated in terms of the validity of its results, the computational complexity and the robustness against noise through some examples including fairly large scale processes. The objective processes are provided by simulation.

The examples we applied are the following four.

(1) Two parallel resistances and a battery (depicted in Fig.1)
(2) Heat conduction at walls of holes[15]
(3) A circuit of photo-meter
(4) Reactor core of power plant

Table 1 is the summary of the specifications of each problem size, complexity and robustness against noise. T_{scf} shows strong dependency on the parameter m and n, i.e., the size of the problem. This is natural, since the algorithm to derive structural canonical forms is NP-hard to the size. In contrast, T_{min} shows very slight dependency on the size of the problem, and the absolute value of the required time is negligible. This observation is also highly consistent with the theoretical view that its complexity should be only $O(n^2)$. The total time T_{tl} does not seem to strongly depend on the size of the problem. This consequence is also very natural, because SDS handles each equation separately. The required time of SDS should be proportional to the number of equations in the model. Instead, the efficiency of the SDS more sensitively depends on the average number of quantities involved in each equation. This tendency becomes clearer by comparing T_{av} with av. The complexity of SDS is known to be around $O(n^2)$.

The last column of Table 1 shows the influence of the noise to the result of SDS2, where Gaussian noise is artificially introduced to the measurements. The noise does not affect the computation time in principle. The result showed that a maximum of 25-35% relative noise amplitude to the absolute value of each quantity was acceptable under the condition that 8 times per 10 trials of SDS2 successfully give the correct structure and coefficients of all equations with statistically acceptable errors. The noise sensitivity dose not increase significantly, because SSF focuses on a complete subset which is a small part of the entire system. Similar discussion holds for SDS. The robustness of SDS2 against the noise is sufficient for practical application.

Finally, the validity of the results are checked. In the example (1), SSF derived the expected structural canonical form. Then SSF gave the following form of

Table 1. Statistics on complexity and robustness

Ex.	m	n	av	T_{scf}	T_{min}	T_{tl}	T_{av}	NL
(1)	4	7	2.5	3	0.00	206	52	35
(2)	8	17	3.9	1035	0.05	725	91	29
(3)	14	22	2.6	1201	0.05	773	55	31
(4)	26	60	4.0	42395	0.11	3315	128	26

m: number of equation, n: number of quantities, av: average number of quantities/equation, T_{scf}: CPU time (sec) to derive structural canonical form, T_{min}: CPU time to derive minimum quantities form, T_{tl}: CPU time to derive all equations by SDS, T_{av}: average CPU time per equation by SDS, NL: limitation of % noise level of SDS.

minimum number of quantities to SDS. Here, each equation is represented by a set of quantities involved in the equation.

$$\{V_e, R_1, I_1\}, \{V_e, R_2, I_2\}, \{V_e, V_1\}, \{V_e, V_2\} \tag{8}$$

As a result, SDS derived the following answer.

$$V_e = I_1 R_1 \, [1], V_e = I_2 R_2 \, [2], V_e = V_1 \, [3] \text{ and } V_e = V_2 \, [4], \tag{9}$$

This is equivalent to Eq.(1) not only in the sense of the invariance theorem but also the quantitativeness. Similarly the original equations could be reconstructed in the other examples, and they have been confirmed to be equivalent to the original in the sense of the invariant theorem and quantitativeness.

6 Discussion and Conclusion

The research presented here characterized under-constrained simultaneous equations in terms of complete subsets, and provided an algorithm to derive their structure through experiments. In addition, the constraints of scale-type and identity are investigated to be applied to the discovery of each complete equation.

SSF can discover a unique structural canonical form of any simultaneous equation system in principle as far as its two basic assumptions noted at the begining of the second section are maintained. It is a generic tool which can be combined with any conventional equation discovery systems not limited to SDS. Moreover, the principle of SSF may be applied in a more generic manner not only to the continous processes but also to some discrete systems as far as the systems have structures to propagate states through simultaneous constraints. Main features of SDS are its low complexity, robustness, scalability and wide applicability to the practical problems. Its bottom up approach to construct a complete equation ensures to derive the simplest solution representation the objective process. However, SDS has some weakness on the class of equation formulae to be discovered. First, the regimes and ensemble formulae must be *read-once formulae*, where each quantity appears at most once in it. Second, the relations among quantities must be *arithmetic*. Third, the formula of every pair of quantities searched in the bivariate test is limited to the relation of a simple *binary operator*. Though these limitations are also reflected to the ability of the SDS2, its performance was shown convincing by applying to examples of fairly large size.

References

1. P.W. Langlay, H.A. Simon, G. Bradshaw and J.M. Zytkow: *Scientific Discovery; Computational Explorations of the Creative Process*, MIT Press, Cambridge, Massachusetts (1987).
2. B. Koehn and J.M. Zytkow: Experimeting and theorizing in theory formation, Proceedings of the International Symposium on Methodologies for Intelligent Systems, ACM SIGART Press, pp. 296–307 (1986).
3. B.C. Falkenhainer and R.S. Michalski: Integrating Quantitative and Qualitative Discovery: The ABACUS System, Machine Learning, Kluwer Academic Publishers, pp. 367–401 (1986).
4. C. Schaffer: A Proven Domain-Independent Scientific Function-Finding Algorithm, Proceedings Eighth National Conference on Artificial Intelligence, AAAI Press/The MIT Press, pp. 828–833 (1990).
5. K.M. Huang and J.M. Zytkow: Robotic discovery: the dilemmas of empirical equations, Proceedings of the Fourth International Workshop on Rough Sets, Fuzzy Sets, and Machine Discovery, pp. 217–224 (1996).
6. M.M. Kokar: Determining Arguments of Invariant Functional Descriptions, Machine Learning, Kluwer Academic Publishers, pp. 403–422 (1986).
7. Y. Iwasaki and H.A. Simon: Causality in Device Behavior, Artificial Intelligence, Elsevier Science Publishers B.V., pp. 3–32 (1986).
8. K. Murota: Systems Analysis by Graphs and Matroids - Structural Solvability and Controllability, Algorithms and Combinatorics, Springer-Verlag, Vol.3 (1987).
9. S. Dzeroski and L. Todorovski: Discovering Dynamics: From Inductive Logic Programing to Machine Discovery, Journal of Intelligent Information Systems", Kluwer Academic Publishers, Vol.3, pp.1–20 (1994).
10. T. Washio and H. Motoda: Discovering Admissible Models of Complex Systems Based on Scale-Types and Identity Constraints, Proc. of Fifteenth International Joint Conference on Artificial Intelligence (IJCAI-97), Vol.2, pp.810–817 (1997).
11. T. Washio and H. Motoda: Discovery of first-principle equations based on scale-type-based and data-driven reasoning, Knowledge-Based Systems, Vol.10, No.7, pp.403–412 (1998).
12. T. Washio and H. Motoda: Discovering Admissible Simultaneous Equations of Large Scale Systems, Proc. of AAAI'98: Fifteenth National Conference on Artificial Intelligence (to appear, accepted) (July, 1998).
13. S.S. Stevens: On the Theory of Scales of Measurement, Science, Vol.103, No.2684, pp.677–680 (1946).
14. MATLAB Reference Guide, The Math Works, Inc. (1992).
15. J. Kalagnanam, M. Henrion and E. Subrahmanian: The Scope of Dimensional Analysis in Qualitative Reasoning, Computational Intelligence, Vol.10, No.2, pp.117–133 (1994).

Discovery of Differential Equations from Numerical Data

Koichi Niijima, Hidemi Uchida*, Eiju Hirowatari **, and Setsuo Arikawa

Department of Informatics, Kyushu University, Fukuoka 812-8581, Japan
E-mail: niijima@i.kyushu-u.ac.jp

Abstract. This paper proposes a method of discovering some kinds of differential equations with interval coefficients, which characterize or explain numerical data obtained by scientific observations and experiments. Such numerical data inevitably involve some ranges of errors, and hence they are represented by closed intervals in this paper. Using these intervals, we design some interval inclusions which approximate integral equations equivalent to the differential equations. Interval coefficients in the differential equations are determined by solving the interval inclusions. Many combinations of interval coefficients can be obtained from numerical data. Based on these interval coefficients, the refutability of differential equations is discussed. As a result, we can identify a differential equation. The efficiency of our discovering method is verified by some simulations.

1 Introduction

By rapid progress of observation and experiment systems, it has become possible to get a large number of numerical data. It is very important to discover some rules from these numerical data. Up to now, many scientists have found such rules in the form of mathematical expressions such as differential and integral equations based on their past experience. However, a large number of numerical data make it difficult. Rules contained in such data need to be discovered automatically with the aid of computer. Langley et al. [4], [5], [6] developed the BACON system for discovering conservation laws in scientific fields. In the paper [2], the ABACUS system was invented to realize quantitative and qualitative discovery. Zembowicz and Żytkow [8] constructed a system for finding equations.

The present paper proposes a method of discovering some kinds of differential equations with interval coefficients, which characterize or explain numerical data obtained by scientific observations and experiments. Such numerical data inevitably involve some ranges of errors, and hence we represent the numerical data by closed intervals. Using these intervals, we design some interval inclusions which approximate integral equations equivalent to the differential equations.

---

* At present, NTT Corporation, Ooita branch, Ooita 870, Japan
** At present, Center for Information Processing Research and Education, Kitakyushu University, Kitakyushu 802-8577, Japan

Interval coefficients in the differential equations are determined by solving the interval inclusions using Hansen's method. We can obtain many combinations of interval coefficients from numerical data. It is desirable to discover a differential equation satisfying all the numerical data. Thus we construct a differential equation whose coefficients consist of intersections of the computed interval coefficients provided that they are not empty. This is just a differential equation identified from all the numerical data. If at least one of the intersection coefficients is empty, then we show that the differential equations, each of which has a combination of interval coefficients, do not possess common solutions. By virtue of this fact, we can refute all the differential equations in the present searching class and proceed to search a larger class of differential equations for a desirable differential equation.

Until now, identification of systems by differential equations has been done using numerical data themselves. Based on these numerical data, parameters occurring in the differential equations are determined by applying the least square methods. Such approaches, however, do not yield the concept of refutability of differential equations. Our identification method can refute the differential equations by checking the emptiness of intersections of computed interval coefficients. This fact enables us to search a larger class of differential equations for a target differential equation.

In Section 2, some notations to be used in this paper are given. Section 3 is devoted to derive interval inclusions from differential equations and to describe Hansen's method. We discuss in Section 4 the refutability of differential equations. In Section 5, we extend the results obtained in Sections 3 and 4 to a system of differential equations. Simulation results are given in Section 6.

2 Preliminary

Numerical data obtained by observation and experiment systems usually contain various kinds of noises such as ones by the systems themselves, noises coming from outside in measurement, and measurement errors. We are often obliged to discover reasonable differential equations from such noisy numerical data. So far, the discovery of differential equations has been carried out using noisy numerical data themselves.

In this paper, we represent the noisy numerical data by closed intervals which may contain true values. Let x be a numerical datum, and $[x_l, x_r]$ a closed interval including x. We denote the interval $[x_l, x_r]$ by $[x]$ which is called an interval number. From now on, the interval $[0, 0]$ is denoted by 0 for simplicity. We give the arithmetic rules for interval addition, subtraction and product as follows:

$$[x_1 + x_3, x_2 + x_4] = [x_1, x_2] + [x_3, x_4],$$

$$[x_1 - x_4, x_2 - x_3] = [x_1, x_2] - [x_3, x_4],$$

$$[\underline{x}, \overline{x}] = [x_1, x_2] * [x_3, x_4],$$

where $\underline{x} = \min\{x_1 x_3, x_1 x_4, x_2 x_3, x_2 x_4\}$ and $\bar{x} = \max\{x_1 x_3, x_1 x_4, x_2 x_3, x_2 x_4\}$. For a vector $v = (v_1, v_2, ..., v_m)$, we define an interval vector $[v]$ by

$$[v] = ([v_1], [v_2], ..., [v_m]),$$

where $[v_i]$ indicates an interval number. Similarly, we can define an interval matrix $[M]$ by

$$[M] = ([M_{ij}]),$$

where the (i, j)-component $[M_{ij}]$ indicates an interval number.

3   Derivation of interval inclusions from differential equations

We consider a single differential equation of the form

$$\frac{dx}{dt} = \sum_{i=0}^{n} a_i \, x^i, \tag{1}$$

where $x = x(t)$. The right hand side is restricted to a polynomial of x and takes a linear form with respect to the parameters a_j. From now on, we call (1) a differential equation of degree n. Although modeling by single differential equations is not so useful, a system of such differential equations has been used for modeling in many application fields. Such systems will be treated in Section 5.

The present problem is to identify the parameters $a_0, a_1, ..., a_n$ in the interval form from the interval numbers $[x(t)]$. We rewrite (1) as

$$\sum_{i=0}^{n} x^i \, a_i = \frac{dx}{dt}. \tag{2}$$

We hit on an idea that approximates (2) by a difference inclusion

$$\sum_{i=0}^{n} [x(t_k)]^i \, [a_i] - \frac{[x(t_{k+1})] - [x(t_k)]}{t_{k+1} - t_k} \supseteq 0, \tag{3}$$

where t_k denote observation points and $[x]^i$ denotes i times products of the interval number $[x]$. However, the difference quotient of interval numbers $([x(t_{k+1})] - [x(t_k)])/(t_{k+1} - t_k)$ causes large numerical errors and the errors propagate in the latter interval arithmetic.

To cope with this problem, we derive an integral equation by integrating both side of (2) from t_0 to t_N:

$$\sum_{i=0}^{n} \int_{t_0}^{t_N} x^i(t) dt \, a_i = x(t_N) - x(t_0). \tag{4}$$

Approximating the integral terms using the trapezoidal rule and replacing both side by interval numbers yield an interval inclusion

$$\sum_{i=0}^{n} [X_i(t_0, t_N)][a_i] - ([x(t_N)] - [x(t_0)]) \supseteq 0, \tag{5}$$

where $[X_i(t_0, t_N)]$ denotes

$$[X_i(t_0, t_N)] = \frac{h}{2}([x(t_0)]^i + [x(t_N)]^i) + h \sum_{k=1}^{N-1} [x(t_k)]^i \tag{6}$$

and the observation points t_k are assumed to be equidistant with mesh size h.

From (5) for $N = k, k+1, ..., k+n$, where $k \geq 1$, we obtain a simultaneous interval inclusion

$$[X^k][a^k] - [r^k] \supseteq 0, \tag{7}$$

where $[X^k]$ is an $(n+1) \times (n+1)$ matrix whose components consist of $[X_i(t_0, t_N)]$ for $0 \leq i \leq n$ and $k \leq N \leq k+n$, and $[a^k]$ and $[r^k]$ denote $([a_0^k], [a_1^k], ..., [a_n^k])$ and $([x(t_k)] - [x(t_0)], [x(t_{k+1})] - [x(t_0)], ..., [x(t_{k+n})] - [x(t_0)])$, respectively. There are many solutions $[a^k]$ of (7) so that the following inclusion is fulfilled:

$$\{ a^k = (a_0^k, a_1^k, ..., a_n^k) \mid X^k a^k = r^k, \quad X^k \in [X^k], \quad r^k \in [r^k] \} \subseteq [a^k]. \tag{8}$$

As a method for obtaining a solution $[a^k]$ of (7) satisfying (8), we know the Gaussian elimination process ([1]). This process, however, contains division in the interval arithmetic, which causes numerical error propagations. We notice here that the left hand side of (8) may be rewritten as

$$\{ a^k = (a_0^k, a_1^k, ..., a_n^k) \mid a^k = (X^k)^{-1} r^k, \quad X^k \in [X^k], \quad r^k \in [r^k] \}.$$

We shall remember Hansen's method [3] which is a technique for computing an interval matrix U such that $\{ (X^k)^{-1} \mid X^k \in [X^k] \} \subseteq U$. The merit of this method is to be able to estimate the bound of an interval matrix H defined by

$$U = \{ (X^k)^{-1} \mid X^k \in [X^k] \} + H.$$

Hansen's method enables us to compute $U[r^k]$ satisfying $[a^k] \subseteq U[r^k]$, and to make the width of the interval vector $[X^k][a^k] - [r^k]$ in (7) as small as possible. In simulations in Section 6, we use this method to compute $U[r^k]$.

4   Refutation of differential equations with interval coefficients

Usually, the number of numerical data are much more than that of parameters contained in the differential equation. Differential equations should be identified from all the numerical data.

We assume that $M + n + 1$ numerical data have been obtained. This means that we can consider M combinations of data of the form $(x(t_0), x(t_k), ..., x(t_{k+n}))$, $k = 1, 2, ..., M$, and can compute $[X^k]$ and $[r^k]$ appearing in (7) for $k = 1, 2, ..., M$. We solve (7) for $k = 1, 2, ..., M$ by Hansen's method and denote the solutions by

$$[a^k] = ([a_0^k], [a_1^k], ..., [a_n^k]), \qquad k = 1, 2, ..., M. \tag{9}$$

Then we consider the differential equations with the interval coefficients $[a^k] = ([a_0^k], [a_1^k], ..., [a_n^k])$

$$\frac{dx}{dt} = \sum_{i=0}^{n} [a_i^k] \, x^i, \qquad k = 1, 2, ..., M. \tag{10}$$

Let us define a set of solutions of (10):

$$S_k = \{ \, x = x(t) \mid \frac{dx}{dt} = \sum_{i=0}^{n} a_i^k x^i, \ t_0 \le t \le T, \ a_i^k \in [a_i^k], \ x(t_0) \in [x(t_0)] \, \},$$

$$k = 1, 2, ..., M. \tag{11}$$

It can be shown by Peano's theorem [7] that the initial value problem appeared in the set S_k has solutions, that is, the set S_k is not empty.

Using the interval coefficients (9), we define an intersection

$$[b_i] = \cap_{k=1}^{M} [a_i^k]. \tag{12}$$

If all $[b_i]$, $i = 0, 1, ..., n$ are not empty, we can say that the differential equations

$$\frac{dx}{dt} = \sum_{i=0}^{n} [b_i] x^i \tag{13}$$

were identified from all the numerical data.

The following theorem holds.

Theorem 1. *Let S_k be defined by (11) and assume that the set $\cap_{k=1}^{M} S_k$ is not empty and does not contain constant solutions. We suppose that all the interval numbers $[b_i]$, $i = 0, 1, ..., n$, defined by (12) are not empty and put*

$$I = \{ \, x = x(t) \mid \frac{dx}{dt} = \sum_{i=0}^{n} b_i x^i, \ t_0 \le t \le T, \ b_i \in [b_i], \ x(t_0) \in [x(t_0)] \, \}. \tag{14}$$

Then we have

$$I = \cap_{k=1}^{M} S_k.$$

Proof. We first prove $I \subseteq \cap_{k=1}^{M} S_k$. Choose any $x \in I$. The function $x = x(t)$ satisfies the differential equation

$$\frac{dx}{dt} = \sum_{i=0}^{n} b_i x^i.$$

Since b_i is in $[b_i] = \cap_{k=1}^{M} [a_i^k]$, it follows that b_i belongs to $[a_i^k]$ for all k. This implies that x is an element of S_k for all k, that is, $x \in \cap_{k=1}^{M} S_k$ which proves $I \subseteq \cap_{k=1}^{M} S_k$.

We next show that $\cap_{k=1}^{M} S_k \subseteq I$. Choose any $x \in \cap_{k=1}^{M} S_k$. Then x belongs to S_k for all k. Therefore, this x satisfies

$$\frac{dx}{dt} = \sum_{i=0}^{n} a_i^k x^i, \qquad a_i^k \in [a_i^k] \tag{15}$$

for all k. It is easily shown that $a_i^k = a_i^\ell$ holds for any $k \neq \ell$ and for all i. Indeed, if $a_i^k \neq a_i^\ell$ for some pair $k \neq \ell$ and for some i, then we have from (15),

$$\sum_{i=0}^{n} (a_i^k - a_i^\ell) x^i = 0.$$

This implies that $x = x(t)$ is a constant solution, which contradicts the assumption.

We put $a_i = a_i^k$. Then the differential equation satisfied by x can be written as

$$\frac{dx}{dt} = \sum_{i=0}^{n} a_i x^i, \qquad a_i \in [a_i^k]$$

for all k. Therefore, we have $a_i \in [b_i] = \cap_{k=1}^{M} [a_i^k]$. This means $x \in I$, which finishes the proof.

This theorem justifies the construction of differential equations by taking intersections of interval coefficients. It is valuable to note that these intersection interval numbers decrease monotonically.

We show the result of Theorem 1 in Figure 1.

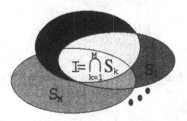

Fig. 1. $I = \cap_{k=1}^{M} S_k$

In the next, we consider the case that some of $[b_i]$ defined by (12) are empty. Then we can prove the following theorem.

Theorem 2. *Suppose that some of $[b_i]$ defined by (12) are empty. We assume that $\cap_{k=1}^{M} S_k$ does not contain constant solutions. Then we have*

$$\cap_{k=1}^{M} S_k = \phi.$$

Proof. Assume that $\cap_{k=1}^{M} S_k$ is not empty. Then there exists $x \in \cap_{k=1}^{M} S_k$. Therefore, this $x = x(t)$ satisfies the differential equations

$$\frac{dx}{dt} = \sum_{i=0}^{n} a_i^k x^i, \quad a_i^k \in [a_i^k], \quad k = 1, 2, ..., M$$

from which we get

$$\sum_{i=0}^{n} (a_i^k - a_i^\ell) x^i = 0. \tag{16}$$

By the first assumption of Theorem 2, we have $[b_{i_0}] = \phi$ for some i_0, where the symbol ϕ denotes the empty set, and hence there exists some pair (k, ℓ) such that $a_{i_0}^k \neq a_{i_0}^\ell$. This fact shows that (16) has only constant solutions, which contradicts the second assumption of Theorem 2.

Theorem 2 shows that if some $[b_i]$ defined by (12) are empty, any differential equation of degree n is not identifiable from the numerical data. This suggests that if at least one of $[b_i]$ defined by (12) is empty, we can refute all the differential equations of degree n, and proceed to search a class of differential equations of degree $n + 1$ for a desirable differential equation.

The result of Theorem 2 is shown in Figure 2.

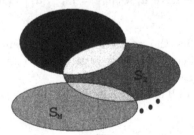

Fig.2. $\cap_{k=1}^{M} S_k = \phi$

Our discovery algorithm of differential equations is as follows:

(i) Set $n = 0$ and $m = 2$.
(ii) Solve the integral inclusions and check whether the intervals $\cap_{k=1}^{m} [a_i^k]$, $0 \le i \le n$, are empty or not.

(iii) If all the intervals are not empty and $m < M$, then replace m by $m + 1$ and
go to (ii). If $m = M$, then stop. If there exists i_0 such that $\cap_{k=1}^{m}[a_{i_0}^k] = \phi$,
then refute all the differential equations of degree n, replace n by $n+1$, reset
$m = 2$ and go to (ii).

5   Extension to a system of differential equations

The results obtained in Section 4 are easily extended to a system of differential
equations. We consider a system of differential equations

$$\frac{dx_\ell}{dt} = \sum_{0 \le |i| \le n_\ell} a_{i,\ell}\, x^i, \qquad \ell = 1, 2, ..., m, \tag{17}$$

where x^i indicates $x_1^{i_1} x_2^{i_2} \ldots x_m^{i_m}$ with $i = (i_1, i_2, ..., i_m)$ and $|i| = \sum_{j=1}^{m} i_j$. In a
similar way as in Section 3, we get an interval inclusion:

$$\sum_{0 \le |i| \le n_\ell} [X_i(t_0, t_N)][a_{i,\ell}] - ([x_\ell(t_N)] - [x_\ell(t_0)]) \supseteq 0, \tag{18}$$

where $[X_i(t_0, t_N)]$ has the same form as (6), but now

$$[x(t_k)]^i = [x_1(t_k)]^{i_1} [x_2(t_k)]^{i_2} \ldots [x_m(t_k)]^{i_m}.$$

Let d_ℓ be the number of i satisfying $0 \le |i| \le n_\ell$, and let $d = \sum_{\ell=1}^{m} d_\ell$. For
each ℓ, we consider the inclusion (18) for $k \le N \le k + d_\ell - 1$. Moreover,
we denote a $d \times d$ matrix $([X_i(t_0, t_N)])_{0 \le |i| \le n_\ell,\ k \le N \le k+d_\ell-1;\ 1 \le \ell \le m}$ again by
$[X^k]$, and a vector $([a_{i,\ell}])_{0 \le |i| \le n_\ell;\ 1 \le \ell \le m}$ again by $[a^k]$. We also denote a vector
$([x_\ell(t_N)] - [x_\ell(t_0)])_{k \le N \le k+d_\ell-1;\ 1 \le \ell \le m}$ by $[r^k]$. Then we get a simultaneous
interval inclusion

$$[x^k][a^k] - [r^k] \supseteq 0$$

which has the same form as (7). Therefore, we can proceed the latter discussion
in the same way as in Sections 3 and 4.

6   Simulations

We carry out some simulations using the discovering system developed previously. Our system works following the discovering algorithm given in Section
4. The higher the degree of polynomials occurring in the differential equations
is, the larger the size of simultaneous inclusions for determining interval coefficients becomes. This causes the propagation of intervals to be found in the
interval arithmetic. So we do not calculate all the interval coefficients simultaneously, but solve subsystems of inclusions with changing the combination of the
interval terms. In our simulations, numerical data are made artificially from the
solutions of differential equations to be identified.

Example 1 (single differential equation).

Let us consider $x(t) = e^t$ in the interval $0 \leq t \leq 10$. We divide the interval into 100 equidistant subintervals and denote the mesh points by $t_k = 0.1k$. Numerical data are given as $x(t_k) = e^{t_k}$ and interval numbers are constructed by adding 1% error to each of the numerical data. Based on these interval numbers, our system first tries to identify the differential equation

$$\frac{dx}{dt} = [a_0].$$

However, this equation is refuted by Theorem 2 because of $[a_0^1] \cap [a_0^2] = \phi$. Next, our system is going to find out the differential equation

$$\frac{dx}{dt} = [a_1]x.$$

In this case, we can obtain the intersection

$$\cap_{k=1}^{99}[a_1^k] = [0.991371, 1.00265].$$

Therefore, the differential equations

$$\frac{dx}{dt} = [0.991371, 1.00265]x$$

were discovered from the given numerical data. From these equations, scientists can guess the differential equation $dx/dt = x$ with high accuracy.

Example 2 (single differential equation).

Next, we consider the logistic curve $x(t) = (1+10e^t)^{-1}$ in the interval $0 \leq t \leq 10$. The division of the interval and the mesh points are the same as in Example 1. Numerical data are given as $x(t_k) = (1 + 10e^{t_k})^{-1}$ and interval numbers are constructed by adding 1% error to each of the numerical data. The searching of a target equation was tried by changing the combination of the terms in the differential equations. As a result, we succeeded to find out the differential equations:

$$\frac{dx}{dt} = [-1.5006, -0.962858]x + [0.930201, 1.06570]x^2.$$

Although the width of the first interval coefficient is not so small, the differential equation

$$\frac{dx}{dt} = -x + x^2$$

can be chosen as a candidate of target differential equations.

Example 3 (system of differential equations).

Finally, we consider two functions $x(t) = -2e^{-t}$ and $y(t) = e^{-t}(\sin t + \cos t)$ in the interval $0 \le t \le 10$. In the same way in Example 1, we divide the interval and denote the mesh points by t_k. Numerical data are given as $x(t_k) = e^{t_k}$ and $y(t_k) = e^{-t_k}(\sin t_k + \cos t_k)$. We construct interval numbers by adding 1% error to each of the numerical data as in Example 1. By changing the combination of the terms in the differential equations, our system succeeded to find out finally the simultaneous differential equations

$$\begin{cases} \dfrac{dx}{dt} = [-2.03367, -1.96269]x + [-2.00822, -1.96263]y, \\[2mm] \dfrac{dy}{dt} = [0.988213, 1.00629]x. \end{cases}$$

Among these equations, scientists can predict the target simultaneous differential equations

$$\begin{cases} \dfrac{dx}{dt} = -2x - 2y, \\[2mm] \dfrac{dy}{dt} = x. \end{cases}$$

7   Conclusion

In this paper, we proposed a method of discovering some differential equations from given numerical data. The feature of our approach lies in constructing interval inclusions from an integral equation equivalent to a differential equation. By solving these interval inclusions, we can obtain various combinations of coefficients in the differential equation in an interval form. If the intersections of these interval coefficients are not empty, we can identify a differential equation having these intersection coefficients. The justification of this was given in Theorem 1. If not the case, we refute all the differential equations in the present searching class and try to search a larger class of differential equations for a target differential equation. This refutation was justified by Theorem 2. Since our method contains interval arithmetic, the accuracy of interval coefficients goes down in proportion to the size of interval inclusions. In the simulation, therefore, the size of interval inclusions was reduced by changing the combination of terms in the differential equation.

There are some problems to be solved in the future. This paper restricts a searching domain to a class of differential equations whose right hand side has a polynomial form. It is a future work to extend the polynomial form to a more general mathematical expression. One more problem is to clarify a relation between the solutions of a discovered differential equation and base interval numbers constructed from the given numerical data. In the simulations, we succeeded the identification of differential equations only for numerical data including small amount of error. We also failed to identify a system of differential equations with larger size. This is also a future work.

References

1. Alefeld, G., Herzberger, J.: Introduction to Interval Computations. Academic Press (1982)
2. Falkenhainer, B. C., Michalski, R. S.: Integrating quantitative and qualitative discovery: the ABACUS system. Machine Learning 1 (1986) 367–401
3. Hansen, E. : Interval arithmetic in matrix computations. part I. SIAM J. Numerical Analysis 2 (1965) 308–320
4. Langley, P. G., Bradshow, L., Simon, H. A.: BACON:5 The Discovery of Conservation Laws. Proceedings of the Seventh International Joint Conference on Artificial Intelligence (1981) 121–126
5. Langley, P. G., Bradshow, L., Simon, H. A: Rediscovering Chemistry with the Bacon System. Machine Learning: An Artificial Intelligence Approach (1983)
6. Langley, P., Zytkow, J., Bradshaw, G.L., Simon, H. A.: Mechanisms for Qualitative and Quantitative Discovery. Proceedings of the International Machine Learning Workshop (1983) 12–32
7. Zeidler, E.: Nonlinear Functional Analysis and its Applications I. Springer-Verlag (1993)
8. Zembowicz, R., Żytkow, J. M.: Discovery of equations: experimental evaluation of convergence. Proceedings Tenth National Conference on Artificial Intelligence (1992) 70–75

Automatic Transaction of Signal via Statistical Modeling

Genshiro Kitagawa and Tomoyuki Higuchi

The Institute of Statistical Mathematics, 4-6-7 Minami-Azabu, Minato-ku, Tokyo
106-8569 Japan

Abstract. The statistical information processing can be characterized
by using the likelihood function defined by giving an explicit form for
an approximation to the true distribution from which the data are generated. This mathematical representation as an approximation, which is
usually called a model, is built based on not only the current data but also
prior knowledge on the object and the objective of the analysis. Akaike
([2] and [3]) showed that the log-likelihood can be considered as an estimate of the Kullback-Leibler information which defines the similarity
between the predictive distribution of the model and the true distribution and proposed the Akaike information criterion (AIC). By the use
of this AIC, it becomes possible to evaluate and compare the goodness
of many models objectively and it enables us to select the best model
among many candidates. In consequence, the minimum AIC procedure
allows us to develop automatic modeling and signal extraction procedures. In this study, we give a simple explanation of statistical modeling
based on the AIC and demonstrate four examples of applying the minimum AIC procedure to an automatic transaction of signals observed
in the earth sciences. In each case, the AIC plays an important role in
making the procedure automatic and objective, and promises to realize
a detail examination of a large amount of data sets, which provides us
with an opportunity to discover new information.

1 Introduction: Statistical Modeling

In statistical information processing, a model is built based on not only the
current data but also prior knowledge on the object and the objective of the
analysis, whereas the conventional data analysis techniques rely on simple manipulation of the current data. To use a proper model for describing the data
makes it possible to combine various knowledge on the object or the information
from other data sets, and can enhance a scientific return from the given data
sets. Namely, necessary information is extracted based on the model (Figure 1).
This is the main feature of statistical information processing.

On the other hand, there is a danger of extracting biased result if an analysis
is made by using improper models. Therefore, in information processing based
on a model, use of proper model is crucial. Further for an automatic statistical

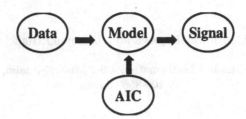

Figure 1   Statistical information processing

information processing procedures, the development of an automatic statistical modeling procedure is necessary. Akaike information criterion AIC ([2]) is an objective criterion to evaluate the goodness of fit of statistical model and facilitates the development of automatic statistical information processing procedures. The authors are also referred to MDL (minimum description length, [13]) for the selection of the orders of the models. Modeling and simplicity in scientific discovery and artificial intelligence are discussed in [15] and [16].

In this paper, we first briefly review the statistical modeling procedure based on information criterion. Then we shall show examples of developing statistical information processing for knowledge discovery in various fields of earth science.

2   Information Criterion and Automatic Selection of Models

The phenomena in real world are usually very complicated and information obtained from the real world is in general incomplete and insufficient. The models which we obtained from and used for such incomplete information is inevitably an approximation to the real world. In modeling, it is expected to describe the complex real world as precise as possible by simpler model. However, if the objective of the modeling is to obtain precise description of the data, it is not obvious why the model should be simple. A clear answer to this basic question was given by Akaike ([2]) from a predictive point of view.

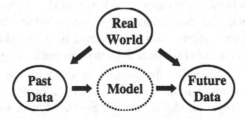

Figure 2   Prediction with statistical model

As shown in Figure 2, in prediction an inference is made on the future data based on the existing data. In statistical prediction, a model is used for prediction and it controls the accuracy of the prediction. If, in the modeling, we adhere to mimic the current data or the phenomenon, then the model will become increasingly more complicated to reproduce the details of the current data. However, by aiming at the improvement of predictive ability, it becomes possible to extract essential information from or knowledge about the object, properly excluding random effects.

Akaike ([2] and [3]) proposed to evaluate the goodness of statistical models by the goodness of the corresponding predictive distributions. Namely, he proposed to evaluate the goodness of the statistical models by the similarity between the predictive distribution of the model and the true distribution that generates the data $Y_N = [y_1, \ldots, y_N]$, and to evaluate its similarity by the Kullback-Leibler information quantity. Here N is the number of data. Under the situation that the true distribution is unknown, it is not possible to compute the Kullback-Leibler information. However, Akaike showed that the log-likelihood

$$\ell(\theta_m) = \log f_m(Y_N|\theta_m) \tag{1}$$

$$= \sum_{n=1}^{N} \log f_m(y_n|Y_{n-1}, \theta_m), \tag{2}$$

that has been used for many years as general criterion for the estimation of parametric models, can be considered as an estimate of the K-L information (precisely, the expected log-likelihood). Here f_m is one of a set of candidate models for a probability density function of the observation, $\{f_m; m = 1, \ldots, M\}$, which is an approximation to the true distribution, and θ_m is the parameter vector of the density f_m. In particular case where y_n is independently and identically distributed, (2) can be given as a very simple form

$$\ell(\theta_m) = \sum_{n=1}^{N} \log f_m(y_n|\theta_m). \tag{3}$$

An optimal parameter estimate, $\widehat{\theta}_m$, is defined by maximizing the log-likelihood function with respect to θ_m.

According to this idea, the maximum likelihood method can be interpreted as an estimation method that aims at minimizing the K-L information. A difficulty arose in the development of automatic modeling procedures, where the log-likelihoods of the models with parameters estimated from data have biases as estimators of the K-L information, and thus the goodness of the models with estimated parameters cannot be compared with this criterion. This bias occurred because the same data set was used twice for the estimation of parameters and for the estimation of the K-L information. Akaike evaluated the bias of the log-likelihood, and defined the information criterion

$$\mathrm{AIC}_m = -2(\text{log likelihood}) + 2(\text{number of parameters})$$

$$= -2\log f_m(Y_N|\widehat{\theta}_m) + 2\,\|\theta_m\| \tag{4}$$

by compensating this bias. Here $\|\theta_m\|$ denotes the dimension of the parameter vector. By the use of this AIC, it becomes possible to evaluate and compare the goodness of many models objectively and it enables us to select the best model among many competing candidates $f_m(\cdot|\theta_m)$; $m = 1,\ldots,M$. As a result, the minimum AIC procedure allows us to develop automatic modeling and signal extraction procedures. This is a breakthrough in statistics and helped the change of statistical paradigm from the estimation within the given stochastic structure to modeling with unknown structure. Using AIC, various data structure search procedures and data screening procedures were developed (see e.g., [1], [14], and [4]).

As mentioned above, a statistical approach to automatic transaction of data relies on using the minimum AIC procedure which is based on the maximum log-likelihood principle. Then the statistical information processing can be characterized by using the likelihood function defined by giving an explicit form for an approximation to the true distribution from which the data are generated. Although the direct application of the AIC is limited to the model with parameter estimated by the maximum likelihood method, its idea can be applied to much wider class of models and estimation procedures and various types of information criteria are developed recently (e.g., [11] and [7]).

3 Least Squares Fit of Regression Models

One of the easiest way to represent random effects in the observation y_n is to adopt an observation model in which an observation error (noise) is assumed to be added to a signal $g(n|\theta)$

$$y_n = g(n|\theta) + \varepsilon_n, \quad \varepsilon_n \sim N(0,\sigma^2), \tag{5}$$

where ε_n is an independently and identically distributed (i.i.d.) Gaussian white noise sequences with mean 0 and unknown variance σ^2. In this case, the log-likelihood can be given by (3) and the maximum likelihood estimates is obtained by minimizing $\sum_{i=1}^{N}[y_n - g(n|\theta)]^2$. Within this framework, the major efforts in developing an automatic procedure are made on preparing a wide variety of candidates for $g(\cdot|\theta)$. We shall show two examples. In each case the determination of the best $g(\cdot|\theta)$ through the AIC plays an important role in making the procedure automatic and objective.

3.1 Automatic Identification of Large-Scale Filed-Aligned Current Structure

The plasma stream from the Sun, solar wind, interacts the Earth's magnetic field and generates three-dimensional current system above the ionosphere. Because conductance along the magnetic field is much higher than that across the magnetic field, the currents flow along magnetic field lines. Such current is called the large-scale field-aligned currents (LSFAC). LSFACs are also related to the dynamics of aurorae. Depending on the number of LSFAC sheets crossed by a satellite and also on the intensity and flow direction (upward/downward) of each LSFAC, a plot of the magnetic fluctuations associated with the LSFAC (as shown in Figure 3), mainly in the east-west (E-W) magnetic component, can have any shape, and we have been depending on visual examination to identify LSFAC systems. We developed a procedure to automatically identify the spatial structure of LSFAC from satellite magnetic field measurements ([5]).

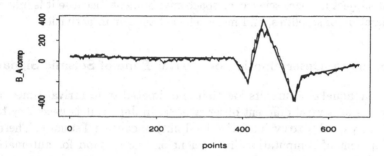

Figure 3 Magnetic field perturbation associated with the LSFAC and fitted polyline.

The required task is to automatically fit the first-order B-spline function with variable node positions, which is sometimes called a polyline or linear spline ([8]). Namely, we adopt a polyline as $g(n|\theta)$ mentioned above. Although node points are fixed in usual spline applications, the benefit of the spline function can be maximized when node points are allowed to move ([6]). We therefore treat a set of node positions and node values as parameters to be estimated. In addition, the number of node points, which determines the number of LSFAC sheets, is one of the fitting parameters. For this modeling, the AIC with J node points is defined by

$$\mathrm{AIC}_J = N \log \widehat{\sigma}_J^2 + 2(2J + 1) + \text{constant}, \qquad (6)$$

where a constant factor is independent of the selection of models, and $\widehat{\sigma}_J^2$ is the maximum likelihood estimates of the variance of the observation error in (5).

We applied the developed procedure to the whole data set of magnetic field measurements made by the Defense Meteorological Satellite Program–F7 (DMSP–F7) satellite during the entire interval of its mission from December 1983 to January 1988. DMSP is a Sun-synchronous satellite with a nearly circular polar orbit at about 835 km in altitude, and thus the orbital period is about 101 minutes. We divide a data file of each polar pass into two parts, dayside and nightside files, by the data point of the highest-latitude satellite position. We have a total of 71,594 data files. Each data file usually contains from 600 to 800 magnetic field vector measurements as well as various geographic and geomagnetic parameters necessary for describing a satellite position at observation time. The sampling interval is 1 second.

The first subject made possible by the developed procedure is to find a four-FAC-sheet structure along dayside passes. Ohtani et al. ([12]) reported only

four events observed by the DMSP–F7. This four-FAC-sheet structure was unexpected phenomena from a viewpoint of the conventional interpretation of the
LSFAC, and happened to be discovered. The developed procedure found 517
northern and 436 southern passes along which the DMSP–F7 observed four LS-
FACs. This discovery allowed us for the first time ever to conduct a statistical
study on what solar wind conditions bring about this peculiar LSFAC. In addition, the developed automatic procedure to identify the structure of LSFAC
systems can be used to conduct space weather forecasting that is becoming an
important subject in space science, as space environment, because it is influential
to the operation of satellites, and more relevant to human activities.

3.2   Automatic Determination of Arrival Time of Seismic Signal

When an earthquake occurs, its location is estimated from arrival times of the
seismic waves at several different observatories. In Japan, it is necessary to determine it very quickly to evaluate the possibility of causing Tsunami. Therefore,
the development of computationally efficient on-line method for automatic estimation of the arrival time of the seismic wave is a very important problem.
At each observatory, three-component seismogram is observed at a sampling
interval of about 0.01 second.

When seismic wave arrives, the characteristics of the record of seismograms,
such as the variances and the spectrum, change significantly. For estimation of
the arrival time of the seismic signal, it is assumed that each of the seismogram
before and after the arrival of the seismic wave is stationary and can be expressed
by an autoregressive model as follows ([17]):

Background model: $y_n = \sum_{i=1}^{m} a_i y_{n-i} + v_n, \quad v_n \sim N(0, \tau_m^2)$

Seismic signal model: $y_n = \sum_{i=1}^{\ell} b_i y_{n-i} + w_n, \quad w_n \sim N(0, \sigma_\ell^2)$

Although the likelihood function of y_n for the AR modeling depends on Y_{n-1},
we can obtain its analytic form as a function of the AR coefficients. Then, given
the observations, the AIC of the locally stationary AR model is obtained by

$$\text{AIC}_k = k \log \widehat{\tau}_m^2 + (N - k) \log \widehat{\sigma}_\ell^2 + 2(m + \ell + 2) \qquad (7)$$

where N and k are the number of data and the assumed arrival time point, and
$\widehat{\tau}_m^2$ and $\widehat{\sigma}_\ell^2$ are the maximum likelihood estimates of the innovation variances of
the background noise model and the seismic signal model, respectively. In this
locally stationary AR modeling, the arrival time of the seismic wave corresponds
to the change point of the autoregressive model. The arrival time of the seismic
signal can be determined automatically by finding the minimum of the AIC_k on
a specified interval.

However, for automatic determination of the change point by the minimum
AIC procedure, we have to fit and compare $K \times (M + 1)^2$ models. Here K is the

number of possible change points and M is the possible maximum AR order of background and signal models. The fitting and finding the minimum AIC model can be realized computationally and efficiently by using the least squares method based on the Householder transformation. By this method, the necessary amount of computation is only twice as much as that for the fitting of single AR model with order M.

Figure 4A shows a portion of a seismogram of a foreshock of Urakawa-Oki Earthquake observed at Moyori, Hokkaido, Japan. 4B shows AIC_k for $k = 850, \ldots, 1150$. From this figure, it can be seen that the AIC becomes the minimum at $k = 1026$. Using the estimated arrival times of seismic signal at several observatories, it is possible to estimate the epicenter of the earthquake automatically. Also, the AIC values shown in Figure 4B can define the likelihood function for the arrival time. Based on that function, it is expected to be able to develop a new maximum likelihood type estimator for the epicenter of the earthquake.

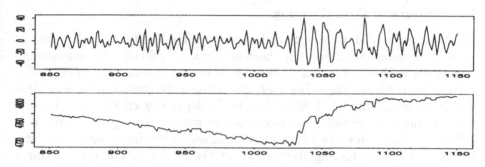

Figure 4   A: Seismic signal. B: AIC value.

4   General State Space Model

The two examples shown above are based on the relatively simpler models which can produce an analytic form of the AIC, because of a simple assumption such that the observation error is linearly added to the regression model and i.i.d. Gaussian white noise sequences. Although the adopted models are suited to each application, general framework for describing the models are available for time series data. This framework is called the general state space model which is a generalization of the state space model ([9]), and is defined by

$$x_n \sim q(\,\cdot\,|x_{n-1}, \theta) \qquad \text{[system model]} \qquad (8)$$
$$y_n \sim r(\,\cdot\,|x_n, \theta) \qquad \text{[observation model]} \qquad (9)$$

where x_n is the state vector at time n. q and r are conditional distributions of x_n given x_{n-1} and of y_n given x_n, respectively. This general state space model can treat the non-Gaussian and non-linear time series model, in contrast to the ordinary state space model.

For state estimation in the general state space model, one step ahead predictor and filter can be obtained by the following recursive formulas called the non-Gaussian filter:

[prediction]

$$p(x_n|Y_{n-1}) = \int_{-\infty}^{\infty} p(x_n|x_{n-1})p(x_{n-1}|Y_{n-1})dx_{n-1} \qquad (10)$$

[filtering]

$$p(x_n|Y_n) = \frac{p(y_n|x_n)p(x_n|Y_{n-1})}{p(y_n|Y_{n-1})}, \qquad (11)$$

where we omit a dependency on θ for simple notation. $p(y_n|Y_{n-1})$ in the filtering is obtained from $\int p(y_n|x_n)p(x_n|Y_{n-1})dx_n$ and then the log-likelihood in the general state space model can be defined by (2).

4.1   Automatic Data Cleaning

In an attempt to predict big earthquake anticipated in Tokai area, Japan, various types of measurement devices have been set since early 80'th. The underground water level is observed in many observation wells at a sampling interval of 2 minutes for almost 20 years. However, the actual underground water level data contains huge amount of (1 % to over 10 %, depending on the year) missing and outlying observations. Therefore, without proper cleaning procedure, it is difficult to fully utilize the information contained in the huge amount of data. We interpolated the missing observations and corrected the outliers by using a non-Gaussian state space model: ([10])

$$t_n = t_{n-1} + w_n, \qquad y_n = t_n + \varepsilon_n, \qquad (12)$$

where $w_n \sim N(0, \tau^2)$. For the observation noise ε_n, we considered Gaussian mixture distribution

$$\varepsilon_n \sim (1 - \alpha)N(0, \sigma_0^2) + \alpha N(\mu, \sigma_1^2) \qquad (13)$$

where α is the rate of contaminated observations, σ_0^2 is the variance of ordinary observations and μ and σ_1^2 are the mean and variance of the outliers. Such a density allows the occurrence of large deviations with a low probability. In this model, a set of $[\tau^2, \alpha, \sigma_0^2, \mu, \sigma_1^2]$ is a parameter vector θ to be optimized. For the filtering and smoothing of the non-Gaussian state space model, we applied the non-Gaussian filter and smoother ([9]). By this Gaussian-mixture modeling of the observation noise, the essential signal t_n is extracted automatically taking account of the effect of the outliers and filling in the missing observations.

4.2   Finding out the Effect of Earthquake in Underground Water Level Data

Even after filling in the missing observations and correcting the outliers, the underground water level is very variable. Further, because the data is affected

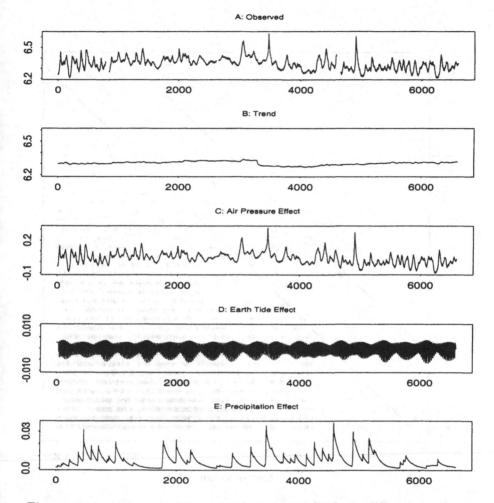

Figure 5   A: A segment of the water level data. B: The extracted seismic effect.

by many other covariates such as barometric air pressure, earth tide and precipitation, it is almost impossible to extract the effect of earthquake by simple manipulation of the data. In an attempt to account for the effect of the covariates on the underground water level, we considered the following model,

$$y_n = t_n + P_n + E_n + R_n + \varepsilon_n, \tag{14}$$

where t_n, P_n, E_n, R_n and ε_n are the trend, the barometric pressure effect, the earth tide effect, the rainfall effect and the observation noise components, respectively ([10]). We assumed that those components follow the models

$$\nabla^k t_n = w_n, \qquad P_n = \sum_{i=0}^{m} a_i p_{n-i}, \tag{15}$$

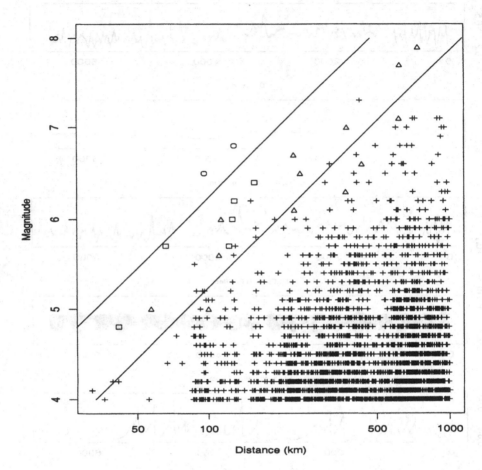

Figure 6 Scatter plot of the earthquakes ([10]).

$$E_n = \sum_{i=0}^{\ell} b_i et_{n-i}, \quad R_n = \sum_{i=1}^{k} c_i R_{n-i} + \sum_{i=1}^{k} d_i r_{n-i} + v_n.$$

Here p_n, et_n and r_n are the observed barometric pressure, the earth tide and the observed precipitation at time n, respectively. These components can be estimated by the state space representation of (8) and (9) and by the use of Kalman filter and the fixed interval smoother.

Figure 5B-E show the extracted coseismic effect, air pressure effect, earth tide effect and precipitation effect obtained by the Kalman smoother. The annual variation of the trend is only about 6cm and the effect of the earthquake with magnitude M=4.8, at a distance D=42 km, is clearly detected. Most of the range of about 45 cm trend variations in Figure 5A can be considered as the effect of barometric pressure, etc..

Figure 6 shows the scatter plot of the earthquakes with log-distance as the horizontal axis and the magnitude as the vertical axis. The earthquakes with $\bigcirc$ -label has detected coseismic effects over 8 cm. Whereas the $\square$- and $\triangle$-labeled events indicates the earthquakes with coseismic effects over 4 and 1 cm. The + labeled events indicate earthquakes without coseismic effects over 1 cm. Two lines in the figure are defined by

$$\bar{M} = M - 2.62 \log_{10} D = C$$

for C=0 and 1.

From the analysis of over 10 years data, we obtained the following important findings: (1) The drop of water level can be seen for most of the earthquakes with magnitude larger than $M > 2.62 \log_{10} D + 0.2$, where D is the hypocentral distance. (2) The amount of the drop can be explained as a function of $M - 2.62 \log_{10} D$. (3) Except for the coseismic effect drop, the trend regularly increases at the rate of about 6cm per year.

5 Summary

In statistical approach to knowledge discovery, a proper modeling of the object is a crucial step. In this paper, we introduced an automatic procedure based on Akaike information criterion, AIC. We demonstrated four applications of the minimum AIC procedure to the actual large data sets observed in the earth science. In each case, an appropriate representation of the signals can enables us to give an explicit form of the AIC and results in the realization of an automatic procedure to handle a large amount of data sets.

References

1. Akaike, H. (1972), "Automatic Data Structure Search by the Maximum Likelihood", *Computers in Biomedicine, A Supplement to the Proceedings of the Fifth Hawaii International Conference on Systems Sciences*, Western Periodicals, California, 99–101.
2. Akaike, H. (1973), "Information Theory and an Extension of the Maximum Likelihood Principle", *2nd International Symposium in Information Theory*, Petrov, B.N. and Csaki, F. eds., Akademiai Kiado, Budapest, pp. 267–281. (Reproduced in *Breakthroughs in Statistics, Vol.I, Foundations and Basic Theory*, S. Kots and N.L. Johnson, eds., Springer-Verlag, New York, 1992, 610–624.)
3. Akaike, H. (1974), "A New Look at the Statistical Model Identification", *IEEE Transactions on Automatic Control*, AC-19, 716-723.
4. Bozdogan, H. (1994), *Proceeding of the First US/Japan Conference on the Frontiers of Statistical Modeling: An Informational Approach*, Kluwer Academic Publishers.

5. Higuchi, T. and Ohtani, S. (1998), "Automatic Identification of Large-Scale Field-Aligned Current Structure," *Research Memorandom*, No. 668, The Institute of Statistical Mathematics. http://www.ism.ac.jp/~ higuchi/AIFACpaper.html
6. Hiragi, Y., Urakawa, H., and Tanabe, K. (1985), "Statistical Procedure for Deconvoluting Experimental Data," *J. Applied Physics*, **58**, No. 1, 5–11.
7. Ishiguro, M., Sakamoto, Y. and Kitagawa, G. (1997), "Bootstrapping log likelihood and EIC, an extension of AIC," *Annals of the Institute of Statistical Mathematics*, Vol. 49, No. 3, 411–434.
8. Ja-Yong, Koo (1997). "Spline Estimation of Discontinuous Regression Functions," *J. Computational and Graphical Statistics*, **6**, No. 3, 266–284.
9. Kitagawa, G. and Gersch, W. (1996), *Smoothness Priors Analysis of Time Series*, Lecture Notes in Statistics, No. 116, Springer-Verlag, New York.
10. Kitagawa, G. and Matsumoto, N. (1996), "Detection of Coseismic Changes of Underground water Level", *Journal of the American Statistical Association*, Vol. 91, No. 434, 521–528.
11. Konishi, S. and Kitagawa, G. (1996), "Generalized Information Criteria in Model Selection ", *Biometrika*, Vol. 83, No. 4, 875–890.
12. Ohtani, S., Potemra, T.A., Newell, P.T., Zanetti, L.J., Iijima, T., Watanabe, M., Blomberg, L.G., Elphinstone, R.D., Murphree, J.S., Yamauchi, M., and Woch, J.G. (1995). "Four large-scale field-aligned current systems in the dayside high-latitude region," *J. Geophysical Research*, **100**, A1, 137–153.
13. Rissanen, J. (1978), "Modeling by shortest data description," *Automatica*, Vol. 14, 465–471.
14. Sakamoto, Y. and Akaike, H. (1978), "Analysis of cross-classified data by AIC", *Annals of the Institute of Statistical Mathematics*, Vol. 30, No.1, 185–197.
15. Simon, H. A. (1977), *Models of Discovery*, D. Reidel Publishing Company, 38–40.
16. Simon, H. A., Veldes-Perez, R. E. and Sleeman, D. H. (1997), "Scientific discovery and simplicity of method," *Artificial Intelligence*, Vol. 91, 177–181.
17. Takanami, T. and Kitagawa, G. (1991), "Estimation of the Arrival Times of Seismic Waves by Multivariate Time Series Model", *Annals of the Institute of Statistical mathematics*, Vol. 43, No. 3, 407–433.

Empirical Comparison of Competing Query Learning Methods

Naoki Abe* , Hiroshi Mamitsuka and Atsuyoshi Nakamura

NEC C& C Media Research Laboratories
4-1-1 Miyazaki, Miyamae-ku, Kawasaki 216-8555 JAPAN
{abe, mami,atsu}@ccm.cl.nec.co.jp

Query learning is a form of machine learning in which the learner has control over the learning data it receives. In the context of discovery science, query learning may prove to be relevant in at least two ways. One is as a method of selective sampling, when a huge set of unlabeled data is available but a relatively small number of these data can be labeled, and a method that can selectively ask valuable queries is desired. The other is as a method of experimental design, where a query learning method is used to inform the experimenter what experiments are to be performed next.

There are mainly two approaches to query learning, the *algorithmic* approach and the *information theoretic* approach. The algorithmic approach to query learning was initiated by Angluin's query learning model [3], and was subsequently explored by many researchers in the area of computational learning theory. The information theoretic approach to query learning dates back to earlier works in statistics and more recent development in the area of neural networks [5, 6]. In this paper, we attempt to compare these two approaches by emprically comparing the performance of learning methods from the two schools in a number of different settings.

As a test bed for our empirical comparison, we mainly consider the problem of learning binary and n-ary relations, in which each dimension can be clustered into a relatively small number of clusters, each consisting of members that are behaviorally indistinguishable. (c.f. [1].) The relation learning problem is especially relevant to discovery science, because most popular databases in the real world are *relational* databases.

The query learning methods we consider are the following: As a method of the algorithmic approach, we employ SDP (Self-Directed-Prediction) algorithm proposed and analyzed in [4]. This algorithm first makes systematical queries to discover the clusters in each dimesion, and eventually reaches a state of perfect generalization. As a method of the information theoretic approach, we employ a variant of 'Query by committee'(QBC) [6]. QBC, in its original form, makes use of many copies of an idealized randomized agent learning algorithm (Gibbs algorithm), and queries the function value of a point at which their predictions

* This author is also affiliated with Department of Computational Intelligence and Systems Sciences, Interdisciplinary Graduate School of Science and Engineering, Tokyo Institute of Technology, 4259 Nagatsuta, Midori-ku, Yokohama 226 JAPAN. This research was supported in part by the Grant-in-Aid of the Ministry of Education, Science, Sports and Culture, Japan.

are maximally spread. Here we combine it with a randomized version of WMP (weighted majority prediction) algorithm proposed and studied in [1].

The empirical results indicate that, with respect to data efficiency, SDP is superior in the absence of noise, and QBC with WMP wins when noise is present. (See Figure 34.) The original QBC is extremely computationally demanding, however, since it assumes an idealized randomized agent algorithm. As a possible remedy for this problem, we have proposed else-where a query learning strategy called 'query by boosting' (QBoost) which combines the idea of QBC with that of performance boosting [2]. With boosting being one of the most important innovations coming from the computational approach to learning, query by boosting is a result of taking advantage of the two approaches. Here we demonstrate, through experiments using both real world data sets and simulation data, that in many situations query by boosting gives the best result overall.

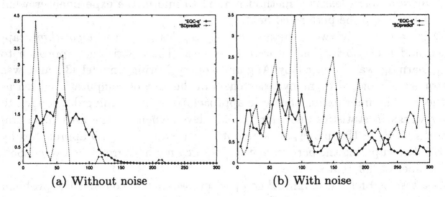

(a) Without noise   (b) With noise

Fig. 1. Learning curves for EQC and SD-Predict.

References

1. N. Abe, H. Li, and A. Nakamura. On-line learning of binary lexical relations using two-dimensional weighted majority algorithms. In *Proc. 12th Int'l. Conference on Machine Learning*, July 1995.
2. N. Abe and H. Mamitsuka. Query learning strategies using boosting and bagging. In *Proceedings of the Fifteenth International Conference on Machine Learning*, 1998.
3. D. Angluin. Learning regular sets from queries and counterexamples. *Inform. Comput.*, 75(2):87–106, November 1987.
4. A. Nakamura and N. Abe. On-line learning of binary and n-ary relations over multi-dimensional clusters. In *Proc. 8th Annu. Workshop on Comput. Learning Theory*, July 1995.
5. G. Paass and J. Kindermann. Bayesian query construction for neural network models. In *Advances in nueral information processing systems 7*, pages 443–450, 1995.
6. H. S. Seung, M. Opper, and H. Sompolinsky. Query by committee. In *Proc. 5th Annu. Workshop on Comput. Learning Theory*, pages 287–294. ACM Press, New York, NY, 1992.

Abstracting a Human's Decision Process by PRISM

Yoshitaka Kameya and Taisuke Sato

Dept. of Computer Science, Tokyo Institute of Technology
2-12-2 Ookayama Meguro-ku Tokyo Japan 152-8552
{kame,sato}@cs.titech.ac.jp

1 Introduction

Everyday we repeatedly make decisions, which are sometimes logical, and sometimes uncertain. In some fields such as marketing, on the other hand, modeling individuals' decision process is indispensable for the prediction of the trend. Logical rules however give us no way to represent uncertainties in the decision process, so new tool is required. PRISM (**PR**ogramming **I**n **S**tatistical **M**odeling) [5] is a programming language for symbolic-statistical modeling, whose theoretical basis is *distributional semantics* and the learning algorithm for *BS programs* [4]. In this paper, we conducted an experiment on a car-evaluation database in UCI ML repository [3], in which attributes of various cars are evaluated by a human. The database was originally developed for the demonstration of DEX [1], whose domain has a hierarchy (Figure 1) on ten multi-valued variables: six attributes of a car, three intermediate concepts and the final evaluation, but intermediate concepts are unobservable. Our purpose is to model the decision process and obtain, from estimated parameters, a probabilistic relation between the evaluation and the intermediate concepts.

2 PRISM programs

Based on the hierarchy in Figure 1 and tuples of the final evaluation and attributes, we modeled a human's evaluation process as a PRISM program. PRISM programs may look like just Prolog programs but they have a built-in probabilistic predicate msw/3, where $\mathtt{msw}(i,n,v)$ holds if the multi-valued probabilistic switch i takes the value v on the trial n (V_i, the set of discrete values of switch i, is assumed to be predefined). Each switch i has a function $\theta_i : V_i \to [0,1]$. $\theta_i(v)$, the probability of switch i taking the value v, is called a *parameter*, where $\sum_{v \in V_i} \theta_i(v) = 1$. In our model, car/2 represents the final evaluation *Car*:

```
car([Buy,Mnt,Drs,Psns,LBt,Sfty],Car):-
    price([Buy,Mnt],Price),tech([Drs,Psns,LBt,Sfty],Tech),
    msw(car(Price,Tech),null,Car).
```

Each parameter of the switches msw(car(Price,Tech),Car) represents an entry of the conditional probability table for $Pr(Car \mid Price, Tech)$. The above clause therefore indicates that *Car* is determined with probability $Pr(Car \mid Price, Tech)$. These probabilistic rules are used to capture the uncertainty in his/her mind.

3 Experimental results

We implemented the EM algorithm [7], an iterative algorithm for maximum likelihood estimation, as a built-in learning routine of PRISM programming system.

Given all tuples, the system estimated the parameters after iterating 1,394 times (It takes about 3 minutes on Pentium II 333 MHz). The estimated conditional probability table for $Pr(Car \mid Price, Tech)$ is shown in Table 1 ("—" stands for 0.0). This table can be seen as a probabilistic relation between the final evaluation (Car) and the intermediate concepts ($Price, Tech$). Intermediate concepts are more abstract than attributes, hence we obtained an *abstract* knowledge of car evaluation. Abstract knowledge helps us get essential ideas. Note that the result is not obvious because we are only given the hierarchy and tuples of the final evaluation and attributes. Our approach is applicable to a situation where given data is not complete (some tuples or some attribute values are missing), or where several humans' data are given. In the latter case, the obtained relation will abstract the *trend* of their evaluation. Furthermore, from the obtained knowledge, a logical rule "**if** $Tech$ is *poor* **then** Car will be *unacc* (unacceptable)" can be extracted. Automatic extraction of such logical rules is the future work.

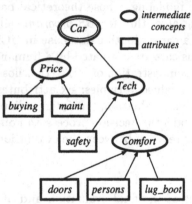

Fig. 1. The concept hierarchy

○ intermediate concepts
□ attributes

Table 1. Estimated conditional probability table

Price	Tech	Car			
		unacc	acc	good	v-good
v-high	good	0.767	0.233	—	—
v-high	satisf	0.993	0.007	—	—
v-high	poor	1.000	—	—	—
high	good	0.107	0.893	—	—
high	satisf	0.975	0.025	—	—
high	poor	1.000	—	—	—
med	good	0.096	0.810	—	0.093
med	satisf	0.963	0.037	—	—
med	poor	1.000	—	—	—
low	good	0.063	0.177	0.432	0.327
low	satisf	0.915	0.058	0.028	—
low	poor	1.000	—	—	—

4   Conclusion

Hierarchical knowledge is also used in the research fields of mining association rules in very large databases [2, 6]. Our model-based approach has a chance to reach more highly understandable and more interesting rules. Besides, if the model is appropriate, the support from statistics makes obtained knowledge statistically reliable. The constructed model itself leaves room for improvement.

References

1. Bohanec, M., Introduction to DEX, Jožef Stefan Institute, Report DP-6240, 1991.
2. Han, J., Cai, Y., and Cercone, N., Knowledge discovery in databases: An attribute-oriented approach, In *Proc. of the 18th VLDB Conf.*, pp.547–559, 1992.
3. Merz, C. J., and Murphy, P. M., *UCI* Repository of machine learning databases, http://www.ics.uci.edu/~mlearn/MLRepository.html, University of California, Irvine, Dept. of Information and Computer Sciences, 1998.
4. Sato, T., A statistical learning method for logic programs with distribution semantics, In *Proc. of Intl. Conf. on Logic Programming*, Tokyo, pp.715–729, 1995.
5. Sato, T., and Kameya, Y., PRISM: A symbolic-statistical modeling language, In *Proc. of the 15th Intl. Joint Conf. on Artif. Intell.*, pp.1330–1335, 1997.
6. Srikant, R., and Agrawal, R., Mining generalized association rules, In *Proc. of the 21st VLDB Conf.*, pp.407–419, 1995.
7. Tanner, M., *Tools for Statistical Inference* (2nd ed.), Springer-Verlag, 1986.

Mechanisms of Self-Organized Renormalizability

Hiroshi H. Hasegawa and Yoshikazu Ohtaki

Department of Mathematical Sciences, Ibaraki University,
Mito, 310-8512, Japan

We construct mathematical models through the analysis of time serise data. But it is not so clear how we can justify the model which really describes the phenomena. It is not sufficient to judge that the models show similar time serises or space patterns. In general, the real systems are complex and the mathematical models are simple. We need more clear reason why the simple mathematical models really describe the complex phenomena.

We try to find the reason of the justification for the mathematical models through the theory of collective coordinates [1] and that of renomalization group [2]. Recently effective theories by integarating out the short timescale variables and renormalize the dynamics for the long time scale variables are constructed [3][4]. This method is known as the theory of renomalization group and is extention of the thermodynamics.

In the construction of the thermodynamics, there exists big separation of the time scale between the short time scale, which characterizes the motions of the micro-scale variables, and the long time scale, which does the macro one. Therefore, we can easily integrate out the micro-scale varibales and obtain the effective theory, the thermodynamics. Recently, this method becomes applicable for systems, which have scaling property from short time scale to long time one.

We are considering classification of systems, in which each class has the same effective theory. We can justify the mathematical model, if the model belongs the same class as the original system.

We expect that the method of the renomalization group is applicable for a system, which has $1/f$ noise[5][6][7]. There are many sytems, which show $1/f$ noise. For instance, some dynamical systems, in which synchronization transites to desynchronization [8]. In complex systems, such as the earthquake, genetic code in DNA and the flow of transportation we can also observe $1/f$ noise, which is known as phenomena of the self-organized criticality or the edge of chaos [9].

The $1/f$ noise behavior of the power spectrum density means the power decay behavior of the correlation function. We can expect self similar behavior for time scale transformation for the system such as $t \to \lambda t$. Therefore we can apply the method of the renomalization group. For some complex systems, we observe self-similarity in space, fractal. We can expect to apply the method of the renomalization group for not only in time but also in space.

In this paper we discuss why the phenomena of the $1/f$ noise are so common. It is pointed out that the $1/f$ noise is related with the transition between synchronization and desynchronization, the self-organized criticality or the edge of chaos. If the transition were only on a point in the parameter space, we could not observe the $1/f$ noise. We expect that there are mechanisms to make it common.

We propose two mechanisms of self-organized renormalizability, that is dynamical appearence of scaling property in macro-scale effective dynamics. One comes from the feedback effects from the underlying micro-scale dynamics. The other does from the cooperative noise effects from the underlying one. These mechanisms would answer the question why $1/f$ noise is common in nature.

Acknowledgement This work is supported by Grant-in-Aid for Science Research from the Ministry of Education, Science and Culture of Japan and also by Ibaraki University Satellite Venture Business Labolatory.

References

1. Y. Abe and T. Suzuki. *Microscopic Theories of Nuclear Collective Motions*, Prog. Theor. Phys. Suppl. **75**(1983)74.
2. K. G. Wilson and J. Kogut. *Renormalization group and ϵ-expansion*, Phys. Repts. **C12**(1974)75.
3. Y. Oono. Butsuri **52**(1997)501.
4. L.-Y. Chen, N. Goldfield and Y. Oono., Phys. Rev. **E54**(1996)376.
5. P. Manneville and Y. Pomeau., Phys. Lett. **77A**(1979)1.
6. J. E. Hirsch, M. Nauenberg and D. J. Scalapino., Phys. Lett. **87A**(1982)391.
7. B. Hu and J. Rudnick. Phys. Rev. Lett. **48**(1982)1645.
8. H. Fujisaka and T. Yamada. Prog. Theor.Phys. **74**(1985)918.
9. P. Bak, C. Tang and K. Wiesenfeld. Phys. Rev. Lett. **59**(1987)381.

An Efficient Tool for Discovering Simple Combinatorial Patterns from Large Text Databases

Hiroki Arimura[1], Atsushi Wataki[1]*, Ryoichi Fujino[1], Shinichi Shimozono[2], and Setsuo Arikawa[1]

[1] Department of Informatics, Kyushu University, Hakozaki 6-10-1, Fukuoka 812–8581, Japan
{arim,wataki,fujino,arikawa}@i.kyushu-u.ac.jp

[2] Dept. of Artificial Intelligence, Kyushu Inst. of Tech., Iizuka 820–8502, Japan
sin@ai.kyutech.ac.jp

Abstract. In this poster, we present demonstration of a prototype system for efficient discovery of combinatorial patterns, called *proximity word-association patterns*, from a collection of texts. The algorithm computes the best k-proximity d-word patterns in almost linear expected time in the total input length n, which is drastically faster than a straightforward algorithm of $O(n^{2d+1})$ time complexity.

Since emerged in early 1990's, data mining has been extensively studied to develop semi-automatic tools for discovering valuable rules from stored facts in large scale databases. A rule which is looked for is an association among attributes that gives a useful property. Mainly a considerable amount of results have been known for well-defined and structured databases, such as relational databases with boolean or numeric attributes [3].

Beside this, recent progress of measuring and sensing technology, storage devices and network infrastructure has been rapidly increasing the size and the species of weakly-structured databases such as bibliographic databases, e-mails, HTML streams and raw experimental results such as genomic sequences. These lead to potential demands to data mining tools for databases where no attributes or structure is assumed in advance. However, there are still a few results in this direction [6]. One difficulty might be that a tool should quickly extract a structure behind the data as well as discover rules of interests. Our aim is to develop an efficient tool for text databases along the lines of data mining.

We consider a data mining problem in a large collection of unstructured texts based on association rules over subwords of texts. A *k-proximity word association pattern* is an expression such as (TATA, TAGT, AGGAGGT; 30) that expresses a rule that if subwords TATA, TAGT, and AGGAGGT appear in a text in this order with distance no more than $k = 30$ letters then a specified property ξ will hold over the text with a probability.

The data mining problem we consider is the *maximum agreement problem* defined as follows. Assume that we are given a collection of documents with an *objective condition*, that is, a binary label ξ over texts in S that indicates if a

---

* Presently working for Fujitsu LTD.

text has a property of interest. A pattern π *agrees with* ξ on s if π matches s if and only if $\xi(s) = 1$. The maximum agreement problem is to find a k-proximity d-word association pattern π that maximizes the number of documents in S on which π agrees with ξ.

The notion of proximity word association patterns extends frequently used *proximity patterns* consisting of two strings and a gap [5]. An algorithm that efficiently solves this problem can be applied in a wide range of practical problems in bioinformatics, e.g., the discovery of a consensus motif from protein sequences in [6]. Further, the maximum agreement problem plays an important role in computational learning theory; it is shown that an algorithm that efficiently solves the problem for a class with moderate complexity will be an efficient learner with the same class in the framework of *Agnostic PAC-learning* [4].

Clearly, the maximum agreement problem by word-association patterns is polynomial-time solvable in $O(n^{2d+1})$ time if the patterns are formed from at most d strings. However, the practical importance of the problem requires more efficient, essentially faster algorithm. Hence, we have devised an algorithm that efficiently solve the maximum agreement problem, which finds a d-words k-proximity word-association pattern with the maximum agreement in expected running time $O(k^{d-1}n\log^{d+1}n)$ and $O(k^{d-1}n)$ space with the total length n of texts, if texts are uniformly random strings [1,2]. Even in the worst case the algorithm runs in time $O(k^d n^{d+1}\log n)$ which is essentially faster than the naive method.

In this poster, we present demonstration of a prototype system based on our algorithm, which is written in C together with suffix tree index structure for managing a text database. From the good average case behavior of the underlying algorithm, we can expect that for the inputs being likely random strings such as the biological sequences the algorithm drastically reduces the computation time for finding the best word-association pattern. In the computational experiments on biological sequences from GenBank database, the prototype system finds the best patterns in less than a minute for an input text of around several tens of kilo bytes.

References

1. Arimura,H., Wataki,A., Fujino, R., Arikawa, S., A fast algorithm for discovering optimal string patterns in large text databases. In Proc. ALT'98, LNAI, Springer, 1998. (To appear.)
2. Arimura, H., Shimozono, S., Maximizing agreement between a classification and bounded or unbounded number of associated words. Proc. ISAAC'98, LNCS, Springer, 1998. (To appear.)
3. Fukuda,T., Morimoto, Y., Morishita, S. and Tokuyama,T., Data mining using two-dimensional optimized association rules. In Proc. SIGMOD'96, 13–23, 1996.
4. Kearns,M. J., Shapire, R. E., Sellie, L. M., Toward efficient agnostic learning. *Machine Learning*, 17(2–3), 115–141, 1994.
5. Manber, U. and Baeza-Yates,R., An algorithm for string matching with a sequence of don't cares. IPL 37, 133–136 (1991).
6. Wang,J. T.-L., Chirn, G.-W., Marr,T. G., Shapiro,B., Shasha,D., Zhang. K., Combinatorial Pattern Discovery for Scientific Data: Some preliminary results. In Proc. SIGMOD'94, (1994) 115–125.

A Similarity Finding System under a Goal Concept

Makoto Haraguchi[1] and Tokuyasu Kakuta[2]

[1] Division of Electronics and Information Engineering, Hokkaido University,
N-13,W-8, Kita-ku, Sapporo 060-8628, JAPAN
E-mail: makoto@db-ei.eng.hokudai.ac.jp
[2] Faculty of Law, Hokkaido University, N-9,W-7, Kita-ku, Sapporo 060-0809, JAPAN
E-mail: kaku@juris.hokudai.ac.jp

1 GDA, an algorithm to find similarities between concepts

We often utilize taxonomic hierarchy of concepts to organize knowledge bases and to reason under them. The taxonomic hierarchy can be in itself a representation of similarities between concepts, depending on our way to define concepts. For instance, placing two concepts A and B under their super concept C means that A and B inherit and share the properties of C. Thus they are considered similar and to be defined as special kinds of C.

However the fact that A is a kind of C is only an aspect of A in the context given by C. We can have another aspect of A involved under a different context. The aim of this Poster and Software Demonstration is to present a software system that can build hypothetical similarities that is useful for analogical reasoning and CBR as well from the viewpoint of variable contexts. A GUI subsystem is also available for novice users at computer operations to analyze the hypothetical similarities and the content of our knowledge base.

For these purposes, we have designed an algorithm called Goal-Dependent Abstraction (GDA) to find a similarity, given a goal formula representing some context and a domain theory including a standard taxonomic hierarchy.

In a word, given a goal G and an order-sorted representation of domain knowledge in which every concept is denoted by a sort symbol, our GDA algorithm tries to find a grouping of sorts such that the original proof of the goal G is preserved, even if we replace one sort in a sort group with another sort in the same group. Any sort in the same group thus shares the explanation of the goal. This means that we need not distinguish two distinct sorts s_1 and s_2 in a group at least in the explanation of G. In this sense we can say that s_1 and s_2 are similar with respect to G. The grouping is now understood as a family of similarity classes of sorts.

2 Object-centered representation

Although the original GDA has been implemented based on a usual order-sorted formalism, the requirement for the similarities is too weak to constrain the search space, the set of all groupings of sort symbols. In addition, a standard order-sorted representation is not appropriate for writing complex knowledge for which we like to examine our algorithm. For these reasons, an object-centered formalism is introduced further into our representation. More exactly speaking, we allow to use a macro language to describe feature structures of sorts and their instance objects as in F-logic. Each macro expression is easily translated into a set of order-sorted formulas. The introduction of feature structures help us not only to write knowledge but also to investigate another constraints in order to reduce our search space. In fact we postulate that a grouping should be consistent with the conceptual structure consisting of subsumption relationships and the feature structures. This means that GDA examines only structural similarities in the sense of classical studies of analogy. Its computational efficiency is therefore improved.

3   Legal Domain to test GDA

Most legal system of legal rules may be incomplete, since there always exists a new case that cannot be covered directly by any rule in the system. The legal rules are thus often used analogically based on some similarities between the condition part of rules and cases in inquiry. To exclude irrelevant similarities from the possible ones and to justify the analogical application, a notion of legal purposes should be considered. The legal purpose coincides with our goal concept. In addition, a complex feature structure is always necessary to write a legal knowledge.

From these facts, we can say that the problem of analogical application of rules in legal domains is adequate to examine the ability of GDA algorithm. The analogy process proceeds as follows. Given a legal case, a legal goal, and a legal domain theory including a legal rule for which analogy is tried, GDA algorithm first checks if the rule and the case share the goal and its proof as well. If the goal is not achieved under both the rule and the case, GDA algorithm fails in applying the rule to the case. Otherwise it starts on finding a grouping, a family of similarity classes, which meets the conditions mentioned in the previous sections. The output grouping is then represented as a hypothetical hierarchy, and is presented to the user graphically. Also an abstract level explanation of the goal is visually shown on one's terminal so that even a lawyer easily checks the appropriateness of hypothetical hierarchy (similarity) and the abstract explanation formed based upon the hierarchy. In case he regards the output inappropriate, GDA examines another possible goal.

The actual example to be demonstrated is a real one: analogical application of Japanese Civil Code 93 to a case of act as agent. The rule is concerned with a notion of declaration of intention. Hence it cannot be applied directly to the case of act as agent. GDA succeeds in forming a hypothetical hierarchy which lawyers agree with, given a goal concept "fairness" of human acts.

Parallel Induction Algorithms for Large Samples

Tohgoroh Matsui, Nobuhiro Inuzuka, Hirohisa Seki, and Hidenori Itoh

Dept. of Intelligence and Computer Science, Nagoya Institute of Technology,
Gokiso-cho Showa-ku Nagoya 466-8555, Japan.
{tohgoroh,inuzuka,seki,itoh}@ics.nitech.ac.jp
http://www-sekilab.ics.nitech.ac.jp/

1 Introduction

We propose three approaches to implement an Inductive Logic Programming (ILP) system FOIL in parallel, partitioning the search space, the training set, and the background knowledge. We experimented on a parallel environment to compare among these approaches and discuss the efficiency of them.

2 ILP and FOIL

ILP, in short, induces a hypothesis H that *explains* target relations from their positive and negative examples E with background knowledge B.

QUINLAN's FOIL [1], known as one of the most successful ILP systems, requires a set of all examples T_R belonging to relation R in B. FOIL starts with the left-hand side of the clause and specializes it keeping a training set T that starts from E to evaluate literals. While a clause C covers negative examples, FOIL adds a literal L_R using R to C, T and T_R are *joined* to T' consisting of tuples covered by $C \wedge L_R$. Positive examples covered by the found clause are removed from E, and FOIL returns the loop with the new training set.

3 Parallel Implementations

Two main approaches to designing parallel algorithms for inductive learning are the search space parallel approach and the data parallel approach. The data of ILP consist of a training set T and background knowledge B, and hence we partition the search space, the training set, and the background knowledge[1].

Search Space Parallel Approach –SSP-FOIL–. Because each candidate for L is independent of another, we can evaluate a candidate without any other candidates, and hence we divide candidates and evaluate each of them.

Training Set Parallel Approach –TSP-FOIL–. Each tuple in T is independent of another in T, so that we can *join* $t_i \in T$ and T_R without $t_j \in T$. Accordingly, we divide T and *join* each of them and T_R, then collect the results to a processor and evaluate the *union* of them.

---

[1] For further details of these algorithms, see [2].

Background Knowledge Parallel Approach –BKP-FOIL–. The converse is logically equivalent, and hence we can *join* $t_{Ri} \in T_R$ and T without $t_{Rj} \in T_R$. We divide T_R and *join* each of them and T, then collect the results to a processor and evaluate the *union* of them.

4   Experimental Results

We implemented the three different algorithms in Java with TCP/IP communication. We experimented on FUJITSU AP3000 with a processor to divide tasks and some processors to compute sub-tasks, using MICHALSKI's `eastbound/1` that contains 216 kinds of cars and 2000 trains.

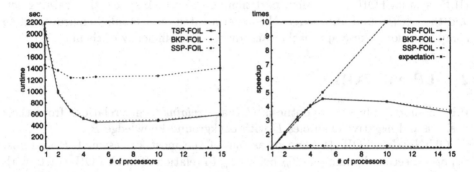

Fig. 1. Run-time and speed-up values of experimental results with MICHALSKI's `eastbound/1` for three different types of partitioning approach.

5   Discussion

Search space parallel approach is worse than the other approaches because the sizes of divided tasks may not be all the same. On the other hand, FOIL and our systems need all positive tuples belonging to relations in background knowledge. If the relations are given as logic programs, we cannot divide the background knowledge efficiently, and hence training set parallel approach is superior to background knowledge parallel approach. However, the experimental results highlighted the problems posed by the amount of communications caused by the strong dependence on FOIL's tuple data representation. Part of our future work includes a reduction of the amount of communications.

References

1. J. R. QUINLAN and R. M. CAMERON-JONES: FOIL: A Midterm Report. In: P. Brazdil (ed.): Proceedings of the 6th European Conference on Machine Learning, Lecture Notes in Artificial Intelligence, Vol. 667, pp. 3-20. Springer-Verlag (1993).
2. Tohgoroh MATSUI, Nobuhiro INUZUKA, Hirohisa SEKI, and Hidenori ITOH: Comparison of Three Parallel Implementations of an Induction Algorithm. Proceedings of the 8th Parallel Computing Workshop. Fujitsu Laboratories Ltd. (1998).

Toward Effective Knowledge Acquisition with First Order Logic Induction

Xiaolong Zhang[1], Tetsuo Narita[1], Masayuki Numao[2]

[1] Solution Center, XI and NII Promotion, IBM Japan, Tokyo 103-8510 Japan
{xzhang,narita}@jp.ibm.com
[2] Department of Computer Science, Tokyo Institute of Technology,
2-12-1 Oh-okayama, Meguro, Tokyo 152, Japan
numao@cs.titech.ac.jp

1 Introduction

Given a set of noisy training examples, an approximate theory probably including multiple predicates and background knowledge, to acquire a more accurate theory is a more realistic problem in knowledge acquisition with machine learning methods. An algorithm called KNOWAR[2] that combines a multiple predicate learning module and a theory revision module has been used to deal with such a problem, where inductive logic programming methods are employed.

Clearly a single information-based heuristic used to select hypothesis (or regularities) and tolerate noise in training data sometimes makes algorithms learn inaccurate theories, due to its own focus on evaluating hypotheses. Two heuristics have been employed in KNOWAR, which are the MDL(minimum description length)-based encoding scheme and the Laplace estimate, aiming at continuously inducing regularities even though a training set is noised.

2 Theory Preference Criterion and Its Applications

From the viewpoint of compression in induction, the MDL principle focuses on the compression of the whole training data. A clause which contributes great compression is of high probability to be selected for further refinement. Suppose a specialized clause C_2 is refined based on a clause C_1 given a training set, the compression produced by C_2 is $Cmp(C_2, C_1)$, whose definition is given as:

$$Cmp(C_2, C_1) = (1 - \frac{e_2}{D_0})E(C_2, D_2) - (1 - \frac{e_1}{D_0})E(C_1, D_1) - C_{mc}(C_2, C_1),$$

where e_1 and e_2 are the number of misclassified examples by clauses C_1 and C_2 respectively. $E(C, D) = P \times (\log_2 \frac{P}{D} - \log_2 \frac{P_0}{D_0})$, $C_{mc}(C\prime_i, C_i) = \log_2(Num)$. $E(C, D)$ refers to the compression of a clauses C, where P_0 and D_0 are the number of positive and total examples in the remaining training set respectively; P and D are the number of the positive examples and total examples covered by the clause C respectively. $C_{mc}(Cl_i, C_i)$ refers to the model complexity for an algorithm to select a clause from the refined ones. C_i is refined into a set of clauses(with a size Num). If a clause Cl has a higher value of $Cmp(Cl, C_1)$, it is added to the beam(A beam search strategy is applied) for further refinement. Details of the encoding scheme can be found in [2].

Another heuristic is the Laplace estimate. Suppose that a clause c covers p positive and n negative examples, the Laplace estimate $Lap(c)$ with respect to c is defined as: $Lap(c) = \frac{p+1}{(p+n)+2}$.

Both two heuristics are employed in refinement, and qualified clauses are selected to beams b_1 and b_2, where b_1 and b_2 are for the MDL and Laplace estimate, respectively. The beam B_1 consists of the clauses with high compression, and b_2 consists of the clauses with high Laplace rate. The clauses in b_1 are firstly refined. If the MDL heuristic is unable to obtain a resultant clause (e.g., b_1 becomes empty), clauses in b_2 are refined further.

The above two heuristics are used in the revising- and learning sessions of KNOWAR (more details refer to [2]). The input data to the algorithm includes an initial theory, a training set probably including noise and background knowledge. The output of the algorithm is a resultant theory that has high predictive accuracy when predicting unseen data. The algorithm uses the revising-clause session to revise the clauses in a given initial theory, and employs the learning-clause session to learn new clauses.

KNOWAR has applied to conduct finite element mesh (FEM) design ([1]). The task is to divide a complex object into a number of finite elements used to structure analysis. There are five objects, a, b, c, d and e, each of them having its corresponding training set. Cross-validation was performed. For each object (e.g., the object a) in turn, the training set was generated from the other objects (c, d, e and f). Learned clauses are tested on all edges. If both heuristics are used, KNOWAR learns a theory with 90.2% accuracy; if only the MDL heuristic is used, the accuracy of the result theory is 88.1%. More experiments in other domains have been done.

3   Conclusions

The main research issues we addressed include the areas of inductive logic programming and theory revision. KNOWAR, with the proposed two heuristics, learns more accurate theories.

References

1. B. Dolšak, I. Bratko, and A. Jezernik. Finite element mesh design: An engineering domain for ILP application. In *Proc. of the Fourth International Workshop on Inductive Logic Programming*, pages 305–320, Germany, 1994.
2. Xiaolong Zhang. *Knowledge Acquisition and Revision with First Order Logic Induction*. PhD thesis, Dept. of Computer Science, Tokyo Institute of Technology, Tokyo, Japan, 1998.

A Logic of Discovery

Jānis Bārzdiņš[1] and Rūsiņš Freivalds[1] and Carl H. Smith[2]

[1] Institute of Math and Computer Science, University of Latvia, Raiņa bulvāris 29, LV-1459, Riga, Latvia
[2] Department of Computer Science, University of Maryland, College Park, MD 20742 USA

A logic of discovery is introduced. In this logic, true sentences are discovered over time based on arriving data. A notion of expectation is introduced to reflect the growing certainty that a universally quantified sentence is true as more true instances are observed. The logic is shown to be consistent and complete. Monadic predicates are considered as a special case.

In this paper, we consider learning models that arise when the goal of the learning is *not* a complete explanation, but rather some facts about the observed data. This is consistent with the way science is actually done. We present a logic of discovery. Standard first order logic is extended so that some lines of a proof have associated expectations of truth. This is used to generalize from $A(a)$ for some set of examples a to a formulae $\forall x A(x)$. We show how to prove all true first order formulae over monadic predicates.

A *expectation level* is a rational number between 0 and 1. Intuitively, a 0 indicates no confidence in the answer and a 1 indicates certainty. The proofs that we construct will start with a finite set of assumptions, and continue with a list of statements, each of which follows by the rules of logic from some prior statements or assumptions. Some of the statements will be paired with an expectation level.

The basic idea is that traditional data such as "$f(3) = 5$" can be taken as true instance of some predicate F, in which case we say "$F(3, 5)$ is true." We consider algorithms that take true instances of predicates and produce, deduce, infer, or whatever, true formulae about the predicate F.

The basic idea is to have the proof develop over infinitely many statements. At first, the idea of an infinite prove may appear blasphemous. However, consider for the moment that induction is really infinitely may applications of monus ponens collected into a single axiom. While in the case of induction, it is easy to see where the infinitely many applications of monus ponens will lead to, we consider cases where the outcome is not so clear. Hence, we explicitly list all the components of our infinite arguments. In a proof of assertion A, A would appear on an infinite subset of the lines of the proof, each time with a larger expectation of truth.

The intuitive meaning of A/e is that, based on the prior steps of the proof, we have expectation e that the formulae A is true. A system of positive weights is employed to calculate the expectations. A *weighting function* is a recursive function w such that $\sum_{n=0}^{\infty} w(n) = 1$. For example, let $w(n) = 1/2^{n+1}$. The idea of the weighting system is that some examples are more relevant then others. The weighting system will be used to determine the expectation of a universally quantified sentence, based on known examples.

Our proof system starts with a standard first order theory. Proofs are built in the standard fashion. We allowed to write "$A(t)$" as a line of a proof if $A(t)$ can be deduced by the usual axioms from prior lines in the proof, or we are given that $A(t)$ is true as an input to the discovery process. For this, an subsequent examples, we consider formulae of a single variable for notational simplicity. However, we envision formulae of several variables. Our target scenario is the discovery of "features" of some observed phenomenon represented (encoded) as a function over the natural numbers. While the data might look like "$f(3) = 5$" we use a predicate symbol F for the function and enter "$F(3,5)$" as a line of the proof we are building.

We add to the standard set of axioms for a first order theory the following e-axiom of expectation axiom for A a formula with n free variables:

$$\frac{A(t_1,\ldots,t_n) \mid (t_1,\ldots,t_n) \in T^{(n)} \subseteq \mathbb{N}^{(n)}}{(\forall x_1,\ldots,x_n)A(x_1,\ldots,x_n)/ \displaystyle\sum_{(t_1,\ldots,t_n)\in T^{(n)}} w(t_1)\cdot \ldots \cdot w(t_n)}$$

The set $T^{(n)}$ denotes the set of values for which we know the predicate A to be true for. We illustrate the use of the e-axiom for the $n = 1$ case. If in some lines of a proof we have $A(2)$ and then later $A(4)$, we would be able to use the e-axiom to obtain $(\forall x)A(x)/e$ where $e = w(2) + w(4)$.

We introduce the t-axiom, or truth axiom that allows the an expectation of 1 to be added to any sentence provable within the standard first order theory.

Definition 1. *A formula A is* epistemically provable *(or eprovable) iff there is an A' that is either A or logically equivalent to A for each $\epsilon > 0$, there is an $e \geq 1 - \epsilon$ such that A'/e is provable using the traditional first order theory augmented with the e-axiom and the t-axiom.*

Theorem 1. *Suppose w and w' are two different weight functions. Then, for any formulae A, A is eprovable with respect to w iff A is eprovable with respect to w'.*

Theorem 2. *First order logic plus the the e-axiom and the t-axiom is sound.*

Assume that all first order formulae are presented in prenex normal form.

Theorem 3. *Suppose Σ is a signature containing only monadic predicates. If f is a first order formula over Σ and I is an interpretation of the predicates in Σ, there is an eproof of f iff f is true according to the interpretation I.*

Theorem 4. *For any $n \geq 1$, for every true Π_n sentence S_1 there is an $(\lceil n/2 \rceil \cdot \omega)$-proof of S_1 and for every true Σ_n sentence S_2 there is an $(\lfloor n/2 \rfloor \cdot \omega + 2)$-proof of S_2.*

A Strong Relevant Logic Model of Epistemic Processes in Scientific Discovery

Jingde Cheng

Department of Computer Science and Communication Engineering
Kyushu University
6-10-1 Hakozaki, Fukuoka 812-8581, Japan

In various mathematical, natural, and social scientific literature, it is probably difficult, if not impossible, to find a sentence form that is more generally used to describe various definitions, propositions, theorems, and laws than the sentence form of 'if ... then ...'. In logic, a sentence in the form of 'if ... then ...' is usually called a conditional proposition or simply conditional which states that there exists a sufficient condition relation between the 'if' part (antecedent) and the 'then' part (consequent) of the sentence. Mathematical, natural, and social scientists always use conditionals in their descriptions of various definitions, propositions, theorems, and laws to connect a concept, fact, situation or conclusion and its sufficient conditions. Indeed, the major work of almost all scientists is to discover some sufficient condition relations between various phenomena, data, and laws in their research fields.

Any scientific discovery must include an epistemic process to gain knowledge of or to ascertain the existence of some empirical and/or logical conditionals previously unknown or unrecognized. As an applied and/or technical science, computer science should provide scientists with some epistemic representation, description, reasoning, and computational tools for supporting the scientists to suppose, verify, and then ultimately discover new conditionals in their research fields. However, no programming paradigm in the current computer science focuses its attention on this issue. The goal of my work are to construct a realistic model of epistemic processes in scientific discovery, and then, based on the model, to establish a novel programming paradigm, named 'Epistemic Programming', which regards conditionals as the subject of computing, takes primary epistemic operations as basic operations of computing, and regards epistemic processes as the subject of programming.

The philosophical observations and/or assumptions on scientific discovery processes and their automation, which underlie my research direction, are as follows:
(1) Any scientist who made a scientific discovery must have worked in some particular scientific field and more specifically on some problem in a particular domain within the field. There is no universal scientist who can make scientific discoveries in every field.
(2) Any scientific discovery must have, among other things, a process that consists of a number of ordered epistemic activities that may be contributed by many scientists in a long duration. Any scientific discovery is nether an event occurring in a moment nor an accumulation of disorderly and disorganized inquisitions.

(3) Any scientific discovery process must include an epistemic process to gain knowledge of or to ascertain the existence of some empirical and/or logical conditionals previously unknown or unrecognized. Finding some new data or some new fact is just an initial step in a scientific discovery but not the scientific discovery.

(4) Any scientific discovery process can be described and modeled in a normal way and therefore it can be simulated by computer programs automatically.

(5) Even if scientific discovery processes can be simulated by computer programs automatically in general, a particular computational process which can certainly perform a particular scientific discovery must take sufficient knowledge specific to the subject under investigation into account. There is no generally organized order of scientific discovery processes that can be applied to every problem in every field.

(6) Any automated process of scientific discovery must be able to assure us of the truth, in the sense of not only fact but also conditional, of the final result produced by the process if it starts from an epistemic state where all facts, hypotheses, and conditionals are regarded to be true.

Generally, for any correct argument in scientific reasoning as well as our everyday reasoning, the premises of the argument must be in some way relevant to the conclusion of that argument. In order to find a fundamental logic system that can satisfactorily underlie relevant reasoning, I have proposed some new relevant logic systems, named 'strong relevant logic', which are free of not only implicational paradoxes in classical mathematical logic but also conjunction-implicational and disjunction-implicational paradoxes in traditional (weak) relevant logics.

I have proposed a strong relevant logic model of epistemic processes in scientific discovery. In comparison with those epistemic process models based on classical mathematical logic, our model has the following features:

(1) We adopt predicate strong relevant logic EcQ as the fundamental logic to underlie epistemic processes. EcQ is paraconsistent and is free of not only implicational paradoxes but also conjunction-implicational and disjunction-implicational paradoxes, and therefore, it allows inconsistent belief set and assure us of the validity of a belief in the form of conditional in any epistemic state, if all premises in the primary epistemic state are true and/or valid.

(2) We use the sentence set itself rather than its deductive closure to represent the current belief state of an agent and distinguishes explicitly known knowledge and implicit consequences (in the sense of strong relevant logic) of the known knowledge.

(3) Since EcQ is paraconsistent, our strong relevant logic model does not require the belief set must be consistent at any time.

(4) We take epistemic deduction, explicitly and implicitly epistemic expansions, and explicitly and implicitly epistemic contractions as primary epistemic operations, and models a belief revision as a process (i.e., models a belief revision as an epistemic process which consists of a sequence of primary epistemic operations) rather than a function.

Four-Fold Table Calculi for Discovery Science

Jan Rauch

Laboratory of Intelligent Systems, Faculty of Informatics and Statistics, University of Economics, W. Churchill Sq. 4, 13067 Prague, Czech Republic rauch@vse.cz

The goal of KDD is to search for interesting and useful patterns in data. Some of these patterns concern relations of two Boolean attributes φ and ψ. Such relations are evaluated on the basis of a four-fold table (FFT for short) $\langle a, b, c, d \rangle$ of φ and ψ. Here a is the number of objects satisfying both φ and ψ, b is the number of objects satisfying φ and not satisfying ψ, c is the number of objects not satisfying φ and satisfying ψ and d is the number of objects not satisfying neither φ nor ψ. An example of such pattern is an association rule [1]. Four-Fold Table Predicate Calculi (FFTPC for short) will be introduced. Formulae of FFTPC correspond to patterns evaluated on the basis of FFT. FFTPC are defined and studied in connection with development of GUHA procedures, see e.g. [2], [3], [4].

A **type of FFTPC** is an integer positive number T. **Formulae of FFTPC** of the type T are of the form of $\varphi \sim \psi$. Here φ and ψ are Boolean attributes derived from basic Boolean attributes $A_1, \ldots, A_T$ using usual propositional connectives $\wedge, \vee, \neg$. Symbol $\sim$ is an FFT quantifier, it corresponds to a $\{0, 1\}$-valued function $F_\sim$ defined for all four-fold tables $\langle a, b, c, d \rangle$. **Models of the FFTPC** of the type T are all $\{0, 1\}$-data matrices with T columns. We consider each such data matrix M with n rows to be a result of an observation of n objects. Basic attribute A_i is interpreted by i-th column of data matrix M for $i = 1, \ldots, T$. **Value** $Val(\varphi \sim \psi, M)$ **of formula** $\varphi \sim \psi$ **in data matrix** M is defined as $F_\sim(a, b, c, d)$ where $\langle a, b, c, d \rangle$ is four-fold table of φ and ψ in data matrix M. We usually write only $\sim (a, b, c, d)$ instead of $F_\sim(a, b, c, d)$.

Several **classes of FFT quantifiers** were defined, see e.g. [2], [4]. An example is a **class of implicational quantifiers**. FFT quantifier $\sim$ is implicational if the condition "if $\sim (a, b, c, d) = 1$ and $a' \geq a \wedge b' \leq b$ then also $\sim (a', b', c', d') = 1$" is satisfied for all FFT $\langle a, b, c, d \rangle$ and $\langle a', b', c', d' \rangle$. Quantifier $\exists^!_{p,\Theta}$ of likely implication with threshold Θ [2] is implicational, an other example of the implicational quantifier is quantifier $\exists_p$ of almost implication with threshold p: $\exists_p (a, b, c, d) = 1$ iff $a \geq p(a + b)$.

Conditions like "$a' \geq a \wedge b' \leq b$" are called **truth preservation conditions** (**TPC** for short) [4]. Condition $a' \geq a \wedge b' \leq b$ is **TPC for implicational quantifiers**. Examples of further classes of FFT quantifiers defined by TPC: **class of double implicational FFT quantifiers**: $a' \geq a \wedge b' \leq b \wedge c' \leq c$, **class of Σ-double implicational FFT quantifiers**: $a' \geq a \wedge b' + c' \leq b + c$, **class of Σ-equivalency FFT quantifiers**: $a' + d' \geq a + d \wedge b' + c' \leq b + c$, **class of FFT quantifiers with F-property**: $a' \geq a \wedge d' \geq d \wedge |b' - c'| \geq |b - c|$.

Deduction rules concerning formulae of FFTPC belong to main reasons why FFTPC can be useful for KDD. The deduction rule is a relation of the form

$\frac{\Phi_1,...,\Phi_n}{\Psi}$, where $\Phi_1,...,\Phi_n,\Psi$ are formulae. This deduction rule is correct if it is satisfied: *If $\Phi_1,...,\Phi_n$ are true in model M, then also Ψ is true in M*. We are interesting in deduction rules of the form $\frac{\varphi\sim\psi}{\varphi'\sim\psi'}$ where $\varphi\sim\psi$ and $\varphi'\sim\psi'$ are formulae of FFTPC. Such deduction rules can be used e.g.:(1) **To reduce an output of a data mining procedure:** If formula $\varphi\sim\psi$ is a part of the output and $\frac{\varphi\sim\psi}{\varphi'\sim\psi'}$ is the correct deduction rule, then it is not necessary to put $\varphi'\sim\psi'$ into the output if the used deduction rule is clear enough. (2) **To decrease number of actually tested formulae:** If formula $\varphi\sim\psi$ is true in the analysed data matrix and $\frac{\varphi\sim\psi}{\varphi'\sim\psi'}$ is the correct deduction rule, then it is not necessary to test $\varphi'\sim\psi'$. Various correct deduction rules are used in the GUHA procedure PC-ASSOC, e.g. [2]. Thus it is reasonable to ask when the deduction rule of the form $\frac{\varphi\sim\psi}{\varphi'\sim\psi'}$ is correct.

An **associated propositional formula** $\pi(\varphi)$ to attribute φ is the same string of symbols as φ but the particular basic attributes are understood as propositional variables. E.g.: $A_1\wedge A_2$ is a derived attribute and $\pi(A_1\wedge A_2)$ is a propositional formula $A_1\wedge A_2$ with propositional variables A_1 and A_2. FFT quantifier $\sim$ is **a-dependent** if there are non-negative integers a, a', b, c, d such that $\sim (a,b,c,d) \neq \sim (a',b,c,d)$, analogously for **b-dependent** quantifier. **Implicational quantifier** $\Rightarrow^*$ **is interesting** if $\Rightarrow^*$ is both a-dependent and b-dependent and if $\Rightarrow^* (0,0) = 0$.

A following theorem is proved in [3]: *If $\Rightarrow^*$ is the interesting implicational quantifier then $\frac{\varphi\Rightarrow^*\psi}{\varphi'\Rightarrow^*\psi'}$ is the correct deduction rule iff at least one of conditions* i), ii) *are satisfied* : (i): *Both* $\pi(\varphi)\wedge\pi(\psi) \to \pi(\varphi')\wedge\pi(\psi')$ *and* $\pi(\varphi')\wedge\neg\pi(\psi') \to \pi(\varphi)\wedge\neg\pi(\psi)$ *are tautologies.* (ii): $\pi(\varphi) \to \neg\pi(\psi)$ *is a tautology* . Similar theorems are proved for further classes of FFT-quantifiers, see e.g. [4].

Further important features of FFT quantifiers are related to the above defined classes of FFT quantifiers. E.g., **tables of critical frequencies** can be used to convert even very complex hypotheses test concerning FFT to a test of very simple inequalities, see e.g. [2], [4]).

This work is supported by grant 47160008 of Ministry of Education and by grant 201/96/1445 of the Grant Agency of the Czech Republic

References

1. Aggraval, R. et al: Fast Discovery of Association Rules. in Fayyad, U. M. et al.: Advances in Knowledge Discovery and Data Mining. AAAI Press, 1996.
2. Hájek, P., Holeňa, M.: Formal logics of discovery and hypothesis formation by machine. Accepted for "Discovery Science 1998".
3. Rauch, J.: Logical foundations of mechanizing hypothesis formation from databases. Thesis, Mathematical institute of Czechoslovak Academy of Sciences Prague, 1986, 133 p. (in Czech)
4. Rauch, J.: Classes of Four-Fold Table Quantifiers. Accepted for "Principles of Data Mining and Knowledge Discovery 1998"

Parallel Organization Algorithm for Graph Matching and Subgraph Isomorphism Detection

Yoshinori Nakanishi and Kuniaki Uehara

Department of Computer and Systems Engineering,
Kobe University
Nada, Kobe, 657–8501 Japan

Graph representation is a flexible and general-purpose tool to represent structured data, it is widely used in various fields, such as circuit design, concept representation, image cognition and so on. It is very significant to detect beneficial information and to retrieve specific substructures from a graph database. However, the computational cost of algorithms using graph representation is huge in general, a quick subgraph isomorphism detection algorithm is required [1], [2].

In the field of artificial intelligence, the RETE algorithm is proposed. The algorithm compiles the RETE network from each element in the conditional part of the rules and achieves matching once at most. In the same way, it is possible to attempt an efficient graph matching based on the idea that each graph corresponds to a production rule and each subgraph corresponds an element in the working memory. For example, if the set of graphs, shown in Fig. 1(a), is given, it is possible to compile the hierarchical network shown Fig. 1(b) by observing the subgraph isomorphism.

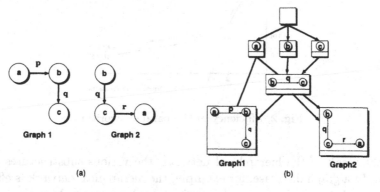

Fig. 1. Graph set and hierarchical network for graph retrival.

We adopt the MDL (Minimum Description Length) principle to detect the subgraph isomorphism out of a graph set. The best model on the MDL principle minimizes the sum of the description length of a subgraph and that of a graph set with respect to the subgraph. In this principle, it is necessary to evaluate all possible subgraphs for efficient description. However, as a matter of fact, the number of the combinations grows huge in a large-scale graph database. Hence we use the heuristics for speed-up detection as follows: once the description

length of a subgraph begins to increase, further combination of the subgraph
will not yield a smaller description length. During the evaluation process, we
reject the candidate subgraph if the description length begins to increase. That
is, our algorithm is a computationally constrained beam search.

Moreover we use the parallelism for speed-up detection as follows: multiple
processors search all subgraphs in given graphs simultaniously and new subgraph
is created if the highest value of combination is found. If the same subgraphs
are detected at the same time, the problem occurs that the hierarchical network
does not reflect the original graphs correctly. This procedure is a critical region
in this algorithm. Fortunately, since the computational cost of this procedure is
small, we think that the influence of the critical region can be neglected.

In order to verify the efficiency of parallelim, we used two kinds of graphs
which have 257 and 1285 vertices respectively. Fig. 2 shows the relation between
the number of processors and the processing time. In Fig. 2, the processing time
is clearly reduced in propotional to the number of processors. In this algorithm,
since the procedure is processed by each processor independently, the evaluation
of the same inter-node may be processed by different processors. If such an
overlap can be removed, it is possible to process them more quickly.

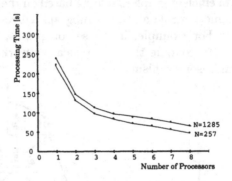

Fig. 2. Efficiency of the parallel algorithm.

As the result of the hierarchical network, the various substructures are discovered. In a graph database, for example, the hierarchical network is efficiently
used for retrieving various sizes of meaningful substructures. Moreover if we use
background knowledge as an evaluation measure, the proper substructures can
be discovered.

References

1. Cook, D. J. and Holder, L. B., "Substructure Discovery Using Minimum Description
   Length and Background Knowledge", *Journal of Artificial Intelligence Research*, 1,
   231–255 (1994).
2. Messmer, B. T. and Bunke, H., "A Network Based Approach to Exact and Inexact
   Graph Matching", Technical Report, LM-93-021, University of Berne, (1993).

Visualizing Semantic Clusters in the Internet Information Space

Etsuya Shibayama[1], Jun Yabe[1], Shin Takahashi[1], and Mitsuru Matsuda[2]

[1] Tokyo Institute of Technology, O-okayama, Meguro, Tokyo, 152-8552, Japan
{etsuya,yabe,shin}@is.titech.ac.jp,
[2] Hitachi Software Engineering Co. Ltd.

1 Introduction

Information resources on the Internet, including Web pages and news articles, constitute a huge, ill-structured, and continuously growing information space. *Knowledge discovery from the Internet* is a challenge. It includes useful knowledge that is difficult to be automatically exploited by the following reasons.

- The Internet is full of junk pages and articles. Conventional techniques rarely tolerate so noisy information as often found on the Internet.
- A single page or an article on the net is often too fine-grained as a unit of knowledge. We need techniques to extract a cluster of inter-related fine-grained pages and/or articles with semantic relationships among them.

In this abstract, we propose new filtering and visualizing techniques, which can be partial solutions to these problems.

2 Filtering by Reference Counting

The current major technique to extract information from the Internet is keyword matching, which, for instance, is employed by the AltaVista search engine. However, Web pages recommended by search engines are often obsolete and sometimes come from untrusted sources.

To alleviate this problem, we propose a filtering technique based on reference counting. We assume that web page creators make links to target pages if the targets are important for them. We have made an experimental system that presents a directed graph of web pages visualized in the following manner.

1. Retrieving Web pages by providing a few keywords to AltaVista.
2. With a force-directed algorithm[1], visualizing a graph whose nodes are collected pages and edges are links among them. In this stage, the pages at the same site are merged and those that have no links to or from any other collected pages are removed.

By this way, the user can easily find pages with a number of links. Furthermore, if several types of pages are mingled, the graph is divided into clusters each of which consists of pages semantically related each other.

In our informal tests, each tester could find *helpful* pages more rapidly with our system than with solely AltaVista. Although the sets of (potentially) presented pages by our system and AltaVista were the same, since our system filtered pages by reference counting and presented semantic structures, the performance could be different.

3   Visualizing Semantic Content and Relationships

We introduce a visualization technique suitable for a thread of news articles and reveal semantic clusters among articles. We follow conventional visualization techniques[2, 3] for document spaces and our technique consists of three phases: keyword extractions, dissimilarity analysis among keywords, layout calculation. However, we design algorithms so that they can tolerate noises produced from the following sources.

- News articles are informal documents and rarely have been proofread.
- Automatic keyword extractions from short articles are difficult and error-prone.

To reduce those noises, we employ the following.

- keyword extraction techniques that are optimized for news articles.
- a two-stage layout algorithm which first calculates the positions of keywords and then those of articles; notice that keyword placements can be more accurate since each keyword occurs frequently in the target newsgroup.
- a force-directed algorithm that takes account of certainty factors.

In our experiences using a prototype system, this technique is often useful to reveal semantic clusters within a thread of news articles even though they share similar keywords.

4   Related Work

Until now, various visualization techniques are proposed for document spaces[4, 2] and retrieved information[3] including Web pages[5]. They provide, however, no effective solutions to junk information that we meet frequently on the net.

References

1. T. M.J.Fruchterman and E. M.Reingold.: Graph Drawing by Force-directed Placement. *Software - Practice and Experience*, 21(11):1129–1164, 1991.
2. J. Tatemura.: Visualizing Document Space by Force-directed Dynamic Layout. *IEEE Symposium on Visual Languages*, 119–120, 1997.
3. S. K. Card.: Visualizing Retrieved Information: A Survey. *IEEE Computer Graphics and Applications*, 16(2):63–67, 1996.
4. M. Chalmers and P. Chitson.: Bead: Explorations in Information Visualization. *ACM SIGIR '92*, 330–337, 1992.
5. P. Pirolli, J. Pitkow, and R. Rao.: Silk from a Sow's Ear: Extracting Usable Structures from the Web.: *ACM CHI '96*, 118–125, 1996.

KN on ZK — Knowledge Network on Network Note Pad ZK *

Sachio Hirokawa[1] and Tsuyoshi Taguchi[2]

[1] Computer Center, Kyushu University , Hakozaki 6-10-1, Fukuoka 812-8581, JAPAN
[2] Department of Electrical Engineering and Computer Science, Kyushu University ,
Hakozaki 6-10-1, Fukuoka 812-8581, JAPAN

1   Links as Knowledge Network

Internet is a great book of knowledge. We are abundantly suplied with information. We can easily access them and can collect a huge amount of data. The problem is the quality of the information, and how to discover the valuable information. Search engines help to decrease the size. However, the results of search have no structure. The authors think that the value is in the connection of several information. Each information can be considered as a segment of knowledge. The value *is* in the network of those segments. To discover a knowledge is to form such a network of knowledge.

2   Practical Experience of Knowledge Network

The final goal of our research is to implement a mechanism to form such knowledge network from massive data in the internet. We plan to achieve this goal in three phases – (a) practical experience, (b) technological experience and (c) formalization and mechanization. Meaningful and successful examples of knowledge network should be the first step. Such examples is valuable itself, even if we do not have many. The system "KN on ZK" is a tool to create such a practical knowledge network. It is not a automatic tool, but a tool to handle manually. We know that a personal bookmark of an specific area is useful, if it is compiled by an appropriate person who knows the area in depth. The information is filtered by his point of view and maintained. We think that a key point of discovery is in the process of growth of such a bookmark. "KN on ZK" is a visual tool to raise the knowledge network. It displays the structure as a directed graph. The graph increases and grows as the users deepen their knowledge.

3   Link Editor and Viewer

"KN on ZK" is an interructive tool to create, edit, view and to publish personal knowledge of URL links. It is implemented in Java and used together with a

* Partially supported by Grant-in-Aid for Scientific Research on Priority Areas No.
10143215 of the Ministry of Education, Science and Culture, Japan.

browser. It is built on top of the network note pad ZK [3, 4] and displays the
links as directed graphs. It is different to other link viewers e.g., Nattoview [5],
WebView [1] and Ptolomaeus [2] in the sense that the link information is the
first citizen in "KN on ZK". We can construct, grow and publish them as our
knowledge.

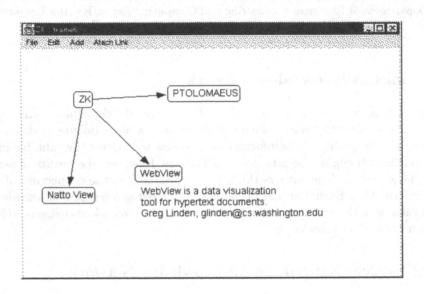

References

1. Linden G.: WebView,
   http://www.cs.washington.edu/homes/glinden/WebView/WebView.html
2. Battista, G.D., Lillo, R., Vernacotola, F.: PTOLOMAEUS,
   http://www.inf.uniroma3.it/v̄ernacot/ptolpage.htm
3. Minami, T., Sazuka H., Hirokawa S., Ohtani T.: Living with ZK - An Approach towards Communication with Analogue Messages, Proc. KES98, Second International
   Conference on Knowledge-based Intelligent Electronic Systems, (1998) 369-374
4. Sazuka H., Minami T., Hirokawa S.: Internet note pad ZK (in Japanese), Proc. DPS
   Workshop, Information Processing Society of Japan (1997) 195-199
5. Shiozawa H., Matsushita Y.:WWW visualization giving meanings to interactive
   manipulations HCI International 97 (1997)
   http://www.myo.inst.keio.ac.jp/NattoView/

Virtual Integration of Distributed Database by Multiple Agents

Takao Mohri and Yuji Takada

Netmedia Lab. Fujitsu Laboratories Ltd.,
2-2-1 Momochi-hama Sawara-ku Fukuoka 814-8588 JAPAN

1 Introduction

Since the Internet becomes popular and widely used, many services including databases are available for users. However, more databases are available, more difficult it becomes to select and use them. One promising way to solve this problem is to use a software agent called *Mediator* (or *Facilitator*) [1]. A mediator has meta-information of databases, and it can select databases that satisfy requests from users, access them in suitable protocols, and summarize their answers. Users have only to know the way to access the mediator. From the view of users, a mediator realizes the virtual integration of distributed databases.

The most simple way to realize a mediator is to implement it as a single program. Such centralized architecture may be enough for a small system of a single organization. However, if the system becomes large and more organizations join in, this simple architecture causes several problems including load concentration, and secure information sharing among organizations.

2 Distributed Architecture of Mediator

To solve those problems of the centralized architecture, we adopt distributed architecture as shown in Figure 1. A medaitor is a distributed system which consists of three kinds of federated agents, *user agents*, *mediation agents*, and *database agents*.

A user agent and a database agent are interfaces to users and databases respectively. A mediation agent forwards requests to suitable agents. Each mediation agent has a condition table which has pairs of condition and destination. It forwards requests to the destination agents if the corresponding condition is satisfied by requests. For efficient selection of databases, more general and stable information of databases is assigned to mediation agents close to user agents, and more specific and frequently changed information is handled by those close to database agents.

This distributed architecture has several advantages over the centralized architecture. For example, loads of a mediator can be balanced over machines because agents that compose a mediator can run in parallel on different machines distributed over the network. Meta-information of databases can be kept within agents. A mediator can be extended by adding agents. User agents and database

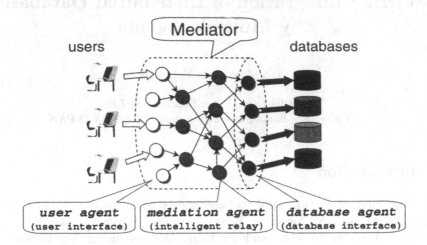

Fig. 1. Mediator Realized by Distributed Multiple Agents

agents can be added for new users and databases respectively. Mediation agents can be added to support more complex networks. Redundant routes can be set between agents to improve robustness. Therefore, our distributed architecture is more scalable, secure, and robust than the centralized architecture.

3   Concluding Remarks

We have already proposed this distributed architecture of a mediator[2], and applied it to an experimental project[3] of CALS(Continuous Acquisition and Lifecycle Support). The advantages of distributed architecture are especially critical for inter-enterprise collaboration such as EDI(Electronic Data Interchange), EC(Electronic Commerce), and CALS.

In the demonstration, we will show how multiple agents are configured for inter-enterprise collaboration, and how they select databases cooperatively and access them in parallel.

References

1. Finin, T., Labrou, Y. and Mayfield, J.: KQML as an agent communication language. Bradshaw, J. (Ed.), Software Agents. MIT Press. (1997) 291–316
2. Takada, Y., Mohri, T. and Fujii, H.: A Multi-Agent Model for Virtual Integration of Distributed Databases. In Proc. of CALS Expo International '97 (1997) Track 2, 99–110
3. Nishiguchi, A., Shida, K., Takada, Y.: Steel Plant CALS Project: Business Process and Information Infrastructure Across Enterprises. In Proc. of CALS Expo International '97 (1997) Track 7, 1–7

Development of Some Methods and Tools for Discovering Conceptual Knowledge

Tu Bao Ho, Trong Dung Nguyen, Ngoc Binh Nguyen, Takuyu Ito

Japan Advanced Institute of Science and Technology (JAIST)
Tatsunokuchi, Ishikawa, 923-1292 JAPAN

1 Objective

This abstract summarizes key points of our ongoing project that aims at *developing* several new methods in knowledge discovery in databases (KDD) and *implementing* them together with related techniques in an interactive-graphic environment. This research lies in several areas if KDD such as supervised classification, conceptual clustering, discovery of association rules, parallel and distributed data mining, genetic programming. Some of them are the continuation of our work done so far [1], [2], [3], and some others have been started recently.

2 Research Topics

2.1 Supervised Classification

CABRO [1] creates and manipulates *decision trees*. It carries out the attribute selection by a new measure for the dependency of attributes (R-measure) stemming from rough set theory. Careful experiments have shown that its performance is comparable to those of several popular methods. We are currently working for new development on model selection, Tree Visualizer, and Interactive Learning Mode of CABRO.

2.2 Conceptual Clustering

OSHAM [2] extracts *concept hierarchies* from hypothesis spaces in the form of Galois lattices. Its key features are a combination of advantages of classical, prototype and exemplar views in concepts, a learning method and a program that allow the integration of human factors and computation. The concept hierarchy can be generated with disjoint or overlapping concepts depending on the user's interest. We are focusing on the development of OSHAM for large datasets.

2.3 Discovery of Association Rules

In many applications, interesting associations often occur at a relatively high concept level, and information about multiple abstraction levels may exist in database organization. We currently do research on mining *multiple-level association rules*. Another topic is to improve the efficiency of mining association rules by using scan reduction and sampling techniques.

2.4   Parallel and Distributed Data Mining Algorithms

We are to design and implement parallel data mining algorithms on the MIMD machines *ncub-3* and *Parsytec* at JAIST. At this step are investigating the parallelization of decision tree induction algorithms and then plan to do so for algorithms of mining multiple-level association rules.

2.5   Genetic Programming

To generate automatically large scale programs is a difficult problem that requires using knowledge. In considering the co-evolution of programs we work for realizing an automatic generation of advanced programs in the frame of the theory of the symbiosis.

2.6   Related Tools for KDD

Some KDD tools have been developed for the pre-processing and post-processing in the KDD process: (1) Tools for sampling and model testing (e.g., random subsampling and automatic *k*-fold stratified cross-validation); (2) Tools for supporting the *interpretation* of unknown instances; (3) Tools for dealing with heterogeneous data such as *discretization* of numerical data or *mixed similarity measures* without discretization.

References

1. Nguyen T.D. and Ho T.B., "An Interactive-Graphic System for Decision Tree Induction", to appear in *Journal of Japanese Society for Artificial Intelligence*, January 1999.
2. Ho T.B., "Discovering and Using Knowledge From Unsupervised Data", *International Journal Decision Support Systems*, Elsevier Science, Vol. 21, No. 1, 27–41, 1997.
3. Ito T., Iba H. and Sato S., "Depth-Dependent Crossover for Genetic Programming", Proceedings of the 1998 *IEEE International Conference on Evolutionary Computation* (ICEC'98), 775–780, IEEE Press, 1998.

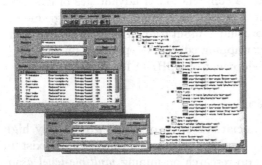
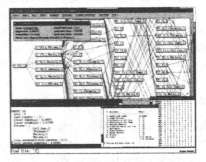

Fig. 1. Decision trees/concept hierarchies produced by CABRO and OSHAM

Reducing the Dimensions of Attributes by Selection and Aggregation

Akiko Aizawa

National Center for Science Information Systems
3-29-1 Otsuka, Bunkyo-ku, Tokyo, 112-8640 Japan
akiko@rd.nacsis.ac.jp

1 Introduction

Given a matrix of objects and attributes, *dual scaling analysis* (DSA) [1] is known to be a promising method of reducing the dimensionality while preserving the underlying structure between objects and attributes [2], [3]. However, due to the computational complexity of matrix inversion, DSA suffers from a scalability problem for data with tens of thousands of attributes, as is often the case in information retrieval applications. The problem thus becomes how to reduce the dimension of the original data at the pre-processing stage, to make analysis feasible. Our study calculates the comparative data losses of two schemes for such dimension reduction, *feature selection* and *feature aggregation*, and proposes a procedure for combining these two schemes. We also evaluate performance using HTTP log data.

2 Theoretical Development

Let D be a $n \times m$ matrix with each cell d_{ij} representing a numerical value associating object i with attribute j. We furthermore assume that a classification hierarchy is given for attributes representing *a priori* knowledge about which subsets of attributes are more likely to be similar to each other.

Assume A is an attribute group defined in the classification hierarchy. *Feature selection* refers to the case where one of the k attributes of A is selected as a representative of the group while the other $(k-1)$ attributes are discarded. On the other hand, *feature aggregation* refers to the case where the k attributes are replaced with a newly generated attribute whose value for each object is a summation of the original k attributes. We calculate the loss for these reduction schemes based on the square error of their equivalent transformations:

$$L_{sel}(A) = \sum_{i=1}^{k-1}\sum_{j=1}^{n}(d_{ij} - \frac{\sum_{i=k}^{m}d_{ij}}{\alpha})^2, \quad where \quad \alpha = \frac{(k-1)\sum_{i=k}^{m}\sum_{j=1}^{n}d_{ij}}{\sum_{i=1}^{k-1}\sum_{j=1}^{n}d_{ij}}, \quad and$$

$$L_{sum}(A) = \sum_{i=1}^{k}\sum_{j=1}^{n}(d_{ij} - \frac{\sum_{i=1}^{k}d_{ij}}{k})^2.$$

The reduction strategy adopted is as follows: (1) start with the root node of the attribute hierarchy; (2) calculate L_{sel} and L_{sum} for the current node; (3) if $L_{sel} < L_{sum}$, expand the node and go to (2); (4) otherwise, stop the expansion and summarize the attributes at the current node; (5) repeat (2)-(4) until all nodes are processed; (6) select the top $l(< m)$ nodes with minimum loss values.

3   Experimental Results

The information source used in our experiment is an HTTP access log obtained at a well-known proxy server site in Japan. We select 12 client domains as objects and 2,400 URLs as attributes and calculate the total access count of each client to each URL. The attribute hierarchy is simply a directory of URLs derived from their notations. The object-attribute matrix is first pre-processed using one of the following: (1) without normalization (RLOG); (2) d_{ij} is set to 1 if there is any access from client (i) to URL (j) or 0 otherwise (UNIQ); (3) each row is normalized to make the row totals equal to 1.0, i.e. $d_{ij} = f_{ij}/\sum_{j=1}^{2400} f_{ij}$ (FREQ).

Feature reduction is performed after normalization to reduce the attribute dimension from 2,400 to 60. We compare the proposed reduction scheme (FSUM) with a naive but commonly practiced method that selects the 60 most frequent URLs as features (FTOP). After the reduction, DSA is applied to each reduced matrix to calculate the similarities for every client pair. Also, as reference data, we directly apply DSA to the original (normalized but not reduced) matrix and calculate the similarities between the client pairs. The performance of FSUM and FTOP is evaluated by the correlation between these similarity value sets using *Pearson's product moment correlation coefficient* for values and *Spearman's rank correlation coefficient* for ranks. The result is shown in Table 1.

Table 1. Performance comparison of FTOP and FSUM

(a) 20-weeks data		Pearson	Spearman	(b) one-week data (20 week average)		Pearson	Spearman
RLOG	FTOP	0.87	0.74	RLOG	FTOP	0.35	0.15
	FSUM	0.90	0.92		FSUM	0.53	0.39
UNIQ	FTOP	–	–	UNIQ	FTOP	–	–
	FSUM	1.00	0.89		FSUM	0.83	0.83
FREQ	FTOP	0.79	0.83	FREQ	FTOP	0.73	0.77
	FSUM	0.92	0.94		FSUM	0.93	0.93

4   Discussion

For all cases, FSUM outperforms FTOP and establishes fairly good correlation with the reference data, which is sufficient for DSA. FTOP does not work with UNIQ since all the features it selects are 1s. For one-week data, the correlation becomes worse in most cases while FSUM applied with FREQ still maintains good performance. This may be because when a client shows relatively low activity throughout the week, most information about the client is lost after reduction. Such information loss can be prevented with FREQ since the values are normalized for each client.

References

1. Nishisato, S.: Analysis of Categorical Data: Dual Scaling and Its Applications, Asakura-Shoten (1982).
2. Deerwester, S., Dumais, S.T., Furnas, G.W., Landauer, T.K. and Harshman, R.: Indexing by Latent Semantic Analysis, *J. Amer. Soc. Info. Sci.*, Vol. 41, No.6, pp. 391-407 (1990).
3. Sugimoto, M., Hori, K. and Ohsuga, S.: A System for Visualizing Viewpoints and Its Application to Intelligent Activity Support, *IEEE Trans. System, Man and Cybernetics*, Vol. 28C, No. 1, pp. 124-136 (1998).

On the Number of Clusters in Cluster Analysis

Atsuhiro Takasu

National Center for Science Information Systems
3-29-1 Otsuka, Bunkyo-ku, Tokyo, 112-8640 Japan
takasu@rd.nacsis.ac.jp

Clustering is a fundamental tool for analyzing the structure of feature spaces. It has been applied to various fields such as pattern recognition, information retrieval and so on. Many studies have been done on this problem and various kinds of clustering methods have been proposed and compared (e.g., [1]). Some clustering algorithms partition the points in the feature space into the given number of clusters, and others generate hierarchical structure of clusters. Either type of clustering algorithms is not able to compare the partitions consisting of different number of clusters and users must specify the *optimal* number of clusters. However, when applying a clustering method, we don't often know how many clusters exist in the given data. This paper discusses the framework for comparing partitions that consist of different number of clusters.

The goodness of partition is measured from two aspects: uniformness in the cluster and difference between the clusters. Typical measures for clustering are the sum of within cluster variances and the trace of within cluster scatter matrix. Clustering algorithms try to find the partition that minimizes these measures. However they are biased about the number of clusters, and we cannot apply them for the comparison of partitions that consist of different number of clusters.

In order to compare partitions consisting of different number of clusters, we introduce a probability model that defines probability distribution on d-dimensional feature space. For the partition we assume that the feature space contains a set of disjoint subspaces $C_1, C_2, \cdots, C_m$ such that the probability density between subspaces is 0 and the probability density inside C_i is p_i/V_i, where V_i stands for the volume of C_i and $\sum_{i=1}^{m} p_i = 1.0$. Then the clustering problem is to find a number m of clusters, subspaces $C_1, C_2, \ldots, C_m$ and parameters $p_1, p_2, \cdots, p_m$ that explain the given data most adequately. In this problem, we can use model selection criteria such as AIC (Akaike Information Criterion) and MDL (Minimum Description Length). Suppose that we select AIC as a criterion. For a set of points and the number m of clusters, let us assume that a set of subspaces $C_i(1 \leq i \leq m)$ contains n_i points. Then, the maximum likelihood estimate of p_i is n_i/n and the MLL (Maximum Log Likelihood) is denoted as $\sum_{i=1}^{m} n_i \log \frac{n_i}{nV_i}$. The AIC is described as the following expression:

$$AIC = -2MLL + 2m = -2\sum_{i=1}^{m} n_i \log \frac{n_i}{nV_i} + 2m . \tag{1}$$

The clustering problem based on AIC is to find m and subspaces that minimize (1).

Let us consider a set P of points and its partition $\{C_1, C_2, \cdots, C_m\}$. If we consider P as one cluster, the MLL is $n \log \frac{1}{V_P}$, while the MLL of the partition $\{C_1, C_2, \cdots, C_m\}$ is $\sum_{i=1}^{m} n_i \log \frac{n_i}{nV_i}$. Since $V_P \geq \sum_{i=1}^{m} V_i$, the following inequalities hold:

$$n \log \frac{1}{V_p} \leq n \log \frac{1}{\sum_{i=1}^{m} V_i} \leq \sum_{i=1}^{m} n_i \log \frac{n_i}{nV_i}$$

where $n = \sum_{i=1}^{m} n_i$. This means that the MLL increases as the space is partitioned into smaller fragments. Simultaneously, the number of parameters m increases. AIC works as a criterion to adjust the overfitting of probability distribution estimation.

As (1) shows, we get smaller AIC for smaller bounding subspace for the given set of points. As the shape of the subspace, we can consider bounding rectangle, bounding sphere, bounding ellipsoid and convex hull. Convex hull is preferable for obtaining small bounding subspace. However, calculation of convex hull in d-dimensional space requires $O(n^{\lfloor d/2 \rfloor + 1})$ [3]. Therefore, it is not feasible to obtain the subspace in the high dimensional feature space and we need to use bounding rectangle or bounding sphere.

It is computationally hard problem to obtain the partition that minimizes (1). However, this framework can be used for selecting the optimal partition created by existing clustering methods. For clustering algorithms that produce tree structure of clusters such as dendrogram of the hierarchical clustering, information criterion is applied in the following ways;

1 for leaves of the tree, merge nodes (points) with siblings until the merged node has volume in the d-dimensional feature space and assign the information criterion to the merged node,
2 apply the following procedure to each node N recursively in the bottom-up way:
   2.1 calculate information criterion IC_s when regarding all points in the subtree of N consist one cluster
   2.2 for the sum of information criterion of N's children (denoted by IC_c), if $IC_s > IC_c$ then assign IC_c to N, otherwise merge the points in the subtree and assign IC_s to N.

References

1. Punj,G. and Stewart, D.W.: "Cluster Analysis in Marketing Research: Review and Suggestions for Applications," Journal of Marketing Research, Vol.XX, pp.134-148 (1983).
2. Akaike,H.: "A New Look at the Statistical Model Identification," IEEE Trans. on Automat. Contr., Vol.AC19, pp.716-723 (1974).
3. Preparata,F.P. and Shamos, M. I.: "Computational Geometry - An Introduction", Springer-Verlag (1985).

Geometric Clustering Models in Feature Space*

Mary Inaba and Hiroshi Imai

Department of Information Science, University of Tokyo, Tokyo, 113-0033 Japan

This paper investigates the use of geometric structures for unsupervised learning from multimedia databases, such as full text databases and image databases. We first provide formulations of the geometric clustering problems by divergences, such as the Kullback-Leibler divergence and the square of Euclidean distance, for the feature space of such databases. Some of our algorithmic results are summarized. Underlying geometric structures of the feature space for the vector-space model for texts and the quadratic form measure for images are revealed to be fully usable from the algorithmic viewpoint.

Our aim in the unsupervised learning is grouping similar objects to generalize them as concepts as well as compress the data amount. Formally, our problem is to find a useful clustering with respect to meaningful criteria. As the meaning criteria, we adopt divergences [1] based on their connection with statistical estimation. Mathematically, for a set S of n vectors t_i in the d-dimensional space, a k-clustering $(S_1, S_2, \ldots, S_k)$ of S is a partition of S into k subsets S_j $(j = 1 \ldots, k)$, i.e., $S_j \cap S_{j'} = \emptyset$ $(j \neq j')$ and $\bigcup_{j=1}^{k} S_j = S$. Each cluster S_j is represented as its centroid denoted by $\bar{t}(S_j)$. Then, for a divergence $D(t, t')$ from t to t', the problem is formulated as an optimization problem

$$\min \left\{ \sum_{j=1}^{k} \sum_{t_i \in S_j} D(\bar{t}(S_j), t_i) \,\middle|\, k\text{-clustering } (S_1, \ldots, S_k) \text{ of } S \right\}$$

Two typical divergences are (1) $D(t, t') = \|t - t'\|^2$, where $\|\cdot\|$ is the L_2 norm, (2) $D(t, t') = \sum_{l=1}^{d} t'_l \log(t'_l/t_l) = D_K(t'\|t)$ for the Kullback-Leibler divergence $D_K(t'\|t)$ where, in this case, $\sum_l t_l = \sum_l t'_l = 1$, $t_l, t'_l > 0$. The case of (1) is often used as a kind of the least square approach, and that of (2) almost coincides with a maximum likelihood method for a certain clustering model based on exponential families of distributions.

For the divergence, the Voronoi diagram can be defined, and it has similar nice properties like the Euclidean Voronoi diagram, for example, see a brief exposition with a video in [4]. By virtue of geometric structures, the following characterization is obtained for the clustering problem: An optimal k-clustering is a partition of n points induced by the Voronoi diagram generated by the centroids of each cluster of the clustering. Based on this characterization, all the possible partitions generated by the Voronoi Diagrams may be enumerated to find an optimal clustering. For the Euclidean case, the number of all possible

---

* This research was supported in part by the Grant-in-Aid on Priority Area, 'Discovery Science', of the Ministry of Education, Science, Sports and Culture, Japan.

partitions is analyzed, and, since this number is large even for moderate k, d, n, a randomized approximate algorithm is developed in [3]. These also hold for the general divergence.

In the sequel the feature spaces of appropriate divergence are described, to which the above-mentioned clustering approach can be applied.

Vector Space Model for Texts In this model, all information items of stored texts are represented by vectors of terms in a space whose dimension is the number d of terms. Counting frequencies of occurrences of terms (keywords), etc., all texts and text queries are represented by vectors t_i in the d-dimensional space. The l-th element in t_i is the 'weight' assigned to the l-th term in the document i. The similarity $\text{sim}(t_i, t_j)$ between two documents represented by t_i and t_j is defined by the inner product of normalized vectors $\text{sim}(t_i, t_j) = (t_i/\|t_i\|)^{\mathrm{T}}(t_j/\|t_j\|)$ [5]. Defining a dissimilarity, $\text{dis}(t_i, t_j)$, between two vectors t_i and t_j, by $\text{dis}(t_i, t_j) = 1 - \text{sim}(t_i, t_j) = \|(t_i/\|t_i\|) - (t_j/\|t_j\|)\|^2/2$. we can impose a natural geometric structure where the square of Euclidean distance is a divergence. Similarly, if we normalize by dividing the sum of total weights instead of normalization by the L_2 norm, the Kullback-Leibler divergence is more natural to use.

Quadratic Form Model for Images In the Query-by-Image-Content (QBIC) system [2] the following measure is used to estimate the closeness of two images by their color frequency vectors. First, quantize the images by $\tilde{d}$ representative colors. Let $B = (b_{pq})$ be a $\tilde{d} \times \tilde{d}$ matrix such that b_{pq} is the negative of some dissimilarity between color i and color j. In the QBIC system, b_{pq} is the negative of squared Euclidean distance between two colors c_p and c_q in the 3-dimensional Luv color space with $\tilde{d} = 64$ or 256. Let t_i and t_j be two normalized frequency vectors of two images with respect to the $\tilde{d}$ colors, normalized by dividing the frequency vector by the total count so that the sum becomes 1. Then, the distance-like function between two images is defined to be $(t_i - t_j)^{\mathrm{T}} B(t_i - t_j)$, which becomes equal to $2\|A(t_i - t_j)\|^2$ where A is a $3 \times \tilde{d}$ matrix whose p-th column is color c_p. Hence, we can reduce the clustering problem by this distance to our geometric clustering problem in the 3-dimensional space.

References

1. S. Amari: *Differential-Geometric Methods in Statistics*. Lecture Notes in Statistics, Vol.28, Springer-Verlag, 1985.
2. C. Faloutsos, R. Barber, M. Flickner, W. Niblack, D. Petkovic and W. Equitz: Efficient and effective querying by image content. *J. Intell. Inf. Syst.*, 3 (1994), 231–262.
3. M. Inaba, N. Katoh and H. Imai: Applications of weighted Voronoi diagrams and randomization to variance-based k-clustering. *Proc. 10th ACM Symp. on Comp. Geom.*, 1994, pp.332–339.
4. K. Sadakane, H. Imai, K. Onishi, M. Inaba, F. Takeuchi and K. Imai: Voronoi diagrams by divergences with additive weights. *Proc. 14th ACM Symp. on Comp. Geom.*, 1998, pp.403–404.
5. G. Salton, J. Allan, C. Buckley and A. Singhal: Automatic analysis, theme generation, and summarization of machine-readable texts. *Science*, 264 (1994), 1421–1426.

Efficient Mining of Association Rules with Item Constraints

Shin-Mu Tseng

Computer Science Division, EECS Department
University of California, Berkeley, CA 94720, USA
Email: tsengsm@cs.berkeley.edu

Discovering association rules is a problem of data mining on which numerous studies have been made. An example of an association rule is: "25 percent of transactions that contain beer also contain diapers; 5 percent of all transactions contain both items". Here, 25 percent is called the confidence of the rule, and 5 percent the support of the rule. Most existing work on this problem focused on finding the association rules between all items in a large database which satisfy user-specified minimum confidence and support. In practice, users are often interested in finding association rules involving only some specified items rather than all items in a query. Meanwhile, based on the searched results in former queries, users might change the minimum confidence and support requirements to obtain suitable number of rules. In this scenario, the users tend to make consecutive queries on interested items with expected quick response rather than wait a long time for getting a lot of association rules between all itemsets in partial which the users are only interested.

Under the item constraints, the existing mining algorithms can not perform efficiently in terms of responding the user's query quickly. The main reason is that the existing mining algorithms mostly need to read the whole database from disks for several passes in each user's query since the database is too large to be resident in memory. Moreover, the existing mining algorithms incur the repeated overhead of reading the whole database several times even a query involves the same interested items as the previous queries but changes only the minimum confidence and support.

In this work, we present a novel mining algorithm which can efficiently discover the association rules between the user-specified items. The principle of the proposed algorithm is as follows. At the first time a database is queried by a user, the whole database is scanned and an index called location index is built for each interested item to record the locations of the transactions the item appears. The location index is designed in a compressed way to reduce the size by recording only the starting and finishing location for continuous occurrences of an item. Another index which records the count of occurrences of each item in the whole database is also built. Once the indexes are built, only the indexes of the user-specified items are used for all subsequent queries to find the association rules bewteen these interested items. To discover the rules, the count index is first scanned to find the items whose numbers of occurrences satisfy the minimum support. Then, the indexes of the qualified items are used to calculate the number of occurrences in same transactions for the subsets of the qualified itemsets. An occurrence for the items of an itemset in a same transaction implies an association between these items. The itemsets which satisfy the user-specified support can thus be obtained by checking the calculated number of occurrences for each itemset. These itemsets are called large itemsets. The proposed algorithm computes the large itemsets by increased size and a pruning method similar to that in [1] is used to determine the further candidate large itemsets with increased size. Finally, the association rules can be found easily by calculating the confidence between the large itemsets.

Compared to the existing mining algorithms, the proposed algorithm has the following main advantages:

First, due to the purpose of mining all association rules between all itemsets, the existing mining algorithms have to scan the whole database several times for each query. In contrast, the proposed algorithm needs at most one scan of the database. For a query, no database read is needed if the interested items have been involved in previous queries; otherwise, the database is scaned once to build the location and count indexes for the new interested items. In either case, only the indexes of interested items rather than the whole database are accessed for all subsequent computation of the large itemsets. Since the size of indexes of interested items is much smaller than that of the whole database, it is clear our algorithm will be more efficient in terms of reducing the overhead of disk access.

Second, the intermediate calculated results in a query can be saved and reused later by further queries to reduce the computation time by the proposed algorithm. For example, in counting the associations among items a, b, c, the number of occurrences for each of (a, b), (a, c) and (b, c) will be calculated by using the location indexes of items a, b and c. If the calculated result of (a, b) is saved as an extended location index, it can be reused by a subsequent query on interested items a, b, d without repeated computation.

Therefore, the proposed algorithm can efficiently discover the association rules among the user-specified items by reducing substantially the disk access overhead for reading the large database and the computation time for counting the associations among the interested items. To our best knowledge, this is the first work which provides quick response time for mining association rules with item constraints from the standpoint of the practical way the users could make queries. Some experiments are in progress to evaluate the performance of the proposed algorithm under various system environments.

References

1. Agrawal, R., Srikant, R.: Fast Algorithms for Mining Association Rules, Proc. 20th VLDB Conference, Santiago, Chile, (1994) 487–499.

GDT-RS: A Probabilistic Rough Induction Approach

Juzhen Dong[1], Ning Zhong[1], and Setsuo Ohsuga[2]

[1] Dept. of Computer Science and Sys. Eng., Yamaguchi University
[2] Dept. of Information and Computer Science, Waseda University

1 Introduction

Over the last two decades, many researchers have investigated inductive methods to learn *if-then* rules and concepts from instances. Among the methods, *version-space* is a typical bottom-up, incremental one[1]. However, it is difficult to handle noisy and incomplete data, and it is weak in mining rules from very large, complex databases. On the other hand, the *rough set* theory introduced by Pawlak is well known for its ability to acquire decision rules from noisy, incomplete data[2].

In this paper, we propose a probabilistic rough induction approach called GDT-RS that is based on the combination of *Generalization Distribution Table (GDT)* and the *Rough Set* theory. Main features of our approach are as follows:

- It can flexibly select biases for search control, and background knowledge can be used as a bias to control the creation of a GDT and the rule induction process;
- It can predict unseen instances and represent explicitly the uncertainty of a rule including the prediction of possible instances in the strength of the rule.

2 Rule Discovery based on the GDT-RS Methodology

The central idea of our methodology is to use a *Generalization Distribution Table (GDT)* as a hypothesis search space for generalization, in which the probabilistic relationships between concepts and instances over discrete domains are represented [3]. The GDT provides a probabilistic basis for evaluating the strength of a rule. It is used to find the rules with larger strengths from possible rules. Furthermore, the rough set theory is used to find minimal relative reducts from the set of rules with larger strengths. The strength of a rule represents the uncertainty of the rule, which is influenced by both unseen instances and noises.

In our approach, the learned rules are typically expressed in $X \rightarrow Y$ *with S*. That is, "a rule $X \rightarrow Y$ has a strength S". Where X denotes the conditions, Y denotes a concept, and S is a "measure of strength" of which the rule holds.

The strength S of a rule $X \rightarrow Y$ in a given as follows:

$S(X \rightarrow Y) = s(X) \times (1 - r(X \rightarrow Y))$.

Where $s(X)$ is the strength of the generalization X, it means that within the possible instances satisfying the generalization X how many of them are observed in the database. r is the rate of noises, it shows the quality of classification, that is, how many instances as the conditions that a rule must satisfy can be classified into some class.

By using our approach, a minimal set of rules with larger strengths can be acquired from databases. There are several possible ways for rule selection in our approach. For example,
- Selecting the rules that contain the most instances;
- Selecting the rules with less attributes;
- Selecting the rules with larger strengths.

We have developed two algorithms called *"Optimal Set of Rules"* and *"Sub Optimal Solution"* respectively for implementing the GDT-RS methodology [4]. We briefly describe here one of the algorithms: *"Sub Optimal Solution"* that is a kind of greedy algorithms as follow:

Step 1. For each instance i, get the generalizations with other instances. Put the generalizations with the instances in an identical class into set G_+, otherwise into set G_-.

Step 2. For the instances existed in both of sets G_+ and G_-, calculate the noise rate (nr) using the following equation:

$$nr = \frac{N_-(g)}{N_+(g) + N_-(g)}, \tag{1}$$

where $N_+(g)$ is the number of generalization g within G_+, $N_-(g)$ is the number of generalization g within G_-.

If the noise rate is greater than a threshold value, g will be deleted from G_+, and the count for g in G_- will be changed.

Step 3. Create the discernibility matrix D according to G_-.

Step 4. Choose a column from D with the maximal number of occurrences in G_+.

Step 5. Delete from D the column chosen in *Step* 4 and all rows marked in this column by non-empty-value.

Step 6. If D is non-empty then goto *Step* 4 else *Step* 7.

Step 7. Go back to *Step* 1 until all of instances are handled.

3   Conclusions

Some of databases such as postoperative patient, meningitis, earthquack, weather, mushroom, cancer have been tested for our approach. By using the greedy algorithm called *"Sub Optimal Solution"* as heuristics, it is possible to obtain more efficient solution with time complexity $O(nm^2 + n^2m)$, where n is the number of instances, m is the number of attributes.

References

1. T.M. Mitchell. "Generalization as Search", *Artif. Intell.*, Vol.18 (1982) 203-226.
2. Z. Pawlak. *ROUGH SETS, Theoretical Aspects of Reasoning about Data*, Kluwer Academic Publishers (1991).
3. N. Zhong, J.Z. Dong, and S. Ohsuga, "Discovering Rules in the Environment with Noise and Incompleteness", *Proc. FLAIRS-97* (1997) 186-191.
4. N. Zhong, J.Z. Dong, and S. Ohsuga, "Data Mining based on the Generalization Distribution Table and Rough Sets", X. Wu et al. (eds.) *Research and Development in Knowledge Discovery and Data Mining*, LNAI 1394, Springer (1998) 360-373.

A Constructive Fuzzy NGE Learning System

Maria do Carmo Nicoletti and Flavia Oliveira Santos

Universidade Federal de Sao Carlos /DC
C.P. 676 - 13565-905 S. Carlos-SP- Brazil
carmo@dc.ufscar.br and flavia@if.sc.usp.br

1 Introduction

Nested Generalized Exemplar (NGE) theory [1] is an incremental form of inductive learning from examples. This paper presents FNGE, a learning system based on a fuzzy version of the NGE theory, describes its main modules and discusses some empirical results from its use in public domains.

2 FNGE Prototype System

The FNGE algorithm was proposed in [2] and can be considered a version of the NGE algorithm (also called EACH) suitable for learning in fuzzy domains. The examples in the training set are described by fuzzy attributes and an associated crisp class. Each attribute is described by a linguistic variable that can assume different linguistic values. Each linguistic value is represented by a fuzzy set. FNGE is an incremental, supervised and constructive learning method. Since its design was substantially based on the NGE theory, we kept this name only as a reference; the FNGE cannot be thought of as a system that induces nested exemplars because that does not mean anything in fuzzy domains. The FNGE algorithm has been implemented as a prototype system also called FNGE, having three main modules, identified as Attribute Definition, Training and Classification Modules. Through the Attribute Definition Module the user provides the system with all the possible attribute linguistic values that exist in the training set, the fuzzy set associated with each possible attribute linguistic value and the number of elements and the elements themselves that constitute the universal set (provided it is finite and discrete). The Training Module is the main module of the FNGE system and implements the FNGE algorithm as described in [2]. It is the module responsible for choosing the closest exemplar to the new example and for generalizing it (when appropriate). It is important to mention that NGE induces hypotheses with the graphical shape of hyperrectangles as a consequence of its generalization process, which "grows" the hyperrectangle when it makes a correct prediction in order to absorb the current training example that lies outside its boundaries. FNGE, due to its fuzzy nature, generalizes hypotheses by (generally) creating new fuzzy values for attributes. In this sense, the FNGE learning phase can be considered a sort of constructive process; however it does not create new attributes, as constructive algorithms usually do, instead, it creates new fuzzy values for the existing attributes. The Classification Module is the responsible for classifying new instances, using the concept(s) learned by the previous module.

3 Experimental Results

Due to the lack of available real-world fuzzy domains, five datasets from the UCI Repository [3] were "transformed" into fuzzy datasets, i.e., datasets where attributes are described by fuzzy sets. It can be seen in Table 1 that FNGE (with weights) has a performance over 70three domains. The performance of FNGE (with weights) was shown to be approximately the same as that of NGE on the Pima Diabetes domain and is slightly superior, on the Postoperative domain. We believe that one of the reasons for the low performance of FNGE (inferior to 50in two domains is the low number of training examples. However, that could be explained as well by a possible inadequacy of the transformation process used, in those domains. By looking at the figures in Table 1 we could risk to say that in average, the FNGE with weights tends to have a better performance than its counterpart; nevertheless, we still believe that we do not have enough data to state that and further investigation needs to be carried on.

Table 1. Average performance of FNGE

Domain	Average Performance of FNGE(%) (with weights)	Average Performance of FNGE(%) (without weights)
Breast Cancer	85.19	95.54
Glass	42.16	23.82
Lung Cancer	30.59	34.51
Pima Diabetes	72.08	56.65
Postoperative	73.08	61.19

Acknowledgments

This work is partially supported by Fapesp and CNPq.

References

1. Salzberg, S.L., "A Nearest Hyperrectangle Learning Method", *Machine Learning* 6, 251–276, 1991
2. Nicoletti and M.C.; Santos, F.O., "Learning Fuzzy Exemplars through a Fuzzified Nested Generalized Exemplar Theory", *Proceedings of EFDAN'96*, Germany, 140–145, 1996
3. Merz, C.J. and Murphy, P.M., "UCI Repository of Machine Learning Databases [http://www.ics.uci.edu/ mlearn/MLRepository.html]. Irvine, CA, 1998

Composing Inductive Applications Using Ontologies for Machine Learning

Akihiro Suyama, Naoya Negishi and Takahira Yamagchi

School of Information, Shizuoka University
3-5-1 Johoku Hamamatsu Shizuoka, 432-8011 Japan
E-mail:{suyama, yamaguti}@cs.inf.shizuoka.ac.jp

1 Introduction

Recently, some community focuses on computer aided engineering inductive applications, as seen in workshops, such as AAAI98/ICML98 workshop on 'The Methodology of Applying Machine Learning' and ECML98 workshop on 'Upgrading Learning to the Meta-level: Model Selection and Data Transformation'. It is time to decompose inductive learning algorithms and organize inductive learning methods (ILMs) for reconstructing inductive learning systems. Given such ILMs, we may invent a new inductive learning system that works well to a given data set by re-interconnecting ILMs.

2 What is CAMLET ?

CAMET is a platform for automatic composition of inductive applications using ontologies, based on knowledge modeling and ontologies technique.

CAMLET has two kinds of ontologies. A process ontology is for ILMs that compose inductive learning systems. An object ontology is for objects manipulated by ILMs from process ontology. In order to specify process and object ontologies, we need to specify conceptual hierarchies and conceptual schemas on two ontologies. In this paper, we describe a method of constructing a process ontology.

In order to specify the conceptual hierarchy of a process ontology, it is important to identify how to branch down processes. Because the upper part is related with general processes and the lower part with specific processes, it is necessary to set up different ways to branch the hierarchy down, depending on the levels of hierarchy. In specifying the upper part of the hierarchy, we have analyzed popular inductive learning systems and then the following five popular and abstract components: 'generating training and test data sets', 'generating a classifier sets', 'estimating data and classifier sets', 'modifying a training data set' and 'modifying a classifier set'. We can place finer components on the upper part, but they seem to make up many redudant composition of inductive learning systems. Thus these five processes have been placed on upper part on the conceptual hierarchy of the process ontology, as shown in Figure 1. In specifying the lower part of the hierarchy, the above abstract component has been divided down using characteristics specific to each.

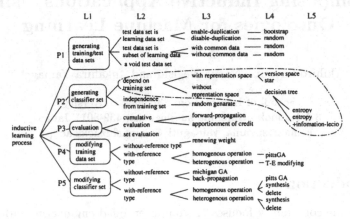

Fig. 1. The Hierarchy of Process Ontology

On the other hand, in order to specify the conceptual scheme of the process ontology, we have identified the learning process scheme including the following roles: 'input', 'output', 'reference', 'pre-process' and 'post-process'.

We apply the basic acrivities [1] to constructing inductive applications using process and object ontologies. When the specification does not go well, it is refined into another one with better competence.

3 Case Studies and Discussions

Based on the basic design, we have implemented CAMLET on UNIX platforms with C language. We did case studies of constructing inductive learning systems for the fourteen different data sets from the UCI Machine Learning Repository. Five complete 5-fold cross-validations were carried out with each data set. For each data set, we compared the performance of CAMLET with popular inductive learning systems, such as C4.5, ID3, Classifier Systems, Neural Networks and Bagged C4.5. As a result, CAMLET constructs inductive learning systems with best competence.

Because the competence of systems constructed by CAMLET is over those of popular inductive learning systems, it turns out for the grain size of process ontology to be proper for the task of composition of inductive applications. CAMLET invents new systems different from popular inductive learning systems.

4 Conclusion

This work comes from inter-discipline between machine learning and ontologies engineering. We put recent efforts on specifications and codes for ontologies and less efficient mechanisms to generate inductive learning systems with best competence.

References

1. Gertjan van Heijst, "The Role of Ontologies in Knowledge Engineering", Dr Thesis, University of Amsterdam, 1995.

TDDA, a Data Mining Tool for Text Databases: A Case History in a Lung Cancer Text Database

Jeffrey A. Goldman[1], Wesley Chu[1], D. Stott Parker[1], and Robert M. Goldman[2]

[1] Computer Science Department, University of California
Los Angeles, California 90095-1596
Email: jeff@ieee.org   URL: http://www.cs.ucla.edu/~goldman
[2] University of Osteopathic Medicine and Health Sciences
Des Moines, Iowa 50312-4198

1 Summary

In this paper, we give a case history illustrating the real world application of a useful technique for data mining in text databases. The technique, Term Domain Distribution Analysis (TDDA), consists of keeping track of term frequencies for specific finite domains, and announcing significant differences from standard frequency distributions over these domains as a hypothesis. In the case study presented, the domain of terms was the pair { right, left }, over which we expected a uniform distribution. In analyzing term frequencies in a thoracic lung cancer database, the TDDA technique led to the surprising discovery that primary thoracic lung cancer tumors appear in the right lung more often than the left lung, with a ratio of 3:2. Treating the text discovery as a hypothesis, we verified this relationship against the medical literature in which primary lung tumor sites were reported, using a standard χ^2 statistic. We subsequently developed a working theoretical model of lung cancer that may explain the discovery.

2 TDDA Applied to a Thoracic Lung Cancer Database

For our text discovery task, we began with a collection of patient thoracic lung tumor radiology reports[1]. In this text base, there are 178 patient histories describing a procedure performed, comparison information to previous exams (if it exists), a clinical history (if given), findings, and a diagnosis. The patients were volunteers from UCLA Medical Center from 1991-1994. Each patient document itself is roughly $1K - 4K$ in size which equates to approximately 250 words. The entire collection is $321K$.

By exploring the first dozen most frequently occurring words, we expect to see words such as "lung," "seen," and "chest," as a part of the general language used by physicians to record information. Similarly, "right," "left," "lobe," "upper," and "lower," appear in reference to a particular location of a tumor. However, the relative distributions of these words differed from our expectations. The ratio of right to left is about $3 : 2$ and upper to lower is about $5 : 3$. This was our first TDDA candidate discovery to be compared against previously published data.

---

[1] The data was obtained as a part of the Knowledge-Based Multimedia Medical Distributed Database System (KMeD) project [2] at UCLA.

Table 1. Summary of Statistical Tests for Tumor Site Location.

Patient Study Group	Primary Right	Primary Left	χ^2 Statistic for 1:1	p-value	χ^2 Statistic for 3:2	p-value
3	696	469	44.2	< 0.00001	0.032	0.858
4	59	46	1.61	0.2045	0.2045	0.4255
1	193	157	3.70	0.0544	3.44	0.0636
5	70	33	13.29	0.0003	2.72	0.0991
6	168	104	15.06	0.0001	0.353	0.5524
Combined Results	1186	809	71.24	< 0.00001	0.253	0.6150
Word Count	886	609	25.66	< 0.00001	0.337	0.5616

In Table 1, we summarize the relevant statistical information for each patient group and our word count. All studies are independent of each other and only primary unilateral tumor cases were included. The χ^2 statistic in the fourth column is calculated assuming an equal distribution between the lungs. The χ^2 statistic in the sixth column is calculated assuming a 3:2 distribution between the lungs[2]. While we have been unable to derive a theory as to why primary tumor site location occurs more frequently in the upper lobe than the lower lobe of the lung, we were able to uncover that fact[3]. In other words, our TDDA was successful in that it revealed this relationship. For left and right, it should be clear that primary lung tumors do not occur equally in both lungs. The published data provides stronger evidence for a 3:2 ratio. In fact, with only two exceptions, the distribution is in line with what we might expect for a randomly distributed 3:2 likelihood ratio. The right lung consists of three branches to each of three lobes while the left lung has only two. This discovery and our model may change how oncologists view the mechanisms of primary lung tumor location.

References

1. Seyhan I. Celikoglu, Talia B. Aykan, Tuncer Karayel, Sabriye Demirci, and Fahir M. Goksel. Frequency of distribution according to histological types of lung cancer in the tracheobronchial tree. *Respiration*, 49:152–156, 1986.
2. Wesley W. Chu, Alfonso F. Cárdenas, and Ricky K. Taira. KMeD: A knowledge-based multimedia medical distributed database system. *Information Systems*, 20(2):75–96, 1995.
3. Gunnar Hillerdal. Malignant mesothelioma 1982: Review of 4710 published cases. *British Journal of Diseases of the Chest*, 77:321–343, 1983.
4. Neil H. Hyman, Jr. Roger S. Foster, James E. DeMeules, and Michael C. Costanza. Blood transfusions and survival after lung cancer resection. *American Journal of Surgery*, 149(4):502–507, 1985.
5. Joel F. Platt, Gary M. Glazer, Barry H. Gross, Leslie E. Quint, Isacc R. Francis, and Mark B. Orringer. CT evaluation of mediastinal lymph nodes in lung cancer. *American Journal of Roentgenology*, 149(4):683–686, October 1987.
6. D. H. Yates, B. Corrin, P. N. Stidolph, and K. Browne. Malignant mesothelioma in south east England: clinicopathological experience of 272 cases. *Thorax*, 52:507–512, 1997.

---

[2] Values for p larger than 0.95 or smaller than 0.05 are considered significant.

[3] This relationship has been observed in [1].

Croww
Classification and Retrieval on WWW

Phyllis Anwyl and Atsushi Itoh and Ayako Oono

Ricoh Ltd.
3-2-3 Shin Yokohama, Kohoku-ku, Yokohama 222-8530, Japan
Phone: 045-477-1570 Fax: 045-477-1563
itoh@ic.rdc.ricoh.co.jp, phyllis@ic.rdc.ricoh.co.jp

A couple of years ago, when we were all facing the problem of how to get the information we want from the ever-spreading web, there were basically two choices for conducting a search – the quantity-oriented gather-all-pages-and-search approach as typified by AltaVista, and the quality-controlled hand-classify-and-search approach as represented by Yahoo. We've since learned that there are significant advantages and drawbacks to both. In response, a number of services and applications, including meta-search engines, personal web robots and cache organizers, have appeared [Lawrence]. For Japanese web documents, the alternatives have often been restricted by the inability to search using Japanese. Our research group has created a system incorporating the better features of the two basic approaches while avoiding their limitations.

Croww, for Classification and Retrieval on the World Wide Web, uses a robot to gather text/html pages and images, automatically categorizes them into an operator-defined classification system, and then allows a full-text search from any category, using Japanese and ASCII characters. Automating the classification of documents, together with relying on the operator's "world knowledge" to define the classification system and direct the retrieval of pages, increased the signal-to-noise ratio of the search results compared to the quantity-oriented approach while reducing the maintenance burden associated with the quality-controlled approach.

The modules in the system have the following features:

Robot – The robot gathers the initial document collection and checks for updates to the documents. It has standard control features, such as the maximum number of sites to visit and timing of the retrieval. The operator uses regular expressions to designate which URLs or groups of URLs to collect or avoid. Thus, the collection can include or exclude specific domains, hosts, directories or filetypes.

Analysis – Each document is given a set of features to be used in the subsequent classification, search and display modules. The set includes title, URL, date stamps, contact address, parent and child links, keywords, description and browsable text. Analysis begins with parsing the HTTP header to extract the bibliographic data. The links to and from the page are catalogued. The HTML document is parsed to extract any META tags and to identify the sections of text meant to be seen by a viewer. When a document does not contain META information for the keywords or description features, its browsable text is analyzed to provide them. Image files contain no META information so the text immediately following the IMAGE tag is used to identify the image.

Classification – Documents and images are categorized using multiple points of view. The primary point of view is that provided by the operator, in the form of a classification tree which associates document keywords with categories. Macros, variables and wildcards are used in category definitions. Secondary points of view are provided using bibliographic data classes from the HTTP header, such as modification date and URL. Thus, the documents can be browsed using any of several criteria, narrowing the target set of documents if a search is intended.

Search – Full-text boolean search in Japanese, including ASCII, is provided for documents and images. The search may target the entire collection or any sub-category. Search results are also classified into categories according to the operator-defined heirarchy.

Display – Accessing a category or conducting a search produces a set of child categories, a list of documents and images as well as the list's parent categories. These are presented to the viewer as an HTML document. The operator can modify the templates used to generate the display pages to customize the output seen by the viewer. Each of the document features is represented as a special variable within an HTML-style template.

Crowww (http://www.ricoh.co.jp/search) is a realistic and practical combination of technologies for classification and searching targeted to the unique qualities of the internet/intranet. It acts as a visualization aid for a difficult-to-grasp set of information by discovering order in a heterogeneous set of documents and giving the user alternative points of view from which to explore the collection. The order is achieved through categorization which places the highest priority on information supplied by the participants in the classification process, whether as META information supplied by authors or as the classification system designed by the operator. The result is a system which maximizes utility and flexibility yet minimizes maintenance burden.

Extracting Knowledge Patterns from Ticket Data

Maria de Fatima Rodrigues, Carlos Ramos and Pedro Rangel Henriques

Polytechnic Institute of Porto, School of Engineering,
Rua Sao Tome, 4200 Porto, Portugal
email: {fr, csr, prh}@dei.isep.ipp.pt

The introduction of bar codes and scanning of those for almost all-commercial products and the computerisation of business transactions (e.g. credit card purchases) have generated an explosive growth in retail data warehouses. Simultaneously, the big competition in this business area has created a significant need for a rigorous knowledge about sales versus clients. The treatment of such volumes of data and the need to understand customer behaviour can only be obtained with a data mining system. We have begin by doing market basket analysis (MBA), however the obtained results have not being so much advantageous, that's why we propose some previous data preparation that have help us to obtain more interesting results.

The goal of MBA is to discover groups of clients who buy similar products and next characterised them demographically. In its original form MBA was defined for a special kind of data, called ticket data. This data describes the contents of supermarket baskets (i.e. collections of items bought together), plus the associated credit card number of the purchaser. This data mining exercise contains two phases: firstly, links between items purchased are discovered using association rule modelling, and secondly, the purchasers of identified product groups are profiled using C5.0 rule induction [1].

The problem of mining association rules over basket data was introduced by Agrawal [2]. However, the algorithms proposed often generate too many irrelevant rules because of the wide variety of patterns in basket data and another problem is that the computations required to generate association rules grows exponentially with the number of items and the complexity of rules being considered.

The solution is to reduce the number of items by generalising them. But, more general items are, usually they are less actionable. One compromise is to use more general items initially and then repeat the rule generation to hone in on more specific items. As the analysis focuses on more specific items, we use only the subset of transactions containing those items. However, this solution does not permit to obtain associations between items being from different sections. Another problem with association rules is that it works best when all items have approximately the same frequency in the data, because items that rarely occur, are in very few transactions, may be pruned.

When applying market basket analysis, it is crucial to use a balanced taxonomy. This involves examine the data to determine the frequency of each item and with this information we can judiciously choose the right level of the taxonomy for groups of items. These generalised items should occur about the same number of times in the data, improving the results of the analysis, because we obtain rules more interesting between products in different levels in the taxonomy.

Another limitation of association rules over ticket data is that they do not provide any information that goes beyond the taxonomy. This can be overtake by using complementary items, that may include information about the transactions themselves, such as whether the purchase was made with cash, a credit card or check, the day of the week or the season. By including a complementary item for the month or the season when a transaction occurred, it's possible to detect differences between seasons and seasonal trends. However, it is not good idea to put in data too many complementary items, because there is a danger, complementary items are a prime cause of redundant rules. We must include only virtual items that could turn into actionable information if found in well-supported, high-confidence association rules.

The system proposed is based on categorising the items purchased in a retail company according to it's buying frequency and the various levels of product categories, and on applying a set of data pre-processing transformations and a set of data mining techniques to mine ticket data. The patterns extracted are actionable because they can suggest new stores layouts; they can determine which products to put on special; they can indicate when to issue coupons, and so on.

References

1. Quilan J.R., "C4.5: Programs for machine learning", Morgan Kaufmann Publishers, 1993.
2. Agrawal, R., S. and Srikant S., "Fast Algorithms for mining association rules, *Proceedings of 20th Very Large Databases (VLDB) Conf*, Santiago, Chile, 1994.

Automatic Acquisition of Phoneme Models and Its Application to Phoneme Labeling of a Large Size of Speech Corpus

Motoyuki Suzuki[1,2], Teruhiko Maeda[2], Hiroki Mori[3] and Shozo Makino[1,2]

[1] Computer Center, Tohoku Univ., 980-8578, Japan.
[2] Graduate School of Information Sciences, Tohoku Univ.
[3] Graduate School of Engineering, Tohoku Univ.

1   Introduction

Research fields such as speech recognition require a large amount of speech data with phoneme label information uttered by various speakers. However, phoneme labeling by visual inspection segmentation of input speech data into corresponding parts of given phoneme by human inspection is a time-consuming job. An automatic phoneme labeling system is required. Currently, several automatic phoneme labeling system based on Hidden Markov Model(HMM) were proposed. The performance of these systems depends on the used phoneme models. In this paper, at first, we propose an acquisition algorithm of accurate phoneme model with the optimum architecture, and then the obtained phoneme models is applied to segment an input speech without phoneme label information into the part corresponding to each phoneme label.

2   The Acquisition algorithm of phoneme model with the optimum architecture

Currently, Hidden Markov Network(HMnet) which is one of a probabilistic automaton is used as phoneme models. It shows higher performance than traditional HMMs. Successive State Splitting algorithm[1] is one of a construction algorithm of HMnet. Their algorithm can simultaneously solve the three factors of the model unit, model architecture and model parameters as a global optimization problem using the same criterion. However, their algorithm is based on the assumption that variation of speech can be described by phoneme contextual factors. It cannot take into account the other factors such as speaker differences. In order to overcome the defect, we propose a new improved algorithm[2]. In this algorithm, each training sample is assigned a path of HMnet independently. It can consider all factors influencing a variation of speech.

3   Phoneme labeling experiments using HMnet

Phoneme models from the HMnet were concatenated according to a given phoneme sequence of input utterance. If the phoneme context not appearing in training

Table 1. Results for labeling experiments.

speaker	HMnet		300		HMM
	100	200	HMnet	others	
male	3.51	3.43	3.31		3.57
			2.94	4.24	
female	4.51	3.88	3.46		4.29
			3.21	4.28	

samples appears in test samples, the context independent HMM is used instead of HMnet. Speech samples uttered by two speakers (a male and a female) were used. 100 sentences uttered by one speaker were used for training of HMnet, and 46 sentences were used as test samples.

Table 1 shows results for phoneme labeling experiments. Difference between the phoneme boundary given by an expert and that with the automatic phoneme labeling system was computed, and the standard deviation is shown in table 1. Moreover, phoneme boundaries were divided to two classes dependent on whether HMnet is used as the preceding and following phoneme models or not, and then standard deviation was calculated in each class when the number of states is set to 300. In the table 1, "HMnet" means the former class and "others" means the latter class.

From these results, HMnet having appropriate number of states showed higher performance than HMM did. Difference between "HMnet" and "others" shows that utilization of HMMs mainly caused degradation of labeling performance. So, if we can use a large amount of speech data, we will construct an automatic phoneme labeling system with higher performance.

4   Conclusion

In order to construct an automatic phoneme labeling system with higher performance, we propose a new HMnet construction algorithm. The algorithm can construct HMnets with the optimum architecture based on the maximum likelihood criterion. We carried out the phoneme labeling experiments. Phoneme labeling system using HMnet showed higher accuracy than that using HMM.

References

1. Takami,J. and Sagayama, S.: A Successive State Splitting Algorithm for Efficient Allophone Modeling. Proc. of ICASSP'92. (1992) 573–576
2. Suzuki,M., Makino,S., Ito,A., Aso,H. and Shimodaira,H.: A New HMnet Construction Algorithm Requiring No Contextual Factors. IEICE Trans. *Inf. & Syst.*, **E78-D**, 6. (1995) 662–668

An Experimental Agricultural Data Mining System

Kazunori Matsumoto

System Integration Technology Center, Toshiba Corp.,
kazunori@sitc.toshiba.co.jp

1 Purposes

Agriculture is an information-intensive industry from an essential point of view. Many factors such as soil, fertilizer, temperature, precipitation, sunray, etc. are all affect harvest, so that information about them is carefully investigated by expert persons in deciding agricultural activities. We thus expect to build an intelligent agricultural information system to assist the experts and to help an improvement on agricultural technologies [7]. Towards this purpose, we firstly need to provide a system which can reveal hidden relations among agricultural factors. Although traditional statistical methods have already applied to this field, we expect recent data mining technologies to bring still more fruitful results. In particular, an expert can easily examines IF - THEN style rules extracted by the typical data mining methods [1, 6], he then may give further investigations around the rules with existing knowledge.

We build an experimental system [5] to establish the necessary technologies for the purpose, which system data mines weather patterns that influence the yield of rice. It runs with the three main steps; (1) collecting all necessary data from distributed databases, (2) making a set of training data from the collected database, and (3) runing data mining algorithms on the training data set.

In the following, we give brief summary for each step.

2 Making Training Data and Mining Them

The current system uses weather data of temperature, sunray, and of precipitaion, and data of rice yielding, all of which are maintained in separate databases. Recent studies on data warehousing [2] provide promising technologies for integrating these databases, however, our application needs more specific database manipulations. In particular, our data are extended in long time period, then effects of agricultural technologies progress must be removed from the yearly yield data. This needs to access an associated knowledge base which stores experts knowledge relating yield evaluation. Further, data at every observatory stations are aggregated by using a geographical knowledge base.

After integrating necessary databases, we make a training set database which is used in the successive data mining step. Most data mining algorithms assume a training data is specified as a set of tuples of feature values, and then they build good rules in terms of features or feature values, which correctly classifies most of the training data. Inadequately selected features cause a searching space explosion and cause decreasing the extracted rule quality [3,4]. Most studies formalize the problem as the feature subset selection, which identifies unrelevant or unimportant features from given initial ones. A direct application of this approach to our case regards every time points as the initial features, then remove unrelevant time points from the initial ones. This direct method is not enough for our case because of the following reasons; (1) this is unaware of a data behavior on an time interval, and (2) agricultural important events, such as seeding, planting, harvesting, are changed year by year, thus a direct comparison of different years data has few agricultural meaning.

We successively enumerate a time interval, as a candidate for a feature, by adjusting each data with the agricultural event database, then check its importance in exploratory manner. Heuristics is used to avoid generating too many intervals.

Finally, we run data mining algorithms [1,6] on the training data set. In case of decision tree algorithm [6], we verify the quality of extracted rules by the usual cross-validation method. Almost 30% of the data are misclassified by the rules, however, suggestive ones can be found in the extracted rules. The analysis of the results will be discussed in a future paper.

3   Conclusion

We find that studies from a database perspective are important issues in developing agricultural data mining systems. Improved version of the experimental system will introduce recent data warehouse technologies to give an unified method for various database manipulation, and performance issues will be also revised.

References

1. R. Agrawal, and R.Srikant: Fast Algorithms for Mining Association Rules, Proc. of VLDB, 1994.
2. R. Kimball: The Data Warehouse Toolkit, John Wiley & Sons,1996
3. R. Kohavi, and D. Sommerfield: Feature Subset Selection using the Wrapper Model: Overfitting and Dynamic Search Space Topology, First Int. Conf. on KDD, 1995.
4. H.Liu, and R.Setiono: A Probabilistic Approach to Feature Selection - A Filter Solution , Proc. of The Thirteenth Int. Conf. on ML, 1996.
5. K. Matsumoto: Exploratory Attributes Search for Time-Series Data: An Experimental System for Agricultural Application, Proc. of PKDD'98.
6. J.R. Quinlan: C4.5: Programs for Machine Learning, Morgan Kaufmann, 1993.
7. R.B.Rao, et.al: Data Mining of Subjective Agricultural Data, Proc. of the Tenth Intl. Conf. Machine Learning, 1993.

Data Mining Oriented System for Business Applications

Yukinobu Hamuro[1], Naoki Katoh[2], and Katsutoshi Yada[1]

[1] Department of Business Administration, Osaka Industrial University, 3-1-1 Nakagaito, Daito, Osaka, 574-8530 Japan, {hamuro, yada}@adm.osaka-sandai.ac.jp
[2] Department of Architecture and Architectural Systems, Kyoto University, Kyoto, 606-8501 Japan, email: naoki@archi.kyoto-u.ac.jp

short abstract

This paper proposes a new concept called *historybase* that helps one to mining data for deriving useful knowledge from a huge amount of data, in particular aiming at business applications. At first we shall point out that data accumulated in databases for daily routine operation is not usually enough for data mining, and that we need to record much more detailed data for the knowledge discovery. Distinguished feature of historybase is to record a stream of events that occur during the course of information processing regardless of the necessity for routine operation, hoping that they are helpful for future data mining. Historybase is already implemented in real business, and has been successfully utilized to improve the business quality.

The necessity to discover useful knowledge from a huge amount of data accumulated in an enterprise has been increasing these days. Recently, for this requirement, much attention has been paid to data warehouse and OLAP[1] as a new technology of database that integrates in a unified manner a huge amount of data obtained through various types of information system for routine operation, so that the collected data can be analyzed from various points of view. However, we claim that in order to carry out data mining process in business applications, the data obtained from the systems for daily business is not sufficient because some important data are lost as shown in the following cases.

Case 1: Suppose one retailer tries to order ten units of some product by using an ordering system of some distribution company. If the available stock is less than ten, the system displays an error message on the terminal. However, the fact that the user tried to order ten units was not recorded.

Case 2: Surprisingly, even nowadays, it is usually the case in retail stores that the receipt number issued by POS register is not accumulated in sales database. Therefore, it is impossible to derive association rules of customer purchasing behavior.

Historybase we propose is designed so as to minimize the loss of such valuable data and to record all operations that occur. This is a major difference from conventional OLAP in which the source of input relies on the output of information system for routine operation that is originally developed for the purpose other than data mining.

In information systems for routine operation, a tremendous amount of data is input and processed and then output, but most of them are not accumulated in databases but has been thrown away although they once existed or were generated in the system. However, we believe that it is essential to the success of data mining in business applications to accumulate all such data since what data will become necessary for future data mining is unpredictable.

In operational system for routine use such as POS system, various kinds of events occur such as scanning the bar code, canceling the registered record, inputting questionnaire and amount of money on deposit, calculating the change, and etc, and types of events may be updated from time to time. Therefore it is difficult to store such series of events in conventional DBMS and OLAP since such systems are constructed based on rigorous model such as entity-relationship model and once they are created it takes long time and costs a lot to update the underlying database schema.

On the other hand, historybase we have designed is constructed based on completely different concept so as to cope with unpredictable requests for system update such as addition of new attributes, and ad hoc and unstructured requests for data mining. More specifically, the historybase system we designed has the following innovative features: (i) all files in the whole system consist of variable-length records and thus are easy to be augmented with new attributes, which greatly enhances the operational flexibility of the system, and (ii) all programs are written by using only UNIX commands and a set of originally developed commands that are a combination of UNIX commands. These features greatly help to reduce time and cost for system development. Data structures that record all such histories as well as the system design concept that support them are referred to as historybase.

We already implemented the historybase in one company that successfully used it for discovering useful business knowledge concerning customer purchasing behavior [4]. Further interesting features of historybase are described in [2,3]. One application system based on historybase will be demonstrated at the conference. The whole sales data used in the system is open to data mining researchers for further investigation by courtesy of the company. Please contact us if you have interest in using it.

References

1. Chaudhuri, S., U. Dayal, An Overview of Data Warehousing and OLAP Technology, SIGMOD Record, Vol.26. No.1, 1997.
2. Hamuro, Y., N. Katoh, Y.Matsuda, and K.Yada, Mining Pharmacy Data Helps to Make Profits, to appear in Data Mining and Knowledge Discovery, 1998.
3. Hamuro, Y., From Database to Historybase, Journal of Osaka Sangyo University (Social Sciences), Vol.108, 1998.
4. Yada, K., N. Katoh, Y. Hamuro, Y. Matsuda, Customer Profiling Makes Profits, Proc. of First International Conference on the PAKeM, 1998.

Moving Object Recognition Using Wavelets and Learning of Eigenspaces

Shigeru Takano, Teruya Minamoto and Koichi Niijima

Department of Informatics, Kyushu University, Kasuga 816-8580, Japan
E-mail: {takano, minamoto, niijima}@i.kyushu-u.ac.jp

1 Introduction

Recognition of moving objects is one of the most important problems in computer vision. This problem has many applications such as individual recognition, gesture recognition and lip reading. Murase and Sakai [1] proposed a method using parametric eigenspace representation to realize efficient recognition of moving objects. This method is independent of the applications and requires a little time to recognize the moving objects. However, it is needed to normalize the position and size of silhouettes extracted from moving images.

In this paper, we present a method combined a wavelet decomposition technique with learning of eigenspaces to recognize moving objects. Our method does not require the above normalization.

2 Feature extraction of moving objects

To extract the feature of moving objects, we adopt a wavelet decomposition method which is often used recently in many application fields of image analysis. We denote an image at time t by $C^1(t) = (c_1^1(t), \cdots, c_N^1(t))^T$, where N is the number of pixels of the image and the symbol T indicates a transpose. For each component $c_i^1(t)$, we compute the high frequencies $\tilde{d}_i(t) = \sum_{k \in \mathbf{Z}} \beta_k c_i^1(2t + k)$ which means a wavelet decomposition. We see from this relation that an odd shift of the time series does not imply the shift of high frequencies $\tilde{d}_i(t)$. Thus we compute other type of high frequencies $\hat{d}_i(t) = \sum_{k \in \mathbf{Z}} \beta_k c_i^1(2t+1+k)$. Using $\tilde{d}_i(t)$ and $\hat{d}_i(t)$, we define $d_i(t)$ as follows: $d_i(t) = \tilde{d}_i([t/2])$ t : even, or $d_i(t) = \hat{d}_i([t/2])$ t : odd, where $[\cdot]$ is the Gaussian symbol. We put $D(t) = (d_1(t), \cdots, d_N(t))^T$. The vector $D(t)$ represents a feature of a moving object at time t.

3 Learning of eigenspace

Using the method in Section 2, we make M high frequency reference images $D^m(t) = (d_1^m(t), \cdots, d_N^m(t))^T, m = 0, 1, \cdots, M - 1$. From these images, we construct a covariance matrix Q whose size is $N \times N$. Parametric eigenspaces can be obtained by computing principle eigenvalues and the corresponding eigenvectors of Q. Generally, the feature of moving objects appears in the first several eigenvectors. This means that $D^m(t)$ can be approximated by these eigenvectors. Since $\dim(D^m(t)) = N$ and $\{e_j\}_{j=1,2,\cdots,N}$ are linearly independent, $D^m(t)$ can be expanded as

$$D^m(t) = \sum_{j=1}^{N} v_j^m(t) e_j.$$

Putting $V^m(t) = (v_1^m(t), \cdots, v_N^m(t))^T$,

$$V^m(t) = (e_1, \cdots, e_N)^T D^m(t)$$

holds. The vector $V^m(t)$ characterizes $D^m(t)$ in the N-dimensional eigenspace. However, since N is usually very large, we approximate $V^m(t)$ by lower dimensional vectors $V^{m,k}(t) = (e_1, \cdots, e_k)^T D^m(t)$. The computation of $V^{m,k}(t)$ means the learning of $D^m(t)$ in the k-dimensional eigenspace.

4   Recognition of moving objects in eigenspace

The vector $V^{m,k}(t)$ forms a trajectory in the k-dimensional eigenspace. The M trajectories obtained by learning have been memorized in advance. Comparing a trajectory for an input moving image with the memorized trajectories, we can recognize the movement of the input image. This is called a method of parametric eigenspace representation. Let $X(t)$ be a high frequency image constructed from an input image. Using V_m and D_m, we define

$$Z^k(t) = (e_1, \cdots, e_k)^T X(t)$$

which represents a projection of $X(t)$ into the k-dimensional eigenspace. To measure the difference between $Z^k(t)$ and $V^{m,k}(t)$, we define a distance of the following form

$$I^{m,k} = \min_{a,b} \sum_{t=0}^{T-1} |Z^k(t) - V^{m,k}(at + b)|,$$

where a and b are used to adjust time stretching and shifting of input images. The parameters a and b are determined depending on $Z^k(t)$.

5   Conclusion

In simulations, we constructed eigenspaces based on video images of human gaits. The experimental results show that our method is effective to extract motion characteristics and to recognize moving objects.

References

1. H.Murase, R.Sakai. (1996). Moving object recognition in eigenspace representation gait analysis and lip reading, Pattern Recognition Letters 17, 155-162.

Search for New Methods for Assignment of Complex Molecular Spectra and a Program Package for Simulation of Molecular Spectra

Takehiko Tanaka and Takashi Imajo

Kyushu University, Department of Chemistry, Faculty of Science,
Hakozaki 6-10-1, Higashiku, Fukuoka 812-8581, Japan
E-mail: tata.scc@mbox.nc.kyushu-u.ac.jp

Recent development of spectroscopic instruments has allowed us to obtain a large amount of spectral data in machine readable forms. High resolution molecular spectra contain abundant information on structures and dynamics of molecules. However, extraction of such useful information necessitates a procedure of *spectral assignment* in which each spectral line is assigned a set of quantum numbers. This procedure has traditionally been performed by making use of regular patterns that are obviously seen in the observed spectrum.

However, we often encounter complex spectra in which such regular patterns may not be readily discerned. The purpose of the present work is to search for new methods which can assist in assigning such complex molecular spectra. We wish to devise computer aided techniques for *picking out regular patterns buried in a list of observed values which look like randomly distributed.* We hope that we may make use of various fruits of information sciences and may depend on great computational power of modern computers.

The traditional method of "Loomis-Wood diagram" [1] which was invented in the pre-computer age has been frequently used as a tool for finding regular patterns in observed spectra. In the present work, we discuss a method which may be taken as an extension of the Loomis-Wood diagram. The procedure is described as follows. Let ν_1, ν_2, and ν_3 be frequencies of three spectral lines arbitrarily chosen from the observed spectrum, and we calculate $\Delta = \nu_1 - \nu_2$ and $\Delta^2 = \nu_1 - 2\nu_2 + \nu_3$. A point corresponding to the calculated values is plotted on a chart with Δ as the vertical axis and Δ^2 as the horizontal axis. The procedure is repeated for every possible set of three spectral lines.

This method is based on the fact that it is a good approximation in most cases to represent the frequencies of spectral lines belonging to a series by a quadratic function of a running number. In such a case, the Δ^2 values calculated from various sets of three consecutive lines in the same series would be almost constant, the Δ values being scattered in a certain range.

We tested the present method on the line frequencies of the OCS molecule for the CO stretching fundamental band. The list of the frequencies included about 200 lines. The resulting plot revealed a characteristic pattern consisting of points concentrated on a narrow line which is almost vertical. We also applied this method to the case of the spectrum of the DCCCl molecule between 1950 and 2005 cm^{-1}. This spectrum is supposed to be composed of the ν_2 bands of the DCC^{35}Cl and DCC^{37}Cl isotopomers and their associated hot bands, which are overlapping with each other. The list of the observed frequencies contained about 1200 lines. We observed concentration of points in the region corresponding to $0.25 \text{ cm}^{-1} < \Delta < 0.45 \text{ cm}^{-1}$ and $-0.0017 \text{ cm}^{-1} < \Delta^2 < -0.0008 \text{ cm}^{-1}$, and random and thinner distribution in the other region. These observations suggest that the present technique may be developed as a useful method for analysis of complex spectra. Discussion of various tests on more complicated spectra will be presented on a poster.

We also report development of a program package for simulation of molecular spectra, with a purpose in part of using calculated spectra for the assessment of the methods for assignment of complex spectra. For this purpose, well characterized synthetic spectra will often be more useful than real spectra with unpredictable noise. The package consists of three main parts. Part I is an assembly of modules, each of which calculates transition frequencies, line strengths, initial state energies etc. for a single band of a molecule. These programs are classified in terms of multiplicity (closed shell, doublet, triplet, etc.), the type of the molecule (linear, symmetric top, asymmetric top, etc.), and the type of the transition ($\Sigma - \Sigma$, $\Sigma - \Pi$, parallel, perpendicular, a-type, etc.).

Part II constitutes the main body of this package. It combines the outputs of appropriate modules in Part I, associates them with linewidths and relative intensities, and synthesizes the spectrum. Part III makes some additional calculation and displays the spectrum.

References

1. Loomis, F. W., Wood, R. W.: Phys. Rev. **32** (1928) 315–333

Computer Aided Hypotheses Based Drug Discovery Using CATALYST^{RTM} and PC GUHA Software Systems (A Case Study of Catechol Analogs Against Malignant Melanoma)

Jaroslava Halova[1], Oldrich Strouf[1], Premysl Zak[2], Anna Sochorova[2], Noritaka Uchida[3], Hiroshi Okimoto[3], Tomoaki Yuzuri[4], Kazuhisa Sakakibara[4], and Minoru Hirota[4]

[1] Institute of Inorganic Chemistry, Academy of Sciences of Czech Republic CZ 250 68 Rez, Czech Republic
halova@iic.cas.cz

[2] Institute of Computer Science, Academy of Sciences of Czech Republic Pod vodarenskou vezi 2, CZ 182 07 Prague 8, Czech Republic
{zak,anna}@uivt.cas.cz

[3] CRC Research Institute , Inc.,2-7-5 Minamisuna, Koto-ku, Tokyo 136-8581 Japan
{n-uchida,h-okimoto}@crc.co.jp

[4] Yokohama National University, Tokiwa-dai 79-5, Hodogaya-ku, Yokohama 240-8501, Japan
T-YUZURI@synchem.bsk.ynu.ac.jp, {mozart,mhirota}@ynu.ac.jp

CATALYST^{RTM} [1] and PC-GUHA [2], [3] software systems have been used in computer aided hypotheses based drug discovery. They are based on quite different principles. CATALYST^{RTM} represents molecular simulation approach based on search for common structure parts of drug molecules responsible for the therapeutic activity (pharmacophores). CATALYST^{RTM} is distributed by Molecular Simulation, Inc . GUHA is acronym of General Unary Hypotheses Automaton. GUHA is academic software distributed by the Institute of Computer Science of the Academy of Sciences of Czech Republic where it is being developed since 1960's. GUHA differs from various statistical packages enabling to test hypotheses that one has formulated, by its explorative character; it automatically generates hypotheses from empirical data by means of computer procedures.

This is the first case when pharmacophoric hypotheses were generated by PC GUHA method on the basis of structure-activity data. The structures of 37 antimelanoma catechol analogs were encoded by unique "fingerprint" descriptors in the same manner as fingerprints are encoded in dactyloscopy. This is the first successful use of fingerprint descriptors for pharmacophoric hypotheses generation. It was enabled by use of PC GUHA in Quantitative Structure-Activity Relationships.

The compounds were classified into five classes according to their therapeutic activity. Quantitative Structure-Activity Relationships (QSAR) have been performed on the basis of structure activity data of catechol analogs against malignant melanoma. Pharmacophoric hypotheses generated by both systems elucidated the influence of pharmacophores on therapeutic activity of catechol analogs against maligant melanoma. The results of PC GUHA confirm the experience that therapeutic activity of catechol analogs against malignant melanoma can be explained by structural characteristics of the benzene ring substituent containing hydrogen bond acceptors (free amino and/or carboxyl groups). The pattern of the hydroxyl substituents of the benzene ring is also important, because the hypotheses concerning the presence of two hydroxyls in ortho position confirm the term catechol analogs against malignant melanoma. The best pharmacophoric hypothesis generated by renowned CATALYST^{RTM} and a hypothesis generated by GUHA are identical: carboxyl and amino group in the side chain with hydroxyl in the para position $\Rightarrow$ high therapeutic activity. The results of both independent methods are consistent. Combining various methods is a useful approach in data analysis. If the results are consistent, then the generated hypotheses are founded more thoroughly. Moreover, GUHA enabled the first successful application of fingerprint descriptors in QSAR. The final GUHA validation will be carried out on recent catechol analogs against malignant melanoma published after the computation was carried out. The GUHA method is widely applicable in discovery science beyond the scope of structure-activity relationships as applied to the computer aided hypotheses based drug discovery.

References

1. CATALYST^{RTM}, Tutorials, April 1996, San Diego : Molecular Simulation, 1996
2. Hájek,P., Sochorova,A., Zvarova, J. : Computational Statistics and Data Analysis 19 (1995) 149-153
3. Hájek,P., Holena, M.: Formal Logics of Discovery and Hypothesis Formation by Machine. In: Motoda, H., Arikawa. S (eds): Proceedings of The First International Conference on Discovery Science, Lecture Notes in Computer Science, LNCS 1532, pp. 291-302, Springer Verlag, Berlin, Heidelberg, New York (1998)

Knowledge Discovery through the Navigation Inside the Human Body

Toyofumi Saito, Jun-ichiro Toriwaki and Kensaku Mori

Nagoya University, Furo-cho, Chikusa-ku, Nagoya 464-8603, JAPAN

1 Introduction

One of remarkable progress in medical image processing is the rapid extension of three-dimensional (3D) images of human body. Imaging technology such as CT and MRI has made it possible to acquire 3D images of patients in fine spatial resolution. Novel usage of such 3D images in both diagnosis and treatment is now being studied actively. Interesting examples include virtualized endoscope system, preoperative surgical simulation, intraoperative surgical aid, and the mass screening of lung cancer. In spite of very active development of new applications, research of these applications from the viewpoint of knowledge processing has not been reported. Although comprehensive surveys were published in [1][2], studies of these kinds of image processing as knowledge processing are not referred.

This paper presents the computer aid to the knowledge acquisition and discovery by human experts through the navigation inside the virtualized human body. We first describe a framework of the knowledge acquisition and discovery from 3D images. Next we show an example using the virtualized endoscope system authors have recently developed [3].

2 Virtualized Human Body and Navigation

We can obtain a set of thin cross section images (called slice) of the human body by using CT and MRI. In computer these cross sectional images are stored in a 3D array, being registered each other. Thus by applying appropriate preprocessings such as segmentation, we can reconstruct the human body of each patient in computer. We call this "the virtualized human body (VHB)". The VHB exactly corresponds to a real human body of a patient. It is different from the Visible Human Data [4] as a general model of the human body in that the VHB is a model of each patient (a patient-specific model).

Thus we can move around inside the VHB, observe it at any viewpoint and any view direction. If high performance graphics machine and carefully designed software are available, doctors will be able to navigate inside the VHB freely, collect necessary information, and reach the diagnosis. We call this way of diagnosis "navigation diagnosis". We already realized parts of this navigation diagnosis as a virtualized endoscope system (VES) [3]. By using VES we can fly through the inside of tubelike organs such as bronchus and vessels.

3 Knowledge Acquisition through Navigation

During the navigation, we can measure various features quantitatively. Accumulating obtained measurements is considered as knowledge acquisition, and further analysis of the acquired knowledge may lead to discovery. This process is represented as the following scheme.

(1) Pictorial data (Slice set)
　　↓ 〈 3D reconstruction 〉
(2) Virtualized human body (3D digital data)
　　↓ 〈 Navigation and measurement 〉

(3) Measurements data / Navigation record (Numerical data and character string)
    ↓ ⟨ Data mining ⟩
(4) Knowledge

Fig.1 is a scene of the virtualized endoscope system [3]. In the main window is seen an endoscopic view of the bronchus. The segment drawn across a branch of the bronchus shows that the diameter of the branch is measured at the location currently seen. By operating the mouse we can fly through inside the bronchus to any part. Also we can stop and measure features in real time at an arbitrary location. All of the routes we have navigated are also recorded as the navigation record. Fig.2 is another view in the neighborhood of tumor in trachea. Medical doctors using this system can collect knowledge they want whenever they detect significant symptom of abnormality.

Similar functions were developed for stomach, vessels and colon, and worked successfully. Before the navigation the target organs are extracted automatically or semiautomatically (segmentation). Segmentation is a typical problem of picture understanding and not always easy to be realized. If it is performed successfully, visualization of 3D structure of organs is efficiently performed by employing the polygon model and the surface rendering, well known in computer graphics. Segmentation is inevitable for feature measurement of organs (Navigation with segmentation).

Navigation without segmentation is possible by utilizing the volume rendering method. In this case we navigate the 3D gray tone picture immediately without any preprocessing like segmentation. This method is applicable to any part of VHB in which segmentation is difficult. However measurement of geometrical features of organs is hardly possible because the surface of the organ is not defined. Human observer will still be able to extract significant knowledge from the rendered image of VHB.

4   Conclusion

It was confirmed experimentally that the knowledge acquisition through the navigation in VHB is useful as a computer aid for human experts knowledge acquisition and discovery. By applying data processing and pattern recognition algorithm including clustering, principal component analysis, and pattern classification to the collected knowledge set, it is expected to realize automated discovery and the advanced discovery aid system. This is the feature problem worth being challenged.

References

1. Special issue on Virtual & Augmented Reality in Medicine, Proc. IEEE, 86, 3 (1998.3)
2. J. Toriwaki : Computer aided diagnosis and treatment with recognition and generation of images, Journal of Computer Aided Diagnosis of Medical Images, 1, 2, pp.1-16 (1997.11)
3. K.Mori et al.:Virtualized endoscope system - an application of virtual reality technology to diagnostic aid, IEICE Trans. Inf.& Sys., E79-D, 6, pp.809-819 (1996.6)
4. M.Ackerman: The visible human project, Proc. of IEEE, 86, 3, pp.504-511 (1998.3)

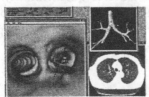   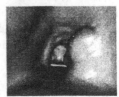

Fig.1 Virtualized endoscope system.   **Fig.2** View of tumor in trachea.

Application of Discovery Science to Solar-Terrestrial Physics

Tohru Araki[1,2], Toshihiko Iyemori[2], Masahito Nose[1], Takaaki Wada[1],
Yoshifumi Futaana[1], Genta Ueno[1], Toyohisa Kamei[2], and Akiniro Saito[1]

[1] Department of Geophysics,
email: araki@kugi.kyoto-u.ac.jp,
WWW home page:http://www-step.kugi.kyoto-u.ac.jp
[2] Data Analysis Center for Geomagnetism and Space Magnetism,
email: iyemori@kugi.kyoto-u.ac.jp,
WWW home page:http://swdcdb.kugi.kyoto-u.ac.jp
Graduate School of Science, Kyoto University,
Kyoto 606-8502, Japan

Direct satellite observations started by Sputnik-I in 1957 has revealed that complex structures exist in the space between the sun and the earth and that various kinds of dynamical processes occur there. The radiation belt, magnetosphere, collision less bow shock and solar wind are the structures which were newly discovered by in-situ satellite observations. Understanding of phenomena such as solar flares, interplanetary disturbances, aurora and magnetic storms has drastically progressed compared with in the pre-satellite era. The space near the earth is now utilized for communications, broadcasting, meteorological and oceanic observations, monitoring of the earth's environment, exploration of natural resources and experiments of physical, chemical and biological sciences. As being symbolized by beginning of construction of the space station in this year, the space has become more important from practical point of view. It is now necessary to nowcast or forecast the state of the space environment by quasi-instantaneous analysis of real time data in order to avoid hazards caused by electromagnetic disturbances.

Now much data have been continuously accumulated from various kinds of three dimensional observations on the ground and in the ionosphere, magnetosphere, interplanetary space and solar atmosphere. It becomes a great problem how to dig out the knowledge of solar terrestrial physics from huge amount of the data. In consideration of these circumstances we show some examples of our trials of application of discovery science to the problems in solar terrestrial physics. In this symposium we will focus on the following topics;

1. Real time detection of Pi2 geomagnetic pulsations using wavelet analysis,
2. Application of neural network analysis to detection of geomagnetic substorms,
3. Automatic detection of interplanetary shocks,
4. Discovery of maximal values of 3-dimensional velocity distribution function.

References

1. Nose, M., Iyemori T., Takeda M., Kamei T., Milling D. K., Orr D., Singer H. J., Worthington E. W. and Sumitomo N.: Automated detection of Pi2 pulsations using wavelet analysis : 1 Method and an application for substorm monitoring, Earth, Planet and Space, 1998 (in press).

Incorporating a Navigation Tool into a WWW Browser

Hiroshi Sawai, Hayato Ohwada and Fumio Mizoguchi

Science University of Tokyo
Noda, Chiba, 278-8510, Japan

1 Introduction

This paper describes a design of Web browser with a navigation tool to find out visually other user's interests in a collection of WWW information. This is accomplished by visualizing the hyperlinks in a homepage as a graph in three dimensional hyper space. This navigation tool is incorporated into this browser, named *HomepageMap*. Functions of *HomepageMap* are realized by extending our *WebMap*[1]. In this paper, we will describe the system overview and an experiment so as to evaluate our proposed navigation tool.

2 System overview

Conventional browsers do not have a tool to visualize hyperlinks, and a user may be in a state of *"lost in hyper space"* for users. *HomepageMap* avoids this state, to adapt a technique of *Hyperbolic tree*[2]. Representing a graph into *Hyperbolic tree*, a user can understand the total structure of a homepage easily. Although *Cone tree*[3] represents a graph on a large scale in three dimension, it can not show the whole graph in one canvas. *HomepageMap* have functions that user can jump to any node in the tree directly, and understand current browsing position easily by rearranging the graph. *"Hyperbolic Tree in Java applet"* by Inxight company[4] have these functions.

HomepageMap have a function like *Graphic History*[5]. While *Graphic history* is a tool which represent a user's browsing history into a tree structure, *HomepageMap* represent a graph in advance and identify whether a node is visited by means of the color of the node.

A significant feature of *HomepageMap* is to indicate interesting documents by showing the access statistics as a bar-chart(see Fig.1). In this feature, user can understand easily where other user's interesting documents exist.

HomepageMap is implemented within our browser. Our browser first parses the hyperlinks in HTML documents, then makes a tree structure for *HomepageMap*. Our system navigates a user by analyzing the interactions between the user, the browser and the tool. A user can open new documents in the browser using *HomepageMap*, and change the graph in *HomepageMap* using the browser.

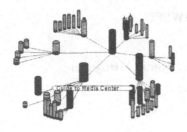

Fig. 1. A graph in *HomepageMap* Each node shows a document, The height of a node shows a frequency of access.

3   Experiment

We conducted an experiment to find out the effectiveness in recognizing interesting WWW documents. In the experiment, we recorded data for two browsing patterns by students; one was not using *HomepageMap* and the other was using *HomepageMap*. There were 83 documents in the homepage. It took about 10 minutes for a student to browse the homepage.

From the result, we made two interesting observations. One is that students could access much more documents using *HomepageMap*, because average number of documents a student accessed were increased from 15 to 21. The other is that students did not need to use backward button frequently in order to revisit documents using *HomepageMap*, because the ratio of revisiting same documents was decreased from 30% to 18%.

We interviewed students regarding impressions of using the *HomepageMap*. Some students were much interesting to use this tool, in particular, to select higher nodes. But some students did not see contents of the homepage sufficiently in spite of the amount of browsing pages were increased. This is because *HomepageMap* shows only a title in each document, and it is necessary to show some keywords in it.

Although most of the students were interesting in other user's interest among the interviews, it is not always that all people have same interests. The author have a idea that the height of a node will show the frequency of a keyword matching for a user.

4   Conclusion

We have designed and demonstrated our visualization tool for navigation in WWW. An experiment showed that the effectiveness has obtained in the revisited documentation search compared with those of conventional browsers. This system will be made generally available in the new future to extract keywords in HTML documents.

References

1. H.Sawai, H.Ohwada, F.Mizoguchi, Designing a browser with the function of *"WebMap"*, *Proc. of the 12th Annual Conference of JSAI*,1998
2. J.Lamping,R.Rao,P.Pirolli, A Focus+Context Technique Based on Hyperbolic Geometry for Visualizing Large Hierarchies, *Proc. of ACM CHI '95*,1995.
3. G.G.Robertson,J.D.Mackinlay,S.K.Card, Cone Trees: Animated 3D Visualizations of Hierarchical Information, *Proc. of ACM CHI '91*, 1991.
4. Inxight Homepage,http://www.inxight.com/
5. E.Ayers,J.Stasko, Using Graphic History in Browsing the World Wide Web, *The 4th International WWW Conference* , 1995.

Author Index

Springer
and the
environment

At Springer we firmly believe that an
international science publisher has a
special obligation to the environment,
and our corporate policies consistently
reflect this conviction.
We also expect our business partners –
paper mills, printers, packaging
manufacturers, etc. – to commit
themselves to using materials and
production processes that do not harm
the environment. The paper in this
book is made from low- or no-chlorine
pulp and is acid free, in conformance
with international standards for paper
permanency.

Springer

Lecture Notes in Artificial Intelligence (LNAI)

Lecture Notes in Computer Science